Methods in Enzymology

Volume 99
HORMONE ACTION
Part F
Protein Kinases

METHODS IN ENZYMOLOGY

EDITORS-IN-CHIEF

Sidney P. Colowick   Nathan O. Kaplan

Methods in Enzymology

Volume 99

Hormone Action

Part F
Protein Kinases

EDITED BY

Jackie D. Corbin

HOWARD HUGHES MEDICAL INSTITUTE
VANDERBILT UNIVERSITY SCHOOL OF MEDICINE
NASHVILLE, TENNESSEE

Joel G. Hardman

DEPARTMENT OF PHARMACOLOGY
VANDERBILT UNIVERSITY SCHOOL OF MEDICINE
NASHVILLE, TENNESSEE

1983

ACADEMIC PRESS

A Subsidiary of Harcourt Brace Jovanovich, Publishers

New York   London
Paris San Diego San Francisco São Paulo Sydney Tokyo Toronto

COPYRIGHT © 1983, BY ACADEMIC PRESS, INC.
ALL RIGHTS RESERVED.
NO PART OF THIS PUBLICATION MAY BE REPRODUCED OR
TRANSMITTED IN ANY FORM OR BY ANY MEANS, ELECTRONIC
OR MECHANICAL, INCLUDING PHOTOCOPY, RECORDING, OR ANY
INFORMATION STORAGE AND RETRIEVAL SYSTEM, WITHOUT
PERMISSION IN WRITING FROM THE PUBLISHER.

ACADEMIC PRESS, INC.
111 Fifth Avenue, New York, New York 10003

United Kingdom Edition published by
ACADEMIC PRESS, INC. (LONDON) LTD.
24/28 Oval Road, London NW1 7DX

Library of Congress Cataloging in Publication Data

Hormone action.

(Methods in enzymology ; v. 36-   )
In pts. C and D Hardman's name is first; pt. F
edited by Jackie D. Corbin and Joel G. Hardman.
Includes bibliographical references and indexes.
CONTENTS: pt. A. Steroid hormones.--pt. B. Peptide
hormones.--[etc.]--pt. F. Protein kinases.
1. Enzymes. 2. Hormones. 3. Cyclic nucleotides.
I. Hardman, Joel G., joint author. II. Title.
III. Series: Methods in enzymology ' v. 36-40.
[DNLM: W1 ME9615K v.36 etc. / WK 102 H8115]
QP601.M49 vol. 36-40   574.1925s [574.19'2]   74-10710
ISBN 0-12-181999-X (v. 99)

PRINTED IN THE UNITED STATES OF AMERICA

83 84 85 86   9 8 7 6 5 4 3 2 1

Table of Contents

Contributors to Volume 99 ix
Preface xiii
Volumes in Series xv

Section I. General Methodology

1. Assays of Protein Kinase	Robert Roskoski, Jr.	3
2. Measurement of Chemical Phosphate in Proteins	Janice E. Buss and James T. Stull	7
3. Removal of Phosphate from Proteins by the Reverse Reaction	David A. Flockhart	14
4. Measurement of Hormone-Stimulated Phosphorylation in Intact Cells	James C. Garrison	20
5. Peptide Mapping and Purification of Phosphopeptides Using High-Performance Liquid Chromatography	Henning Juhl and Thomas R. Soderling	37

Section II. Purification and Properties of Specific Protein Kinases

A. Cyclic Nucleotide-Dependent Protein Kinases

6. Catalytic Subunit of cAMP-Dependent Protein Kinase	Erwin M. Reimann and Richard A. Beham	51
7. Regulatory Subunits of Bovine Heart and Rabbit Skeletal Muscle cAMP-Dependent Protein Kinase Isozymes	Stephen R. Rannels, Alfreda Beasley, and Jackie D. Corbin	55
8. cGMP-Dependent Protein Kinase	Thomas M. Lincoln	62
9. Insect (cAMP–cGMP)-Dependent Protein Kinase	Alexander Vardanis	71
10. Preparation of Partially Purified Protein Kinase Inhibitor	Keith K. Schlender, Jennifer L. Tyma, and Erwin M. Reimann	77
11. Inhibitor Protein of the cAMP-Dependent Protein Kinase: Characteristics and Purification	Susan Whitehouse and Donal A. Walsh	80

12. Use of NMR and EPR to Study cAMP-Dependent Protein Kinase — ALBERT S. MILDVAN, PAUL R. ROSEVEAR, JOSEPH GRANOT, CATHERINE A. O'BRIAN, H. NEAL BRAMSON, AND E. T. KAISER — 93

13. Synthesis of Oligopeptides for the Study of Cyclic Nucleotide-Dependent Protein Kinases — DAVID B. GLASS — 119

14. Affinity Labeling of cAMP-Dependent Protein Kinases — SUSAN S. TAYLOR, ANTHONY R. KERLAVAGE, AND MARK J. ZOLLER — 140

15. Photoaffinity Labeling of the Regulatory Subunit of cAMP-Dependent Protein Kinase — ULRICH WALTER AND PAUL GREENGARD — 154

16. Use of 1,N^6-Etheno-cAMP as a Fluorescent Probe to Study cAMP-Dependent Protein Kinase — HILLARY D. WHITE, STEPHEN B. SMITH, AND EDWIN G. KREBS — 162

17. Using Analogs to Study Selectivity and Cooperativity of Cyclic Nucleotide Binding Sites — STEPHEN R. RANNELS AND JACKIE D. CORBIN — 168

18. Reversible Autophosphorylation of Type II cAMP-Dependent Protein Kinase: Distinction between Intramolecular and Intermolecular Reactions — JACK ERLICHMAN, RAPHAEL RANGEL-ALDAO, AND ORA M. ROSEN — 176

19. Use of Immunological Approaches to Identify a Brain Protein Kinase Isozyme — DWIJEN SARKAR, JACK ERLICHMAN, NORMAN FLEISCHER, AND CHARLES S. RUBIN — 187

20. Using Mutants to Study cAMP-Dependent Protein Kinase — MICHAEL GOTTESMAN — 197

21. Protein Modulation of Cyclic Nucleotide-Dependent Protein Kinases — GORDON M. WALTON AND GORDON N. GILL — 206

22. Substrate-Directed Regulation of cAMP-Dependent Phosphorylation — M. R. EL-MAGHRABI, T. H. CLAUS, AND S. J. PILKIS — 212

23. Use of Microinjection Techniques to Study Protein Kinases and Protein Phosphorylation in Amphibian Oocytes — JAMES L. MALLER — 219

24. Determination of the cAMP-Dependent Protein Kinase Activity Ratio in Intact Tissues — JACKIE D. CORBIN — 227

25. Radiolabeling and Detection Methods for Studying Metabolism of Regulatory Subunit of cAMP-Dependent Protein Kinase I in Intact Cultured Cells — ROBERT A. STEINBERG — 233

B. Calcium-Dependent Protein Kinases

26. Phosphorylase Kinase from Rabbit Skeletal Muscle	PHILIP COHEN	243
27. Cardiac Phosphorylase Kinase: Preparation and Properties	HEI SOOK SUL, BERNADETTE DIRDEN, KAREN L. ANGELOS, PATRICK HALLENBECK, AND DONAL A. WALSH	250
28. Separation of the Subunits of Muscle Phosphorylase Kinase	K.-F. JESSE CHAN AND DONALD J. GRAVES	259
29. Use of Peptide Substrates to Study the Specificity of Phosphorylase Kinase Phosphorylation	DONALD J. GRAVES	268
30. Smooth Muscle Myosin Light Chain Kinase	MICHAEL P. WALSH, SUSAN HINKINS, RENATA DABROWSKA, AND DAVID J. HARTSHORNE	279
31. Calcium-Activated, Phospholipid-Dependent Protein Kinase (Protein Kinase C) from Rat Brain	USHIO KIKKAWA, RYOJI MINAKUCHI, YOSHIMI TAKAI, AND YASUTOMI NISHIZUKA	288
32. Liver Calmodulin-Dependent Glycogen Synthase Kinase	M. ELIZABETH PAYNE AND THOMAS R. SODERLING	299

C. Cyclic Nucleotide and Calcium-Independent Protein Kinases

33. Casein Kinase I	GARY M. HATHAWAY, POLYGENA T. TUAZON, AND JOLINDA A. TRAUGH	308
34. Casein Kinase II	GARY M. HATHAWAY AND JOLINDA A. TRAUGH	317
35. Pyruvate Dehydrogenase Kinase from Bovine Kidney	FLORA H. PETTIT, STEPHEN J. YEAMAN, AND LESTER J. REED	331
36. Glycogen Synthase Kinase-3 from Rabbit Skeletal Muscle	BRIAN A. HEMMINGS AND PHILIP COHEN	337
37. Double-Stranded RNA-Dependent eIF-2α Protein Kinase	RAY PETRYSHYN, DANIEL H. LEVIN, AND IRVING M. LONDON	346
38. Rhodopsin Kinase	HITOSHI SHICHI, ROBERT L. SOMERS, AND KATSUHIKO YAMAMOTO	362

39. Polyamine-Dependent Protein Kinase and Phosphorylation of Ornithine Decarboxylase in *Physarum polycephalum* — VALERIE J. ATMAR AND GLENN D. KUEHN — 366

D. Tyrosine-Specific Protein Kinases

40. Characterization of the Abelson Murine Leukemia Virus-Encoded Tyrosine-Specific Protein Kinase — JEAN YIN JEN WANG AND DAVID BALTIMORE — 373

41. Purification of the Receptor for Epidermal Growth Factor from A-431 Cells: Its Function as a Tyrosyl Kinase — STANLEY COHEN — 379

42. Detection and Quantification of Phosphotyrosine in Proteins — JONATHAN A. COOPER, BARTHOLOMEW M. SEFTON, AND TONY HUNTER — 387

43. Base Hydrolysis and Amino Acid Analysis for Phosphotyrosine in Proteins — TODD M. MARTENSEN AND RODNEY L. LEVINE — 402

AUTHOR INDEX 407

SUBJECT INDEX 421

Contributors to Volume 99

Article numbers are in parentheses following the names of contributors.
Affiliations listed are current.

KAREN L. ANGELOS (27), *Department of Biological Chemistry, University of California School of Medicine, Davis, California 95616*

VALERIE J. ATMAR (39), *Department of Chemistry, New Mexico State University, Las Cruces, New Mexico 88003*

DAVID BALTIMORE (40), *Whitehead Institute for Biomedical Research and Department of Biology, Massachusetts Institute of Technology, Cambridge, Massachusetts 02139*

ALFREDA BEASLEY (7), *Howard Hughes Medical Institute, Vanderbilt University School of Medicine, Nashville, Tennessee 37232*

RICHARD A. BEHAM (6), *Department of Biochemistry, Medical College of Ohio, Toledo, Ohio 43699*

H. NEAL BRAMSON (12), *Laboratory of Bioorganic Chemistry and Biochemistry, Rockefeller University, New York, New York 10021*

JANICE E. BUSS (2), *Department of Medicine, School of Medicine, University of California, San Diego, La Jolla, California 92093*

K.-F. JESSE CHAN (28), *Department of Pharmacology, Yale University School of Medicine, New Haven, Connecticut 06510*

T. H. CLAUS (22), *Lederle Laboratories, Pearl River, New York 10965*

PHILIP COHEN (26, 36), *Department of Biochemistry, University of Dundee, Dundee DD1 4HN, Scotland*

STANLEY COHEN (41), *Department of Biochemistry, Vanderbilt University, Nashville, Tennessee 37232*

JONATHAN A. COOPER (42), *Molecular Biology and Virology Laboratory, The Salk Institute, San Diego, California 92138*

JACKIE D. CORBIN (7, 17, 24), *Howard Hughes Medical Institute, Vanderbilt University School of Medicine, Nashville, Tennessee 37232*

RENATA DABROWSKA (30), *Department of Biochemistry of Nervous System and Muscle, Nencki Institute of Experimental Biology, Polish Academy of Sciences, 02-093 Warsaw, Poland*

BERNADETTE DIRDEN (27), *Department of Biological Chemistry, University of California School of Medicine, Davis, California 95616*

M. R. EL-MAGHRABI (22), *Department of Physiology, Vanderbilt University, Nashville, Tennessee 37232*

JACK ERLICHMAN (18, 19), *Departments of Medicine and Biochemistry, Albert Einstein College of Medicine, Bronx, New York 10461*

NORMAN FLEISCHER (19), *Department of Medicine, Albert Einstein College of Medicine, Bronx, New York 10461*

DAVID A. FLOCKHART (3), *Laboratory of Cellular and Molecular Physiology, Howard Hughes Medical Institute and Department of Physiology, Vanderbilt University School of Medicine, Nashville, Tennessee 37232*

JAMES C. GARRISON (4), *Department of Pharmacology, University of Virginia Medical School, Charlottesville, Virginia 22908*

GORDON N. GILL (21), *Department of Medicine, University of California, San Diego, La Jolla, California 92093*

DAVID B. GLASS (13), *Department of Pharmacology, Emory University School of Medicine, Atlanta, Georgia 30322*

MICHAEL GOTTESMAN (20), *Laboratory of Molecular Biology, National Cancer Institute, National Institutes of Health, Bethesda, Maryland 20205*

JOSEPH GRANOT (12), *Department of Physics, Purdue University School of Science at Indianapolis, Indianapolis, Indiana 46223*

DONALD J. GRAVES (28, 29), *Department of Biochemistry and Biophysics, Iowa State University, Ames, Iowa 50011*

PAUL GREENGARD (15), *The Rockefeller University, New York, New York 10021*

PATRICK HALLENBECK (27), *Department of Biological Chemistry, University of California School of Medicine, Davis, California 95616*

DAVID J. HARTSHORNE (30), *Muscle Biology Group, Departments of Biochemistry and Nutrition and Food Science, University of Arizona, Tucson, Arizona 85721*

GARY M. HATHAWAY (33, 34), *Biochemistry Department, University of California, Riverside, California 92521*

BRIAN A. HEMMINGS (36), *Friedrich-Miescher Institut, CH-4002 Basel, Switzerland*

SUSAN HINKINS (30), *Muscle Biology Group, Department of Nutrition and Food Science, University of Arizona, Tucson, Arizona 85721*

TONY HUNTER (42), *Molecular Biology and Virology Laboratory, The Salk Institute, San Diego, California 92138*

HENNING JUHL (5), *Department of Medicine, Marselisborg Hospital, DK-8000 Arhus, Denmark*

E. T. KAISER (12), *Laboratory of Bioorganic Chemistry and Biochemistry, The Rockefeller University, New York, New York 10021*

ANTHONY R. KERLAVAGE (14), *Department of Chemistry, University of Pennsylvania, Philadelphia, Pennsylvania 19104*

USHIO KIKKAWA (31), *Department of Biochemistry, Kobe University School of Medicine, Kobe 650, Japan*

EDWIN G. KREBS (16), *Howard Hughes Medical Institute Laboratories, Department of Pharmacology, University of Washington, Seattle, Washington 98195*

GLENN D. KUEHN (39), *Department of Chemistry, New Mexico State University, Las Cruces, New Mexico 88003*

DANIEL H. LEVIN (37), *Harvard-MIT Division of Health Sciences and Technology, Cambridge, Massachusetts 02139*

RODNEY L. LEVINE (43), *Laboratory of Biochemistry, National Heart, Lung, and Blood Institute, National Institutes of Health, Bethesda, Maryland 20205*

THOMAS M. LINCOLN (8), *Department of Pharmacology, School of Medicine, University of South Carolina, Columbia, South Carolina 29208*

IRVING M. LONDON (37), *Harvard-MIT Division of Health Sciences and Technology and Department of Biology, Massachusetts Institute of Technology, Cambridge, Massachusetts 02139*

JAMES L. MALLER (23), *Department of Pharmacology, University of Colorado School of Medicine, Denver, Colorado 80262*

TODD M. MARTENSEN (43), *Laboratory of Biochemistry, National Heart, Lung, and Blood Institute, National Institutes of Health, Bethesda, Maryland 20205*

ALBERT S. MILDVAN (12), *Departments of Physiological Chemistry and Chemistry, The Johns Hopkins University School of Medicine, Baltimore, Maryland 21205*

RYOJI MINAKUCHI (31), *Department of Biochemistry, Kobe University School of Medicine, Kobe 650, Japan*

YASUTOMI NISHIZUKA (31), *Department of Biochemistry, Kobe University School of Medicine, Kobe 650, Japan*

CATHERINE A. O'BRIAN (12), *Laboratory of Bioorganic Chemistry and Biochemistry, Rockefeller University, New York, New York 10021*

M. ELIZABETH PAYNE (32), *Laboratory of Molecular Cardiology, National Institutes of Health, Bethesda, Maryland 20205*

RAY PETRYSHYN (37), *Harvard-MIT Division of Health Sciences and Technology, Cambridge, Massachusetts 02139*

FLORA H. PETTIT (35), *Clayton Foundation Biochemical Institute, The University of Texas at Austin, Austin, Texas 78712*

S. J. PILKIS (22), *Department of Physiology, Vanderbilt University, Nashville, Tennessee 37232*

RAPHAEL RANGEL-ALDAO (18), *Centro de Investigaciones Biomedicas, Universidad De Carabobo, Nucleo Aragua, Maracay, Venezuela*

STEPHEN R. RANNELS (7, 17), *Howard Hughes Medical Institute, Vanderbilt University School of Medicine, Nashville, Tennessee 37232*

LESTER J. REED (35), *Clayton Foundation Biochemical Institute and Department of Chemistry, The University of Texas at Austin, Austin, Texas 78712*

ERWIN M. REIMANN (6, 10), *Department of Biochemistry, Medical College of Ohio, Toledo, Ohio 43699*

ORA M. ROSEN (18), *Department of Molecular Pharmacology, Albert Einstein College of Medicine, Bronx, New York 10461*

PAUL R. ROSEVEAR (12), *Department of Physiological Chemistry, The Johns Hopkins University School of Medicine, Baltimore, Maryland 21205*

ROBERT ROSKOSKI, JR. (1), *Department of Biochemistry, Louisiana State University Medical Center, New Orleans, Louisiana 70119*

CHARLES S. RUBIN (19), *Department of Molecular Pharmacology, Albert Einstein College of Medicine, Bronx, New York 10461*

DWIJEN SARKAR (19), *Department of Molecular Pharmacology, Albert Einstein College of Medicine, Bronx, New York 10461*

KEITH K. SCHLENDER (10), *Department of Pharmacology, Medical College of Ohio, Toledo, Ohio 43699*

BARTHOLOMEW M. SEFTON (42), *Molecular Biology and Virology Laboratory, The Salk Institute, San Diego, California 92138*

HITOSHI SHICHI (38), *Institute of Biological Sciences, Oakland University, Rochester, Michigan 48063*

STEPHEN B. SMITH (16), *School of Pharmacy, University of Southern California, Los Angeles, California 90065*

THOMAS R. SODERLING (5, 32), *Howard Hughes Medical Institute and Department of Physiology, Vanderbilt University, Nashville, Tennessee 37232*

ROBERT L. SOMERS (38), *Laboratory of Vision Research, National Eye Institute, National Institutes of Health, Bethesda, Maryland 20205*

ROBERT A. STEINBERG (25), *Biological Sciences Group, University of Connecticut, Storrs, Connecticut 06268*

JAMES T. STULL (2), *Department of Pharmacology and Moss Heart Center, University of Texas Health Science Center at Dallas, Dallas, Texas 75235*

HEI SOOK SUL (27), *Department of Molecular Pharmacology, Albert Einstein College of Medicine, Bronx, New York 10461*

YOSHIMI TAKAI (31), *Department of Biochemistry, Kobe University School of Medicine, Kobe 650, Japan*

SUSAN S. TAYLOR (14), *Department of Chemistry, University of California, San Diego, La Jolla, California 92093*

JOLINDA A. TRAUGH (33, 34), *Department of Biochemistry, University of California, Riverside, California 92521*

POLYGENA T. TUAZON (33), *Department of Biochemistry, University of California, Riverside, California 92521*

JENNIFER L. TYMA (10), *Department of Pharmacology, Medical College of Ohio, Toledo, Ohio 43699*

ALEXANDER VARDANIS (9), *Agriculture Canada, London Research Centre, London, Ontario N6A 5B7, Canada*

DONAL A. WALSH (11, 27), *Department of Biological Chemistry, University of California School of Medicine, Davis, California 95616*

MICHAEL P. WALSH (30), *Department of Medical Biochemistry, Faculty of Medicine, University of Calgary, Calgary, Alberta T2N 4N1, Canada*

ULRICH WALTER (15), *Departments of Physiological Chemistry and Medicine, University of Würzburg, D-8700 Würzburg, Federal Republic of Germany*

GORDON M. WALTON (21), *Department of Medicine, University of California, San Diego, La Jolla, California 92093*

JEAN YIN JEN WANG (40), *Whitehead Institute for Biomedical Research and Center for Cancer Research, Massachusetts Institute of Technology, Cambridge, Massachusetts 01239*

HILLARY D. WHITE (16), *Department of Microbiology and Immunology, University of North Carolina, Chapel Hill, North Carolina 27514*

SUSAN WHITEHOUSE (11), *Veterans Administration Medical Center, 3801 Miranda Avenue, Palo Alto, California 94304*

KATSUHIKO YAMAMOTO (38), *Institute of Biological Sciences, Oakland University, Rochester, Michigan 48063*

STEPHEN J. YEAMAN (35), *Department of Biochemistry, The University, Newcastle-upon-Tyne NE1 7RU, England*

MARK J. ZOLLER (14), *Cold Spring Harbor Laboratory, Cold Spring Harbor, New York 11724*

Preface

It is clear that nature has selected protein phosphorylation as an important regulatory process in cell function. Since interest in protein kinases has increased enormously over the past few years, the publication of this book in 1983 is timely in view of the fact that so many new investigators are entering the field.

The book has been divided into two broad sections. The first covers general methodology which should be useful to all investigators in the field, and the second includes individual chapters on those specific protein kinases which had been identified and purified when the articles for this publication were being written. The latter section has been subdivided into four groups based either on the type of activator or on the amino acid which serves as phosphate acceptor.

A small amount of redundancy could not be avoided. For example, it is seen that a certain protein kinase which had been first studied using one protein substrate is found to be identical to a protein kinase which had been studied using another protein substrate. In some such cases, the real physiological substrate(s) of the enzyme is not known. Lack of absolute substrate specificity seems to be a general property of protein kinases and may result in the existence of "silent" phosphorylation sites in intracellular proteins, particularly when the protein is present at high concentrations. Lack of absolute specificity can be explained in part at least by the fact that many protein kinases are related in evolution. To our knowledge, the first prediction of this phenomenon was the homology between cAMP- and cGMP-dependent protein kinases. Homology has also been found between such apparently diverse protein kinases as the catalytic subunit of cAMP-dependent protein kinase and certain tyrosine-specific protein kinases. The recent breakthroughs in protein chemistry and genetic engineering should allow the classification of protein kinases into families and superfamilies in the near future. It is obviously useful to be informed about the evolution of protein kinases since knowledge accumulated in one system can be gainfully applied in another.

In this volume there may appear to be unusual emphasis on cAMP-dependent protein kinase, and the reason is not entirely due to the bias of one of the volume editors. This particular protein kinase is one of the earliest to be described and has probably been the most thoroughly characterized. For the reasons outlined above it is hoped that the knowledge accumulated on the cAMP-dependent protein kinase will be wisely applied to the studies of other protein kinases.

Not only did we think it appropriate to conform with the *Methods in Enzymology* policy of using 1 μmol per minute as an enzyme unit, but we also believed that interested persons might find it easier and more informative, particularly when specific protein kinase chapters are contained in the same volume, to compare turnover rates using the same unit of activity.

We apologize to those researchers who, in a rapidly moving field, have recently contributed excellent work which would have been appropriate for this book but appeared too late to be included.

JACKIE D. CORBIN
JOEL G. HARDMAN

METHODS IN ENZYMOLOGY

EDITED BY

Sidney P. Colowick and Nathan O. Kaplan

VANDERBILT UNIVERSITY
SCHOOL OF MEDICINE
NASHVILLE, TENNESSEE

DEPARTMENT OF CHEMISTRY
UNIVERSITY OF CALIFORNIA
AT SAN DIEGO
LA JOLLA, CALIFORNIA

I. Preparation and Assay of Enzymes
II. Preparation and Assay of Enzymes
III. Preparation and Assay of Substrates
IV. Special Techniques for the Enzymologist
V. Preparation and Assay of Enzymes
VI. Preparation and Assay of Enzymes (*Continued*)
 Preparation and Assay of Substrates
 Special Techniques
VII. Cumulative Subject Index

METHODS IN ENZYMOLOGY

EDITORS-IN-CHIEF
Sidney P. Colowick   Nathan O. Kaplan

VOLUME VIII. Complex Carbohydrates
Edited by ELIZABETH F. NEUFELD AND VICTOR GINSBURG

VOLUME IX. Carbohydrate Metabolism
Edited by WILLIS A. WOOD

VOLUME X. Oxidation and Phosphorylation
Edited by RONALD W. ESTABROOK AND MAYNARD E. PULLMAN

VOLUME XI. Enzyme Structure
Edited by C. H. W. HIRS

VOLUME XII. Nucleic Acids (Parts A and B)
Edited by LAWRENCE GROSSMAN AND KIVIE MOLDAVE

VOLUME XIII. Citric Acid Cycle
Edited by J. M. LOWENSTEIN

VOLUME XIV. Lipids
Edited by J. M. LOWENSTEIN

VOLUME XV. Steroids and Terpenoids
Edited by RAYMOND B. CLAYTON

VOLUME XVI. Fast Reactions
Edited by KENNETH KUSTIN

VOLUME XVII. Metabolism of Amino Acids and Amines (Parts A and B)
Edited by HERBERT TABOR AND CELIA WHITE TABOR

VOLUME XVIII. Vitamins and Coenzymes (Parts A, B, and C)
Edited by DONALD B. MCCORMICK AND LEMUEL D. WRIGHT

VOLUME XIX. Proteolytic Enzymes
Edited by GERTRUDE E. PERLMANN AND LASZLO LORAND

VOLUME XX. Nucleic Acids and Protein Synthesis (Part C)
Edited by KIVIE MOLDAVE AND LAWRENCE GROSSMAN

VOLUME XXI. Nucleic Acids (Part D)
Edited by LAWRENCE GROSSMAN AND KIVIE MOLDAVE

VOLUME XXII. Enzyme Purification and Related Techniques
Edited by WILLIAM B. JAKOBY

VOLUME XXIII. Photosynthesis (Part A)
Edited by ANTHONY SAN PIETRO

VOLUME XXIV. Photosynthesis and Nitrogen Fixation (Part B)
Edited by ANTHONY SAN PIETRO

VOLUME XXV. Enzyme Structure (Part B)
Edited by C. H. W. HIRS AND SERGE N. TIMASHEFF

VOLUME XXVI. Enzyme Structure (Part C)
Edited by C. H. W. HIRS AND SERGE N. TIMASHEFF

VOLUME XXVII. Enzyme Structure (Part D)
Edited by C. H. W. HIRS AND SERGE N. TIMASHEFF

VOLUME XXVIII. Complex Carbohydrates (Part B)
Edited by VICTOR GINSBURG

VOLUME XXIX. Nucleic Acids and Protein Synthesis (Part E)
Edited by LAWRENCE GROSSMAN AND KIVIE MOLDAVE

VOLUME XXX. Nucleic Acids and Protein Synthesis (Part F)
Edited by KIVIE MOLDAVE AND LAWRENCE GROSSMAN

VOLUME XXXI. Biomembranes (Part A)
Edited by SIDNEY FLEISCHER AND LESTER PACKER

VOLUME XXXII. Biomembranes (Part B)
Edited by SIDNEY FLEISCHER AND LESTER PACKER

VOLUME XXXIII. Cumulative Subject Index Volumes I–XXX
Edited by MARTHA G. DENNIS AND EDWARD A. DENNIS

VOLUME XXXIV. Affinity Techniques (Enzyme Purification: Part B)
Edited by WILLIAM B. JAKOBY AND MEIR WILCHEK

VOLUME XXXV. Lipids (Part B)
Edited by JOHN M. LOWENSTEIN

VOLUME XXXVI. Hormone Action (Part A: Steroid Hormones)
Edited by BERT W. O'MALLEY AND JOEL G. HARDMAN

VOLUME XXXVII. Hormone Action (Part B: Peptide Hormones)
Edited by BERT W. O'MALLEY AND JOEL G. HARDMAN

VOLUME XXXVIII. Hormone Action (Part C: Cyclic Nucleotides)
Edited by JOEL G. HARDMAN AND BERT W. O'MALLEY

VOLUME XXXIX. Hormone Action (Part D: Isolated Cells, Tissues, and Organ Systems)
Edited by JOEL G. HARDMAN AND BERT W. O'MALLEY

VOLUME XL. Hormone Action (Part E: Nuclear Structure and Function)
Edited by BERT W. O'MALLEY AND JOEL G. HARDMAN

VOLUME XLI. Carbohydrate Metabolism (Part B)
Edited by W. A. WOOD

VOLUME XLII. Carbohydrate Metabolism (Part C)
Edited by W. A. WOOD

VOLUME XLIII. Antibiotics
Edited by JOHN H. HASH

VOLUME XLIV. Immobilized Enzymes
Edited by KLAUS MOSBACH

VOLUME XLV. Proteolytic Enzymes (Part B)
Edited by LASZLO LORAND

VOLUME XLVI. Affinity Labeling
Edited by WILLIAM B. JAKOBY AND MEIR WILCHEK

VOLUME XLVII. Enzyme Structure (Part E)
Edited by C. H. W. HIRS AND SERGE N. TIMASHEFF

VOLUME XLVIII. Enzyme Structure (Part F)
Edited by C. H. W. HIRS AND SERGE N. TIMASHEFF

VOLUME XLIX. Enzyme Structure (Part G)
Edited by C. H. W. HIRS AND SERGE N. TIMASHEFF

VOLUME L. Complex Carbohydrates (Part C)
Edited by VICTOR GINSBURG

VOLUME LI. Purine and Pyrimidine Nucleotide Metabolism
Edited by PATRICIA A. HOFFEE AND MARY ELLEN JONES

VOLUME LII. Biomembranes (Part C: Biological Oxidations)
Edited by SIDNEY FLEISCHER AND LESTER PACKER

VOLUME LIII. Biomembranes (Part D: Biological Oxidations)
Edited by SIDNEY FLEISCHER AND LESTER PACKER

VOLUME LIV. Biomembranes (Part E: Biological Oxidations)
Edited by SIDNEY FLEISCHER AND LESTER PACKER

VOLUME LV. Biomembranes (Part F: Bioenergetics)
Edited by SIDNEY FLEISCHER AND LESTER PACKER

VOLUME LVI. Biomembranes (Part G: Bioenergetics)
Edited by SIDNEY FLEISCHER AND LESTER PACKER

VOLUME LVII. Bioluminescence and Chemiluminescence
Edited by MARLENE A. DELUCA

VOLUME LVIII. Cell Culture
Edited by WILLIAM B. JAKOBY AND IRA PASTAN

VOLUME LIX. Nucleic Acids and Protein Synthesis (Part G)
Edited by KIVIE MOLDAVE AND LAWRENCE GROSSMAN

VOLUME LX. Nucleic Acids and Protein Synthesis (Part H)
Edited by KIVIE MOLDAVE AND LAWRENCE GROSSMAN

VOLUME 61. Enzyme Structure (Part H)
Edited by C. H. W. HIRS AND SERGE N. TIMASHEFF

VOLUME 62. Vitamins and Coenzymes (Part D)
Edited by DONALD B. MCCORMICK AND LEMUEL D. WRIGHT

VOLUME 63. Enzyme Kinetics and Mechanism (Part A: Initial Rate and Inhibitor Methods)
Edited by DANIEL L. PURICH

VOLUME 64. Enzyme Kinetics and Mechanism (Part B: Isotopic Probes and Complex Enzyme Systems)
Edited by DANIEL L. PURICH

VOLUME 65. Nucleic Acids (Part I)
Edited by LAWRENCE GROSSMAN AND KIVIE MOLDAVE

VOLUME 66. Vitamins and Coenzymes (Part E)
Edited by DONALD B. MCCORMICK AND LEMUEL D. WRIGHT

VOLUME 67. Vitamins and Coenzymes (Part F)
Edited by DONALD B. MCCORMICK AND LEMUEL D. WRIGHT

VOLUME 68. Recombinant DNA
Edited by RAY WU

VOLUME 69. Photosynthesis and Nitrogen Fixation (Part C)
Edited by ANTHONY SAN PIETRO

VOLUME 70. Immunochemical Techniques (Part A)
Edited by HELEN VAN VUNAKIS AND JOHN J. LANGONE

VOLUME 71. Lipids (Part C)
Edited by JOHN M. LOWENSTEIN

VOLUME 72. Lipids (Part D)
Edited by JOHN M. LOWENSTEIN

VOLUME 73. Immunochemical Techniques (Part B)
Edited by JOHN J. LANGONE AND HELEN VAN VUNAKIS

VOLUME 74. Immunochemical Techniques (Part C)
Edited by JOHN J. LANGONE AND HELEN VAN VUNAKIS

VOLUME 75. Cumulative Subject Index Volumes XXXI, XXXII, and XXXLIV–LX
Edited by EDWARD A. DENNIS AND MARTHA G. DENNIS

VOLUME 76. Hemoglobins
Edited by ERALDO ANTONINI, LUIGI ROSSI-BERNARDI, AND EMILIA CHIANCONE

VOLUME 77. Detoxication and Drug Metabolism
Edited by WILLIAM B. JAKOBY

VOLUME 78. Interferons (Part A)
Edited by SIDNEY PESTKA

VOLUME 79. Interferons (Part B)
Edited by SIDNEY PESTKA

VOLUME 80. Proteolytic Enzymes (Part C)
Edited by LASZLO LORAND

VOLUME 81. Biomembranes (Part H: Visual Pigments and Purple Membranes, I)
Edited by LESTER PACKER

VOLUME 82. Structural and Contractile Proteins (Part A: Extracellular Matrix)
Edited by LEON W. CUNNINGHAM AND DIXIE W. FREDERIKSEN

VOLUME 83. Complex Carbohydrates (Part D)
Edited by VICTOR GINSBURG

VOLUME 84. Immunochemical Techniques (Part D: Selected Immunoassays)
Edited by JOHN J. LANGONE AND HELEN VAN VUNAKIS

VOLUME 85. Structural and Contractile Proteins (Part B: The Contractile Apparatus and the Cytoskeleton)
Edited by DIXIE W. FREDERIKSEN AND LEON W. CUNNINGHAM

VOLUME 86. Prostaglandins and Arachidonate Metabolites
Edited by WILLIAM E. M. LANDS AND WILLIAM L. SMITH

VOLUME 87. Enzyme Kinetics and Mechanism (Part C: Intermediates, Stereochemistry, and Rate Studies)
Edited by DANIEL L. PURICH

VOLUME 88. Biomembranes (Part I: Visual Pigments and Purple Membranes, II)
Edited by LESTER PACKER

VOLUME 89. Carbohydrate Metabolism (Part D)
Edited by WILLIS A. WOOD

VOLUME 90. Carbohydrate Metabolism (Part E)
Edited by Willis A. Wood

VOLUME 91. Enzyme Structure (Part I)
Edited by C. H. W. HIRS AND SERGE N. TIMASHEFF

VOLUME 92. Immunochemical Techniques (Part E: Monoclonal Antibodies and General Immunoassay Methods)
Edited by JOHN J. LANGONE AND HELEN VAN VUNAKIS

VOLUME 93. Immunochemical Techniques (Part F: Conventional Antibodies, Fc Receptors, and Cytotoxicity)
Edited by JOHN J. LANGONE AND HELEN VAN VUNAKIS

VOLUME 94. Polyamines (in preparation)
Edited by HERBERT TABOR AND CELIA WHITE TABOR

VOLUME 95. Cumulative Subject Index Volumes 61–74, 76–80 (in preparation)
Edited by EDWARD A. DENNIS AND MARTHA G. DENNIS

VOLUME 96. Biomembranes [Part J: Membrane Biogenesis: Assembly and Targeting (General Methods; Eukaryotes)] (in preparation)
Edited by SIDNEY FLEISCHER AND BECCA FLEISCHER

VOLUME 97. Biomembranes [Part K: Membrane Biogenesis: Assembly and Targeting (Prokaryotes, Mitochondria, and Chloroplasts)] (in preparation)
Edited by SIDNEY FLEISCHER AND BECCA FLEISCHER

VOLUME 98. Biomembranes [Part L: Membrane Biogenesis (Processing and Recycling)] (in preparation)
Edited by SIDNEY FLEISCHER AND BECCA FLEISCHER

VOLUME 99. Hormone Action (Part F: Protein Kinases)
Edited by JACKIE D. CORBIN AND JOEL G. HARDMAN

VOLUME 100. Recombinant DNA (Part B)
Edited by RAY WU, LAWRENCE GROSSMAN, AND KIVIE MOLDAVE

VOLUME 101. Recombinant DNA (Part C)
Edited by RAY WU, LAWRENCE GROSSMAN, AND KIVIE MOLDAVE

VOLUME 102. Hormone Action (Part G: Calmodulin and Calcium-Binding Proteins) (in preparation)
Edited by ANTHONY R. MEANS AND BERT W. O'MALLEY

VOLUME 103. Hormone Action (Part H: Neuroendocrine Peptides) (in preparation)
Edited by P. MICHAEL CONN

VOLUME 104. Enzyme Purification and Related Techniques (Part C) (in preparation)
Edited by WILLIAM B. JAKOBY

VOLUME 105. Oxygen Radicals in Biological Systems (in preparation)
Edited by LESTER PACKER

Section I

General Methodology

[1] Assays of Protein Kinase

By ROBERT ROSKOSKI, JR.

Assay Method

$$\text{ATP} + \text{protein} \rightarrow \text{phosphoprotein} + \text{ADP}$$

Principle. Protein kinases catalyze the transfer of the γ-phosphoryl group of ATP to an acceptor protein substrate. Their activity is conveniently measured by using [γ-^{32}P]ATP and an appropriate acceptor substrate. Unreacted ATP and its metabolites are then resolved from the radioactive protein substrate by a variety of techniques. One widely used procedure to achieve this involves protein precipitation in cellulose strips by trichloroacetic acid followed by extensive washing.[1] We developed an alternative procedure for resolving phosphohistone from ATP and its metabolites based upon adsorption of phosphohistone onto phosphocellulose strips.[2] ATP and its metabolites, on the other hand, fail to bind to phosphocellulose. Using the synthetic Ser-peptide (Leu Arg Arg Ala Ser Leu Gly) as substrate, Glass and co-workers used 30% acetic acid to effect binding of the phosphopeptide to phosphocellulose discs.[3] We have subsequently developed a general method for studying both peptide and protein phosphorylation using phosphoric acid to convert the acceptor peptide or protein substrates into positively charged forms which bind to the phosphocellulose paper.[4] The phosphoric acid, moreover, more effectively displaces ATP (and its metabolites) from the phosphocellulose strips than does water or 30% acetic acid.

In addition to measuring the activity of cyclic nucleotide-dependent protein kinases, the phosphocellulose method can be used for phosphorylase kinase, myosin light chain kinase, and other protein kinase reactions. To document specificity, phosphate incorporation should be dependent upon added peptide or protein substrate.

Reagents

Morpholinopropanesulfonic acid (MOPS), pH 7.0, 500 mM; 100 mM MgCl$_2$, bovine serum albumin, 2.5 mg/ml

[1] J. D. Corbin and E. M. Reimann, this series, Vol. 38C, p. 287.
[2] J. J. Witt and R. Roskoski, Jr., *Anal. Biochem.* **66**, 253 (1975).
[3] D. B. Glass, R. A. Masaracchia, J. R. Feramisco, and B. E. Kemp, *Anal. Biochem.* **87**, 566 (1978).
[4] P. F. Cook, M. E. Neville, K. E. Vrana, F. T. Hartl, and R. Roskoski, Jr., *Biochemistry* **21**, 5794 (1982).

Histone (e.g., Sigma histone type II A mixture), 10 mg/ml, or Ser-peptide (e.g., Peninsula Laboratories, Leu Arg Arg Ala Ser Leu Gly), 1.0 mM (or other acceptor protein substrate)
[γ-^{32}P]ATP, 1.0 mM, 100 cpm/pmol
Cyclic AMP, 100 μM
Protein kinase
Phosphoric acid (5 ml/liter of 85% phosphoric acid, 75 mM)

Procedure. The protein kinase reaction mixture contains 50 mM MOPS (pH 7.0), 10 mM MgCl$_2$, 0.25 mg/ml bovine serum albumin, 1.0 mg/ml histone or 100 μM Ser-peptide (or other acceptor protein), 100 μM ATP, and 10 μM cAMP (if desired), H$_2$O and enzyme to give a final volume of 50 μl. The reaction is initiated by enzyme addition. Following incubation (2–10 min as desired), portions (25 μl) are withdrawn, spotted onto 1 × 2-cm phosphocellulose strips (e.g., Whatman P81) and immersed in 75 mM phosphoric acid (10 ml per sample) to terminate the reaction. Samples lacking enzyme (blanks) are added last since the desorption of labeled ATP is time dependent.

The strips are swirled gently for 2 min, the phosphoric acid is decanted, and the phosphocellulose strips are washed twice more (2 min each) in phosphoric acid with gentle agitation. The stirring may be performed with a stirring rod or with a magnetic stirrer. The agitation must be carefully monitored and gentle because the phosphocellulose strips are fragile. After drying (optional) in air, with a hair dryer, or in an oven (100°, 5 min), the radioactivity is measured by liquid scintillation spectrometry with Budget-Solve or other appropriate scintillant or by measuring Cerenkov radiation (in the absence of scintillant). After application of 250,000 cpm, blanks are less than 250 cpm. With a fourth 2-min wash, blanks of less than 100 cpm are obtained.

Spectrophotometric Assay

$$ATP + peptide \rightarrow phosphopeptide + ADP$$
$$ADP + phosphoenolpyruvate \rightarrow ATP + pyruvate$$
$$pyruvate + DPNH \rightarrow lactate + DPN^+$$

Principle. With this technique, the formation of ADP in the protein kinase reaction is coupled to the pyruvate kinase reaction to produce pyruvate which is, in turn, coupled to the lactate dehydrogenase reaction with the concomitant oxidation of DPNH to DPN$^+$.[4] The decrease in absorbance at 340 nm is used to determine the reaction rate using an extinction coefficient of 6220 M^{-1} cm^{-1}. The optimum concentrations of lactate dehydrogenase and pyruvate kinase were determined according to Cleland[5] and validated by varying the protein kinase concentration.

Reagents

    0.5 M MOPS (pH 7.0), 0.5 M KCl, 50 mM MgCl$_2$
    10 mM ATP
    10 mM PEP
    1 mM Ser-peptide
    3000 units/ml pig heart lactate dehydrogenase
    1400 units/ml rabbit muscle pyruvate kinase
    Cyclic AMP-dependent protein kinase catalytic subunit (0.017 units = 890 ng)
    100 mM DPNH (stored at 0° for less than 10 days)

Procedure. The final 1-ml incubation mixture contains the following components: 100 mM MOPS, 15 U lactate dehydrogenase, 7 U pyruvate kinase, 100 mM KCl, 1 mM phosphoenolpyruvate, 10 mM MgCl$_2$, 1 mM ATP, 100 μM Ser-peptide, 200 μM DPNH. Contaminant ADP is converted to ATP instantaneously. After recording the minor ATPase activity of the components, the reaction is initiated by addition of enzyme (usually 2 μl). The background rate (ATPase activity) is subtracted from the rate obtained in the presence of C subunit. The rates are monitored at 340 nm with a signal output of 0.05 to 0.1 full-scale absorbance. A plot of velocity against concentration of C subunit is linear up to at least 0.08 units (4 μg/ml).

Discussion

    The radioisotopic method for cAMP-dependent protein kinase can be used with tissue homogenates and purified enzyme. Because crude homogenates contain ATPase activity and the heat-stable protein kinase inhibitor,[6] activities obtained following DEAE-cellulose chromatography are greater than that of the initial homogenate for brain, heart, skeletal muscle, and liver.

    The method is generally applicable for all acceptor proteins except those acidic proteins which are not positively charged at pH 1.8. Small basic peptides such as Ser-peptide can also be employed. The initial peptide substrate should have a charge of +3 at pH 1.8 for the method to work. *N*-Acetyl Ser-peptide, for example, is incompletely recovered on the phosphocellulose under these conditions. The phosphoric acid procedure is preferable to 30% acetic acid for several reasons. First, the non-

---

[5] W. W. Cleland, *Anal. Biochem.* **99,** 142 (1979).

[6] D. A. Walsh, C. D. Ashby, C. Gonzalez, D. Calkins, E. H. Fischer, and E. G. Krebs, *J. Biol. Chem.* **246,** 1977 (1971).

specific absorption is less. This may be due to the displacement of ATP and its metabolites by the higher concentration of phosphate. Furthermore, it is more rapid, economical and the strips are easier to handle in dilute phosphoric acid than in 30% acetic acid.

The phosphoric acid procedure can be used to monitor autophosphorylation of the type II R subunit and to measure incorporation of labeled substances into the C subunit. The same stoichiometries are obtained with this procedure as with trichloroacetic acid precipitation. The capacity of the paper is much greater than the amount of acceptor substrates commonly employed (up to 10 mg/ml protein or 250 μM peptide) per 25-μl portion. If very high concentrations are used, then the capacity of the paper should be checked to ensure that binding is complete.

The spectrophotometric assay is valuable in enzyme kinetic studies because it is continuous. That the experimental conditions yield initial velocities is apparent immediately and not after a series of assays have been performed as with the radioisotopic method. Any nucleoside which participates in the pyruvate kinase couple can be used and synthesis of radiolabeled nucleotide is not required. Phosphorylation of synthetic peptides which fail to bind completely to the phosphocellulose (N-acetyl Ser-peptide) can be measured with the spectrophotometric assay. The assay can be used to measure activity in preparative DEAE-cellulose chromatographic fractions. Most of the ATPase activity observed is Ser-peptide and cAMP-dependent. The radioisotopic assay, however, is much more rapid for processing the large numbers of fractions associated with enzyme purification.

The cAMP-dependent protein kinase exhibits a broad pH optimum. Buffers other than MOPS can be effectively employed. The enzyme is sensitive to conditions of ionic strength and particularly Mg^{2+} concentration. Increasing the free Mg^{2+} from 1 to 10 mM decreases the K_m fivefold (50 to 10 μM) and also decreases the V_{max} fivefold.[4] When using low concentrations of high specific activity ATP, the 10 mM MgCl$_2$ decreases the K_m to give higher incorporation, but decreases the V_{max}.

The synthetic acceptor peptide obtained from commercial sources (Sigma Chemical Co., Boehringer-Manheim, and Peninsula Laboratories) contains 25–50% nonpeptide material by weight. This was determined both spectrophotometrically and by the radioisotopic procedure using limiting acceptor peptide. In the radioisotopic procedure, a 20-fold excess of [γ-^{32}P]ATP can be used to make this determination.

Acknowledgments

This work was supported by Grants from the U.S. Public Health Service (NS-15994) and the Muscular Dystrophy Association.

[2] Measurement of Chemical Phosphate in Proteins

By JANICE E. BUSS and JAMES T. STULL

Analytical procedures for measuring the phosphate content in proteins have suffered in general from a lack of sensitivity. This problem has required preparation of large amounts of a purified protein for phosphate analysis, a task which cannot easily be accomplished for many phosphoproteins. Such preparation is particularly difficult in cases in which the protein must be purified from tissue biopsy samples obtained for investigations of protein phosphorylation *in vivo*. The sensitivity of the procedure described below is 30 times greater than the standard Fiske–SubbaRow procedure for measuring inorganic phosphate and measures as low as 0.2 nmol phosphate. This procedure incorporates two methods. First, the purified protein sample is ashed to convert protein-bound phosphate to inorganic phosphate. Second, the inorganic phosphate is measured after complexation of phosphomolybdate with the triphenylmethane dye, malachite green.

Preparation of Proteins

The direct chemical measurement of phosphate covalently bound to a specific protein can be attempted only on proteins that are free of other phosphate-containing compounds. Therefore, it is essential that during the early stages of investigation both the nature of the phosphate bond to be characterized and the extent of nonprotein phosphate contamination be determined. This applies both to proteins isolated in such abundance that the fortunate investigator may monitor purification by simple protein assays and to proteins that require identification or assay with radioactive phosphate probes.

Noncovalent association of phosphate-containing compounds with proteins may be of three general types: low-molecular-weight cofactors or regulatory groups such as nucleotides, pyridoxal phosphate, or phosphosugars, nucleic acids, or phospholipids.[1] The fact that the association is noncovalent does not prevent the compound from interacting with the protein with an affinity sufficient to confound assays of protein-bound phosphate. The affinity of 3′ : 5′-cyclic adenosine monophosphate (cyclic AMP) for the type I regulatory subunit of cyclic AMP-dependent protein kinase is sufficiently high that dialysis against 4 M urea is necessary for

---

[1] G. Taborsky, *Adv. Protein Chem.* **28**, 1 (1974).

complete removal of the nucleotide.[2] Because 1 mole of a nucleotide contains several phosphates, contamination of only a small fraction of a protein may lead to significant errors in the determination of covalent phosphate content. If the protein under study interacts with nucleic acids, the potential for such phosphate contamination is even greater. Ribosomal proteins obviously are at risk, as are most proteins found in the cell nucleus, if not from design then through exposure during isolation. Nucleic acids can also act as specific modulators of enzyme activity, as in the activation of the interferon-inducible protein kinase by double-stranded RNA.[3] The 67,000 M_r form of this kinase may, in addition, be covalently modified by phosphorylation. Thus, as is true with many enzymes, both covalently and noncovalently attached phosphate may be present, and the extent of either type of modification may vary with experimental conditions or interventions.[4] The rigor with which the protein sample must be washed will depend upon the extent and affinity of such noncovalent associations.

Phosphate covalently attached to amino acid residues in protein has been found in three forms: as acyl phosphate attached to the carboxyl function of aspartic or glutamic acid; as a phosphoamidate with either of the two imidazole nitrogens of histidine or with the epsilon amino group of lysine; and as an O-monoester with the hydroxyl group of serine, threonine, or tyrosine.[1,4]

The most common form of phosphorylated amino acid is the O-phosphomonoester. In mammalian cells, phosphoserine constitutes approximately 90%, phosphothreonine approximately 10%, and phosphotyrosine less than 0.1% of the acid-stable phosphoamino acids.[5] The recent discovery of a class of protein kinases which modify proteins through phosphorylation of tyrosine residues emphasizes the importance of determining the type of phosphate bond associated with the protein being studied.

Trichloroacetic acid precipitation of the protein often suffices for removal of low-molecular-weight cofactors such as phosphosugars. The removal of phosphate by extraction of the protein with organic solvents indicates the presence of a phospholipid. The chemical lability of phosphate bonds can be used to differentiate various types of phosphate associated with protein. Glycosidic phosphate esters and acyl phosphates are acid labile, and oligonucleotides or nucleic acids are destroyed by heating in acid. Precipitation of the protein with trichloroacetic acid followed by treatment of the precipitate for 20 min at 90° in 10% trichloroacetic acid is

[2] S. E. Builder, J. A. Beavo, and E. G. Krebs, *J. Biol. Chem.* **255**, 230 (1980).
[3] H. Grosfeld and S. Ochoa, *Proc. Natl. Acad. Sci. U.S.A.* **77**, 6526 (1980).
[4] M. Weller, in "Protein Phosphorylation" (J. R. Lagnado, ed.), p. 1. Pion, London, 1979.
[5] T. Hunter and B. M. Sefton, *Proc. Natl. Acad. Sci. U.S.A.* **77**, 1311 (1980).

sufficient to remove RNA from ribosomal proteins.[6] Such treatment will also hydrolyze phosphohistidine or phospholysine and must be avoided if these phosphoamino acids are present. However, the phosphoamidates are stable to alkali treatment and survive treatment at 100° in 1.5 M potassium hydroxide for 13 hr.[4] Acyl phosphates are selectively destroyed by treatment for 30 min at 37° in 0.8 M hydroxylamine.[4] Free O-phosphomonoesters are stable to base treatment but when present on residues in a polypeptide chain are surprisingly labile. Phosphoserine residues are the most sensitive to hydrolysis by alkali with phosphothreonine less sensitive, and phosphotyrosine being resistant. A comparison of the alkali lability of the phosphoserine in histone H2b to the phosphotyrosine of an immunoglobulin in Rous sarcoma virus tumor-bearing rabbit serum[7] is presented in the table. The lability of a phosphorylated amino acid to either acid or alkali will probably vary in different proteins depending on the particular environment of the modified residue.[8] The relative alkali stability of the phosphotyrosine bond has been used to detect phosphotyrosine-containing proteins in two-dimensional acrylamide gels despite the overwhelming presence of the other phosphomonoesters. It has so far not been possible to devise a technique to selectively hydrolyze phosphate from serine while maintaining quantitative recovery of phosphotyrosine in the same sample (J. E. Buss, unpublished observations). However, a recent method describes rapid base hydrolysis of proteins to quantitatively recover tyrosine-phosphate, which is measured on an amino acid analyzer with a fluorometric detection system.[9]

Reagents

> Trichloroacetic acid solutions: 20%, 16%, 5% (store at 0–4°). Since dilute trichloroacetic solutions decompose, these solutions should be stored no longer than 1 week.
> Sodium hydroxide: 0.1 N

Procedure

Once the type of phosphate bond to be measured has been determined, appropriate washing procedures may be devised to ensure removal of contaminating, phosphate-containing compounds. The following technique was developed for the assessment of phosphomonoesters and will

---

[6] L. Bitte and D. Kabat, this series, Vol. 30, p. 563.
[7] J. E. Kudlow, J. E. Buss, and G. N. Gill, *Nature* (*London*) **290**, 519 (1981).
[8] D. B. Bylund and T.-S. Huang, *Anal. Biochem.* **73**, 477 (1976).
[9] T. M. Martensen and R. L. Levine, this volume [43].

EFFECT OF ALKALI TREATMENT ON HYDROLYSIS OF
PHOSPHOTYROSINE AND PHOSPHOSERINE[a]

Treatment	Phosphotyrosine		Phosphoserine	
	pmol ^{32}P remaining	%	pmol ^{32}P remaining	%
0', 4°	0.072	100	43.4	100
5', 23°	0.068	94	38.1	88
5', 55°	0.061	85	29.5	68
60', 55°	0.042	58	11.3	26
60', 55°, 1 N NaOH	0.016	22	1.4	3

[a] Phosphorylation of tyrosine residues in plasma membranes and the added substrate, anti-src serum from Rous sarcoma virus tumor-bearing rabbits, was produced by the epidermal growth factor-stimulated protein kinase endogenous to the membranes.[7] Membranes were solubilized and treated with 0.3 μM epidermal growth factor, anti-src serum was added, and phosphorylation was initiated by the addition of ATP to a final concentration of 0.5 μM, with 20 μCi [γ-^{32}P]ATP in 300 μl. The reaction was allowed to proceed for 30 min at 4°. Tyrosine is the only amino acid phosphorylated under these reaction conditions.[7] Phosphorylation of serine residues in histone fraction H2b (Sigma) was produced by the catalytic subunit of the cAMP-dependent protein kinase. Phosphorylation was initiated by the addition of ATP to a final concentration of 50 μM, with 10 μCi [γ-^{32}P]ATP in 100 μl. The reaction was allowed to proceed for 30 min at 32°. Both reactions were terminated by the addition of trichloroacetic acid to a final concentration of 10%. After centrifugation, the precipitates were washed with 5% trichloroacetic acid, drained, quickly solubilized in ice-cold 0.1 N NaOH, and divided into aliquots. Tubes were incubated for the indicated times at 4, 23, or 55°. NaOH was also added to final concentration of 1 N to a phosphotyrosine and phosphoserine sample incubated for 60 min at 55°. The alkali treatments were terminated by neutralization of the base with an equal volume of 0.1 or 1 N HCl. Triplicate aliquots of the protein solution were transferred to cellulose filter papers; the protein was precipitated with trichloroacetic acid and washed as described.[7] Residual radioactivity as protein-bound ^{32}P was then measured.

remove ribosomal RNA from native tropomyosin and pyridoxal phosphate cofactor from phosphorylase b. The procedure should not be altered without verification that such changes do not lead to variable or inaccurate results from inadequate washing. To ensure adequate protein recovery during the trichloroacetic acid precipitations, 100 μg of crystallized bovine serum albumin is routinely added to samples that contain less than 25 μg of protein. If more than 250 μg of protein are used, more ashing reagent may be needed (see below). Because the heat treatment with trichloroacetic acid has the potential to decrease protein recovery by 10–20%, the retention of the sample protein may be quantified with duplicate samples carried through the entire washing procedure (without addition of bovine serum albumin), or separation of an aliquot from the final sodium hydroxide step. The fluorescamine assay for protein[10] is particularly useful for small amounts of sample because it easily detects less than 1 μg of protein. With small sample volumes the sodium hydroxide-solubilized protein may be used directly to circumvent recovery problems in the final trichloroacetic acid precipitation.

The protein sample is placed in a nondisposable, thick-walled 6 × 50-mm borosilicate glass tube. The tubes should be prewashed with 6 N HCl to remove contaminating phosphate. Disposable 6 × 50-mm tubes usually cannot withstand the intense heat during the ashing procedure. Please note that if the procedure is modified to include an incubation with hot alkali that phosphate may be leached from the glass. If alkali-treated blanks indicate that such leaching is a problem, the sample should be precipitated in heat- and alkali-resistant plastic tubes, and then transferred to the glass tubes for ashing.

Samples should contain 0.2 to 1.5 nmol of phosphate. All solutions are to be at 4° unless otherwise noted. Trichloroacetic acid is added to give a final concentration of 10%. Some proteins may require a higher concentration of trichloroacetic acid to be precipitated and other proteins, especially those with a basic charge, may require a 5% trichloroacetic acid–0.25% tungstate solution, pH 1.6, for quantitative precipitation.[11] The sample should be mixed rapidly upon addition of trichloroacetic acid to prevent the formation of large aggregates. The precipitate is collected by centrifugation at low speed to avoid packing the pellet so tightly that subsequent resuspension is difficult (a setting of number 3 for 5 min on a table-top IEC clinical centrifuge is routinely used). The supernatant is carefully decanted or aspirated with a finely drawn Pasteur pipet, and the pellet is mixed with 125 μl of 0.1 N NaOH at 4°. If the sample does not quickly and completely dissolve, another 25 μl of 0.1 N NaOH is added. This step should be completed in 1 minute or less, particularly if the protein-bound phosphate is phosphoserine. If a longer time is required, then the protein pellet was probably packed too tightly. The solubilized protein is then immediately precipitated by the addition of 100 μl of 20% trichloroacetic acid. The sample is centrifuged and the supernatant is removed as described above. If phospholipids contaminate the sample, the precipitate may be washed with 200 μl of ethanol:diethyl ether (1:1, v/v) and again collected by centrifugation. A further wash with 200 μl of

[10] S. Udenfriend, S. Stein, P. Bohlen, W. Dairman, W. Leimgruber, and M. Weigeke, *Science* **178,** 871 (1972).

[11] To prepare the 5% trichloroacetic acid-0.25% tungstate solution, add 20 ml of 100% trichloroacetic acid to 200 ml of water. Adjust to pH 1.6 with solid NaOH and then add 1.0 g of tungstic acid, sodium salt. Readjust to pH 1.6 with 1 N HCl. Dilute the solution to 400 ml with water and store at 4°.

chloroform:methanol (2:1, v/v) may be included, but may cause problems as the protein will often float after centrifugation. The washed precipitate is resuspended in 200 μl of 16% trichloroacetic acid and heated for 20 min at 90° to destroy nucleic acids. This is most conveniently done with the sample tubes in a heat-resistant rack in a metal pan. The tubes should be capped to prevent loss of volume or contamination. A size 1X Caplug (Protective Closures, Inc., 7310 La Cienega Boulevard, Inglewood, CA 90302) pierced with a needle to release pressure works well. The tubes are cooled on ice and the protein is again collected by centrifugation. The resolubilization and precipitation steps are repeated. The precipitated protein is resuspended in 200 μl of 5% trichloroacetic acid and collected by centrifugation.

Measurement of Phosphate

Reagents

Care must be used in selection of chemical reagents because of phosphate contamination and interference with the assay blank. The importance of high-quality reagents is discussed in detail by Penny.[12,13]

Phosphate standards: Dried KH_2PO_4 (100° for 3 hr) is used to prepare a 10 mM stock solution that is normally stored frozen. Solutions containing 5 to 100 μM KH_2PO_4 are prepared as standards and are also stored frozen.

Ammonium molybdate: 10% (w/v) $(NH_4)_6Mo_7O_{24} \cdot 4H_2O$ in 4 N HCl. Store at room temperature.

Malachite green: 0.2% malachite green stored at room temperature.

Phosphate reagent: 1 volume of the ammonium molybdate solution is mixed with 3 volumes of the malachite green solution. After stirring for 30 min, the mixture is filtered into a plastic bottle. Filter paper and funnel are prewashed with 6 N HCl and drained well before filtering the phosphate reagent. Do not rinse filter paper with H_2O after the 6 N HCl. The phosphate reagent may be stored for several weeks at room temperature although a small amount of precipitate may form during this time. This precipitate usually causes no problem if it does not get into the assay tubes. Either refilter before use or carefully decant the solution.

HCl: 1.2 N

Ashing reagent: 10% $MgNO_3 \cdot 6H_2O$ in 95% ethanol. Store at room temperature.

Procedure. The precipitated protein pellet in the 6 × 50-mm borosilicate glass tubes should contain between 0.2 and 1.5 nmol of protein-bound

[12] C. L. Penny, *Anal. Biochem.* **75**, 201 (1976).
[13] C. L. Penny, *Anal. Biochem.* **89**, 297 (1978).

phosphate. Standards should be added to separate glass tubes at this time. Add 25 µl of ashing reagent to each tube; mix and dry the solution by gentle heating. Evaporating the alcohol solution to dryness can be difficult in these small tubes because of the tendency for the contents to pop out if heated too quickly. A heating block or oven provides best results. If an attempt is made to dry the tubes over a flame, *do not point the open end of the tubes at your face or at your friends*. The best place to try this procedure is in a hood.

After the tubes are dried, heat each one over an intense flame until the brown fumes disappear and a white residue remains. No black particles should be present. After the tubes have cooled, add 150 µl of 1.2 N HCl to dissolve the white residue. Then add 50 µl of phosphate reagent and mix. After several minutes, measure the developed color at 660 nm in microcuvettes. The cuvettes should also be prewashed in 6 N HCl. It is best to have a set of microcuvettes to be used only for phosphate assays to avoid contamination by phosphate from routine laboratory use. The cuvette should be checked before use by adding undiluted phosphate reagent and determining by visual inspection if any green color develops after 10 min.

A typical assay has a blank value of 0.05 A or less when compared to H_2O (Fig. 1). The assay is generally linear up to 1.5 nmol of phosphate (Fig. 1). Recovery of protein-bound phosphate (phosphoserine or phos-

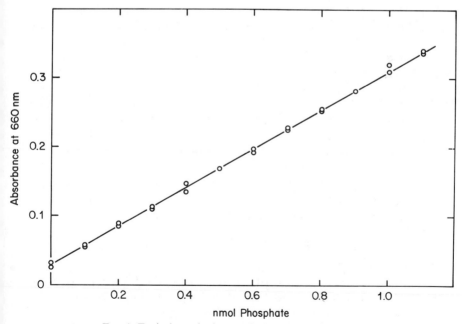

FIG. 1. Typical standard curve for inorganic phosphate.

phothreonine) is generally 90% or greater. However, it may be useful to measure protein-bound phosphate in a standard phosphoprotein such as phosphorylase a. The most common problem encountered by investigators in initially performing this procedure is a large blank value. Values greater than 0.1 A can seriously compromise the sensitivity of the assay procedure. Therefore, it is essential to be scrupulous in eliminating contaminating phosphate from test tubes, microcuvettes, etc. It is also important to obtain reagent grade chemicals that are not significantly contaminated with phosphate.

[3] Removal of Phosphate from Proteins by the Reverse Reaction

By DAVID A. FLOCKHART

Covalently bound phosphate can be removed from proteins by simple chemical means or by enzymatically catalyzed reactions. Chemical means include mild alkali treatment to remove phosphate from serine and threonine residues through β-elimination reactions.[1] However, such reactions are only one of the many effects of treating proteins with mild alkali.[1] The lack of specificity of such methods therefore makes them unsuitable for studying the effect of removing phosphate from specific sites in proteins, in particular those sites which are not alkali-labile. Enzymatic methods include dephosphorylation stimulated by nucleophilic reagents,[2] or catalyzed by protein phosphatases or by protein kinases which can catalyze the *removal* of phosphate from phosphorylated substrates in the presence of high concentrations of Mg · ADP. These methods have the advantage that they allow the possibility of activity measurement after a relatively mild and specific treatment to remove phosphate.

The reversal of protein kinase activity is a useful biochemical tool which can be used to address a number of specific problems. These include (1) the removal of phosphate from sites whose phosphorylation has been catalyzed by a known protein kinase, but where a sufficiently pure or active protein phosphatase is unavailable; (2) the removal of phosphate from discrete sites in multiply phosphorylated proteins where a sufficiently specific phosphatase is unavailable; (3) the determination of the

[1] J. R. Whitaker, *in* "Chemical Deterioration of Proteins" (J. R. Whitaker and M. Fujimaki, eds.), p. 145. American Chemical Society, Washington, D.C., 1980.
[2] P. P. Layne and V. A. Najjar, *J. Biol. Chem.* **250,** 966 (1975).

free energy of hydrolysis of protein-bound phosphate by calculation of the equilibrium constant of a protein kinase reaction; and (4) the determination of effector mechanisms through the study of the influence of activators or inhibitors upon the reverse reaction of a protein kinase.

Protein phosphatase activities have been reported which are able to catalyze the removal of phosphate from serine, threonine,[3,4] or tyrosine[5-7] residues. However, relatively few protein phosphatases have been purified to homogeneity and of these most have low protein specificity. For example, a number of phosphorylase phosphatase activities have been reported, all of which are active to varying degrees upon other phosphorylated protein substrates such as histones, caseins, phosvitin, pyruvate kinase, phosphorylase kinase, or glycogen synthase.[3,4] The paucity of information about homogeneous and specific phosphatases means that the reverse reaction might often be the tool of choice in situations where one would like to remove phosphate from a particular site in order to observe, for example, a greater effect of phosphorylation on activity or in order to determine a detailed relationship between phosphate content and activity.

The reversal of protein kinase activity in the presence of high concentrations of Mg · ADP has been observed when phosphorylase kinase,[8] cAMP-dependent protein kinase,[9-11] cGMP-dependent protein kinase,[12] or phosvitin kinase[13,14] is used to catalyze the reaction. As a result researchers have been able to study the effect of removal of phosphate from sites in protein substrates of cAMP-dependent protein kinase. El-Maghrabi et al.[11] used the reversal of cAMP-dependent protein kinase to catalyze the removal of phosphate from pyruvate kinase and to establish a more detailed correlation between the phosphorylation state and the

---

[3] E. Y. C. Lee, S. R. Silberman, M. K. Ganapathi, S. Petrovic, and H. Paris, *Adv. Cyclic Nucleotide Res.* **13,** 95 (1980).

[4] P. Cohen, *Nature (London)* **296,** 613 (1982).

[5] D. L. Brautigan, P. Bornstein, and B. Gallis, *J. Biol. Chem.* **256,** 6519 (1981).

[6] J. G. Foulkes, R. F. Howard, and A. Ziemiecki, *FEBS Lett.* **130,** 197 (1981).

[7] G. Swarup, K. V. Speeg, Jr., S. Cohen, and D. L. Garbers, *J. Biol. Chem.* **257,** 7298 (1982).

[8] Y. Shizuta, R. L. Khandelwal, J. L. Maller, J. R. Vandenheede, and E. G. Krebs, *J. Biol. Chem.* **252,** 3408 (1977).

[9] Y. Shizuta, J. A. Beavo, P. J. Bechtel, F. Hofmann, and E. G. Krebs, *J. Biol. Chem.* **250,** 6891 (1975).

[10] O. M. Rosen and J. Ehrlichman, *J. Biol. Chem.* **250,** 7788 (1975).

[11] M. R. El-Maghrabi, W. S. Haston, D. A. Flockhart, T. H. Claus, and S. J. Pilkis, *J. Biol. Chem.* **255,** 668 (1980).

[12] D. A. Flockhart and J. D. Corbin, unpublished observations, 1978.

[13] M. Rabinowitz and F. Lipmann, *J. Biol. Chem.* **235,** 1043 (1960).

[14] K. Lerch, L. W. Muir, and E. H. Fischer, *Biochemistry* **14,** 2015 (1975).

activity of that enzyme. Rosen and Erlichman[10] used the reverse reaction to study interaction of cAMP and polyarginine with the protein kinase and to investigate the relationship between the phosphorylation state of the isozyme II regulatory subunit and kinase activity.

There is also precedent for the reversal of protein kinase activity being used to calculate the free energy of hydrolysis of particular phosphate bonds. Thus, it has been shown that the phosphate incorporated into pyruvate kinase under the action of cAMP-dependent protein kinase has a free energy of hydrolysis of −6.6 kcal/mol.[11] The equivalent value for casein has been determined to be −6.5 kcal/mol.[9]

The particular methods and conditions described in this chapter have been designed for work with the cyclic nucleotide-dependent protein kinases. The principles involved, however, should also be applicable to many other purified protein kinases[15] which might effectively remove phosphate from proteins under appropriately designed conditions.

Principle

The method involves the use of high concentrations of Mg · ADP and a cyclic nucleotide-dependent protein kinase to catalyze the removal of phosphate from sites of interest under appropriate conditions.

$$\text{Protein-P} + \text{Mg} \cdot \text{ADP} \rightarrow \text{Protein} + \text{Mg} \cdot \text{ATP} \tag{1}$$

Materials and Methods

Purified Proteins. The catalytic subunit of cAMP-dependent protein kinase was prepared according to Sugden *et al.*[16] and Flockhart *et al.*[17] and stored in a 1 : 1 solution of glycerol:350 mM potassium phosphate, 0.1 mM dithiothreitol, pH 6.8. Homogeneous cGMP-dependent protein kinase was prepared by the procedure of Lincoln *et al.*[18] The isozyme II of the regulatory subunit from bovine heart was prepared as described previously.[17,19] Yeast hexokinase was obtained from Boehringer Mannheim.

Reagents. The solutions below are made up in 50 mM 2-[N-morpholino]ethanesulfonic acid (MES)-Cl, pH 6.25 in advance and stored at 0–4°. The pH of this buffer is important since the reverse reaction of cAMP-dependent protein kinase has been reported to have a sharper and more

[15] D. A. Flockhart and J. D. Corbin, *CRC Crit. Rev. Biochem.* **12,** 133 (1982).
[16] P. H. Sugden, L. A. Holladay, E. M. Reimann, and J. D. Corbin, *Biochem. J.* **159,** 409 (1976).
[17] D. A. Flockhart, D. M. Watterson, and J. D. Corbin, *J. Biol. Chem.* **255,** 4435 (1980).
[18] T. M. Lincoln, W. L. Dills, Jr., and J. D. Corbin, *J. Biol. Chem.* **252,** 4269 1977).
[19] J. D. Corbin, P. H. Sugden, D. A. Flockhart, T. M. Lincoln, and D. McArthy, *J. Biol. Chem.* **253,** 3997 (1978).

acidic pH optimum than the forward reaction,[9,10] and this also appears to be the case for phosvitin kinase[13] and phosphorylase kinase.[8]

 200 mM 2-mercaptoethanol.
 50 mM magnesium acetate.
 10 mM Na$_2$ · ADP.
 0.01 mM Na$_2$ · cAMP.
 0.01 mM Na$_2$ · cGMP.

The following solutions are made up fresh on the day of the experiment in the same buffer:

 100 units/ml yeast hexokinase made up in 100 mM glucose.
 0.1–0.2 U/ml of catalytic subunit of cAMP-dependent protein kinase in 50 mM potassium phosphate, pH 6.8 containing 0.2 mg/ml bovine serum albumin.
 0.1–0.2 units/ml cGMP-dependent protein kinase in 0.2 mg/ml bovine serum albumin.
 Phosphorylated protein substrate in 0.2 mg/ml bovine serum albumin.

Protein kinase solutions are always made up immediately prior to use, since both of these enzymes are unstable at dilute protein concentrations (<0.2 mg/ml) in the absence of substrate. One unit of protein kinase activity = 1 μmol/min ^{32}P incorporated into histone.[16,18]

Preparation of Phosphorylated Protein Substrate. If it is possible to incorporate radioactive label into a substrate using cAMP-dependent protein kinase, then one can incubate the substrate under suitable conditions, such as those that have been described for phosphorylation of pyruvate kinase,[11] or the isozyme II of the regulatory subunit,[17] and stop the reaction by the addition of an excess of EDTA. The nucleotide can then be removed by addition of 1 mg/ml bovine serum albumin followed by precipitation of the incubation mixture (0.1 ml) with 60% ammonium sulfate (0.9 ml in water) and centrifuging for 15 min at 10,000 g in a bench microfuge. The pellet should then be resuspended in 60% ammonium sulfate and washed twice before dialysis against 10 volumes of 50 mM MES-Cl, 1 mM EDTA, pH 6.25, to remove ammonium sulfate and any remaining traces of ATP. The same procedure can be used to remove ADP after the reverse reaction incubation if one intends to attempt rephosphorylation.

Procedure

For Reverse Reaction with cAMP-Dependent Protein Kinase. The reaction is carried out in 75 × 12-mm polystyrene tubes placed in a circulating water bath maintained at 30°. To the reaction tube are added 50 μl of MES buffer solution followed by 10 μl mercaptoethanol, 10 μl 50 mM magnesium acetate, and 10 μl 20 mM ADP. The substrate, protein kinase, and hexokinase-glucose solutions should then be diluted to the appropriate concentrations in MES buffer containing 0.2 mg/ml bovine serum albumin. The substrate (10 μl) and hexokinase-glucose (10 μl) solutions are added and the reaction started immediately with 10 μl of 0.1–0.2 units/ml catalytic subunit. The incubation should be continued for up to an hour and then stopped with 10 μl 200 mM EDTA, pH 7.0 before assays are performed.

For Reverse Reaction with cGMP-Dependent Protein Kinase. This protocol is similar to that for the cAMP-dependent enzyme. Cyclic GMP (10 μl of 0.01 mM) is included since it is essential for the activity of the enzyme. Other effectors do not appear to be necessary unless they are substrate-directed.

Substrate-Directed Effectors. The reverse reaction may not be possible if the substrate is not in the appropriate conformation. For example it has been shown that glucose, a phosphorlyase effector, is a requirement for the removal of phosphate from phosphorylase a catalyzed by the reversal of the phosphorylase kinase reaction. Glucose appears to act by dissociating the tetrameric structure of phosphorylase a into the dimeric form. In the same manner, cAMP is necessary if phosphate is to be removed from the regulatory subunit of cAMP-dependent protein kinase under the action of the catalytic subunit[10,12] or cGMP-dependent protein kinase.[12] Caution must be exercised in such experiments to keep the cAMP concentration low (<0.05 mM) since the nucleotide can bind to the catalytic site and inhibit the activity of the catalytic subunit under these conditions.[20,21]

Controls. Control incubations in the absence of ADP, in the absence of magnesium, in the absence of substrate, in the absence of protein kinase, and in the absence of hexokinase trap should all be run. In some situations such as the removal of phosphate from pyruvate kinase,[11] the hexokinase trap appears important. It seems less important when the phosphorylated regulatory subunit is used as a substrate for the reverse reaction.[10] Many proteases are particularly active under these incubation conditions (30° and 5 mM magnesium) and so it is important to perform controls to check for the reversibility of any effects which appear to be brought about by the reaction.

Assay. Removal of phosphate from protein can be followed by using

---

[20] J. Hoppe, W. Freist, R. Marutzky, and S. Shaltiel, *Eur. J. Biochem.* **90**, 427 (1978).
[21] D. A. Flockhart, W. Freist, J. Hoppe, and J. D. Corbin, *Fed. Proc. Fed. Am. Soc. Exp. Biol.* **39** (1980). (Abstr.)

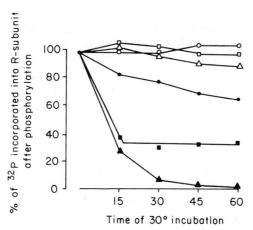

FIG. 1. Removal of ^{32}P from R-subunit. Isozyme II regulatory subunit prepared from bovine heart (5.6 μg) was incubated for 10 min at 0–4° in 0.1 ml containing 50 mM MES pH 6.25, 5 mM magnesium acetate, 20 mM 2-mercaptoethanol, 10^{-6} M cAMP, 0.002 units/ml of the catalytic subunit of cAMP-dependent protein kinase (C-subunit), and 0.04 mM [γ-^{32}P]ATP (300 cpm/pmol). Aliquots (20 μl) were taken to determine ^{32}P incorporation into protein and then test solution (20 μl) was added and the tubes placed immediately in a water bath at 30°. Incubations were carried out for the times indicated and then aliquots (20 μl) taken for determination of ^{32}P incorporation into protein. Test solutions were made up in 20 mM MES, pH 6.25 with the following additions: nothing (△), hexokinase and glucose solution (□), hexokinase and glucose solution with 0.2 units/ml of C-subunit (○), 10 mM ADP (●), 10 mM ADP with C-subunit (■), 10 mM ADP with C-subunit and hexokinase and glucose (▲).

the filter paper assay of Corbin and Reimann,[22] if the substrate is labeled with ^{32}P. An example of such an experiment, including pertinent controls, is shown in Fig. 1. Although this kind of analysis is simple to perform, caution should be exercised in the interpretation of data obtained in this way. The specific activity of phosphate is difficult to determine in such situations and chemical measurements of protein bound phosphate by a procedure such as that of Ames[23] as adapted for protein-bound phosphate by Buss and Stull[24] should be performed when there is sufficient protein to do so. We have found[11,17] that these procedures can detect as little as 0.5 nmol of protein-bound phosphate.

Calculation of Free Energy of Hydrolysis ($\Delta G°$). If the amount of chemical phosphate in a protein can be determined, then it is possible to take the reverse reaction to equilibrium using a known concentration of

---

[22] J. D. Corbin and E. M. Reimann, this series, Vol. 38, p. 287.
[23] B. N. Ames, this series, Vol, 8, p. 116.
[24] J. E. Buss and J. T. Stull, this volume [2].

ADP, to measure the ATP formed and hence to calculate an equilibrium constant.[9-11] From the equilibrium constant (K_{eq}) the free energy of hydrolysis can be calculated according to Eq. (2),

$$\Delta G° = -RT \ln K_{eq} \qquad (2)$$

The free energy of the γ-phosphate bond (normally assumed to be −8.4 kcal/mol) can then be subtracted in order to calculate the free energy of hydrolysis of the phosphoprotein bond.

[4] Measurement of Hormone-Stimulated Protein Phosphorylation in Intact Cells

By JAMES C. GARRISON

The use of radioactive phosphate to measure the phosphorylation state of enzymes in intact cells was begun in 1956 by Rall, Sutherland, and Wosilait, who used this approach to demonstrate that epinephrine and glucagon stimulated the phosphorylation of phosphorylase in dog liver slices.[1] The recent development of sodium dodecyl sulfate (SDS)–polyacrylamide gel systems capable of resolving very large numbers of proteins has increased the utility of this technique and it is now being applied to a number of tissues responsive to hormones or neurotransmitters.

The advantages of measuring changes in protein phosphorylation in intact cells include (1) The protein kinases and substrates are in their native state during the phosphorylation reaction, obviating arguments about the relevance of the phosphorylation events. (2) Changes can be observed in the phosphorylation states of proteins when the hormone message and/or the protein kinase mediating the reaction are not clearly defined. This property has been useful in studying insulin and Ca^{2+}-dependent responses in a number of cells. (3) The technique displays all of the phosphorylation changes that occur in the cell following stimulation with a given agent. This feature allows assessment of the overall response of the cell to a stimulus and has helped to uncover important sites of regulation by covalent modification. (4) In some systems, a change in the phosphorylation state of certain proteins may be the most sensitive indicator of hormone action.

The major disadvantages of the intact cell phosphorylation approach include (1) The identity and/or function of the proteins phosphorylated in response to a given agonist are not known in many cases. (2) The tech-

[1] T. W. Rall, E. W. Sutherland, and W. D. Wosilait, *J. Biol. Chem.* **218**, 483 (1956).

nique does not identify the kinase(s) which have participated in the response. (3) It is often difficult to correlate the phosphate content of a substrate and a change in enzyme activity when the experiments are performed in intact cells. Nevertheless, the technique of measuring protein phosphorylation in intact cells has been advantageous in a number of systems. Table I presents a representative list of cells in which the tech-

TABLE I
SELECTED EXAMPLES OF TISSUES STUDIED BY THE METHOD OF
INTACT CELL PROTEIN PHOSPHORYLATION

Tissue	Agonist used	References
Brain		
Cortical slices	K^+ ion, veratridine, cyclic AMP	a
Synaptosomes	K^+ ion, veratridine, A23187	b
Cultured cells		
C_6 glioma	Norepinephrine, isoproterenol	c
S-49	Cyclic AMP, isoproterenol	d
Secretory cells		
Adrenal medulla	Nicotine, veratridine, ionomycin	e
Mast cells	A23187, Ca^{2+} ion, compound 48/80	f
Platelets	Collagen, A23187, PGE_1, thrombin	g, h
Parotid	Isoproterenol, dibutyryl cyclic AMP, carbachol	i, j
Other cells		
Avian erythrocytes	Isoproterenol, PGE_1	k
Fat cells	Insulin, epinephrine, ACTH, glucagon	l, m
Hepatocytes	Insulin, glucagon, α-agonists, vasopressin	n, o
Adrenal cortical cells	ACTH	p

[a] J. Forn and P. Greengard, *Proc. Natl. Acad. Sci. U.S.A.* **75**, 5195 (1978).
[b] B. K. Kreuger, J. Forn, and P. Greengard, *J. Biol. Chem.* **252**, 2764 (1977).
[c] V. E. Groppi, Jr. and E. T. Browning, *Mol. Pharmacol.* **18**, 427 (1980).
[d] R. A. Steinberg, *Cold Spring Harbor Conf. Cell Proliferation* **8**, 179 (1981).
[e] C. M. Amy and N. Kirshner, *J. Neurochem.* **36**, 847 (1981).
[f] W. Seighart, T. Theoharidies, S. L. Alper, W. W. Douglas, and P. Greengard, *Nature (London)* **275**, 329 (1978).
[g] R. J. Haslam, J. A. Lynham, and J. E. B. Fox, *Biochem. J.* **178**, 397 (1979).
[h] R. M. Lyons and J. O. Shaw, *J. Clin. Invest.* **65**, 242 (1980).
[i] B. J. Baum, J. M. Freiberg, H. Ito, G. S. Roth, and C. R. Filburn, *J. Biol. Chem.* **256**, 9731 (1981).
[j] R. Jahn and H. D. Soling, *Proc. Natl. Acad. Sci. U.S.A.* **78**, 6903 (1981).
[k] S. A. Rudolph and P. Greengard, *J. Biol. Chem.* **255**, 8534 (1980).
[l] W. B. Benjamin and I. Singer, *Biochemistry* **14**, 3301 (1975).
[m] J. Avruch, G. R. Leone, and D. B. Martin, *J. Biol. Chem.* **251**, 1511 (1976).
[n] J. Avruch, L. A. Witters, M. C. Alexander, and M. A. Bush, *J. Biol. Chem.* **253**, 4754 (1978).
[o] J. C. Garrison, *J. Biol. Chem.* **253**, 7091 (1978).
[p] T. M. Koroscil and S. Gallant, *J. Biol. Chem.* **256**, 6700 (1981).

nique has been used. The sections below present protocols for performing these experiments based on the author's experience with isolated hepatocytes. The commentary provided in each section is intended to help the reader transfer these protocols to other systems. Interested readers may also wish to consult a similar article by Rudolph and Krueger.[2]

Preparation of ^{32}P-Labeled Cells and Extraction of Proteins

Labeling the Cells. Isolated rat hepatocytes are prepared by perfusing the liver with collagenase according to a modification of the method originally described by Berry and Friend.[3,4] The cells are resuspended in a Krebs–Ringer bicarbonate buffer with a reduced (0–0.1 mM) phosphate content in order to increase the specific activity of the added ^{32}PO$_4$. The final cell concentration is adjusted to about 20–30 mg protein/ml (2–3 × 10^7 cells/ml). No albumin is added to the cell suspensions because its presence would distort the polyacrylamide gels eventually used to resolve the phosphoproteins. Carrier-free ^{32}PO$_4^{3-}$ (New England Nuclear) is added to a concentration of 10–20 μCi/mg cell protein and 2–5 ml of cells are incubated in 25-ml plastic flasks under 95% O$_2$–5% CO$_2$ for 45 min to allow the specific activity of the [^{32}P]ATP in the cell to reach a constant value.[5] At 45 min, the hepatocytes are stimulated with the agent(s) of choice for the desired period (usually 3–15 min) and the incubations are stopped as described below.

Comments on Cell Labeling Protocols. The conditions chosen for labeling hepatocytes were designed with the following considerations in mind. (1) Since only 10–15% of the added ^{32}PO$_4^{3-}$ is incorporated into the cells, the phosphate content of the Krebs–Ringer bicarbonate buffer was reduced to obtain the highest possible specific activity of radioactive phosphate. Even under these conditions most experiments require about 0.5 mCi of ^{32}P per flask to label the cells adequately. No obvious problems with cell integrity or viability have been noted with hepatocytes in the reduced phosphate medium. However, this point should be considered when applying these protocols to other systems. (2) The choice of the equilibration time is essentially a compromise between labeling the cells' ATP pool and labeling total cell proteins. The hepatocyte membrane is fairly permeable to phosphate and the ATP pools equilibrate within 30–40

[2] S. A. Rudolph and B. K. Krueger, *Adv. Cyclic Nucleotide Res.* **10**, 107 (1979).
[3] J. C. Garrison and R. C. Haynes, Jr., *J. Biol. Chem.* **250**, 2769 (1975).
[4] M. N. Berry and D. S. Friend, *J. Cell Biol.* **43**, 506 (1969).
[5] J. C. Garrison, M. K. Borland, R. D. Moylan, and B. J. Ballard, *Cold Spring Harbor Conf. Cell Proliferation* **8**, 529 (1981).

min, a situation that may not exist in other cell types.[6,7] However, the proteins in the hepatocyte have *not* equilibrated in this time and continue to gain phosphate for at least 2 hr.[5,8] Since most of the phosphoproteins in hepatocytes and other cells are *not* responsive to hormones,[6,9] incubations beyond the time necessary to equilibrate the ATP pool increase the total background of radioactive proteins in the cell and on the autoradiographs used to visualize the labeled proteins. In contrast, most hormone-sensitive phosphoproteins equilibrate quickly with the [^{32}P]ATP pool because their phosphorylated sites are constantly subject to the action of kinases and phosphatases. For these reasons, we have elected to stimulate the cells with hormones shortly after the specific activity of the ATP pool reaches equilibrium.

Extraction of Phosphorylated Proteins. Once the cells have been stimulated with the appropriate agent, the next step is to extract the relevant proteins without their further modification by proteases, phosphatases, protein kinases, or other enzymes. The problem with this goal is that the investigator does not know the true state of the proteins in the intact cell. With this caveat in mind, two protocols used in the author's laboratory to extract cytosolic proteins are outlined below.

Homogenization of Cells in a Dounce Homogenizer. Incubations of labeled cells are stopped by a 10- to 20-fold dilution of the cells into an ice-cold medium containing 10 mM TES, pH 7.4, 20 mM sucrose, 100 mM NaF, 15 mM EDTA, and 2 mM EGTA. This dilution is performed by pouring 2–3 ml of cells into 35 ml of ice-cold buffer in a 50-ml plastic centrifuge tube. The cells are pelleted at 1000 g for 2 min, the supernatant is removed and discarded as radioactive waste, and the cells are homogenized at 0° in 0.5–1.5 ml of the above buffer with 30 strokes of a glass on glass Dounce homogenizer. The homogenizer is rinsed three times with buffer between each sample and the rinses are discarded as radioactive waste. The homogenate can be fractionated by a variety of methods (see below). Since we are interested in the effects of hormones on the phosphorylation of proteins in the cytoplasm, the homogenate is centrifuged at 100,000 g for 1 hr in a Beckman 50 Ti rotor to obtain a whole cell particulate fraction and the supernatant proteins. These fractions are then prepared for one or two-dimensional gel electrophoresis as described below.

Fractionation of Cells with Digitonin. Janski and Cornell have described a very useful method for fractionating liver cells into cytoplasmic

[6] V. E. Groppi, Jr. and E. T. Browning, *Mol. Pharmacol.* **18,** 427 (1980).
[7] S. A. Rudolph and P. Greengard, *J. Biol. Chem.* **255,** 8534 (1980).
[8] J. Avruch, L. A. Witters, M. C. Alexander, and M. A. Busch, *J. Biol. Chem.* **253,** 4754 (1978).
[9] J. C. Garrison and J. D. Wagner, *J. Biol. Chem.* **257,** 13135 (1982).

proteins and a particulate fraction in 5–10 sec.[10] In this procedure, 90–95% of the cytoplasmic proteins are released by lysing the plasma membrane with digitonin, then the rest of the cell structures are removed by rapid centrifugation at 13,000 g through a hydrocarbon layer in an Eppendorf microfuge. This method is superior because the proteins can be boiled in SDS within 1 min of fractionation, minimizing the chances of dephosphorylation or proteolysis. Moreover, cleanup of radioactive waste is simplified. The digitonin causes minimal interference with one-dimensional electrophoresis provided the sample is diluted at least sixfold into the sample buffer. Digitonin does not interfere at all with two-dimensional separations. This procedure is performed as follows. Before the experiment, the caps are removed from 2-ml Eppendorf tubes and they are loaded with 800 μl of a medium containing 4 mg/ml digitonin, 10 mM TES, pH 7.4, 50 mM NaF, 10 mM EDTA, 5 mM EGTA, and 200 mM sucrose. A hydrocarbon layer with a density of $\rho = 1.05$ is layered under the fractionation medium (90:110 bromodecane:bromododecane from Aldrich and Sigma, respectively). Usually 150 μl is used, although other volumes may be necessary with other systems. Since the degree of digitonin fractionation is very temperature sensitive,[10] the tube is maintained at 27° in a water bath. A new cap is made to seal the Eppendorf tube from a Fisher microfuge cap (Catalog No. 04-978-145) by making an X-shaped cut in the bottom of the cap with a razor. The X-shaped cut allows the insertion of the tip of an adjustable micropipet (Pipetman or Finnpipet) and the natural resilience of the plastic closes the cap upon withdrawal of the tip.

Hepatocytes are incubated as described above except that only 0.5 ml of cells is needed because the fractionation procedure uses only 200 μl of cell suspension. To perform the fractionation, the Eppendorf tube is placed in the centrifuge opposite a balance tube. The cells are removed from the incubation with a micropipet, the tip inserted through the X-shaped cut in the cap of the Eppendorf tube, and 200 μl of cell suspension is injected into the fractionation medium taking care *not* to disturb the hydrocarbon layer. The fractionation is allowed to proceed for 6–8 sec to ensure release of 90–95% of the cytoplasmic proteins and then the centrifuge is run for 30 sec. When the centrifuge stops, the supernatant proteins and/or the cell pellet are removed from the Eppendorf tube and prepared for electrophoresis as described below. The tube is discarded in the radioactive waste.

Preparation of Cells without Fractionation. A number of investigators have elected not to fractionate the cells before preparing the samples for

---

[10] A. M. Janski and N. W. Cornell, *Biochem. J.* **186**, 423 (1980).

gel electrophoresis.[7,11,12] This approach may be the method of choice when one is working with cultured cells or other systems containing small amounts of protein. To stop an incubation one can spin the cells for 30–60 sec in a microfuge, remove the supernatant medium, rapidly resuspend the cells to a final concentration of 0.5–1 mg protein/ml in a buffer containing 50 mM Tris, pH 6.8, 10% glycerol (v/v), 1% SDS (w/v), 1–2% 2-mercaptoethanol (v/v), and boil the mixture for 3–5 min. Alternatively, a concentrated stock solution of Tris, SDS, and 2-mercaptoethanol can be added directly to the cells followed by boiling. The latter method has the disadvantages that all of the $^{32}PO_4^{3-}$ in the incubation vessel is carried into the sample and the protein content is more dilute. Samples prepared by either method can be subjected to electrophoresis directly (see below).

Comments on the Fractionation Procedure. Our fractionation buffers were designed to be isotonic in an attempt to maintain the integrity of intracellular organelles. The high concentrations of NaF and EDTA are used to inhibit the actions of phosphatases and kinases, respectively, following disruption of the cell. A large number of control experiments have demonstrated that hormone-induced changes in the phosphorylation state of cytoplasmic proteins are stable in this medium for up to 24 hr at 5° and at least 15–30 min at 37°.[13,14]

Proteolysis is another potential problem to consider in the preparation of cell fractions for electrophoresis.[13] We have used the proteolysis inhibitors phenylmethylsulfonyl fluoride (PMSF), leupeptin, pepstatin, benzamidine, and others in the fractionation media. In most cases we have not been able to demonstrate clear effects of these compounds.

If cell fractions such as mitochondria, microsomes, or plasma membranes are of interest, it is important to realize that most subcellular fractionation schemes are performed in a medium of low ionic strength. The inclusion of 100 mM NaF and 15 mM EDTA in the buffers may cause particulate material to aggregate and to pellet at lower centrifugal forces than expected. Fortunately, some investigators have found that large amounts of F$^-$ and EDTA may not be necessary during fractionation procedures and have resolved phosphorylated proteins in mitochondrial, microsomal, nuclear, and membrane fractions.[6,15–17] It is difficult to rec-

---

[11] B. J. Baum, J. M. Freiberg, H. Ito, G. S. Roth, and C. R. Filburn, *J. Biol. Chem.* **256**, 9731 (1981).
[12] C. M. Amy and N. Kirshner, *J. Neurochem.* **36**, 847 (1981).
[13] J. C. Garrison, *J. Biol. Chem.* **253**, 7091 (1978).
[14] J. C. Garrison, M. K. Borland, V. A. Florio, and D. A. Twible, *J. Biol. Chem.* **254**, 7147 (1979).
[15] J. Avruch, G. R. Leone, and D. B. Martin, *J. Biol. Chem.* **251**, 1505 (1976).
[16] J. Avruch, G. R. Leone, and D. B. Martin, *J. Biol. Chem.* **251**, 1511 (1976).
[17] J. E. B. Fox, A. K. Say, and R. J. Haslam, *Biochem. J.* **184**, 651 (1979).

ommend any one scheme for maintaining the phosphorylation state of individual proteins while fractionating cells. Each tissue should be approached on an individual basis.

Finally, it should be noted that while the method of boiling the entire cell in SDS to prepare samples for electrophoresis eliminates the questions associated with cell fractionation, the approach does create some unique problems. For example, some of the inorganic phosphate added to the incubation is incorporated into lipids and nuclear material. If these compounds are not removed prior to SDS–polyacrylamide gel electrophoresis they tend to cause an increase in background radioactivity during the separation. This problem can obscure hormone-induced changes in minor proteins especially in the high-molecular-weight region of the gel.[6,11,12]

Important Controls

It is important to perform controls to ensure that (1) the specific activity of the [^{32}P]ATP in the cells is not changed by the addition of hormones, drugs or neurotransmitters; (2) the radioactive phosphate observed in the gels is incorporated into proteins; (3) the changes observed in phosphorylation state did not occur after homogenization of the cell; (4) the net phosphate content of the proteins is increased by hormone treatment. Methods for performing these controls have been published by a number of investigators.[6,8,12–15,18]

Specific Activity. The specific activity of the [^{32}P]ATP can be measured by a number of methods, but the most accurate and convenient is high-pressure liquid chromatography (HPLC). ATP samples can be prepared for this assay by incubating about 1 ml of hepatocytes with 0.1 μCi ^{32}PO$_4^{3-}$/mg cell protein, pelleting the cells for 15–30 sec in an Eppendorf microfuge, removing the supernatant, and adding 0.75 ml of 0.6 M perchloric acid to the cell pellet. After thorough mixing, the precipitated protein is removed by centrifugation, the supernatant placed in ice-cold test tubes and neutralized by addition of 75 μl of ice-cold 3.0 M KHCO$_3$. The solution is centrifuged to clarify it and the supernatant containing the ATP is used for the HPLC analysis. In our experiments the [^{32}P]ATP was purified on a Whatman Partisil-10 strong anion-exchange column by isocratic elution with 0.5 M KH$_2$PO$_4$ at pH 4.5,[19] however other columns and eluants may be suitable. The amount of ATP eluted from the column is monitored by its absorbance at 254 nm, the entire peak collected and the

[18] B. K. Krueger, J. Forn, and P. Greengard, *J. Biol. Chem.* **252**, 2764 (1977).
[19] W. L. Terasaki, G. Brooker, J. de Vellis, D. Inglish, C. Hsu, and R. D. Moylan, *Adv. Cyclic Nucleotide Res.* **9**, 33 (1977).

^{32}P content determined by liquid scintillation counting. The ratio of these two determinations gives the specific activity of the ATP in the α, β, and γ positions.

The specific activity of the γ-phosphate of the ATP can be determined by quantitative enzymatic transfer of the terminal phosphate to glucose with hexokinase.[20] Briefly, a 1- to 2-ml aliquot of the [^{32}P]ATP purified by HPLC is brought to pH 7.5 with KOH; then concentrated stock solutions are added to bring ATP to a concentration of 100 μM, glucose to 10 mM, MgCl$_2$ to 10 mM, and hexokinase to 3 U/ml. The reaction mixture is allowed to incubate at 25° to effect a quantitative conversion of ATP to glucose-6-PO$_4$. The extent of conversion can be monitored spectrophotometrically.[20] The radioactive glucose-6-^{32}PO$_4$ is separated from the [^{32}P]ATP by adding the reaction mixture to a small column containing 1 ml of Bio-Rad AG-1 and eluting the glucose-6-^{32}PO$_4$ with 2 ml of H$_2$O. The specific activity of the γ-phosphate in the ATP can be determined from the counts in the glucose-6-^{32}PO$_4$ and the concentration of ATP present in the original sample purified by HPLC.

Controls for the Nature of the Protein-Phosphate Bond. A number of investigators have described procedures for ensuring that the ^{32}PO$_4$ observed in gels is not in phospholipids, is covalently linked to protein, and has the chemical characteristics of the phosphoserine or phosphothreonine bond.[6,8,12–15,18] To test for the presence of labeled phospholipids, the cell extracts are precipitated with 5% trichloroacetic acid (TCA) and washed three times with a solution of 2:1 chloroform:methanol or 90% acetone. Following the extraction, the proteins are washed in ice-cold 1 M NaCl by resuspension and precipitation to remove traces of TCA and solvents. The protein precipitate can then be dissolved in the one-dimensional sample buffer as described above, heated, and subjected to electrophoresis. To test for acyl-bound phosphate, the proteins can be precipitated in 5% TCA at 0 and 95°, the TCA removed by washing with 1 M NaCl, and the proteins prepared for electrophoresis. The phosphoserine bond can be detected by treating the proteins with 0.5 N NaOH at 0° for 5 min or 90° for 1 min, precipitating the protein with excess TCA, washing to remove the TCA as above, and subjecting the proteins to electrophoresis. The serine–phosphate bond should be sensitive only to hot NaOH. Since acrylamide gels are stable to the conditions outlined above, it is also possible to perform these tests with the proteins fixed in the gels.[6]

Phosphorylation Occurring after Homogenization of the Cell. To determine if the observed phosphorylation pattern is occurring following homogenization of the cells, exogenous protein kinase and [γ-^{32}P]ATP are

[20] A. B. Leiter, M. Weinberg, F. Isohashi, M. F. Utter, and T. Linn, *J. Biol. Chem.* **253**, 2716 (1978).

added to an *unlabeled* cell extract and processed through the sample preparation, gel electrophoretic, and autoradiographic procedures. Any radioactivity incorporated into proteins following this control will have occurred posthomogenization. Since most protein kinases require Mg^{2+} ion, the inclusion of large amounts of EDTA in the extraction buffers should ensure that posthomogenization phosphorylation does not occur.[8,13]

Charge Shifts. It is possible to demonstrate increases in protein phosphorylation by changes in the position of the protein in the isoelectric focusing dimension of a two-dimensional gel because an increase in phosphate content causes the protein to focus at a new, more acidic isoelectric point.[21,22] Therefore, if a protein moves to a more acidic isoelectric point following hormone stimulation, its net phosphate content has increased. If the protein is in high enough concentration in the cell to be stained with Coomassie brilliant blue or silver, it is possible to demonstrate these charge shifts by comparing the stained protein patterns from control and hormone-treated cells.[9] In this situation, autoradiographs from ^{32}P-labeled cells are needed only to provide the initial location of the phospho-form of the protein.

Separation and Resolution of Proteins

Virtually every recent study utilizing the technique of intact cell protein phosphorylation has employed some type of gel electrophoretic technique to resolve the proteins in the cell extracts. While a detailed description of this powerful separation technique is beyond the scope of this article, a brief discussion of the advantages and disadvantages of various gel systems is included below.

One-Dimensional Gels. The most useful gel technique for studies of intact cell protein phosphorylation is the discontinuous one-dimensional SDS–polyacrylamide slab gel run by the method of Laemmli.[23] The theory of this technique has been published,[24-26] and an excellent practical description of the method can be found in a paper by Ames.[27] The advantages of this system are (1) it separates proteins by their subunit molecular

---

[21] P. H. O'Farrell, *J. Biol. Chem.* **250**, 4007 (1975).
[22] R. A. Steinberg, P. H. O'Farrell, U. Friedrich, and P. Coffino, *Cell* **10**, 381 (1977).
[23] U. K. Laemmli, *Nature (London)* **227**, 680 (1970).
[24] L. Ornstein, *Ann. N.Y. Acad. Sci.* **121**, 321 (1964).
[25] B. J. Davis, *Ann. N.Y. Acad. Sci.* **121**, 404 (1964).
[26] A. Chrambach, T. M. Jovin, P. J. Svenden, and D. Rodbard, in "Methods of Protein Separation" (N. Catsimpoolas, ed.), Vol. 2, p. 27. Plenum, New York, 1976.
[27] G. F. L. Ames, *J. Biol. Chem.* **249**, 634 (1974).

weights; (2) as many as 40 different samples can be compared in one run with the commonly used slab gel units; (3) proteins from most fractions of the cell will enter and run in the gel system; (4) quantitative comparisons of the changes in protein phosphorylation can be made easily with a spectrophotometer following autoradiography of the dried gel (see below). The major disadvantage of the one-dimensional gel system is that its resolving power is limited as compared to two-dimensional gel systems (see below).

Sample Preparation. We routinely prepare cytoplasmic protein samples for one-dimensional electrophoresis in the following manner. (1) The protein concentration of each sample is measured and adjusted to equality. It is useful to have the protein concentration in the range of 1–7 mg/ml. (2) The protein samples are diluted at least 1 part sample plus 4 parts of a buffer containing 62 mM Tris, pH 6.8, 12.5% glycerol (v/v), 1.25% SDS (w/v), 2.5% 2-mercaptoethanol (v/v), and 0.0025% bromophenol blue (w/v) and heated to 100° in a stoppered glass tube for 4–5 min. Twenty microliters of this sample is loaded into the sample well of the stacking gel with a Hamilton syringe for a total protein load in the range of 4–28 μg of protein per well. Equal amounts of protein are loaded in each well.

Conditions for Gel Electrophoresis. Gels are run in a Hoefer Model 500 or 600 slab gel apparatus under the following conditions: (1) 0.75-mm separating gels of 8–12% acrylamide are used, depending on the subunit molecular weights of the proteins of interest; (2) a 3% acrylamide stacking gel that extends at least 1 cm below the sample well is used; (3) gels are run at a constant current of 15–20 mA/slab and the gel plates are cooled with tap water at 10–15°; (4) upper and lower tank buffers are 0.025 M Tris, 0.1% SDS (w/v), and 0.190 M glycine, pH 8.3. We routinely use 0.75-mm gels because they stain, destain, and dry much more rapidly than thicker gels.

The finished gel is stained by immersion in a solution of 10% acetic acid (v/v), 25% isopropyl alcohol (v/v), 0.05% Coomassie brilliant blue (w/v) for at least 2 hr.[28] The staining solution is slowly depleted with use and should be replaced accordingly. If silver staining is desired, the method of Morrissey offers a number of advantages.[29] Coomassie blue-stained gels are destained by charcoal absorption of the stain in a commercial destainer (Hoefer Model 530) for about 1.5 hr, treated with 1% glycerol (v/v), 10% acetic acid (v/v) for 30–45 min, and vacuum dried onto filter paper using a commercial gel drier (Hoefer Model 540). The dried gel is used for autoradiography (see below). The results of a typical experiment are shown in Fig. 1.

[28] G. Fairbanks, T. L. Steck, and D. F. H. Wallach, *Biochemistry* **10**, 2606 (1971).
[29] J. H. Morrissey, *Anal. Biochem.* **117**, 307 (1981).

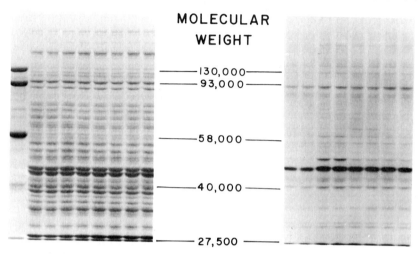

FIG. 1. A one-dimensional Coomassie blue-stained gel and its corresponding autoradiograph. Duplicate samples of cytosolic proteins from ^{32}P-labeled hepatocytes treated with vehicle (CON), glucagon (GLUC), norepinephrine (NOREPI), and the Ca^{2+} ionophore A23187 were run in adjacent wells. The autoradiograph was made by exposing the dried gel to Kodak Min-R film for 7 days. [Data from J. C. Garrison, J. Biol. Chem. **253**, 7091 (1978).]

Comments on the Gel Procedures. Consistency and optimal resolution in each gel run will be aided by using fresh solutions, washing the top of the separating gel about 30 min after polymerization with 0.1% SDS to remove unpolymerized acrylamide, and cleaning out the sample wells in the stacking gel before loading the samples. The latter steps can be quickly accomplished by loading a 10-ml glass syringe with 0.1% SDS, fitting a 26- or 27-gauge needle, and washing the gel surface with a stream of fluid.

Two-Dimensional Gels. Two-dimensional gels have the capability to resolve thousands of proteins.[21] While the increased complexity of this separation method hinders its routine use in protein phosphorylation studies, it does provide the following advantages: (1) The high resolution may display phosphorylation changes in major or minor proteins that are not apparent on one-dimensional gels; (2) the resolution of the system is great enough to allow the positive identification of the proteins in a crude cell extract without further purification; (3) when a protein's phosphorylation state is increased, the protein focuses at a more acidic position in the gel demonstrating a net gain of phosphate by the protein. Aside from its complexity the major disadvantages of the two-dimensional technique are that only one sample can be run per gel and that quantitation of the resultant autoradiographs usually requires computer techniques.

Sample Preparation. We routinely prepare samples for two-dimensional electrophoresis by the method of Anderson and Anderson,[30] using the following protocol: (1) The cytoplasmic proteins from either of the extraction procedures are adjusted so that the protein content is equal (in the range of 3–7 mg/ml), (2) 100 μl of sample is added to 200 μl of a solution containing 2% SDS (w/v), 10% glycerol (v/v), and 5% 2-mercaptoethanol (v/v) and heated to 100° in a stoppered glass tube for 4–5 min, and (3) the sample is cooled to 25° and 1 mg of urea/μl of sample is added (300 mg urea/300 μl sample). We use Schwarz-Mann Ultrapure urea. (4) Stock NP-40 (the detergent Nonidet P40, Particle Data Laboratories, Elmhurst, Illinois) is added to a final concentration of 3–4% (v/v) and LKB ampholines are added to a final concentration of 2% (v/v). (5) The ampholine distribution can be varied as needed. We usually add a mix of ampholines in the ratio of 1 part pH 3.5–10, 1.5 parts pH 5–7, and 2.5 parts pH 6–8. The same ampholine ratio is also used in the isoelectric focusing gel. (6) Equal amounts of protein (25–50 μg of protein in up to 50 μl) are loaded on each isoelectric focusing gel with a Hamilton syringe, overlayed with 20 μl of 8 M urea containing 1% ampholines, and run according to O'Farrell or Garrels.[21,31]

Conditions for Gel Electrophoresis. Isoelectric focusing gels are prepared and run in tubes with a 2.2 mm i.d. as described by O'Farrell with the modification that the acrylamide concentration is reduced to 3% and the NP-40 concentration raised to 3%. Gels are run in the isoelectric focusing dimension at 500 V for 16 hr and 1000 V for 1–2 hr. SDS–polyacrylamide slabs (0.75 mm) of 10% acrylamide are prepared and run as described above for the one-dimensional technique. We usually run four to six slabs at a time in two or three Hoefer Model 500 slab gel units. It is our usual practice to run the dye front off the bottom of the gel for 3–5 min to allow ampholines to exit from the gels. Two-dimensional gels are stained, destained, and dried for autoradiography as described above. The results of a typical experiment are shown in Fig. 2.

Comments on the Two-Dimensional Gel Technique. It is very important that the isoelectric focusing gels be properly polymerized; therefore, we use fresh TEMED and ammonium persulfate. Another point at which care is required is in the transfer of the first dimension focusing gel to the second dimension slab gel. If the O'Farrell system is used, best results are obtained if the stacking gel is high enough in the plates that the focusing

---

[30] L. Anderson and N. G. Anderson, *Proc. Natl. Acad. Sci. U.S.A.* **74**, 5421 (1977).
[31] J. I. Garrels, *J. Biol. Chem.* **254**, 7961 (1979).

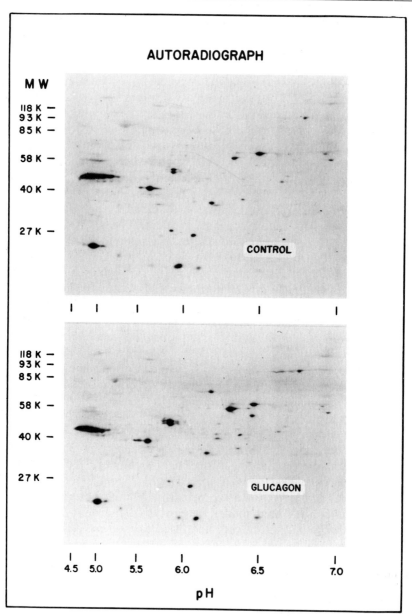

FIG. 2. Autoradiographs from hepatic cytosolic proteins separated by two-dimensional electrophoresis. The proteins were isolated from control cells and those treated with 100 nM glucagon for 4 min. The dried gels were exposed to Dupont Chronex-4 film for 5 days. [Data from J. C. Garrison and J. D. Wagner, *J. Biol. Chem.* **257**, 13135 (1982).]

gel can sit *directly* on the stacking gel, allowing very little of the agarose used in this step to come between the focusing gel and the stacking gel. We accomplish this by pouring the stacking gel to the top of the plates and overlaying it with a Teflon rod. This leaves a semicircular depression in the top of the stacking gel, ensuring close contact between the focusing gel and the stacking gel. An excellent alternative method for making this transfer *without* agarose has been published by Garrels.[31] In this system, the focusing gel is of the correct size to be forced down between the glass plates forming the separating gel. Many descriptions of the particulars of the two-dimensional technique are available.[21,31–33] Recently an entire issue of *Clinical Chemistry* was devoted to two-dimensional electrophoresis.[33a]

Identification of Phosphoproteins

Identification of the functional role of the proteins phosphorylated in the intact cell in response to various stimuli is an important goal. The main strategy is to make reasonable guesses as to the possible identity and function of a protein whose phosphorylation changes with the agonist studied. Usually, the major tools available are the subunit molecular weight of the phosphoprotein and knowledge of the biochemistry of the tissue under study. One can then use the appropriate purification techniques to determine if the protein of interest is, in fact, coincident with the phosphoprotein in the gel system. We have used one or more of the following criteria to identify hepatic proteins in the gel patterns. (1) The purified protein of interest migrates with the phosphoprotein in the one-dimensional and/or two-dimensional gel patterns. (2) The phosphoprotein can be purified from labeled cell extracts via the known purification scheme for that protein. (3) The phosphoprotein of interest can be precipitated from the cell extract with a specific antibody to that protein. The combined use of one of these criteria and the resolving power of a two-dimensional gel provides convincing identifications. A more detailed description of strategies for protein identification can be found in a discussion by Anderson and Anderson.[33a]

Quantitation of Changes in Phosphorylation

Quantitation of the changes that occur in the phosphorylation state of the proteins resolved on the SDS–polyacrylamide gels can be accomplished by slicing and counting the gels or by autoradiography.

[32] N. G. Anderson and N. L. Anderson, *Anal. Biochem.* **85**, 331 (1978).
[33] N. L. Anderson and N. G. Anderson, *Anal. Biochem.* **85**, 341 (1978).
[33a] *Clinical Chemistry* **28**, 737–1092 (part II) (1982).

Slicing and Counting Gels. The lanes in the stained one-dimensional slab gel can be cut from the gel with a razor blade and the strip sliced into small pieces with one of the commercial gel slicing devices (i.e., Hoefer Model 280). This will produce a series of 100–200 slices about 1 mm wide that can be counted in a liquid scintillation counter. The only advantage of this method is that it provides a direct measurement of counts per minute. The disadvantages are that it is tedious, it destroys the gel, and it gives very poor resolution as compared to autoradiography. It would be an exhausting task to perform this analysis on a two-dimensional gel.

Autoradiography. The ^{32}P-labeled proteins resolved in one- or two-dimensional gels can be detected easily by autoradiography with a number of medical X-ray films. This method is very sensitive (0.5–1 cpm can be observed), provides excellent resolution, and can be quantitative provided care is taken not to saturate the film with overly long exposures.

There is an extremely wide range of X-ray films that are suitable for detection of ^{32}PO$_4$. It is most convenient to use films that can be developed in the automatic processors used by Radiology Departments.

Intensifying screens can be used to increase the sensitivity of the exposure up to 10-fold with the penalty of reducing resolution.[34] We regularly use films of different speeds with or without intensifying screens depending on the sensitivity and/or resolution required. Our usual procedures are as follows: (1) The dried gel is exposed to a fast film such as Kodak X-OMAT AR (XAR) for 12–30 hr to get a quick assessment of the results. (2) The gel is then exposed to a slower, higher resolution film such as Kodak X-OMAT K (XK-1) or Min R for 3–7 days to obtain an image suitable for quantitation. (3) Intensifying screens may be used for the first exposure but they are not used for quantitative work because screens cause the film response to deviate from linearity (see below). We presently use standard medical X-ray cassettes for autoradiography because they provide a light-tight frame that gives excellent film/gel contact and are easily stored and transported. Table II provides the characteristics of the X-ray films that we have found useful. Investigators with no experience with autoradiography may wish to obtain Kodak's pamphlets M3-508 and M5-15 on biomedical imaging and M3-138 on the speeds of different film/screen combinations.

Densitometry of Autoradiographs. Before describing these techniques, it should be stressed that the human eye is extremely sensitive to shades of density. It is rare that quantitation of the phosphorylation patterns reveals a density change that is not obvious on visual inspection. All X-ray films we have tested respond to ^{32}P with a linear relationship between counts per minute and optical density up to a density of about 2.5 OD. Therefore it is possible to scan an autoradiograph in a spectropho-

---

[34] R. Swanstrom and P. R. Shank, *Anal. Biochem.* **86**, 184 (1978).

TABLE II
CHARACTERISTICS OF SOME MEDICAL X-RAY FILMS

Film	Single or double emulsion	Speed[a]	Usual exposure time (days)	Resolution	Used with intensifying screens	Optical background at 550 nm
Kodak X-OMAT AR	Double	Very fast	0.5–1.5	Poor	Yes[b]	0.50 OD
Kodak X-OMAT K Dupont Chronex 4 3M XUD	Double	Medium	3–6	Good	Yes[b]	0.25–0.35 OD
Kodak Min R Kodak Ortho M	Single	Slow	7–15	Excellent	Yes[b]	0.20 OD

[a] The speeds of these films measured with ^{32}P and expressed relative to Kodak X-OMAT AR (the fastest) are: X-OMAT AR = 1.00; Chronex 4, X-OMAT K, XUD ≅ 0.24; Ortho M = 0.18; Min R = 0.092.

[b] Intensifying screens are useful only when the film is sensitive to the wavelength of light emitted by the screens. See Kodak Bulletin M3-138 for proper film/screen combinations.

tometer to quantitate the effects of hormones on the phosphorylation of proteins.[2,13] We quantitate autoradiographs from one-dimensional gels in a Gilford Model 2520 20-cm gel scanner mounted on a Gilford Model 250 spectrophotometer. The X-ray film is scanned at a wavelength of 550 nm and a slit width of 0.1 mm is used in the scanning attachment. The lane of interest is cut from the X-ray film with a paper cutter so that straight edges are obtained. Care is taken to mount the film in the scanning attachment so that the light beam travels through the center of the bands. (The bands tend to taper at the edges leading to an artifactually low density reading.) It should be noted that the most accurate scans are made using high-resolution single-emulsion X-ray films exposed to sharply resolved proteins. No optical tricks will improve the resolution of an overexposed film from a poorly run acrylamide gel. The optical density on the X-ray film is recorded as the film moves across the light path and the hormonally induced changes can be compared with the control by overlaying the records on a light box. Peaks of interest should be compared by aligning the valleys on either side of the peak. This method compensates for changes in film background along different areas of the lane.[2,12,13] Since the peaks compared are essentially equalateral triangles with the same base and differing heights, the changes in phosphorylation are directly related to changes in the peak heights. The linear relationship between peak height and peak area has been verified.[2,13]

The quantitation of autoradiographic data from two-dimensional gels is a much more complex task. While it is possible to perform this job manually,[6] quantitation of more than a few films requires a high-speed, rotating drum densitometer such as an Optronics P-1000 and computer techniques to manage the tremendous amount of data generated from the autoradiographs. While this subject is beyond the scope of this article, computer software has been developed to accomplish these tasks and the interested reader is referred to the original references.[31,33a,35-38]

Photography of Gels and Autoradiographs

Faithful photographic reproduction of Coomassie blue-stained gels and autoradiographs can be obtained by using the following conditions as starting points.

Gels Dried on Filter Paper. Coomassie blue stained gels are photographed using Kodak Contrast Process Pan film (4 × 5 in.). The filters are K-2, O-2, or A-2, depending on the contrast needed. The film is processed in Kodak HC-110 developer for 6.5 min at 68°F. Prints are made with Kodak Kodabrome paper using Grades 1–5, again depending on the contrast desired.

Autoradiographs. X-Ray film is photographed using Kodak Tri-X Orthochromatic film (4 × 5 in.) with white fluorescent back-lighting from a lightbox. K-2 or O-2 filters are used, depending on the contrast desired. The film is developed in HC-110 developer for 13 min at 68°F and printed as described above.

Wet Gels. Wet gels are photographed on a lightbox modified to give dark-field illumination, on a white background using Kodak Tri-X Orthochromatic Film with K-2 or Y-2 filters depending on the contrast desired. The film is developed in Kodak HC-110 for 13 min at 68°F and printed on Kodak Polycontrast RC paper.

Acknowledgments

This work was supported by NIH Grant No. AM-19952 and Career Development Award AM-00491. The author thanks Pat Barnett of Eastman Kodak Company for helpful discussions about X-ray films and Anne Russell and Mike Pittard of the University of Virginia Department of Biomedical Communications for sharing their methods for photography of gels and autoradiographs.

[35] J. Bossinger, M. J. Miller, K. Vo, E. P. Geiduschek, and N. Xuong, *J. Biol. Chem.* **254**, 7986 (1979).
[36] P. Lemkin, C. Merril, L. Lipkin, M. Van Keuren, W. Oertel, M. Shapiro, M. Wade, M. Schultz, and E. Smith, *Comput. Biomed. Res.* **12**, 517 (1979).
[37] J. Taylor, N. L. Anderson, B. P. Coulter, A. E. Scandora, and N. G. Anderson, *in* "Electrophoresis '79" (B. J. Radola, ed.), p. 329. De Gruyter, Berlin, 1980.
[37a] N. L. Anderson, J. Taylor, A. E. Scandora, B. P. Coulter, and N. G. Anderson, *Clin. Chem.* **27**, 1807 (1981).
[38] J. C. Garrison and M. L. Johnson, *J. Biol. Chem.* **257**, 13144 (1982).

[5] Peptide Mapping and Purification of Phosphopeptides Using High-Performance Liquid Chromatography

By HENNING JUHL *and* THOMAS R. SODERLING

Protein phosphorylation is now firmly established as an important mechanism for regulation of diverse cellular functions, such as metabolism, membrane transport, muscle contraction, protein synthesis, neuronal activity, and cellular transformation by viruses.[1] Many phosphoproteins have multiple phosphorylation sites. One of the most complicated of these is skeletal muscle glycogen synthase, which contains at least seven phosphorylation sites per subunit (Fig. 1). These multiple phosphorylations can be catalyzed by a number of different protein kinases.[2,3] In order to establish the specificities of the kinases for the several phosphorylation sites, we developed a HPLC methodology which is rapid and generally gives good recoveries of phosphopeptides.

Materials

Skeletal muscle glycogen synthase and the catalytic subunit of cAMP-dependent protein kinase were purified from rabbit muscle as described by Soderling *et al.*[4] Casein kinase II was a gift from Drs. Traugh and Hathaway (University of California, Riverside) and was also purified from rabbit skeletal muscle.[3] Phosphorylase kinase was purified from rabbit skeletal muscle.[5] Calmodulin (CaM)-dependent synthase kinase from rabbit liver (pp. 413–424 of Ref. 1) and cAMP-independent glycogen synthase kinase were purified from rabbit liver. Synthase kinase 3 was a gift from Dr. Philip Cohen (University of Dundee, Scotland) and synthase kinase F_A was a gift from Dr. J. R. Vandenheede (Katholieke Univ. Leuven). *N*-Tosyl-L-phenylalanine (TPCK) trypsin was purchased from Worthington.

Phosphorylation of glycogen synthase was conducted using 50 mM Tris (pH 7.5), 25 mM NaF, 10 mM magnesium acetate, 0.1 mM [γ-^{32}P]ATP (100–1000 cpm/pmol) or 0.4 mM [γ-^{32}P]GTP (2000–5000 cpm/

---

[1] O. M. Rosen and E. G. Krebs, *Cold Spring Harbor Conf. Cell Proliferation* **8** (1981).
[2] T. R. Soderling and B. S. Khatra, *in* "Calcium and Cell Function" (W. Y. Cheung, ed.), Vol. III. Academic Press, New York, 1982.
[3] P. J. Roach, *Curr. Top. Cell. Regul.* **20**, 45 (1981).
[4] T. R. Soderling, M. F. Jett, N. J. Hutson, and B. S. Khatra, *J. Biol. Chem.* **252**, 7517 (1977).
[5] T. Hayakawa, J. P. Perkins, D. A. Walsh, and E. G. Krebs, *Biochemistry* **12**, 567 (1973).

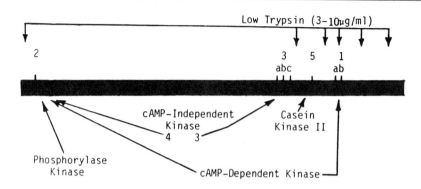
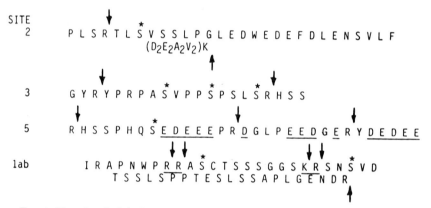

Fig. 1. Phosphorylation sites in skeletal muscle glycogen synthase. The approximate location of the phosphorylation sites in the synthase subunit as well as the sites of limited tryptic cleavage and specificities of synthase kinases are shown in the top half. In the bottom part the amino acid sequences[8] of the phosphorylation sites (marked by asterisk) and the sites of extensive tryptic hydrolysis are given.

pmol), 0.2 to 2 mg/ml glycogen synthase, and the appropriate kinase. In the absence of added kinase there was no detectable phosphorylation of synthase.

Proteolysis of ^{32}P-Labeled Synthase

In most experiments the ^{32}P-labeled synthase was digested for 5 hr with 1 mg/ml trypsin. In some instances advantage was taken of the known sensitivity of glycogen synthase to limited proteolysis by trypsin. Trypsin at 3–10 µg/ml rapidly (less than 15 min) reduces the subunit size of synthase from 90,000 to about 73,000 by hydrolysis of several phospho-

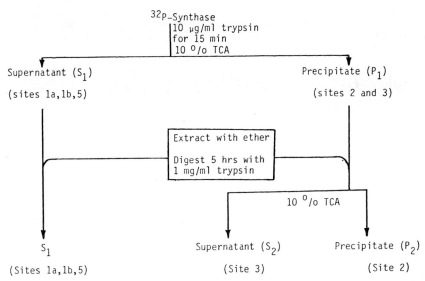

FIG. 2. Scheme for the differential trypsinization of ^{32}P-labeled synthase. See text for details.

peptides from the carboxy-terminus[6] (Fig. 1). At this very low trypsin concentration less than 15% of site 3 was solubilized.[7] The 10% trichloroacetic acid (TCA) precipitate (P_1), containing sites 2 and 3, and the TCA supernatant (S_1), containing sites 1a, 1b, and 5, were washed with ether to extract the TCA and were further digested with 1 mg/ml trypsin for 5 hr. The redigested P_1 was again treated with 10% TCA which precipitated site 2 (P_2) leaving site 3 in the supernatant (S_2). This procedure is summarized in Fig. 2. The second trypsinization for 5 hr at 1 mg/ml was found to be optimal since more prolonged digestion (e.g., 12 hr) gave rise to further degradation of sites 1b and 2, perhaps due to slight chymotryptic activity in the trypsin. Digestion for less than 3 hr resulted in incomplete digestion of site 5, probably due to the large number of proline residues.

HPLC Separation of Phosphorylation Sites

Several different gradient systems were tested to optimize separation and recoveries of the various ^{32}P-labeled peptides. Either 1-propanol or acetonitrile gradients in 0.1% trifluoroacetic acid (TFA) were found to be satisfactory. Reversed phase HPLC was performed using a Beckman Model 421 instrument and a Beckman Ultrasphere-ODS (C_{18}) column

---

[6] T. R. Soderling, *J. Biol. Chem.* **251**, 4359 (1976).
[7] P. J. Parker, N. Embi, F. B. Caudwell, and P. Cohen, *Eur. J. Biochem.*, in press (1982).

(0.46 × 25 cm) at flow rates of 1 ml/min. Recoveries of ^{32}P-labeled peptides were in the range of 70–85% except for site 2, which was variable between 50 and 70%. Slightly better recoveries of site 2 were obtained using 1-propanol than acetonitrile.

Figures 3 and 4 present illustrative HPLC chromatograms of ^{32}P-labeled peptides from synthase phosphorylated by several different kinases. Analogous chromatograms were obtained using other kinases to

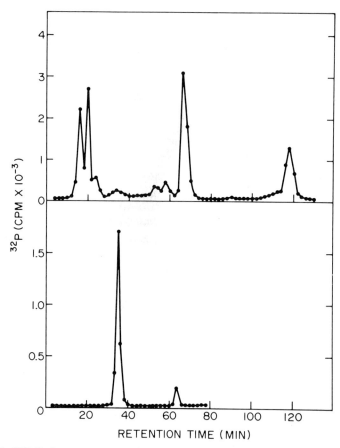

FIG. 3. HPLC chromatograms of ^{32}P-labeled peptides from glycogen synthase. ^{32}P-labeled synthase prepared using cAMP-dependent kinase (top) or casein kinase II (bottom) was digested 5 hr with 1 mg/ml trypsin and chromatogramed on reverse phase HPLC in 0.1% TFA with an acetonitrile gradient of 0–5% (10 min), 5–38% (100 min), and 38–50% (5 min). Peaks at 18–22 min represent site 1a; 35 min, site 5; 65–68 min, site 1b; and 118–122 min, site 2.

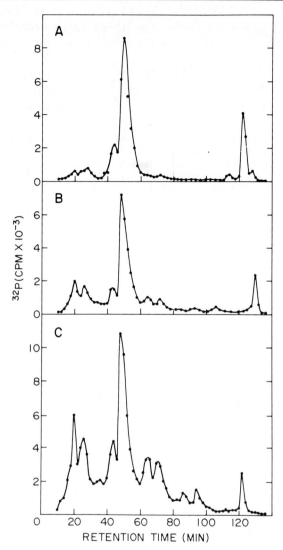

FIG. 4. HPLC chromatograms of ^{32}P-labeled peptides from glycogen synthase. Synthase was phosphorylated using [γ-^{32}P]GTP and three different partially purified preparations of synthase kinase 3: (A) kinase 3 from Dr. P. Cohen; (B) kinase F_A from Dr. Vandenheede; (C) kinase 3 prepared in our laboratory. The ^{32}P-labeled synthase was digested withs trypsin and chromatogrammed as in Fig. 3.

TABLE I
RETENTION TIMES OF SYNTHASE PHOSPHORYLATION
SITES FROM HPLC[a]

	Phosphorylation sites				
Gradient	1a	5	3	1b	2
Acetonitrile	20, 24	35	40–55	68	120
1-Propanol	24	40	44–62	70	116

[a] Acetonitrile gradient (1 ml/min) was 0–5% (10 min), 5–38% (100 min), 38–50% (5 min) in 0.1% TFA. Propanol gradient (0.5 ml/min was 0–25% (100 min), 25–50% (10 min) in 0.1% TFA.

determine the elution positions of the phosphorylation sites (Table I), and the specificities of the protein kinases are shown in Table II.

The identity of each major ^{32}P-labeled peptide, except site 3, was established by amino acid composition after further purification to homogenity on HPLC as well as their elution positions on Sephadex G-50 (not shown). Site 1a eluted from HPLC as two peaks since trypsin can cleave site 1a either between the arginine–alanine or arginine–arginine residues. In the latter case, the N-terminal arginine is quite resistant to further cleavage, thereby giving rise to two peptides containing site 1a.

Identification of site 3 is still somewhat tentative due to the lack of a highly purified preparation of kinase 3. All kinase 3 preparations that we tested contained significant amounts of kinase 4 and the catalytic subunit

TABLE II
SPECIFICITIES OF PROTEIN KINASES FOR THE PHOSPHORYLATION SITES
IN GLYCOGEN SYNTHASE AS DETERMINED BY HPLC[a]

Protein kinase	mol ^{32}P/subunit	Glycogen synthase phosphorylation site				
		1a	1b	2	3	5
cAMP-dependent	2.0	+	+	+		
Casein kinase II	0.4					+
Phosphorylase kinase	0.5			+		
cAMP-independent	0.5			+		
CaM-dependent	1.0		+	+		
Synthase kinase 3	0.3				+	

[a] Specificities for the kinases were determined from HPLC chromatograms analogous to those shown in Figs. 3 and 4.

of cAMP-dependent protein kinase. The identity of the phosophopeptides eluting between 40 to 55 min as site 3 was based on the fact that (1) they were only seen using preparations containing kinase 3 activity, (2) they were not released from glycogen synthase by 2–10 µg/ml trypsin but were solubilized using trypsin at 1 mg/ml, and (3) they were the only major peptides detected using kinase 3 preparations and [γ-^{32}P]GTP. Thus, all available data are consistent with the phosphopeptides between 40 and 55 min representing site 3. However, definitive proof will require the purification of these peptides for amino acid analysis. One would expect up to three phosphopeptides since site 3 can exist as mono-, di-, and triphosphopeptides. However, with *in vitro* phosphorylation one sees predominantly the triphosphopeptide, some diphosphopeptide, and very little monophosphopeptide.[9]

This HPLC method was utilized to map the sites phosphorylated *in vivo* using synthase purified from normal rabbits. This synthase, which contained about 2.8 moles of phosphate per mole of subunit,[10] was subjected to differential trypsinization as outlined in Fig. 2. Samples S_1, S_2, and P_2 were run on HPLC, and the fractions were assayed for inorganic phosphate. As shown in Fig. 5, S_1 had peaks corresponding to sites 1a (22 min), 5 (35 min), and 1b (66 min). Sample S_2 had 3 peaks between 40 and 60 min that are probably the phosphopeptides containing site 3. Fraction P_2 had a peak at 80 min representing site 2. Thus, it appears that all of these sites are phosphorylated *in vivo* and can be separated using this HPLC methodology.

Selective Purification of Phosphopeptides

It is often of interest to know the amino acid sequence surrounding the phosphorylation site since such data can give information about the specificity determinants of protein kinases. The following procedure was developed to give selective purification of peptides containing [^{32}P]serine.

Alkaline β-elimination of phosphoserine in proteins followed by nucleophilic addition to the unsaturated dehydroalanine residue [see Eq. (1)] has been used to identify such modified residues.[11,12] Since the conversion of a phosphoserine to a dehydroalanine residue by β-elimination would be expected to have a considerable influence on the hydrophobicity of a

[8] C. Picton, A. Aitken, T. Bilham, and P. Cohen, *Eur. J. Biochem.*, in press (1982).
[9] B. A. Hemmings, D. Yellowlees, J. C. Kernohan, and P. Cohen, *Eur. J. Biochem.* **119**, 443.
[10] V. S. Sheorain, B. S. Khatra, and T. R. Soderling, *J. Biol. Chem.* **257**, 3462 (1982).
[11] D. L. Simpson, J. Hranisavljevic, and E. A. Davidson, *Biochemistry* **11**, 1849 (1972).
[12] J. R. Whitaker, *Am. Chem. Soc. Symp. Ser.* **123**, 145 (1980).

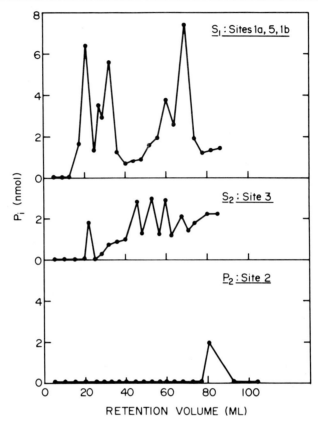

FIG. 5. HPLC chromatogram of synthase phosphorylated *in vivo*. Synthase (2.8 mg), purified from control rabbits and containing 2.8 mol P_i/mol subunit, was subjected to differential trypsinization as outlined in Fig. 2. Each sample (S_1, S_2, P_2) was subjected to reverse phase HPLC, and fractions were assayed for inorganic phosphate. The HPLC gradients for S_1 and S_2 were as in Fig. 3 and for P_2 was 0–30% (70 min), 30–50% (3 min) of 1-propanol in 0.1% TFA.

small peptide, this procedure seemed well suited for the selective purification of [32P]-labeled peptides using reverse phase HPLC. A protein containing [32P]serine would be proteolytically hydrolyzed (e.g., trypsin), and the peptide mixture subjected to reverse phase HPLC. Those fractions containing the [32P]-labeled peptide, together with contaminating peptides, would be subjected to alkaline β-elimination and rechromatography on HPLC. The contaminating peptides would elute in the same position as in the first chromatogram. However, the peptide which now contained the dehydroalanine residue rather than the phosphoserine would be more

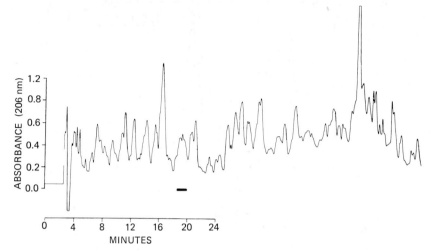

FIG. 6. HPLC chromatogram of ^{32}P-labeled Kemptide plus tryptic digest of BSA. A mixture of 35 nmol each of ^{32}P-labeled Kemptide and tryptic digest of BSA was chromatogrammed on reversed phase HPLC in 0.1% TFA with an acetonitrile gradient of 5–15% (15 min), 15–25% (20 min), 25–50% (20 min). Fractions 19–20 (bar) contained the radioactivity and were pooled. Reproduced by permission from Soderling and Walsh.[12a]

hydrophobic and would elute with a longer retention time. The purified dehydroalanine containing peptide could be reacted with [^{35}S]sulfite which converts the dehydroalanine to [^{35}S]cysteic acid, thereby identifying the original site of phosphorylation.

$$\begin{array}{c}\text{O}\\\ominus\text{O}-\overset{\|}{\text{P}}-\text{O}^{\ominus}\\|\\\text{O}\\|\\\text{CH}_2\\|\\-\text{NH}-\text{CH}-\overset{\|}{\underset{\text{O}}{\text{C}}}-\end{array}\xrightarrow{\ominus\text{OH}}\begin{array}{c}\text{CH}_2\\\|\\-\text{NH}-\overset{}{\underset{}{\text{C}}}-\overset{\|}{\underset{\text{O}}{\text{C}}}-\end{array}\xrightarrow{\text{SO}_3^{\ominus}}\begin{array}{c}\text{O}\\\ominus\text{O}-\overset{\|}{\text{S}}-\text{O}^{\ominus}\\|\\\text{CH}_2\\|\\-\text{NH}-\text{CH}-\overset{\|}{\underset{\text{O}}{\text{C}}}-\end{array}\quad(1)$$

This procedure was tested by adding to a tryptic digest of bovine serum albumin an equimolar amount (35 nmol) of the synthetic peptide L-R-R-A-S-L-G (Kemptide, Ref. 13) which had been phosphorylated at the serine residue using the cAMP-dependent protein kinase. This mixture was subjected to reverse phase HPLC (Fig. 6), and the fractions were counted for radioactivity. The ^{32}P-labeled Kemptide eluted at 17–18% acetonitrile (19–20 min), and these fractions were pooled and lyophilized.

[12a] T. R. Soderling and K. Walsh, *J. Chromatogr.* **253**, 243 (1982).
[13] B. E. Kemp, D. J. Graves, E. Benjamini, and E. G. Krebs, *J. Biol. Chem.* **252**, 4888 (1977).

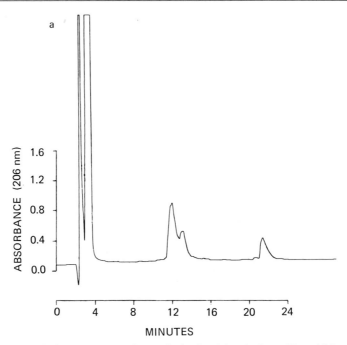

FIG. 7. HPLC chromatograms after β-elimination (a) and after sulfite addition (b). The radioactive fractions from Fig. 6 were subjected to alkaline β-elimination (see text) and chromatogrammed (left). The peak at 22 min was pooled, subjected to sulfite addition (see text), and chromatogrammed (right). Both chromatograms used a 10–30% acetonitrile gradient in 0.1% TFA over 30 min. Reproduced by permission from Soderling and Walsh.[12a]

The peptides were dissolved in 0.1 ml of 0.1 N NaOH, and $CaCl_2$ was added to 0.1 M. It has been shown that the β-elimination reaction is greatly accelerated in the presence of calcium.[14] The β-elimination was accomplished by incubation at 40° for 2 hr, and the sample was again run on HPLC. The contaminating peptides eluted at about 18% acetonitrile (12–14 min) and the dehydroalanine containing peptide eluted at about 25% acetonitrile (22 min) (Fig. 7a). This latter peak was lyophilized, dissolved in 0.1 ml of 0.1 N NaOH plus 0.1 M Na_2SO_3, and incubated for 2 hr at 37° to achieve the sulfite addition. This mixture was again run on HPLC (Fig. 7b), and two peaks eluting at about 13 min were pooled separately, lyophilized, and subjected to amino acid analysis. Both peaks had the amino acid content expected for the cysteic acid derivative of Kemptide (not shown), including 1 mole of cysteic acid, and probably represent the

[14] T. A. Sundararajan, K. S. Kumar, and P. S. Sarma, *Biochem. Biophys. Acta* **28,** 148 (1958).

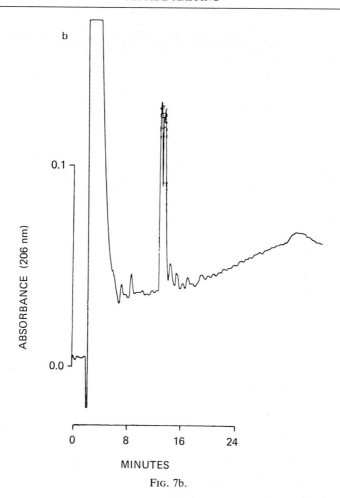

Fig. 7b.

diastereoisomers formed by the sulfite addition reaction. Recoveries of peptides from the HPLC were 50–80%. The sulfite addition was time, temperature, and concentration dependent. The addition reaction should be complete within 2 hr at 40° using 0.1 M Na_2SO_3, but may vary depending on the nature of the peptide.

Certain precautions must be kept in mind to avoid potential problems (see Ref. 12). The phosphopeptides should be reduced and subjected to performic acid oxidation to convert cysteine to cysteic acid since cysteine can readily undergo β-elimination in alkali to form dehydroalanine. Exposure to alkali should be kept to the minimum required time since a number of reactions such as conversion of arginine to ornithine, hydrolysis of certain peptide bonds (e.g., G–G and G–S), and β-elimination of O-glycosidic bonds can occur on prolonged alkaline conditions. Our experience with phosphopeptides indicates that β-elimination is usually complete in 2–4 hr at 40° using 0.1 N NaOH and 0.1 M $CaCl_2$. Another potential problem is reaction of adjacent lysine residues with the dehydroalanine to form lysinoalanine. Sequence analysis of the purified cysteic acid-containing peptide would be best accomplished using a solid phase sequenator since the polar thiazolinone derivative of cysteic acid does not extract satisfactorily from the liquid sequenator.

A potentially important application of this β-elimination and [^{35}S]-sulfite addition technique would be to peptide map *in vivo* phosphorylation sites. One could purify the phosphoprotein after *in vivo* phosphorylation. The purified protein would be proteolytically digested, performic acid oxidized, subjected to alkaline β-elimination and [^{35}S]sulfite addition. The elution times of ^{35}S-labeled peptides would be mapped on HPLC. Since the cysteic acid derivized peptide has almost the same retention time as the phosphoserine peptide, labeled peptides from the *in vivo* phosphorylation can be readily related to elution times of peptides derived from the ^{32}P-labeled protein *in vitro*.

Acknowledgments

This work was supported in part by NIH Grant AM 17808. The authors wish to thank Ms. Martha Bass for her excellent technical assistance and Dr. Kenneth Walsh for his consultation on certain aspects of this project.

Section II

Purification and Properties of Specific Protein Kinases

A. Cyclic Nucleotide-Dependent Protein Kinases
Articles 6 through 25

B. Calcium-Dependent Protein Kinases
Articles 26 through 32

C. Cyclic Nucleotide and Calcium-Independent Protein Kinases
Articles 33 through 39

D. Tyrosine-Specific Protein Kinases
Articles 40 through 43

[6] Catalytic Subunit of cAMP-Dependent Protein Kinase

By Erwin M. Reimann and Richard A. Beham

The cAMP-dependent protein kinase has been shown to phosphorylate several proteins.[1] Some enzymes, such as glycogen synthase and phosphorylase kinase, can be phosphorylated in more than one site and by more than one kinase.[2] Highly purified preparations of the cAMP-dependent protein kinase are particularly useful in studies on proteins which are phosphorylated at multiple sites and/or by multiple kinases. Preparations of the catalytic subunit of the cAMP-dependent protein kinase are used frequently because the catalytic subunit is easy to prepare and because the regulatory subunit has no known influence on catalytic activity when the concentration of cAMP is optimal. Preparation of the catalytic subunit involves dissociation of the subunits in the presence of cAMP:

$$R_2C_2 + 4\ cAMP \rightarrow R_2cAMP_4 + 2C$$

where R and C refer to regulatory and catalytic subunits, respectively. It has been shown that each mole of R dimer binds 4 mol of cAMP.[1] Isolation of pure C is facilitated by the fact that the catalytic subunit is a basic protein whereas R_2 and R_2C_2 are quite acidic.[3] If R_2C_2 is adsorbed to DEAE-cellulose, C can be selectively eluted by washing the resin with low concentrations of cAMP.[4,5] Because the stability of C is increased at ionic strengths >0.1, this method is more suitable for the preparation of C from Type II R_2C_2, which binds to DEAE-cellulose at this ionic strength. To minimize elution of unwanted protein during the elution of C, the ionic strength of the cAMP wash is decreased slightly relative to the preceding buffer wash. The eluted C is collected on a hydroxylapatite column to concentrate and further purify the enzyme. The elution of C from the DEAE column occurs in ~1 bed volume rather than a sharp peak because cAMP binds to the resin. The elution volume can be reduced by the use of cAMP derivatives which have diaminoalkyl substituents since these are

---

[1] D. B. Glass and E. G. Krebs, *Annu. Rev. Pharmacol. Toxicol.* **20**, 363 (1980).
[2] E. G. Krebs and J. A. Beavo, *Annu. Rev. Biochem.* **48**, 923 (1979).
[3] L.-J. Chen and D. A. Walsh, *Biochemistry* **10**, 3614 (1971).
[4] J. Erlichman, A. H. Hirsh, and O. M. Rosen, *Proc. Natl. Acad. Sci. U.S.A.* **68**, 731 (1971).
[5] P. H. Sugden, L. A. Holladay, E. M. Reimann, and J. D. Corbin, *Biochem. J.* **159**, 409 (1976).

bound less tightly to DEAE-cellulose, but even with such derivatives, the enzyme still elutes in >0.5 bed volume.

Materials

DEAE-cellulose (DE23) is obtained from Whatman and hydroxylapatite (BioGel HTP) from Bio-Rad. Histone (Type IIA) is obtained from Sigma or Aldrich. The stock histone solution (30 mg/ml) is heated at 100° for 10 min. In some histone preparations this increases the rate of phosphorylation and inactivates contaminating kinases. [γ-^{32}P]ATP is prepared by the method of Walseth and Johnson.[6] The biochemical purity of the ATP is tested in a kinase reaction mixture (see below) which contains 0.01 kinase units/ml and 10 μM [γ-^{32}P]ATP. Under these conditions, the ^{32}P is quantitatively transferred from [γ-^{32}P]ATP to histone in less than 30 min. Some commercial preparations of [γ-^{32}P]ATP have been found to contain large amounts of ^{32}P-labeled material which adsorbs to charcoal but does not serve as substrate for the kinase. The results of this test can be used to calculate moles ^{32}P incorporated into protein substrates based on a corrected specific activity.

All P_i buffers contain equal parts of mono- and dibasic potassium phosphate, 0.1 mM dithiothreitol, and other additives as indicated.

Protein Kinase Assay

Protein kinase is assayed as described previously[7] or more frequently by a modification[8] of the method of Witt and Roskoski.[9] In the latter modification, the reaction mixture contains 50 mM morpholinoethanesulfonic acid, 6 mg histone/ml, 10 mM magnesium acetate, 200 μM [γ-^{32}P]ATP, and varying amounts of kinase (0.01–0.5 mU/ml) diluted in 50 mM P_i containing 1 mM EDTA and 1.5 mg gelatin/ml. After incubation for 10 min, 40-μl aliquots are withdrawn and spotted onto 2 × 2-cm pieces of P81 cellulose and dropped into a beaker containing 50 mM NaCl. The papers are subjected to four washes of 50 mM NaCl for 5 min each and then a wash in acetone for 5 min followed by drying and counting in liquid scintillation solution. The unit is defined as that amount of kinase which catalyzes the transfer of 1 μmol of ^{32}P from ATP to histone in 1 min under these conditions.

---

[6] T. F. Walseth and R. A. Johnson, *Biochim. Biophys. Acta* **562**, 11 (1979).
[7] J. D. Corbin and E. M. Reimann, this series, Vol. 38, p. 287.
[8] K. K. Schlender and E. M. Reimann, *J. Biol. Chem.* **252**, 2384 (1977).
[9] J. J. Witt and R. Roskoski, *Anal. Biochem.* **66**, 253 (1975).

Preparation of C

All steps are carried out at 4°. Two fresh beef hearts (~800 g each) are homogenized in 10 mM P_i containing 1 mM EDTA. The homogenate is centrifuged at 10,000 g for 30 min and the supernatant is filtered through cheesecloth and glass wool. Supernatant is applied to a DEAE-cellulose column (7.5 × 20 cm) and washed overnight with 5–10 liters of 55 mM P_i containing 1 mM EDTA. Elution of C is accomplished by washing the column with 400 ml of 45 mM P_i containing 10–100 μM cAMP followed by ~2 liters of 45 mM P_i buffer. Approximately one bed volume after initiating the cAMP wash, the absorbance at 260 nm (or 280 nm) drops by about a factor of 2 due to the decrease in ionic strength. This change in absorbance signals the beginning of elution of C (Fig. 1). At this point, the column outlet is connected directly to a hydroxylapatite column (2.6 × 4 cm) equilibrated in 45 mM P_i. After ~1000 ml have been collected, the absorbance at 260 nm of the DEAE effluent increases due to the emergence of cAMP, signaling the end of the C peak. This is not evident in Fig. 1 since 10 μM cAMP has little absorbance at 280 nm. The hydroxylapatite

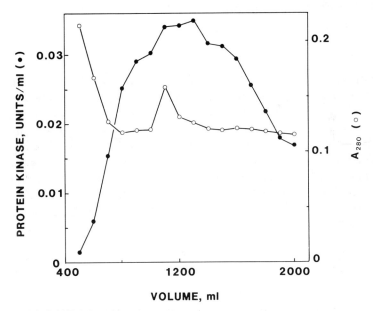

FIG. 1. Elution of C from DEAE-cellulose. Bovine heart extract was applied to DEAE-cellulose and the catalytic subunit was eluted as described in the text. The connection between the DEAE-cellulose column and the hydroxylapatite column was momentarily interrupted at 100-ml intervals to obtain aliquots for assay.

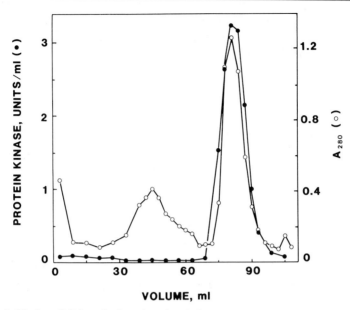

FIG. 2. Elution of C from hydroxylapatite. A linear P_i gradient was used to elute C as described in the text.

column is disconnected from the DEAE column and developed with a linear gradient using 60 ml of 45 mM P_i in the mixing chamber and 60 ml of 500 mM P_i in the other chamber of the gradient maker. The kinase elutes as a single protein peak at about 200 mM P_i (Fig. 2). This procedure can be easily completed in two working days. The preparation can be stored for months at 4° or it can be quick frozen after adding sucrose to a concentration of 20% (w/w). The frozen enzyme can be stored indefinitely at −70°, but repeated freezing and thawing progressively inactivates the enzyme. This procedure yields 10–15 mg of electrophoretically pure enzyme having a specific activity of 3–4 units/mg, representing a purification of ~1000-fold with a yield of ~30%.

Concluding Remarks

With slight modifications, this method has been applied to a variety of other tissues including beef liver,[5] pig gastric mucosa,[10] and rabbit kidney.[11] The tissue of choice appears to be the beef heart because it is available in large quantity, it contains predominantly Type II protein

[10] E. M. Reimann and J. D. Corbin, *Fed. Proc. Fed. Am. Soc. Exp. Biol.* **35,** 1384 (1976).
[11] K. K. Schlender and E. M. Reimann, *Proc. Natl. Acad. Sci. U.S.A.* **72,** 2197 (1975).

kinase, and larger amounts of tissue extract can be adsorbed to a given amount of DEAE-cellulose without overloading. The purification procedure can be scaled down by a factor of approximately 40 if only small amounts of C are needed.

Acknowledgments

This work was supported in part by U.S.P.H.S. Grants AM-15611 and AM-19231, by U.S.P.H.S Research Career Development Award AM-00446 to E.M.R., and by a Grant-in-Aid from the American Heart Association and with funds contributed in part by the AHA Northwestern Ohio Chapter, Inc.

[7] Regulatory Subunits of Bovine Heart and Rabbit Skeletal Muscle cAMP-Dependent Protein Kinase Isozymes

By STEPHEN R. RANNELS, ALFREDA BEASLEY, and JACKIE D. CORBIN

Cyclic AMP-dependent protein kinase exists in two major isozymic forms, Types I and II, which are present in nature in varying relative concentrations, depending upon the tissue and species studied.[1,2] Although the importance of cAMP-dependent protein kinase in metabolic regulation is well established,[3] definitive roles for each of the two isozymes are not understood. The overall mechanism of activation for both forms of the enzyme is similar:

$$R_2C_2 + 4\ cAMP \rightleftharpoons R_2(4\ cAMP) + 2C$$

where the regulatory subunit dimer (R_2) binds cAMP and the active catalytic subunits (C) dissociate from R.[4-7] The differences between isozymes I and II reside with the regulatory moieties, as the C subunits appear to be identical.[8] Thus, a complete understanding of the importance of each

[1] J. D. Corbin, S. L. Keely, and C. R. Park, *J. Biol. Chem.* **250**, 218 (1975).
[2] J. D. Corbin and S. L. Keely, *J. Biol. Chem.* **252**, 910 (1977).
[3] J. A. Beavo and E. G. Krebs, *Ann. Rev. Biochem.* **48**, 923 (1979).
[4] F. Hofmann, J. A. Beavo, P. J. Bechtel, and E. G. Krebs, *J. Biol. Chem.* **250**, 7795 (1975).
[5] O. M. Rosen and J. Erlichman, *J. Biol. Chem.* **250**, 7788 (1975).
[6] J. D. Corbin, P. H. Sugden, L. West, D. A. Flockhart, T. M. Lincoln, and D. McCarthy, *J. Biol. Chem.* **253**, 3997 (1978).
[7] W. Weber and H. Hilz, *Biochem. Biophys. Res. Commun.* **90**, 1073 (1979).
[8] J. D. Corbin, S. L. Keely, T. R. Soderling, and C. R. Park, *in* "Advances in Cyclic Nucleotide Research" (G. I. Drummond, P. Greengard, and G. A. Robinson, eds.), Vol. 5, p. 265. Raven, New York, 1975.

isozyme in the cell requires a detailed functional and structural study of their regulatory subunits. A rapid and simple method for the isolation of homogeneous regulatory subunits from each isozyme (R_I and R_{II}) is presented.

Materials and Methods

Millipore filtration binding assays are done as described by Sugden and Corbin.[9] Binding mixtures consist of 10 to 20 μl of sample and 50 μl of 50 mM potassium phosphate buffer (pH 6.8) containing 1 mM EDTA, 2 M NaCl, 0.5 mg/ml of Sigma Type II-A histone, and 1 μM [^{3}H]cAMP. In the case of cAMP binding to holoenzyme, filtration can be performed immediately after the sample is added. For R subunits containing bound cAMP, an incubation of 45 min at 30° is sufficient for complete nucleotide exchange. In both cases, 1 ml of cold 10 mM potassium phosphate (pH 6.8) containing 1 mM EDTA (phosphate/EDTA) is added to the reaction mixture and the entire solution is passed through a premoistened Millipore filter. The reaction tube is rinsed with an additional 1 ml of buffer and the filter is washed twice with 4 ml each. The filters are dried and counted by liquid scintillation. Protein is determined by the method of Lowry et al.[10] or Bradford[11] using bovine serum albumin as the standard. SDS–polyacrylamide gel electrophesis (10%) is performed using the method of Weber and Osborn.[12]

The [^{3}H]cAMP is obtained from ICN. DEAE-cellulose (DE-11 or DE-22), histone (Type II-A), ammonium sulfate (grade 1), and cAMP are from Sigma. DEAE-Sephadex A-50 is from Pharmacia, and the cAMP-Sepharose 4B affinity gels are purchased from P-L Biochemicals, Inc.

Preparation of Isozyme II Regulatory Subunit

Isozyme II regulatory subunit (R_{II}) is isolated from bovine heart following modification of the procedures of Dills et al.[13] and Corbin et al.[6] Fresh bovine hearts are obtained from a local slaughterhouse and immediately cooled to 4°. All subsequent steps are performed at 4°. The number of hearts used (~ 0.8–1.0 kg each), can be based on an approximate yield of 5–10 mg purified R_{II} per kg tissue. The trimmed tissue is ground in a meat grinder and homogenized (3 ml/g) in phosphate/EDTA for three

[9] P. H. Sugden and J. D. Corbin, *Biochem. J.* **159**, 423 (1976).
[10] O. H. Lowry, N. J. Rosebrough, A. L. Farr, and R. J. Randall, *J. Biol. Chem.* **193**, 265 (1951).
[11] M. Bradford, *Anal. Biochem.* **72**, 248 (1976).
[12] K. Weber and M. Osborn, *J. Biol. Chem.* **244**, 4406 (1969).
[13] W. L. Dills, J. A. Beavo, P. J. Bechtel, and E. G. Krebs, *Biochem. Biophys. Res. Commun.* **62**, 70 (1975).

bursts of 30 sec each using a large Waring Blendor. After centrifugation at 10,000 g for 30 min the supernatant is filtered through glass wool and mixed batchwise with settled DEAE-cellulose, which is slowly stirred for 2 hr. Approximately 1 liter of settled DEAE-cellulose is used for each 5 kg of starting tissue for large preparations. R subunit can be isolated from one heart or less, but larger preparations are desirable because the stability of R is enhanced throughout the preparation if the overall protein concentration is high. Following sedimentation, the supernatant is decanted and discarded and the DEAE-cellulose is washed batchwise with 10 liters of phosphate/EDTA and again with 10 liters of phosphate/EDTA containing 0.15 M NaCl. The latter wash removes many contaminating proteins including R_I. The DEAE-cellulose is then poured into an 8 × 90 cm column. The enzyme is eluted with 4–5 liters of buffer containing 0.35 M NaCl, and 500-ml fractions are collected manually and assayed for [^{3}H]cAMP binding activity. Those fractions containing binding activity, usually fractions 3 and 4, are pooled and the protein concentrated by precipitation with $(NH_4)_2SO_4$ (400 g/liter). After centrifugation at 10,000 g for 30 min, the precipitate is gently resuspended in 100–300 ml of phosphate/EDTA and dialyzed for 2 hr. The dialyzed solution is centrifuged again to remove precipitated proteins and applied to a column of 8(6-aminohexylamino)-cAMP-Sepharose 4B equilibrated in phosphate/EDTA. The affinity gel contains a minimum of 1–2 μmol of covalently bound AHA-cAMP/ml of packed material (P-L Biochemicals, Type 3). To ensure complete binding of R_{II} to the column, the amount of gel used should contain approximately twice the concentration of covalently bound AHA-cAMP than the total estimated binding sites in the protein solution (2–4 ml gel/1–2 μmol cAMP binding capacity of the sample). The protein solution is applied slowly (~0.25 ml/min) and the column is washed with 50–100 ml of phosphate/EDTA containing 2 M NaCl. The column is then washed with 10 ml of phosphate/EDTA, 3 ml of phosphate/EDTA containing 10 mM 5'-AMP (adjusted to pH 6.8), and again with several ml of buffer. R_{II} is eluted by equilibrating the column with phosphate/EDTA containing 10 mM cAMP (pH adjusted to 6.8 after adding the nucleotide). One- to 2-milliliter fractions are collected after several hours and again 24 and 48 hr later. The purity of each fraction is assessed by SDS–polyacrylamide disc gel electrophoresis using 10% gels. All fractions are dialyzed for 3 days against several changes of phosphate/EDTA to remove free cAMP. A typical preparation is summarized in Table I.[6]

Preparation of Isozyme I Regulatory Subunits

Isozyme I R subunit is prepared from either fresh or frozen (Pel-Freez) rabbit skeletal muscle. Approximately 1–2 mg purified R_I can be isolated

TABLE I
PURIFICATION OF ISOZYME II REGULATORY SUBUNIT FROM BOVINE HEART

Step	Volume (ml)	Protein (mg)	cAMP bound (nmol)	Specific activity (nmol cAMP/mg)	Yield (%)
10,000 g supernatant	5500	64075	769	0.012	100
0.3 M NaCl elution of DEAE-cellulose	1700	2202	439	0.199	57
cAMP-Sepharose affinity column	3	10.75	351	33	46

from each kilogram of starting tissue. New Zealand rabbits (3.5–5 kg) are anesthetized by an ear-vein injection of 5 ml of Nembutal, exsanguinated by throat incision, and fresh leg and back muscles are removed and immediately ground and homogenized in ice-cold phosphate/EDTA (3 ml/g) as described above for heart. If frozen muscle is used, it is partially thawed at 4° and used immediately. Following centrifugation at 10,000 g for 30 min, the supernatant (from 8 kg muscle) is filtered and mixed batchwise with 1.5–2 liters of DEAE-cellulose. Alternatively, at this stage this supernatant can be replaced by the pH 6.1 supernatant fraction obtained in a standard phosphorylase kinase purification.[14] Again, slow stirring or occasional mixing for 2 hr is adequate for quantitative binding to DEAE-cellulose. The supernatant is decanted and the DEAE-cellulose is washed batchwise with 20 liters of phosphate/EDTA and then poured into a column (8 × 90 cm). After the phosphate/EDTA reaches the top of the DEAE-cellulose, cyclic AMP binding activity is eluted with phosphate/EDTA containing 0.10 M NaCl (in 500 ml fractions), and pooled. Pooled samples are concentrated by $(NH_4)_2SO_4$ precipitation (400 g/liter), resuspended in phosphate/EDTA, dialyzed, and cleared by centrifugation as described above for R_{II}. The resulting solution (200–300 ml) is applied to a cAMP-Sepharose 4B affinity column. To obtain the native cAMP-saturated R_I, N^6-(2-aminoethylamino)-cAMP-Sepharose 4B should be used (P-L Biochemicals, Type 2). If 8(6-aminohexylamino)-cAMP-Sepharose 4B is used, R_I is not readily removed by cAMP elution but can be eluted with phosphate/EDTA containing 8 M urea. In this case, a short equilibration period (2 hr) and elution with phosphate/EDTA containing 10 mM cAMP will remove potentially contaminating R_{II} and its fragments and nonspecific nucleotide binding proteins before rapid R_I elution with urea.

[14] R. J. DeLange, R. G. Kemp, W. D. Riley, R. A. Cooper, and E. G. Krebs, *J. Biol. Chem.* **243**, 2200 (1968).

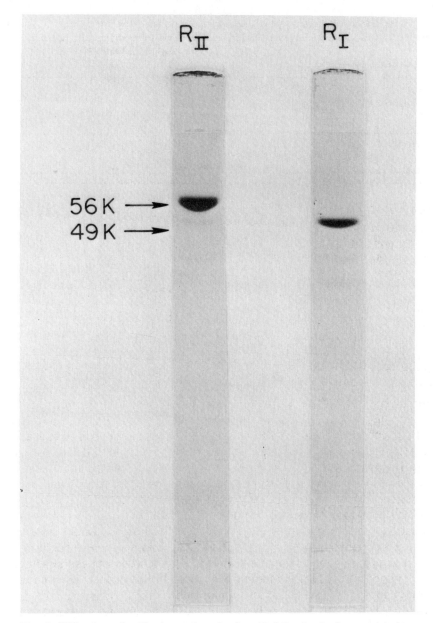

FIG. 1. SDS-polyacrylamide electrophoresis of purified R subunits from protein kinase isozymes I and II.

TABLE II
Physical Parameters of R_I and R_{II}[15-19]

	M_r from							
	$s_{20,w}$ and Stokes radius	SDS gels	Amino acid sequence determination	$s_{20,w}$	Stokes radius (nm)	f/f_0	Axial ratio	pI	cAMP binding sites monomer
R_I	81,500	49,000	—	4.7	4.20	1.47	8.5	5.45, 5.57	2
R_{II}	100,300	56,000	45,084	4.5	5.40	1.74	14	5.34, 5.40	2

For both methods, following application of the sample, the affinity column is washed with 100 ml of phosphate/EDTA containing 2 M NaCl. If urea elution (15 min equilibration) is used, the protein is dialyzed immediately to remove urea and free cAMP. Subsequent urea elutions following a 24 hr equilibration should be taken and dialyzed. If cAMP elution of R_I from N^6-(2-aminoethylamino)-cAMP-Sepharose 4B is used, a final urea elution (after several with cAMP) will yield additional R_I.

SDS–polyacrylamide gels of purified R_I (M_r = 49,000) and R_{II} (M_r = 56,000) are shown in Fig. 1, and a partial list of their physical characteristics is shown in Table II.[15-19] In addition to these data, R_{II} possesses a serine which is rapidly phosphorylated by the C subunit,[20] and other sites which also contain covalently-bound phosphate.[16,21] Although R_I is not readily phosphorylated by C subunit, it can be phosphorylated by cGMP-dependent protein kinase,[22] and it contains other endogenous sites of phosphorylation.[23]

Discussion

The procedures described above for purification of R_I and R_{II} are based on relatively large amounts of starting material. If scaling down is

---

[15] S. R. Rannels and J. D. Corbin, *J. Cyclic Nucleotide Res.* **6**, 201 (1980).
[16] D. A. Flockhart, D. M. Watterson, and J. D. Corbin, *J. Biol. Chem.* **255**, 4435 (1980).
[17] R. Rangel-Aldao, J. W. Kupiec, and O. M. Rosen, *J. Biol. Chem.* **254**, 2499 (1979).
[18] R. L. Geahlen and E. G. Krebs, *J. Biol. Chem.* **255**, 9375 (1980).
[19] K. Takio, S. B. Smith, E. G. Krebs, K. A. Walsh, and K. Titani, *Proc. Natl. Acad. Sci. U.S.A.* **79**, 2544 (1982).
[20] J. Erlichman, R. Rosenfeld, and O. M. Rosen, *J. Biol. Chem.* **249**, 5000 (1974).
[21] B. A. Hemmings, A. Aitken, P. Cohen, M. Rymond, and F. Hofmann, *Eur. J. Biochem.* **127**, 473 (1982).
[22] R. L. Geahlen and E. G. Krebs, *J. Biol. Chem.* **255**, 1164 (1980).
[23] R. A. Steinberg, P. H. O'Farrell, U. Friedrich, and P. Coffino, *Cell* **10**, 381 (1977).

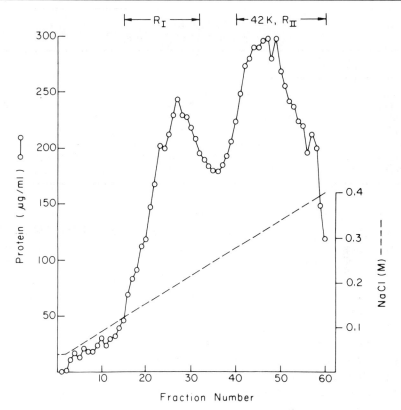

FIG. 2. Separation of R_I, R_{II}, and fragments by chromatography on DEAE-Sephadex.

appropriate, the final volume of cAMP-binding material which is eluted from DEAE-cellulose may be small enough (<1 liter) that concentration by $(NH_4)_2SO_4$ before applying it to the affinity matrix is unnecessary. It is still advisable to centrifuge all material to prevent blockage of the column. Omitting the $(NH_4)_2SO_4$ precipitation step saves at least 6 hr and makes it possible to go from intact tissue to the cAMP-affinity column in <48 hr.

After purification, if an R subunit isozyme is found to contain trace amounts of either the other isozyme or proteolytic fragments as determined by SDS–gel electrophoresis, a simple additional purification step can be used. Most often, R_I eluted from the N^6-ethane-affinity gel by cAMP contains small amounts of R_{II} and fragments. In Fig. 2 is shown the chromatography on DEAE-Sephadex A-50 of a sample of R_I which had been determined to be highly contaminated with R_{II} and a fragment of $M_r = 42,000$. The protein solution (~8 mg) is slowly applied to a small column of DEAE-Sephadex A-50 (2 ml) equilibrated with phosphate/

EDTA containing 40 mM NaCl. Protein is eluted with a 50 ml 40–400 mM NaCl gradient and 25-drop (0.8 ml) fractions are collected. R_I elutes approximately between 12 and 26 ml, whereas R_{II} and the $M_r = 42,000$ fragment elute between 35 and 60 ml.

Regulatory subunits from both isozymes can be isolated from a single tissue source when both types are present in approximately equal amounts, such as in porcine skeletal muscle.[24] Nucleotide-free regulatory subunits are most often prepared by elution from cAMP-affinity columns with 8 M urea followed by dialysis or by treatment of the R · cAMP complex with urea and removal of urea by Sephadex G-25 chromatography.[25] Care should be taken to remove the urea from the R subunits as rapidly as possible, and even rapid removal does not entirely protect against irreversible nucleotide-binding site perturbation.[26] However, both cAMP-eluted and urea-eluted R subunits readily recombine with C subunit and inhibit its activity. A method for the purification of C subunit is presented in this volume.[27] The choice of tissue and method of elution ultimately depends upon the quantity of R_I or R_{II} subunits needed and the nature of the studies for which they are to be used.

[24] M. J. Zoller, A. R. Kerlavage, and S. S. Taylor, *J. Biol. Chem.* **254**, 2408 (1979).
[25] S. E. Builder, J. A. Beavo, and E. G. Krebs, *J. Biol. Chem.* **255**, 3514 (1980).
[26] J. D. Corbin and S. R. Rannels, *J. Biol. Chem.* **256**, 11671 (1981).
[27] E. M. Reimann and R. A. Beham, this volume [6].

[8] cGMP-Dependent Protein Kinase

by THOMAS M. LINCOLN

Since the discovery of cGMP-dependent protein kinase in various arthropod tissues[1] and the mammalian cerebellum,[2] several attempts had been made to purify the enzyme. These procedures, namely ion-exchange chromatography[1] and gel filtration,[3] successfully separated the cGMP-dependent enzyme from the cAMP-dependent protein kinase. The small amount of cGMP-dependent protein kinase present in most animal tissues, however, meant that conventional preparative techniques would yield only small amounts of enzyme from large quantities of starting mate-

[1] J. F. Kuo and P. Greengard, *J. Biol. Chem.* **245**, 2493 (1970).
[2] F. Hofmann and G. Sold, *Biochem. Biophys. Res. Commun.* **49**, 1100 (1972).
[3] J. F. Kuo, *Proc. Natl. Acad. Sci. U.S.A.* **71**, 4037 (1974).

rial. Thus, the application of cyclic nucleotide affinity chromatography seemed to be a plausible alternative to conventional techniques since the cGMP-dependent protein kinase was known to bind its ligand with high affinity.

Three immobilized ligands have been successfully used for the purification of mammalian cGMP-dependent protein kinase: 8-2-aminoethylamino-cAMP-Sepharose 4B,[4] 8-6-aminohexylamino-cAMP-Sepharose 4B,[5] and 8-2-aminoethylthio-cGMP-Sepharose 4B.[6,7] The preliminary preparative procedures leading up to the affinity chromatographic steps and the methods used for the elution of the cGMP-dependent protein kinase from the three resins in principle are the same. The 8-2-aminoethylamino-cAMP-Sepharose 4B resin is commercially available, making the cGMP derivative somewhat disadvantageous unless one undertakes its synthesis.[7,8] The purification procedure for cGMP-dependent protein kinase from bovine lung described below has been performed using all three ligands and has yielded similar results.

Experimental Methods

Cyclic GMP-Dependent Protein Kinase Assay. The standard assay for cGMP-dependent protein kinase was conducted in a volume of 100 μl containing 20 mM Tris–Cl, pH 7.5, 100 mM magnesium acetate, 10 μM [γ-^{32}P]ATP (200 to 300 cpm/pmol), 10 μg of either mixed histone (type II-A) or F2b histone, and 0.1 μM cGMP when present. The reaction was initiated with the addition of enzyme and incubations were carried out at 30° for 5 min. The reaction was terminated by pipetting an aliquot of the mixture onto filter paper discs. The discs were immersed in 10% trichloroacetic acid and processed as described previously.[9] One unit of enzyme activity is the amount of enzyme required to transfer 1 μmol of phosphate from ATP to histone in 1 min.

Other Procedures. Eight percent polyacrylamide gel electrophoresis in sodium dodecyl sulfate (SDS–PAGE) was performed as described by Laemmli[10] using Coomassie blue G-250 to stain the protein bands. Protein

---

[4] G. N. Gill, K. E. Holdy, G. M. Walton, and C. B. Kanstein, *Proc. Natl. Acad. Sci. U.S.A.* **73,** 3918 (1976).
[5] J. D. Corbin, personal communication (1982).
[6] T. M. Lincoln, W. L. Dills, and J. D. Corbin, *J. Biol. Chem.* **252,** 4269 (1977).
[7] V. Flockerzi, N. Speichermann, and F. Hofmann, *J. Biol. Chem.* **253,** 3395 (1978).
[8] W. L. Dills, J. A. Beavo, P. J. Bechtel, K. R. Meyers, L. J Sakai, and E. G. Krebs, *Biochemistry* **15,** 3724 (1976).
[9] J. D. Corbin and E. Reimann, this series, Vol. 38, p. 287.
[10] U. K. Laemmli, *Nature (London)* **227,** 680 (1970).

was determined by the Lowry method,[11] and in some cases by the modified procedures of Ross and Schatz.[12] 8-2-Aminoethylthio-cGMP was synthesized and purified as described previously.[8] Coupling to CNBr-Sepharose 4B was as described earlier.[6]

Materials. Histone was purchased from either Sigma Co. (type II-A) or Worthington Diagnostics (F2b). CNBr-Sepharose 4B was from Pharmacia while 8-2-aminoethylamino-cAMP-Sepharose 4B was obtained from P-L Biochemicals. DE-11 and DE-52 cellulose were from the Whatman Co.

Results and Discussion

Purification Procedures: Step 1. Preparation of the Extract. Freshly obtained bovine lungs (2–3 kg) were obtained from a local slaughterhouse. The lungs were transported to the laboratory on ice to keep them as cold as possible. All further procedures were performed at 4°. Dissected lung tissue (1.5 to 2 kg) was removed with a knife leaving behind the trachea, large bronchiolar tubes, and fat tissue. The tissue was ground with a meat grinder into a fine, pulpy mass and suspended in 2 volumes of 20 mM sodium phosphate–2 mM EDTA–25 mM 2-mercaptoethanol, pH 7.0 (PEM buffer) in a 4-liter Waring blender (approximately 3/4 capacity). The tissue was homogenized in 10-sec bursts at each speed (low, medium, and high), and the procedure was repeated twice. The viscous homogenate was centrifuged in a 3-liter capacity rotor at 12,000 g for 20 min. The supernatant was poured through glass wool into a 4-liter flask, and the extract was diluted in an equal volume of PEM buffer and mixed thoroughly.

Step 2. DEAE-Celluose Chromatography. The entire extract (approximately 6 liters) was gently mixed for 1 hr with 4 liters of DE-11 (or DE-52) previously equilibrated in the same buffer. The DEAE-extract suspension was then poured into a large column (8-liter capacity) and the extract was passed through the settling DEAE. This procedure, in contrast to batch absorption followed by the removal of the extract before filling the column, yielded a greater recovery of cGMP-dependent protein kinase from the DEAE-cellulose column upon salt elution. After the extract had passed into the DEAE-cellulose bed, the column was washed with 20 liters of PEM containing 50 mM NaCl. This fraction contained a large amount of cAMP-dependent protein kinase type I but less than 1% of the cGMP-dependent protein kinase from the original extract. The cGMP-

---

[11] O. H. Lowry, N. J. Rosebrough, A. L. Farr, and R. J. Randall, *J. Biol. Chem.* **193**, 265 (1951).

[12] E. Ross and G. Schatz, *Anal. Biochem.* **54**, 304 (1973).

dependent protein kinase was then eluted with 12 liters of PEM containing 150 mM NaCl, although 80 to 90% of the enzyme could be recovered from the column in the second through the sixth liter of the 150 mM NaCl wash and used as the source for the next purification step.

Step 3. Ammonium Sulfate Precipitation. The NaCl wash was mixed with solid ammonium sulfate (1 kg/2 liters of effluent) for 1 hr with occasional stirring. The suspension was centrifuged at 12,000 g for 20 min and the supernatant was discarded. The pellets were resuspended in a minimum volume of PEM buffer (usually about 300–400 ml for every 2 kg of lung tissue) and placed in dialysis sacks. The suspensions were then dialyzed overnight against 24 liters of PEM. At this point, the cloudy suspension cleared considerably, even though a significant amount of ammonium sulfate was still present in the dialysis sack. This residual amount of salt did not affect the recovery of cGMP-dependent protein kinase at this step nor at the following affinity chromatographic step. Thus, further dialysis was not deemed necessary and only prolonged the purification. In fact, in more recent preparations a 2-hr dialysis was used without noticeable differences in results.[5] The dialysate was collected and centrifuged at 12,000 g for 20 min to remove insoluble material and the supernatant passed through glass wool into a flask. The volume of the suspension had approximately doubled during the dialysis.

Step 4. Affinity Chromatography. The entire ammonium sulfate fraction was passed through a 3-ml column of either 8-2-aminoethylamino-cAMP, 8-6-aminohexylamino-cAMP, or 8-2-aminoethylthio-cGMP Sepharose 4B (approximately 2–4 μmol of nucleotide per ml of packed gel) at about 25 ml/hr. Greater than 90% of the cGMP-dependent protein kinase was bound to the column during the loading procedure using either immobilized ligand. Faster flow rates of up to 100 ml/hr did not decrease the binding of the enzyme to the 8-2-aminoethylthio-cGMP-Sepharose 4B presumably due to its extremely high affinity for the cGMP derivative.[6] Elution of the cGMP-dependent protein kinase was performed as follows: The column was first washed with 30 ml of PEM followed by 30 ml of PEM containing 1 M NaCl. These fractions contained only minute amounts of cGMP-dependent protein kinase and were subsequently discarded. The column was then washed with 1 ml of PEM then with 0.5 ml of PEM containing 10 mM cGMP and the flow was halted. The column was removed from the cold and allowed to stand at room temperature for 30 min. This enabled the free cGMP to displace the cGMP-dependent protein kinase from the immobilized ligand. The column was then washed with 3 ml of PEM plus 10 mM cGMP at room temperature and fractions were collected in tubes placed on ice. This procedure was repeated four times. A typical profile for elution of cGMP-dependent protein kinase

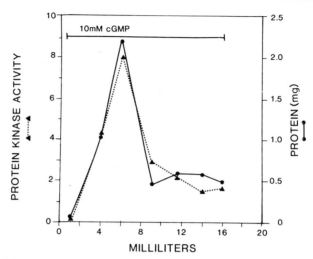

FIG. 1. Elution of cGMP-dependent protein kinase from 8-2-aminoethylthio-cGMP-Sepharose 4B.

from 8-2-aminoethylthio cGMP-Sepharose 4B is shown in Fig. 1. Greater than 60% of the protein was eluted in the first 6 ml. Continued washing eluted residually bound protein and kinase activity even up to 16 ml, attesting to the high affinity of the enzyme for the immobilized cGMP ligand. An alternate procedure for the elution of residual cGMP-dependent protein kinase was adopted as follows after the first two elutions: the immobilized ligand was removed from the column and incubated overnight in 3 ml of PEM plus 10 mM cGMP in a centrifuge tube. The beads were then removed by centrifugation and the supernatant was saved for cGMP-dependent protein kinase determination. An extra 3 mg of protein could usually be recovered by this procedure. A summary of the purification procedure from 1 kg of bovine lung using the cGMP-Sepharose 4B affinity column is shown in the table. Greater than a 4000-fold purification was obtained through the affinity chromatography step. The enzyme had a specific activity of approximately 0.3 units/mg protein using mixed histone as the substrate or approximately 1 unit/mg protein using the preferred phosphate acceptor, F2b histone.

Step 5. Sepharose 6B Chromatography (Optional). The protein eluted by the affinity chromatography step was 80–90% pure as judged by SDS–PAGE (Fig. 2). Most of the protein migrated as a single band with an M_r = 80,000. One major contaminant migrating at 50,000 to 55,000 M_r was usually seen and could be removed by gel filtration. The eluted cGMP-

PURIFICATION OF cGMP-DEPENDENT PROTEIN KINASE FROM BOVINE LUNG
BY 8-2-AMINOETHYLTHIO-cGMP AFFINITY CHROMATOGRAPHY[a]

Step	Volume (ml)	Protein (mg)	Protein kinase[b] (units)	Recovery (%)
12,000 g supernatant	3,000	69,000	5.64	—
DEAE-cellulose	7,000	11,700	2.44	43
Ammonium sulfate	150	4,950	1.64	29
cGMP-Sepharose	12	3.75	1.40	25

[a] Two kilograms of bovine lung was used as the starting material, and the preparation was divided into two 1-kg parts: part one (Fig. 1) was chromatographed on 8-2-aminoethylthio-cGMP-Sepharose 4B and part two (not shown) was chromatographed on 8-2-aminoethylamino-cAMP-Sepharose 4B.

[b] Histone (type II-A) was used as the substrate.

dependent protein kinase was concentrated by applying the sample to a 0.5-ml DEAE-cellulose column (conveniently poured into a pasteur pipet) equilibrated in PEM. The enzyme was eluted with 1 ml of PEM containing 500 mM NaCl and applied to a Sepharose 6B column (0.9 × 60 cm) equilibrated with PEM buffer. The cGMP-dependent protein kinase eluted as a single peak of protein corresponding to a $M_r = 160,000$; while the minor contaminant(s) eluted as a lower molecular weight species. The kinase was pooled, concentrated on a 0.5 ml DEAE-cellulose column as described above, and dialyzed exhaustively against PEM. Since the holoenzyme M_r was estimated at 160,000, the enzyme probably consists of a dimer of identical subunits.[6] This hypothesis has been upheld in experiments from several laboratories and now appears incontrovertible.[4,7]

Comments on the Purification

The purification of the cGMP-dependent protein kinase is simple and rapid using either affinity ligand. Typical purifications can be accomplished in 5 to 7 days following the above procedures. In many cases, the time for purification can be abbreviated even more. For example, the use of a porous fritted disc funnel attached to a vacuum flask rather than a gravity-operated column can decrease dramatically the time required for DEAE-cellulose chromatography. As mentioned earlier, smaller elution volumes from the DEAE-cellulose column can also be used with only minor losses in yield. This not only reduces the elution time required but also reduces considerably the amount of material to be handled at the

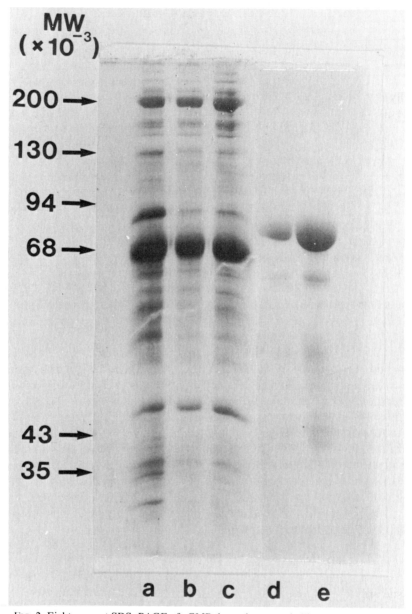

FIG. 2. Eight percent SDS–PAGE of cGMP-dependent protein kinase purified by affinity chromatography. Lane a, 20.0 µg protein from 12,000 g supernatant; Lane b, 8.0 µg protein from DEAE-cellulose elution; Lane c, 8.0 µg protein from 60% ammonium sulfate precipitate; Lane d, 4 µg cGMP-dependent protein kinase from 8-2-aminoethylthio-cGMP-Sepharose 4B; Lane e, 20 µg cGMP-dependent protein kinase from 8-2-aminoethylthio-cGMP-Sepharose 4B. Molecular weight standards are myosin (200,000), β-galactosidase (130,000), phosphorylase (94,000), bovine serum albumin (68,000), ovalbumin (43,000), and glyceraldehyde-3-phosphate dehydrogenase (35,000).

ammonium sulfate step. Finally, the Sepharose 6B chromatography step may be omitted depending on the purity of enzyme desired. As will be mentioned below, one of the contaminants usually seen is the regulatory subunit of the cAMP-dependent protein kinase and this can be successfully removed by eluting the affinity column with 0.1 mM cAMP prior to the cGMP elution. It has also been observed that there is less contaminating regulatory subunit in the earlier fractions eluted from the immobilized cAMP affinity column.

It can be observed from Table I that there is a significant loss of total activity of cGMP-dependent protein kinase between the crude extract and the DEAE-cellulose step. On the other hand, there is very little loss of cGMP binding between these two steps (not shown). The discrepancy is due to the presence in the crude extract of a cGMP-independent protein kinase activity present which is assayed together with cGMP-dependent protein kinase activity using 100 mM Mg^{2+} and 10 μM ATP. Thus, the overall recovery of protein kinase activity is artificially low due to the fact that it was not possible to assay *specifically* the cGMP-dependent protein kinase activity in the crude extract using histone as the substrate.

Potential Problems and Pitfalls

Perhaps the most common problem encountered during the purification of the cGMP-dependent protein kinase is loss of the enzyme through proteolysis. It has been observed that limited tryptic proteolysis of cGMP-dependent protein kinase produces three major protein bands detected in 7.5% SDS–polyacrylamide gels at M_r = 75,000, 50,000, and 30,000.[6] Originally, considerable variability in yield was encountered in the purification when the starting material was obtained from local slaughterhouses. It was subsequently learned that some abattoirs treat living animals with proteolytic enzymes to give the meat a tender texture. This could have had an influence on the yield of cGMP-dependent protein kinase, although this cannot be proven. It has been noticed, however, that previously frozen tissue did not yield appreciable amounts of enzyme so that the fresher the source of tissue the better the yield will be. This is not surprising since lung and the large amounts of blood contained in the tissue are rich in proteolytic enzymes. Aging or freeze-thawing the tissue probably releases and activates numerous proteases which could affect enzyme yields. Likewise, longer purification times (particularly at the ammonium sulfate step) encourage proteolytic breakdown. EDTA undoubtedly inhibits much cation-dependent proteolysis, and the inclusion of this compound is recommended throughout the purification.

Another problem associated with the purification is contamination by cAMP-dependent protein kinase regulatory subunits and fragments of regulatory subunits. This is not surprising since both affinity ligands bind cAMP and cGMP receptors. The contamination of cGMP-dependent protein kinase with regulatory subunits can be minimized in two ways: first, the elution of the DEAE column with 50 mM NaCl removes practically all of the type I cAMP-dependent protein kinase (but unfortunately not the free type I regulatory subunit), which makes up approximately half of the cAMP-dependent protein kinase activity in bovine lung. Furthermore, elution of the cGMP-dependent protein kinase with 150 mM NaCl does not elute a large amount of the type II cAMP-dependent enzyme although some contamination undoubtedly occurs. Second, elution of the immobilized cyclic nucleotide affinity column with small amounts of cAMP (i.e., 0.1 mM) prior to cGMP elution effectively removes the type II R subunit.[7] Of course, if either of these methods fails to yield a pure product, gel filtration or sucrose gradients may be the final answer.

The storage of the cGMP-dependent protein kinase does not usually pose a serious problem. Pure enzyme has been stored up to 6 months or more at 4° in PEM buffer or at $-20°$ in the same buffer containing 50% glycerol with no significant loss of activity or proteolytic breakdown. The enzyme has been frozen in small aliquots by adding glycerol (10%) followed by quick-freezing in dry ice-acetone. It should be stated that the glycerol is absolutely necessary and repetitive freezing and thawing must be avoided. The storage of the immobilized cyclic nucleotide-Sepharose 4B likewise does not present a problem. After elution of the enzyme, the column can be "cleaned" by washing with several volumes of 8 M urea plus 1% Triton X-100 followed by reequilibration in PEM or any other buffer. The Sepharose 4B beads can be stored in 50% glycerol at $-20°$ or at 4° in 0.2% sodium azide. The former storage procedure is recommended. The beads have been reused for several years with no significant loss in binding properties.

A final problem encountered with the purification of the cGMP-dependent protein kinase is the removal of bound cGMP from the enzyme. Because cGMP is bound with such high affinity ($K_D < 1 \times 10^{-8}$ M), ordinary dialysis is insufficient to remove the nucleotide. One approach to this problem is to treat the enzyme with charcoal to bind free cGMP.[6] Typically, pure cGMP-dependent protein kinase is diluted 1 : 100 in PEM containing 2 mg/ml bovine serum albumin. The latter stabilizes the dilute enzyme. Activated charcoal (such as Norite A) is added to a final concentration of 2 to 5 mg/ml charcoal. Higher amounts of charcoal inhibit total enzyme activity presumably by binding to the enzyme itself. The enzyme-charcoal mixture is then incubated for 20 min at room temperature with occasional mixing and the suspension is centrifuged to remove the charcoal. The enzyme treated in this manner can be stimulated some 5- to

10-fold by cGMP. It is important to dilute the cGMP-dependent protein kinase prior to charcoal treatment in order to lower the probability that dissociated cGMP will rebind to the enzyme. In many instances, however, it is desirable to have a nondiluted enzyme free of cGMP. In this instance, the concentrated enzyme is incubated for 1 hr at room temperature with 1 mM cAMP followed by another hour at 4°. Charcoal is then added as described above and the suspension is incubated for 20 min at room temperature followed by centrifugation. The enzyme can then be dialyzed against PEM buffer to remove traces of cAMP. A 5-fold stimulation of the cGMP-dependent protein kinase by cGMP is observed for enzyme treated in this fashion. An alternate procedure was described by Flockerzi et al.[7] for the removal of cGMP. In this case, cGMP-dependent protein kinase is adsorbed to DEAE-cellulose and washed at room temperature with PEM buffer containing 0.1 mM cAMP. The cAMP is then removed from the enzyme by washing the column with at least 10 volumes of PEM buffer at 4°. The cGMP-free cGMP-dependent protein kinase can then be eluted from the column with NaCl and dialyzed. In both instances, little loss of *total* cGMP-dependent protein kinase activity is observed.

[9] Insect (cAMP–cGMP)-Dependent Protein Kinase

By Alexander Vardanis

This enzyme is one of a number of cyclic nucleotide-dependent protein kinases found in insects. It has the unusual property of similar affinity for either cyclic AMP or cyclic GMP.[1] It is present in relatively high titer only in insects that have just molted, and is associated with insect epidermis.[2] It is purified from the thoracic integument and legs of newly emerged (0–24 hr) adults of the grasshopper *Melanoplus sanquinipes*.

Insect Culture

A colony of a nondiapausing strain of the grasshopper is grown in a controlled environment room programmed to reproduce a typical summer day. A continuous supply of eggs is obtained from an adult population of approximately 400 females and 200 males. The time from egg to adult is

[1] A. Vardanis, *J. Biol. Chem.* **255**, 7238 (1980).
[2] A. Vardanis, *Insect Biochem.* **12**, 399 (1982).

approximately 6 weeks. The insects are fed a dry diet consisting of a mixture of 200 g wheat bran, 200 g alfalfa, 20 g dry brewer's yeast, and 45 ml of corn oil. This diet is supplemented by fresh young barley seedlings (1–2 leaf stage). The emerging adults are recognized by their fully developed wings. The insect cages are inspected at the beginning and end of each day and the adult insects are removed and dissected.

Dissection

The head, abdomen, and wings are removed and all adhering tissues are teased off from the thoracic cavity. The tissues remaining are the integument of the thorax and legs and the flight and leg muscles. The dissection takes approximately 15 sec per insect and the material obtained weighs an average of 80 mg. The material can be stored in the deep freeze ($-20°$), for long periods (up to 4 months) before homogenization.

Standard Assay Procedure

Protein kinase activity is assayed by measuring incorporation of ^{32}P from [γ-^{32}P]ATP into histone. The 75-μl incubation mixture contains 10 mM potassium phosphate (pH 6.8), 0.24 mM [γ-^{32}P]ATP (8,000 to 20,000 cpm/nmol), 4 mM magnesium acetate, 50 mM NaF, 6.6 mg/ml of histone II or IIA (Sigma), cyclic nucleotide as indicated, and enzyme. Incubation is for 10 min at 30°. The reaction is stopped by spotting an aliquot (50 μl) of the mixture on a filter paper disc and immersing in ice-cold 10% trichloroacetic acid as described by Corbin and Reimann.[3] The usual washing procedure consists of six 15-min washes: four in trichloroacetic acid followed by 95% ethanol and ether (10 ml of liquid per filter paper disc). The discs are placed in counting vials containing 10 ml of a solution of 4 g of Omnifluor/liter of toluene, and counted in a liquid scintillation spectrometer. One unit of activity is defined as that amount of enzyme that catalyzes the incorporation of 1 μmol of ^{32}P per minute in the described assay. Under the conditions of this assay there is a blank value, representing nonenzymic adsorption of radioactivity onto the histone, that usually amounts to 0.1% of the introduced radioactivity. Occasionally, depending on the histone type and lot of [^{32}P]ATP used, a blank higher than the above amount is obtained. In that instance, acceptable zero time and boiled enzyme blanks are obtained by extending the duration of the washing procedure. We wash four times with 10% trichloroacetic acid, with at least one overnight wash, and three 30 min washes.

---

[3] J. D. Corbin and E. M. Reimann, this series, Vol. 38, p. 287.

Purification

Step 1. Preparation of Crude Extract

All handling of enzyme preparations is done at 0–4°. A 20% (w:v) homogenate of the insect tissue is made in 10 mM potassium phosphate (pH 6.8), 1 mM EDTA (buffer A), by treating in a Virtis homogenizer at half speed for 1.5 min. This homogenate is spun twice at 5000 g for 5 min and the final supernatant solution is then spun at 100,000 g for 60 min. The high-speed supernatant solution is strained through glass wool and is used as the starting material for subsequent purification steps.

Step 2. Batch Treatment with DEAE-Cellulose

The high-speed supernatant solution is added to DEAE-cellulose (DE-52, Whatman) that has been previously equilibrated with buffer A, in the proportion of 2 g wet weight cellulose/g of initial insect tissue. The mixture is allowed to stand for 30 min and the solution is recovered by filtration. (The cAMP–cGMP)-dependent protein kinase does not adsorb to DEAE at the pH and ionic strength of the buffer used. Two cAMP-dependent enzymes originating from the flight muscle in the preparation do adsorb and can be recovered separately from the cellulose by elution with a linear gradient of 0–0.4 M NaCl in buffer A.[1,2]

Step 3. Batch Treatment with CM-Sephadex C-50

The filtrate from the previous step is treated with CM-Sephadex C-50 preswollen and equilibrated with buffer A, in the proportion of 0.1 g dry weight of Sephadex/g of initial insect tissue. The solution is again recovered by filtration, and concentrated to approximately 5 ml using PM10 Amicon Diaflo Ultrafiltration membranes.

Step 4. Gel Filtration on Sephadex G-200

The concentrated filtrate from the previous step is applied on to a Sephadex G-200 column (2.5 × 100 cm), packed, and equilibrated with 10 mM potassium phosphate (pH 6.8), 50 mM NaCl, 1 mM EDTA, 5 mM 2-mercaptoethanol (buffer B). The elution pattern of the enzyme from the column is illustrated in Fig. 1. The active fractions are pooled as shown in the figure, concentrated to 2–5 ml as above, and used for further purification.

Step 5. Binding to and Elution from a cAMP Affinity Column

The column is 0.9 × 7 cm of cAMP-agarose (Sigma). It consists of cAMP attached through its N^6-amino group to an 8-carbon spacer bound

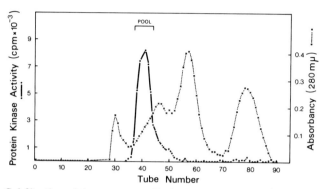

FIG. 1. Gel filtration of the enzyme on Sephadex G-200. The starting material was the filtrate of a high-speed supernatant solution from 1-day-old insects after a batch treatment with DEAE-cellulose followed by a batch treatment with CM-Sephadex C-50. It was applied on a 2.5 × 86-cm column of Sephadex G-200 and eluted with buffer B at a flow rate of 17 ml/hr. Fraction volume was 4.8 ml. Aliquots (30 μl) of each fraction were assayed for protein kinase activity in the presence of 1 μM cAMP. Peak fractions were pooled as shown, concentrated to 2–5 ml, and used as enzyme source for experiments or further purification. Taken from A. Vardanis, *J. Biol. Chem.* **255**, 7238 (1980).

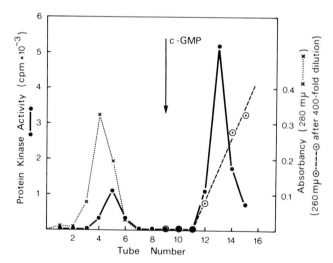

FIG. 2. Adsorption and elution of the enzyme on an affinity column. The enzyme source was 2 ml of the pooled concentrated preparation from the G-200 step. It was applied to a 0.9 × 7-cm column of cAMP-agarose (Sigma). Elution was effected with a 20 mM cGMP solution in buffer B. Flow rate, 5 ml/hr; fraction volume, 2 ml. All fractions after the cGMP addition were dialyzed overnight with at least four changes of buffer B. Aliquots (30 μl) of fractions were assayed for kinase activity in the presence of 1 μM cGMP.

to agarose. The concentrated eluate from the previous step is applied to this column and the enzyme is eluted with 20 mM cGMP in buffer B. All fractions after cGMP addition are dialyzed overnight against four changes of buffer B and assayed. The elution profile is shown in Fig. 2. The portion of activity that does not bind to the affinity column represents enzyme that is no longer dependent on cyclic nucleotide. The bound enzyme appears together with the cyclic nucleotide in the eluate. The three peak tubes contain homogeneous enzyme. A summary of the purification is presented in the table.

Stability of the Enzyme

The enzyme is extremely stable up until the G-200 step with no loss of activity for periods up to 1 week at 0° or several months at −80°. After elution from the G-200 column, however, there were two separable losses. Total activity losses (50% decline in 48 hr at 0°) and a small but steady loss in cyclic nucleotide-binding ability accompanied by an increase in cyclic nucleotide-independent activity. It is therefore advisable to go through steps 3 and 4 as quickly as possible.

PURIFICATION SUMMARY[a,b]

Fraction	Total volume (ml)	Total protein (mg)	Specific activity (units/mg protein × 10^6)	Total activity (units × 10^6)
High-speed supernatant	60	420	234.7[c]	98,574[c]
DEAE eluate	55	146	146.7	21,418
CM-Sephadex eluate	55	88	238.0	20,944
G-200 eluate after concentration	3.8	9.5	1,159.0	11,010
Affinity column eluate[d]	0.5	0.22	18,940.0	4,166

[a] Taken from A. Vardanis, *J. Biol. Chem.* **255,** 7238 (1980).
[b] The preparation was from 20 g of 1-day-old insects (thoraces and legs). Assay in the presence of 1 μM cAMP.
[c] This activity includes the two cAMP-dependent enzymes.
[d] Dialyzed and concentrated after elution.

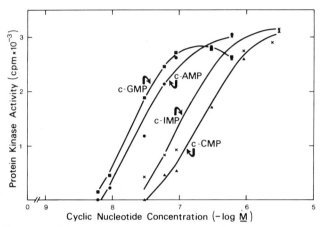

FIG. 3. Response of the enzyme to activation by cyclic nucleotides. Enzyme source was the pooled concentrated preparation from the G-200 step. Activity in the absence of cyclic nucleotides was 214 cpm and has been subtracted from the figures shown. The standard assay was used with additions of ■—■, cGMP; ●—●, cAMP; ×—×, cIMP; and ▲—▲, cCMP, at the concentrations shown. Taken from A. Vardanis, *J. Biol. Chem.* **255**, 7238 (1980).

Characteristics of the Enzyme

The enzyme has similar affinity for cAMP and cGMP (see Fig. 3) with K_a values of 43 and 25 nM, respectively, and lower affinity for cIMP and cCMP (K_a of 160 and 340 nM, respectively). The K_m for ATP is 86 μM and among several types of histones, the slightly lysine-rich subgroup f_{2a} is the best phosphate acceptor. Protamine and casein do not act as acceptor proteins. Maximal activities (at 0.24 mM ATP) are obtained with 1 mM Mg^{2+} while Mn^{2+} at concentrations up to 6 mM is completely ineffective. The natural acceptor substrate for this enzyme is not known. The native enzyme has a molecular weight of 180,000 estimated by gel filtration. The enzyme does not dissociate upon activation with cyclic nucleotide, and gives a single protein upon treatment with SDS whose mobility corresponds to a molecular weight of 90,000. The enzyme is associated with an active epidermis and its titer declines dramatically in the 7 days following the insect molt.[2]

[10] Preparation of Partially Purified Protein Kinase Inhibitor

By KEITH K. SCHLENDER, JENNIFER L. TYMA, and ERWIN M. REIMANN

The heat-stable inhibitor of cAMP-dependent protein kinase specifically inhibits kinase activity by binding to the free catalytic subunit.[1] The inhibitor can be used as a tool to elucidate cellular processes which are mediated by cAMP-dependent protein kinases.[2] There are numerous reports of experiments which utilized crude inhibitor (purified through the heating, trichloroacetic acid precipitation, and dialysis steps[1]) to inhibit cAMP-dependent protein kinases. More purified preparations of the inhibitor are useful since the crude inhibitor is often too dilute[3] and it contains protein contaminants which may serve as substrates for protein kinases.[4] We describe here a simple and convenient modification[3] of the procedure of Walsh and co-workers[1,5] for the partial purification of the heat-stable inhibitor of cAMP-dependent protein kinases.

Materials

Histone (Type II-A) obtained from Sigma was heated at 100° for 10 min as described in this volume.[6] [γ-^{32}P]ATP was prepared by the method of Walseth and Johnson.[7] The catalytic subunit of cAMP-dependent protein kinase (spec. act. 3 to 4 units/mg protein) was prepared as described in this volume.[6] DEAE-cellulose (DE-52) and phosphocellulose paper (P-81) were obtained from Whatman. Frozen rabbit skeletal muscle was purchased from Pel-Freeze.

Assay

The inhibitor assay is based on the inhibition of histone kinase activity of purified catalytic subunit of cAMP-dependent protein kinase.[3] The kinase reaction contained 50 μl of test mixture [70 mM morpholinoethane-

---

[1] C. D. Ashby and D. A. Walsh, this series, Vol. 38, p. 350.
[2] J. A. Traugh, C. D. Ashby, and D. A. Walsh, this series, Vol. 38, p. 290.
[3] K. K. Schlender and E. M. Reimann, *J. Biol. Chem.* **252**, 2384 (1977).
[4] J. R. Kanter and L. L. Brunton, *J. Cyclic Nucleotide Res.* **7**, 259 (1981).
[5] D. A. Walsh, C. D. Ashby, C. Gonzalez, D. Calkins, E. H. Fischer, and E. G. Krebs, *J. Biol. Chem.* **246**, 1977 (1971).
[6] E. M. Reimann and R. A. Beham, this volume [6].
[7] T. F. Walseth and R. A. Johnson, *Biochim. Biophys. Acta* **562**, 11 (1979).

sulfonic acid (pH 6.5), 30 mM KF, 0.4 mM EGTA, 3 mM theophylline, 8.5 mg histone/ml, 285 μM [γ-^{32}P]ATP 10 to 20 cpm/pmol, and 14 mM magnesium acetate], 10 μl of the appropriately diluted inhibitor and 20 μl of catalytic subunit (50 μunits in 25 mM KH$_2$PO$_4$, 25 mM K$_2$HPO$_4$, 1 mM EDTA, 0.1 mM dithiothreitol, and 1.5 mg gelatin/ml). After incubation for 10 min at 30°, 40-μl aliquots were withdrawn and spotted onto 1 × 2-cm pieces of phosphocellulose filter paper and dropped into a beaker containing 50 mM NaCl (>5 ml/paper). The solution was slowly stirred and was changed at 5-min intervals for a total of four washes. The papers were then washed in acetone for 5 min, dried under a heat lamp, and placed in vials for liquid scintillation counting.

Purification

Frozen rabbit skeletal muscle (2.5 kg) is placed in plastic bags and thawed under running tap water. Unless otherwise noted all subsequent operations are conducted at 0 to 4°. The tissue is ground with a power meat grinder and homogenized in 5 liters of 4 mM EDTA, pH 6.8. The homogenate is centrifuged at 10,000 g for 35 min and the supernatant is filtered through glass wool. Alternatively, the pH 6.1 supernatant from a glycogen pellet preparation[8] can be used. The supernatant is placed in two 3-liter beakers and heated with stirring to 90° (requires approx. 35 min). The beakers are then placed in ice to chill the suspension and the material is filtered through several layers of cheesecloth. The cheesecloth is squeezed to remove fluid. The remaining particulates are removed by filtering through filter paper. The clear supernatant solution is made to 5% (w/v) trichloroacetic acid by the rapid addition with stirring of 100% (w/v) trichloroacetic acid. After standing for 10 min, the preparation is collected by centrifuging at 10,000 g for 15 min. The supernatant is discarded and the drained pellets are dissolved in 40 ml of 1 M K$_2$HPO$_4$. The high concentration of phosphate neutralizes the trichloroacetic acid precipitate and the protein is rapidly dissolved. The solution is placed in a dialysis bag and dialyzed twice (4 and 12 hr) against 4 liters of glass distilled water and then 6 hr against 1 liter of 50 mM Tris · HCl, 1 mM EDTA, pH 7.5 at 4°. Insoluble material is removed from the dialyzate by centrifuging at 20,000 g for 10 min. The inhibitor fraction may be stored at this step for months at −70° with little change in activity.

The supernatant is applied to a DEAE-cellulose column (0.9 × 8 cm) equilibrated with 50 mM Tris · HCl, 1 mM EDTA, pH 7.5 at 4°. The column is washed with approx. 5 bed volumes of equilibrating buffer and the inhibitor is eluted with equilibrating buffer containing 100 mM NaCl.

[8] E. G. Krebs, this series, Vol. 8, p. 543.

The inhibitory activity which is eluted in two or three 1-ml fractions is stable when stored for months at −70°.

Figure 1 shows an inhibition curve of catalytic subunit of the cAMP-dependent protein kinase using the purified inhibitor. At the highest concentration of inhibitor, protein kinase activity was inhibited 99%.

Comments about the Preparation

This procedure yields an inhibitor preparation which is purified 500- to 1000-fold in yields of >50%. The DEAE-cellulose column in addition to purification concentrates the inhibitor so that <1 µl is needed to completely inhibit cAMP-dependent protein kinase in a typical assay of extracts from a variety of tissues.[3] Although the inhibitor is not electrophoretically pure, it contains very little protein substrate for several

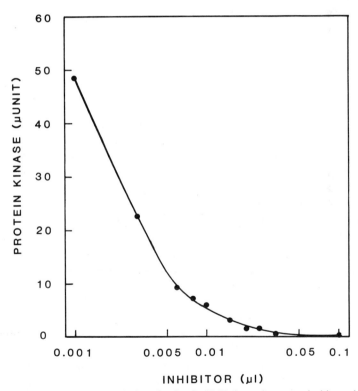

FIG. 1. Inhibition of the catalytic subunit of cAMP-dependent protein kinase by heat-stable inhibitor. Purified inhibitor (4.8 mg/ml based on $A_{280}:A_{260}$ ratio) was assayed as described in the text. In the absence of any added inhibitor, there were 50.9 µunits of histone kinase activity per assay.

cAMP-independent protein kinases.[9] In addition, it has little if any inhibitory activity toward several cAMP-independent glycogen synthase kinases, histone kinases, phosvitin kinases, and casein kinases.[3,9-11] The partially purified inhibitor described here should be useful for many studies of protein kinases. The methods of preparation and characterization of more purified inhibitor are also presented in this volume.[12]

Acknowledgments

This research was supported in part by U.S.P.H.S. Grants AM-14273 and AM-19231 and by U.S.P.H.S. Research Career Development Awards AM-00337 to K.K.S. and AM-00446 to E.M.R.

[9] Unpublished results.
[10] K. K. Schlender, S. J. Beebe, J. C. Willey, S. A. Lutz, and E. M. Reimann, *Biochim. Biophys. Acta* **615**, 324 (1980).
[11] K. K. Schlender, S. J. Beebe, and E. M. Reimann, *Cold Spring Harbor Conf. Cell Proliferation* **8**, 389 (1981).
[12] D. A. Walsh, this volume [11].

[11] Inhibitor Protein of the cAMP-Dependent Protein Kinase: Characteristics and Purification

By SUSAN WHITEHOUSE and DONAL A. WALSH

The literature contains descriptions of a number of apparently different heat-stable inhibitor proteins of the cAMP-dependent protein kinase.[1-5] One potential explanation for this is the recent observation that the inhibitor protein, initially partially purified and characterized by Walsh and co-workers,[2] can exhibit a number of different charge and size/shape species.[6-8] On ion-exchange chromatography, three charge forms, $Inhibitors_{1,2,3}$ (denoting their order of elution), are readily detected and separated. Each charge form may exhibit two size/shape species, named I and I'. Based on the physicochemical data to date, it has been proposed

[1] D. A. Walsh, C. D. Ashby, C. Gonzalez, D. Calkins, E. H. Fischer, and E. G. Krebs, *J. Biol. Chem.* **246**, 1977 (1971).
[2] J. G. Demaille, K. A. Peters, and E. H. Fischer, *Biochemistry* **16**, 3080 (1977).
[3] J. S. Tash, M. J. Welsh, and A. R. Means, *Cell* **21**, 57 (1980).
[4] A. Szmigielski, A. Guidotti, and E. Costa, *J. Biol. Chem.* **252**, 3848 (1977).
[5] H. Weber and O. M. Rosen, *J. Cyclic Nucleotide Res.* **3**, 415 (1977).

that I and I' are related as shape conformers.[6] Although I is the major, if not only, size/shape form detected in initial tissue extracts, it may be converted to I' at later stages of the purification of the protein. I' to I conversion, however, has never been observed. Until all the factors involved in the conversion are understood it cannot be fully avoided, and hence, partially purified and purified preparations may contain mixtures of the I and I' forms. Such preparations may be expeditiously monitored for the presence of the two forms by SDS–gel electrophoresis[9] and methods are available for their separation. The preparation described here may be used to purify all inhibitor forms that have been detected (I_1, I'_1, I_2, I'_2, I_3, and I'_3).

Assay Principle

Assay of the inhibitor protein is based on its ability to inhibit the phosphorylation of histone as catalyzed by the purified catalytic subunit of the cAMP-dependent protein kinase.

Reagents

Assay Buffer. 100 mM MES buffer (pH 6.8)–20 mM magnesium chloride–15 mM 2-mercaptoethanol.

Substrate Solution. Histone (Sigma IIA) dissolved in 10 mM MES buffer (pH 6.8) to a final concentration of 30 mg/ml. Prior to use, the solution is adjusted to pH 6.8 and dialyzed for 12 hr at 4° against 10 mM MES buffer (pH 6.8).

Dilution Buffer. BSA dissolved in 10 mM MES buffer (pH 6.8) to a final concentration of 0.5 mg/ml. The pH is adjusted to 6.8 and the solution is dialyzed prior to use, as described above for the preparation of substrate.

[γ-^{32}P]ATP. 1 mM solution (specific activity ~100 cpm/pmol) in water. [γ-^{32}P]ATP may be obtained commercially or prepared by a modification of the method of Glynn and Chappell.[10,11]

---

[6] J. M. McPherson, S. Whitehouse, and D. A. Walsh, *Biochemistry* **18**, 4835 (1979).
[7] S. Whitehouse, J. M. McPherson, and D. A. Walsh, *Arch. Biochem. Biophys.* **203**, 734 (1980).
[8] S. Whitehouse and D. A. Walsh, *J. Biol. Chem.* **257**, 6028 (1982).
[9] Abbreviations used: MES, 2-(*N*-morpholino)ethanesulfonic acid; BSA, bovine serum albumin; TCA, trichloroacetic acid; EDTA, ethylenediaminetetraacetic acid; SDS, sodium dodecyl sulfate.
[10] I. M. Glynn and J. B. Chappell, *Biochem. J.* **90**, 147 (1964).
[11] D. A. Walsh, J. P. Perkins, C. O. Brostrom, E. S. Ho, and E. G. Krebs, *J. Biol. Chem.* **246**, 1968 (1971).

Protein Kinase. The catalytic subunit of the cAMP-dependent protein kinase is purified from bovine cardiac muscle by the method of Beavo *et al.*[12] The enzyme is stored at $-20°$ in 50% glycerol–5 mM MES buffer (pH 6.8)–50 mM sodium chloride–7 mM 2-mercaptoethanol.

Procedure

The inhibitor fraction to be assayed is diluted in dilution buffer so that a 30-μl aliquot added to the assay will give between 10 and 50% inhibition. Assay buffer (20 μl), and inhibitor (30 μl) are added to disposable (1 ml) plastic tubes in an ice bath. Immediately prior to use, the catalytic subunit is diluted in the substrate solution such as would give in the assay an incorporation of 49 ± 7 pmol of phosphate per minute, and the solution is maintained on ice. Aliquots (20 μl) of this mixture are added to the assay tubes at timed intervals. The assay tubes are then transferred to a water bath at 30°. Following a 10-min preincubation, the reaction is initiated by the timed addition of 10-μl aliquots of [γ-^{32}P]ATP prewarmed to 30°. The reaction is terminated after 10 min by the transfer of 50-μl aliquots of the reaction mixtures to 1 × 1-cm squares of Whatman ET31 filter paper which are immersed immediately in cold (0°) 10% TCA. The papers are washed and [^{32}P]phosphate incorporation is determined as described by Corbin and Reimann.[13] Included in each assay are (a) a standard inhibitor curve, using sufficient inhibitor protein to produce inhibition in the range of 10–90%; (b) a blank (to determine nonspecific [γ-^{32}P]ATP binding) containing all the components of the assay with the exception of the inhibitor protein and the catalytic subunit, and (c) a control, containing all the components with the exception of the inhibitor protein.

Unit of Activity

One unit of inhibitor has been defined as that amount which will decrease a control velocity of 49 ± 7 pmol of phosphate incorporated per minute by 7 pmol of phosphate incorporated per minute in the standard assay. This unit of activity is essentially identical to that initially defined[14] using casein as the substrate for the protein kinase, but histone is a better substrate than casein and so the assay for the inhibitor protein with histone has greater sensitivity.

---

[12] J. Beavo, P. Bechtel, and E. G. Krebs, this series, Vol. 38, p. 299.
[13] J. D. Corbin and E. M. Reimann, this series, Vol. 38, p. 287.
[14] C. D. Ashby and D. A. Walsh, *J. Biol. Chem.* **247,** 6637 (1972).

General Comments on the Assay

In order to quantify inhibitor activity accurately, both the level of control activity and the specific activity of the protein kinase should be maintained within narrow limits. In the experiment illustrated in Fig. 1a, the extent of inhibition is depicted as a function of increasing concentrations of catalytic subunit. The degree of inhibition is dependent on the concentration of both the catalytic subunit and the inhibitor protein, increasing concentrations of the catalytic subunit requiring increasing concentrations of inhibitor to effect the same degree of inhibition. This illustrates the necessity of assaying the inhibitor protein with a specific concentration of protein kinase (i.e., 49 ± 7 pmol/min). This is because the inhibitor protein is a tightly bound inhibitor[2,6] and, as a consequence, in the reaction $C + I \rightleftharpoons CI$ the extent of CI formed is very dependent on the concentration of either C or I in the reaction mixture. In addition to this concern, prolonged storage of the protein kinase, which may result in loss of catalytic activity, is not necessarily associated with an equivalent decrease in catalytic subunit-inhibitor interaction. This is depicted in Fig. 1b; standard curves were constructed employing catalytic subunit prepa-

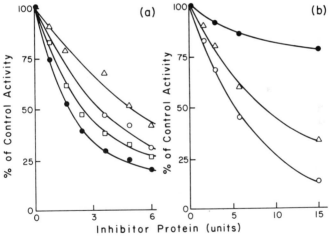

FIG. 1. (a) The dependency of the extent of inhibition by the inhibitor protein on the amount of protein kinase. Initial protein kinase activities were ●—●, 20.3 μunits; □—□, 43.75 μunits; ○—○, 70 μunits; and △—△, 162 μunits, respectively. From C. D. Ashby and D. A. Walsh, *J. Biol. Chem.* **247,** 6637 (1972). (b) Effect of inhibitor protein on catalytic subunit activity of preparations of differing specific activities induced by storage at 4° over an 8-month period. The specific activities of the protein kinase used were ○—○, 3.2 units/mg; △—△, 1.2 units/mg; and ●—●, 0.1 units/mg. In each case sufficient protein kinase was added to give a control activity of 49 pmol/min. From J. M. McPherson, S. Whitehouse, and D. A. Walsh, *Biochemistry* **18,** 4835 (1979).

rations of 3.2, 1.2, and 0.1 units/mg with decreased activity a consequence of age-dependent denaturation. In each case, sufficient protein kinase was added by increasing the amount of protein with the aged enzyme to give a control activity of 49 pmol/min. As is noted, a greater increment of protein inhibitor was required to effect an equivalent decrease in protein kinase activity for a partially denatured enzyme (i.e., protein kinase that had a diminished catalytic activity retained a more normal capacity to bind inhibitor protein). Thus, the concentration of the catalytic subunit in inhibitor assays should be (a) maintained within a narrow range, and (b) either purified preparations of catalytic subunit with specific activities of no less than 2.5 units/mg or, if partially purified preparations are to be employed, care should have been taken to avoid partial denaturation of the kinase.

Purification of the Inhibitor Protein

The purification procedure takes advantage of the heat and acid stability of the protein. Purification to homogeneity is achieved by affinity chromatography with the catalytic subunit of the protein kinase as the immobilized ligand. The affinity resin is prepared and stored as described by Demaille and co-workers.[2]

Step 1: Tissue Extraction. Eight New Zealand White female rabbits, each weighing 6–8 lb, are sacrificed by intravenous injection of Nembutal (60 mg/kg body weight) and exsanguination. The muscle (about 4 kg) from the hind legs and back is iced, ground in a chilled meat grinder, and homogenized in 2.5 volumes (ml/g) of cold 4 mM EDTA (pH 7.0) in a Waring blender at top speed for 1 min at 4°. After centrifugation at 7000 g for 40 min, the supernatant is decanted through glass wool. Note: At this point, the extract may also be employed for the preparation of phosphorylase kinase, glycogen synthase, or phosphofructokinase, which are precipitated by lowering the pH to 6.0 ± 0.1 by the dropwise addition, with stirring, of 7% (v/v) acetic acid. The solution is stirred for 10 min at 4° and then centrifuged at 7000 g for 40 min. The supernatant is the source of the inhibitor protein. This step may be omitted if the inhibitor is the only protein to be purified.

Step 2: Heat Treatment. The supernatant is neutralized with 6 N NH$_4$OH and then heated to 90° (with rigorous overhead stirring) in a stainless-steel bucket using two Fischer burners. As soon as the temperature has been reached, the bulk of denatured protein is removed by filtering through a double layer of cheesecloth. The solution is then transferred to a bucket in ice and cooled to 40°, with overhead stirring. Centrifugation

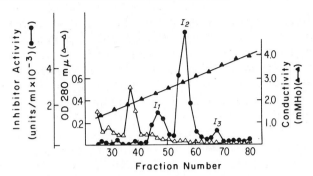

FIG. 2. Separation of the inhibitor charge isomers by chromatography on DEAE at pH 5.0. Conditions are presented in the text.

for 30 min at 7000 g in a refrigerated Sorvall RC-3 centrifuge (4°) results in further cooling and the removal of remaining denatured protein.

Step 3: Trichloroacetic Acid Precipitation. This and all subsequent operations are performed at 4°. Trichloroacetic acid (100% w/v) is added rapidly,[15] with vigorous overhead stirring, to the supernatant obtained from Step 2 to a final concentration of 5%. After stirring on ice for 30 min, the precipitate is collected by centrifugation at 7000 g for 30 min. The precipitate is resuspended in approximately 80 ml of deionized distilled water. The pH is adjusted to 7.0 with 6 N NH$_4$OH, and the solution is then dialyzed in Spectrapor 1 dialysis tubing against two changes (10 liters each) of 5 mM potassium phosphate buffer (pH 7.0), containing 1 mM EDTA, until the conductivity is below 1 mmho.

Step 4: DEAE-Cellulose Chromatography. At this step, the charge forms of the inhibitor protein are separated. The solution is adjusted to pH 5.0 with 7% (v/v) acetic acid and stirred for 30 min. The supernatant obtained following centrifugation for 30 min at 34,000 g is applied to a column (1.5 × 30 cm) of DE-52 equilibrated with 5 mM sodium acetate (pH 5.0). The column is washed with 250 ml of this buffer and subsequently developed with a linear (1-liter) sodium acetate gradient from 5 to 350 mM (pH 5.0). A typical elution profile is shown in Fig. 2. The fractions containing the individual inhibitor peaks are pooled, neutralized with 1 N NaOH, and dialyzed, in Spectrapor 1 dialysis tubing, for 12 hr against two changes (4 liters each) of Buffer A (30 mM KH$_2$PO$_4$–0.1 mM EDTA–15 mM 2-mercaptoethanol, adjusted to pH 6.7 with 1 M KOH) containing 0.15 M KCl.

---

[15] If the trichloroacetic acid is added slowly there is a lower yield of inhibitor protein.

TABLE I
PURIFICATION DATA FOR THE INHIBITOR PROTEIN

Fraction number	Fraction	Specific activity[a,b] (units/mg of protein)	Yield[c] (%)	Purification (n-fold)
I	Initial extract[c]	6.2–11.6		1
II	Heat filtrate	$(3.0–4.2) \times 10^2$	100	43[d]
III	5% TCA precipitate	$(0.4–1.1) \times 10^3$	30–80	76
IV	DE-52 eluate, Inhibitor$_2$	$(3.5–6.7) \times 10^4$	8–22	6,000
V	Affinity chromatography	$(3.5–6.2) \times 10^5$	8–20	25,000

[a] Ranges indicated for three typical preparations.
[b] Protein was determined by the method of Lowry et al.[16] employing BSA as standard.
[c] Preparation from 3.7 kg of rabbit skeletal muscle.
[d] Accuracy for determination of the activity in the initial extract is compromised by the presence of protein kinase substrates and trypsin insensitive inhibitors; this value is based on the assumption of 100% yield at the initial step.

Step 5: Affinity Chromatography. The dialysate is loaded (at 20 ml/hr) onto a 0.9 × 3-cm bed of catalytic subunit affinity resin equilibrated with Buffer A containing 0.15 M KCl. The column is washed in this buffer until the 254 nm absorbance of the eluate is at background level. The inhibitor protein is eluted with 30 ml of Buffer A containing 1 M KCl and 0.3 M guanidinium hydrochloride (pH 6.7). The eluate is dialyzed (in Spectrapor 3 dialysis tubing) for 24 hr, at 4°, against four changes (4 liters each) of 1 mM MES buffer, pH 6.8. The dialysate is concentrated by lyophilization, the powder suspended in a minimal volume of water, and the protein stored at −20° in 50-μl aliquots. A summary of the purification data for the *Inhibitor$_2$* charge isomer, purified from 3.7 kg of rabbit skeletal muscle, is presented in Table I.[16] The protein is homogeneous by the criterion of Coomassie Blue R$_{250}$ staining following electrophoresis on both 9% nondenaturing and 12% SDS gels. *Inhibitor$_1$* and *Inhibitor$_3$* charge isomers may also be purified by this procedure.

General Comments on the Purification

1. With some preparations of affinity resin, it has been found necessary to modify the washing procedure. Although inhibitor activity has never been detected in the breakthrough fraction, it has occasionally been found to "leak off" during the washing of the resin with Buffer A containing 0.15 M KCl. For those preparations where this occurs (repeated use of

[16] O. H. Lowry, N. J. Rosebrough, A. L. Farr, and R. J. Randall, *J. Biol. Chem.* **193**, 265 (1951).

the resin may also result in such a situation), inhibitor protein and affinity resin are equilibrated in Buffer A: the sample is loaded as above, and the resin then washed with a linear (200 ml) gradient of Buffer A containing 0–200 mM KCl at pH 6.7. The inhibitor protein is subsequently eluted and concentrated as described above.

2. Due to the tendency for inhibitor to convert from the I to the I' species, preparations should be monitored for the presence of the two forms. Several methods are available:

    a. *SDS Gel Electrophoresis*. This method is sufficiently sensitive to use when inhibitor is present in low concentrations (minimum requirement, approximately 400 units/ml) and consequently it may be used to assess the composition of both pure preparations and of fractions at any stage of the preparation following the heat treatment (Step 2). The migration of the protein is determined by elution from the gels and assay of activity. Immediately following electrophoresis of the inhibitor protein (12% polyacrylamide gels, 0.1% SDS, method according to Weber and Osborn[17]), the gels are sliced into 2-mm sections which are placed into tubes containing 200 μl of deionized, distilled water. The tubes are sealed and shaken for 12 hr at 25° on an Orbital shaker. Aliquots (10–30 μl) are assayed for inhibitor activity. The recovery of inhibitor from the gel ranges from 60 to 90%. The mobility of I (relative to the tracking dye bromphenol blue) is 0.56 ± 0.02 while that of I' is 0.73 ± 0.02. As an example of the use of this method, the presence of different proportions of I and I' in fractions at various stages of the preparation is depicted in Fig. 3. Only one size/shape form, I, is detected in the heat supernatant (Fig. 3a). Following chromatography on DE-52, however, different preparations of the *Inhibitor*$_2$ charge isomer may contain differing proportions of the I and I' forms (b–d). The composition of the purified preparation following affinity chromatography is identical to that of the preparation applied. That is, on chromatography of a preparation containing only the I form (Fig. 3b) the purified preparation is I (Fig. 3e). However, if preparations containing mixtures of the two forms are applied to the affinity column (Fig. 3c and d), they are not separated and the purified preparation contains the two forms in the same relative proportions as in the preparation that was applied (Fig. 3f and g, respectively).

    b. *Gel Exclusion Chromatography*. The I and I' forms of the inhibitor protein are readily separated on Sephadex G-75, I eluting with a V_e/V_0 of 1.5 while that of I' is 1.95 (Fig. 4a). As easily distinguished markers, I coelutes with soybean trypsin inhibitor while I' elutes a few fractions after cytochrome c. Typically, concentrated samples (containing approxi-

---

[17] K. Weber and M. Osborn, *J. Biol. Chem.* **244**, 4406 (1969).

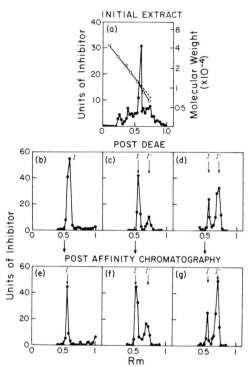

FIG. 3. Determination by SDS gel electrophoresis of the composition of inhibitor preparations at different stages of the purification. The electrophoresis system was that of Weber and Osborn[17] with 12% SDS–polyacrylamide gels. The location of the inhibitor protein was determined as described in the text. (a) heat filtrate (fraction II); (b–d) three different preparations at the DE-52 eluate stage (Fraction IV) (as described in detail elsewhere,[8] preparations at this stage exhibit a variable composition of *Inhibitor I* and *I'*); (e–g) composition of inhibitor protein preparations following affinity chromatography (Fraction V). The preparations indicated in b, c, and d yielded those indicated in e, f, and g, respectively.

mately 10,000 units/ml) can be readily analyzed by this method using a 0.9 × 50 cm column of Sephadex G-75. If dilute samples are to be analyzed (e.g., the crude cell extracts) inhibitor activity cannot be assayed directly, but may be measured following concentration of fractions with a V_e/V_0 of 1.3–1.7 for I and 1.8–2.2 for I' as described previously; an example of such a use is presented in McPherson et al.[6]

c. *Other Methods to Distinguish I and I' Forms.* In addition to differences in size/shape characteristics, the I and I' inhibitor forms also differ slightly in net charge, and can be separated either on ion-exchange chromatography on DE-52 at pH 7.0 (Fig. 4b) or by electrophoresis at pH 8.9

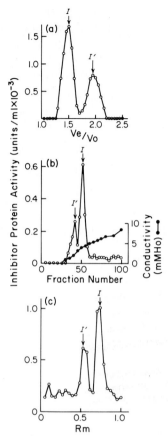

FIG. 4. Separation of the I and I' inhibitor forms of $Inhibitor_2$ charge isomer. In (a) inhibitor protein (1×10^5 units) was chromatographed on Sephadex G-75 (0.9×50 cm) equilibrated in 5 mM Tris chloride, 20 mM NaCl at pH 7.5; 0.6 ml fractions were collected. In (b) inhibitor protein (2×10^5 units) was equilibrated with 5 mM MES buffer, pH 7.0, and applied to a column of DE-52 (0.9×12 cm) equilibrated with the same. The column was washed with 100 ml of buffer and then eluted with a 300 ml linear gradient of 0–500 mM NaCl in 5 mM MES buffer, pH 7.0. In (c) inhibitor protein (6×10^3 units) was electrophoresed on 9% nondenaturing polyacrylamide gels according to the procedure of Orstein[18] and Davis.[19] At the termination of the electrophoresis, the gels were sliced into 2-mm sections and the inhibitor protein eluted by shaking overnight at 26° in 200 μl of 200 mM MES buffer, pH 6.5. In both (a) and (b) the identity of the inhibitor fractions can be readily confirmed by SDS–gel electrophoresis as in Fig. 3.

in a nondenaturing gel system[14,15] (Fig. 4c).[18,19] In the latter, I migrates with an R_m (relative to the tracking dye bromphenol blue) of 0.74 ± 0.02 while that of I' is 0.58 ± 0.03. The differing electrophoretic mobility of the two forms in this system is predominantly due to differences in charge rather than size, with the I and I' forms of the *Inhibitor*$_2$ exhibiting an identical dependency of mobility on gel concentration when analyzed by the procedure of Hedrick and Smith.[20]

3. As depicted by the electrophoretic profiles shown in Fig. 3e–g, purified inhibitor preparations may, depending on the degree of conversion of I and I' occurring throughout the preparation, contain either only *I-Inhibitor* or a mixture of both the I and I' inhibitor forms. Each may be obtained in a homogeneous preparation by separation of the two forms by preparative Sephadex G-75 chromatography.

Properties of the Inhibitor Protein

The presence of multiple charge forms of the inhibitor protein is not limited to rabbit skeletal muscle, and multiple forms have also been observed with both bovine cardiac muscle[7] and bovine brain.[21] They are also present in preparations of the protein which have not involved the rather harsh heat and TCA treatments.[7] The best studied size forms of the protein are the I and I' forms of the *Inhibitor*$_2$ charge isomer, the major charge isomer isolated by the procedure described. The properties of these forms are presented in Table II. The molecular size characteristics of the I and I' forms of the *Inhibitor*$_2$ and *Inhibitor*$_3$ charge forms appear identical to those described for the *Inhibitor*$_2$ size/shape forms.[7] Detailed kinetic analyses of the I forms of *Inhibitor*$_1$ and *Inhibitor*$_3$ have not been performed, but the K_i of the I' form of *Inhibitor*$_3$ has been reported to be fourfold higher than that of the I' forms of *Inhibitor*$_1$ and *Inhibitor*$_2$.[22]

Conversion of I and I' during protein purification can be minimized by performing the purification as rapidly as possible and by avoiding either extensive periods of dialysis and/or lyophilization from dilute solutions. Dialysis and subsequent lyophilization in ammonium acetate or bicarbonate buffers has been shown[6] to potentially promote conversion of I to I'. However, treatments generally considered harsher or denaturing such as heating to 65° in the presence of 1% SDS and 1% Triton X-100, treatment

---

[18] L. Ornstein, *Ann. N.Y. Acad. Sci.* **121,** 321 (1964).
[19] B. Davis, *Ann. N.Y. Acad. Sci.* **121,** 404 (1964).
[20] J. L. Hedrick and A. J. Smith, *Arch. Biochem. Biophys.* **126,** 155 (1968).
[21] J. G. Demaille, K. A. Peters, T. P. Standjord, and E. H. Fischer, *FEBS Lett.* **86,** 113 (1978).
[22] C. Ferraz, J. G. Demaille, and E. H. Fischer, *Biochimie* **61,** 645 (1979).

TABLE II
PROPERTIES OF THE I AND I' FORMS OF THE $Inhibitor_2$ CHARGE ISOMER

	I	I'
Molecular size		
Sephadex G-75	22,000	11,000
SDS gel electrophoresis	10,000–11,000	6,500–8,000
Nondenaturing gel electrophoresis	10,000–11,000	10,000–11,000
Amino acid analysis	8,300–8,700	8,900–9,100[a]
Stokes radius (Å)	21	15
f/f_0	1.3	1.0
Axial ratio	4	1
pI	4.1	4.1
K_i (pM)	~500	~500
Mode of inhibition		
With respect to protein or peptide substrate	Competitive	Competitive
With respect to Mg^{2+}-ATP	Uncompetitive	Uncompetitive

[a] Taken from C. Ferraz, J. G. Demaille, and E. H. Fischer, *Biochimie* **61**, 645 (1979).

with 15% TCA or 0.3 M guanidinium hydrochloride, or heating to 95° do not cause conversion of I to I' (6 and unpublished observations). To date, studies on the use of inhibitors of proteolytic enzymes to potentially block I to I' conversion have yielded only negative results.

Physiological Role of the Inhibitor Protein

Quantitation of Inhibitor Protein in Crude Extracts. The quantitation of inhibitor activity in crude extracts is compromised by factors which interfere with the assay. Contaminating kinases and protein substrates may be removed by first boiling the extract for 5 min (followed by centrifugation to remove denatured protein). Since the heat supernatant may contain, in addition, nonprotein factors which inhibit the catalytic subunit, it is essential to assess the trypsin sensitivity of the inhibition. To do so, inhibitor fraction (50 µl) is incubated for 30 min at 30° with 30 µl of 100 mM MES buffer (pH 6.8) in the presence and absence of 10 µl of trypsin (1.3 mg/ml). The proteolytic reaction is terminated by the addition of 10 µl of soybean trypsin inhibitor (4 mg/ml) and the samples assayed for inhibitor activity. Trypsin-sensitive (i.e., inhibitor protein) inhibitory activity is then quantitated from [inhibitor activity without trypsin treatment] minus [inhibitor activity with trypsin treatment]. Although this is the best method available to date, results should be analyzed with caution and a fully reliable method, not compromised by the presence of forms with potentially different stabilities, remains to be developed.

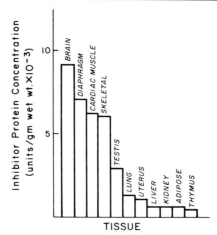

FIG. 5. Distribution of the inhibitor protein in rabbit tissues. Tissues were prepared as described by Ashby and Walsh.[14]

Tissue Distribution

The distribution of inhibitor protein in rabbit tissues is indicated in Fig. 5. The concentration of the protein varies widely between tissues, whereas, in contrast, cAMP-dependent protein kinase activity in most tissues is approximately the same (within 2-fold). While there is sufficient inhibitor protein in heart and skeletal muscle to inhibit approximately 20% of protein kinase activity, in those tissues containing 10-fold less inhibitor it is questionable whether the protein could, under normal circumstances, play a role in regulating the activity of the kinase.

Regulation of Tissue Levels of the Inhibitor Protein

1. In the Sertoli cell-enriched rat testis, inhibitor activity was increased by intraperitoneal injection of follicle-stimulating hormone. The stimulation was specific and dependent on both the age of the animal, and continued protein synthesis. The increase in inhibitor was paralleled by a decrease in protein kinase activity.[23]

2. Inverse relationships between inhibitor and protein kinase activities have also been demonstrated in rat brown adipose tissue during perinatal and postnatal development[24,25] and in confluent nutrient-deprived Chinese hamster ovary cells stimulated to proliferate.[26]

[23] J. S. Tash, J. R. Dedman, and A. R. Means, *J. Biol. Chem.* **254,** 1241 (1979).
[24] J. P. Skala, G. I. Drummond, and P. Hahn, *Biochem. J.* **138,** 195 (1974).
[25] J. P. Skala and B. L. Knight, *J. Biol. Chem.* **252,** 1064 (1977).
[26] M. Costa, *Biochem. Biophys. Res. Commun.* **78,** 1311 (1977).

Although the assessment of inhibitor activity in crude extracts is uncertain, these studies suggest that in some systems, the protein may be under hormonal and nutritional control and indicate that the inhibitor protein may be an important physiological regulator of cAMP-dependent protein phosphorylation.

Acknowledgment

This work was supported by Grant AM 21019 from the National Institutes of Health.

[12] Use of NMR and EPR to Study cAMP-Dependent Protein Kinase

By ALBERT S. MILDVAN, PAUL R. ROSEVEAR, JOSEPH GRANOT, CATHERINE A. O'BRIAN, H. NEIL BRAMSON, and E. T. KAISER

Magnetic resonance methods, such as EPR and NMR, have long been applied to the study of metal-activated and metalloenzymes and their substrate complexes.[1-5] The application of these powerful methods to bovine heart cAMP-dependent protein kinase represents their extension to a more complex system. In this system there are two dissociable divalent cations per enzyme complex[6] and two enzyme-bound flexible substrates with many possible conformations, ATP[7-9] and a large peptide.[9,10] Additionally, the interaction of two catalytic subunits with another protein, the regulatory dimer[11] and the interaction of four allosteric effector

[1] A. S. Mildvan and M. Cohn, *Adv. Enzymol.* **33**, 1 (1970).
[2] A. S. Mildvan, *Adv. Enzymol.* **49**, 103 (1979).
[3] A. S. Mildvan and J. L. Engle, this series, Vol. 26, p. 654.
[4] A. S. Mildvan and R. K. Gupta, this series, Vol. 49, p. 322.
[5] A. S. Mildvan, *Philos. Trans. R. Soc. Ser. B* **293**, 65 (1981).
[6] R. N. Armstrong, H. Kondo, J. Granot, E. T. Kaiser, and A. S. Mildvan, *Biochemistry* **18**, 1230 (1979).
[7] J. Granot, H. Kondo, R. N. Armstrong, A. S. Mildvan, and E. T. Kaiser, *Biochemistry* **18**, 2339 (1979).
[8] J. Granot, A. S. Mildvan, E. Brown, H. Kondo, H. N. Bramson, and E. T. Kaiser, *FEBS Lett.* **103**, 265 (1979).
[9] J. Granot, A. S. Mildvan, H. N. Bramson, and E. T. Kaiser, *Biochemistry* **19**, 3537 (1980).
[10] J. Granot, A. S. Mildvan, H. N. Bramson, N. Thomas, and E. T. Kaiser, *Biochemistry* **20**, 602 (1981).
[11] J. Granot, A. S. Mildvan, K. Hiyama, H. Kondo, and E. T. Kaiser, *J. Biol. Chem.* **255**, 4569 (1980).

cAMP molecules with the regulatory dimer were investigated.[12] Because of the presence of two divalent cations per active site of the catalytic subunit, one an activator, the other an inhibitor, the magnetic resonance studies were greatly facilitated by the use of substitution inert trivalent metal ion complexes of ATP and AMPPCP.[9,10] When a diamagnetic metal–nucleotide complex was required, Co(III) was used. When a paramagnetic complex was needed, Cr(III), which has three unpaired electrons, was used. The preparation, isolation, and characterization of these substitution inert M(III)–nucleotide complexes, have been described by Cleland and co-workers.[13–15] Detailed reviews of the structural and mechanistic conclusions based on the magnetic resonance studies of protein kinase have been published elsewhere.[16,17] This chapter will concentrate on the principles and practical aspects of the magnetic resonance methods employed.

Magnetic Resonance Studies of the Catalytic Subunit

Mn^{2+} and Mg^{2+} Binding Studies

Two independent micromethods, which yielded complementary results, were used to study the binding of Mn^{2+} to the catalytic subunit and to its nucleotide complex: electron paramagnetic resonance (EPR) spectroscopy which observes free Mn^{2+} and the longitudinal proton relaxation rate (PRR) of water which detects bound Mn^{2+}.

Principles and Use of the EPR Method[18–22]

EPR is a form of spectroscopy which makes use of the fact that atoms, molecules, or ions such as Mn^{2+} containing unpaired electrons, when placed in a magnetic field, absorb microwaves. A plot of microwave ab-

---

[12] P. R. Rosevear, A. S. Mildvan, C. O'Brian, and E. T. Kaiser, unpublished observations (1982).
[13] R. D. Cornelius, P. A. Hart, and W. W. Cleland, *Inorg. Chem.* **16**, 2799 (1977).
[14] D. Dunaway-Mariano and W. W. Cleland, *Biochemistry* **19**, 1496 (1980).
[15] W. W. Cleland and A. S. Mildvan, *Adv. Inorg. Biochem.* **1**, 163 (1979).
[16] J. Granot, A. S. Mildvan, and E. T. Kaiser, *Arch. Biochem. Biophys.* **205**, 1 (1980).
[17] E. T. Kaiser, R. N. Armstrong, D. W. Bolen, H. N. Bramson, J. Stingelin, N. Thomas, J. Granot, and A. S. Mildvan, *Cold Spring Harbor Conf. Cell Proliferation* **8**, 67 (1981).
[18] M. Cohn and J. Townsend, *Nature (London)* **173**, 1090 (1954).
[19] A. S. Mildvan and M. Cohn, *Biochemistry* **2**, 910 (1963).
[20] J. A. Fee, this series, Vol. 49, p. 512.
[21] G. H. Reed and W. J. Ray, *Biochemistry* **10**, 3190 (1971).
[22] G. Palmer, *Adv. Inorg. Biochem.* **2**, 153 (1980).

sorption versus microwave frequency would constitute an EPR spectrum. Actually, for ease of instrumentation, the derivative of microwave absorption is plotted against the applied magnetic field. High field corresponds to low frequency and conversely.

EPR has long been a valuable method for detecting and quantitatively studying complexes of Mn^{2+}.[18,19] This is because micromolar levels of the free manganous ion [$Mn^{2+}(H_2O)_6$] in solution are detectable by EPR, and because the EPR spectrum of Mn^{2+} is altered when Mn^{2+} is complexed, usually diminishing by an order of magnitude. The amplitude of the EPR spectrum in a mixture of free and bound Mn^{2+} provides, to a good approximation, a direct measure of the concentration of free Mn^{2+}.

Qualitatively, the reason for this decrease in amplitude is discussed here. More detailed and quantitative treatments can be found elsewhere.[20,21] The Mn^{2+} ion has five unpaired d electrons yielding a net electron spin S of 5/2. Hence, in a magnetic field Mn^{2+} has $2S + 1$ or six energy levels between which electronic transitions may be induced by the absorption of microwave energy. Six energy levels allow five transitions. In a symmetric complex such as $Mn^{2+}(H_2O)_6$ there is no preferred orientation of the net electron spin in the absence of a magnetic field. In the presence of a magnetic field, each of the five electron spin transitions occurs at approximately the same energy or frequency, resulting in an intense, narrow EPR spectrum. Because the naturally occurring isotope of Mn^{2+}(^{55}Mn) has a nuclear spin $I = 5/2$, the EPR spectrum of $Mn^{2+}(H_2O)_6$ is split by the electron–nuclear hyperfine interaction into $2I + 1$ or six intense resonances of very similar amplitude.

If Mn^{2+} binds a small ligand such as acetate or ATP, its observed EPR spectrum markedly decreases in amplitude and broadens due to an increase in the rate of electron spin relaxation (i.e., dephasing and reorientation of the magnetic vectors of the unpaired electrons). If Mn^{2+} binds to a large molecule, the amplitude of its EPR spectrum also decreases by an order of magnitude. This may be due in part to relaxation effects but is predominantly due to the combined effects of the asymmetry of its ligands and slowed molecular tumbling. The asymmetry of the ligands around the Mn^{2+} makes the energies of the five electronic transitions unequal and the rotation of the large complex is too slow to average these inequalities. Hence, in macromolecular complexes of Mn^{2+} the six intense resonances observed in $Mn^{2+}(H_2O)_6$ are each split into five resonances yielding a total of thirty much weaker signals.

For Mn^{2+} binding studies, samples of 30–50 μl volume were made, with Mn^{2+} concentrations varying from 5 μM to 5 mM and constant levels of ligand. The ligand (ADP, AMPPCP, the catalytic subunit or its nucleotide complex) was present at a concentration greater than or equal to the dissociation constant of Mn^{2+}, typically 30–100 μM, so that Mn^{2+} binding could be detected. Control samples contained Mn^{2+} but no ligand. Appropriate nonliganding or weakly liganding buffers such as 50 mM Tris–Cl, pH 7.5 or Good buffers[23] were used. The samples were drawn up into quartz capillary tubes, as previously described.[3] A single 4-min sweep was made on an X-band EPR spectrometer with temperature controlled by flowing nitrogen gas. Other spectrometer settings were typically spectral width, 2000 G; field set at 3200 G; modulation amplitude, 10 G; microwave power, 50 mW; and time constant, 1 sec. The relative amplitude of the Mn^{2+} EPR spectrum in the presence and absence of ligand provides to a reasonable approximation a direct measure of the fraction of Mn^{2+} which is free. Titrations based on this principle are best analyzed by Scatchard plots of $[Mn^{2+}]_b/[Ligand]_T[Mn]_f$ on the y axis versus $[Mn^{2+}]_b/[Ligand]_T$ on the x-axis, where the subscripts b, f, and T refer to bound, free, and total. The x intercept yields the number (n) of Mn^{2+} binding sites per ligand and the y intercept yields the ratio n/K_D where the dissociation constant K_D is defined as $[Mn]_f$ [ligand sites]$_f$/$[Mn]_b$.

The binding of Mg^{2+} or of other diamagnetic cations to Mn^{2+}-binding sites on ligands can also be studied by EPR in competition with Mn^{2+}. For such studies, the dissociation constant of Mn^{2+} is first determined. An EPR titration measuring the displacement of Mn^{2+} by varying levels of Mg^{2+} at constant concentrations of Mn^{2+} and ligand can be analyzed to yield the dissociation constant of Mg^{2+}.[24]

Principles and Use of PRR Method

The PRR method for Mn^{2+} binding studies is based on a change in the paramagnetic effect of Mn^{2+} on the longitudinal relaxation rate ($1/T_1$) of water protons when a Mn^{2+} complex is formed. The PRR method has been discussed in detail in a previous volume of this series.[3] The same samples used for EPR may be used for PRR measurements as well. It is most convenient to prepare the samples in PRR tubes to measure the PRR first, followed by the EPR measurements. As with the EPR method, the PRR method may also be used to study binding of Mg^{2+} to Mn^{2+} binding sites on ligands by competitive Mn^{2+} displacement studies.[6,24]

The stoichiometry and dissociation constants of ternary and higher complexes containing enzyme, Mn^{2+}, and substrates may also be determined by appropriate PRR titrations, provided that a change in the effects of Mn^{2+} on the relaxation rate of water protons occurs as the higher

---

[23] N. E. Good, G. D. Winget, W. Winter, T. N. Connolly, S. Izawa, and R. M. M. Singh, *Biochemistry* **5**, 467 (1966).
[24] A. S. Mildvan and M. Cohn, *J. Biol. Chem.* **240**, 238 (1965).

complex is formed. In such studies typical sample volumes were 30–40 μl containing 100 μM catalytic subunit, 400 μM MnCl$_2$, 40 mM Tris–Cl buffer pH 7.5, 0.1 M KCl, and 0.1 mM dithiothreitol. To avoid dilution of the enzyme or the Mn^{2+}, such solutions were titrated with otherwise identical solutions containing a substrate (e.g., ADP) over a concentration ranging from 10 to 300 μM.[6] The observed paramagnetic effect of Mn^{2+} on $1/T_1$ of water protons may be expressed as an enhancement factor ε^* defined as

$$\varepsilon^* = \frac{1/T_1^* - 1/T_1^{0*}}{1/T_1 - 1/T_1^0} \quad (1)$$

In Eq. (1), $1/T_1^*$ is the longitudinal relaxation rate measured in the presence of Mn^{2+}, enzyme (and other ligands). $1/T_1$ is measured in the presence of the same concentration of Mn^{2+} in absence of enzyme or ligands, while the $1/T_1^{0*}$ and $1/T_1^0$ values are diamagnetic corrections measured in the absence of Mn^{2+} but in the presence and absence, respectively, of enzyme. The ε^* value, which is the factor by which ligand binding alters the relaxivity of Mn^{2+}, is the weighted average of the enhancement factors of all forms of Mn^{2+}.

In such titrations, free Mn^{2+}, binary Mn^{2+} complexes of enzyme and substrate, binary enzyme–substrate complexes, ternary and higher complexes containing enzyme Mn^{2+}, and substrate are present. Hence, such titrations are best analyzed by computer-generated titration curves, which take into account all components and all separately measured binary dissociation constants.[3,6,25] The new information provided by these titrations are the stoichiometry and the dissociation constants of the ternary and higher complexes. Here again, the dissociation constant of Mg^{2+} from higher complexes can be determined by competition studies with Mn^{2+}.

Using both the EPR and PRR methods,[6] only very weak and nonstoichiometric binding of Mn^{2+} to the catalytic subunit of protein kinase was detected. However, the presence of nucleotide substrates [ADP, the Δ-isomer of β,γ-bidentate Co(NH$_3$)$_4$ATP] or nucleotide substrate analogs [AMPPCP, the Λ-isomer of β,γ-bidentate Co(NH$_3$)$_4$ATP] induced the tight and stoichiometric binding of Mn^{2+} to the binary enzyme–nucleotide complex. The stoichiometry of nucleotide binding to the catalytic subunit was found to be 1:1. The stoichiometry of Mn^{2+} binding to the enzyme–ADP and to the enzyme–AMPPCP complex was 2:1, in the latter case with unequal affinity. The tighter of the two binding sites for Mn^{2+} disap-

---

[25] A. S. Mildvan, J. Granot, G. M. Smith, and M. N. Liebman, *Adv. Inorg. Biochem.* **2**, 211 (1980).

peared in the enzyme complex of Co(NH$_3$)$_4$ATP where only one Mn^{2+} binding site was detected. This finding and the range of values of the dissociation constants (6–19 μM) suggested that the tight Mn^{2+}-binding site was on the enzyme-bound nucleotide while the weaker site was on the enzyme, but near the nucleotide. Kinetic studies of the activator and inhibitor constants of Mn^{2+} indicated the tight site to be the activating site while the weaker site was inhibitory. Competitive binding studies with Mg^{2+} indicated that this cation binds with lower affinity (K_D ~2 mM) at both Mn^{2+} sites, while kinetic studies indicated both activation and inhibition by Mg^{2+} at millimolar levels.[6]

Conformations and Arrangement of Enzyme-Bound Substrates

Two NMR methods have been used to study the conformations and arrangement of the nucleotide[7,26,27] and peptide substrates[10] on the catalytic subunit of protein kinase. The paramagnetic probe T_1 method measures distances up to ~20 Å from a bound paramagnetic center, such as Mn^{2+}, to NMR-detectable nuclei (^{1}H, ^{31}P) of the bound substrates. The nuclear Overhauser effect (NOE) may be used to measure interproton distances up to ~5 Å and has been especially useful in refining the conformation of bound nucleotide substrates on several kinases including pyruvate kinase,[28] creatine kinase,[26] and protein kinase.[26,27]

Paramagnetic Probe-T_1 Method

The theory and applications of the paramagnetic probe-T_1 method have been described in detail in this series[3,4] and elsewhere.[1,25] Those aspects unique to protein kinase will therefore be stressed here.

Briefly, the effect of an enzyme-bound paramagnetic metal on the longitudinal relaxation rates ($1/T_{1p}$) of the nuclei of a substrate which is exchanging into a site near the metal, is a function of five measurable parameters: the lifetime (τ_M) of the complex, the stoichiometry (q) of bound metal and substrate, the outer-sphere contribution to the relaxation rate ($1/T_{o.s.}$), the correlation time (τ_c) for dipolar interaction, and the metal–nucleus distance (r). Solving the relevant equations in terms of the distance (r) yields[25]

---

[26] P. R. Rosevear, P. Desmeules, G. L. Kenyon, H. N. Bramson, E. T. Kaiser, and A. S. Mildvan, *A.C.S. Meet. Biol. Div. U. Minn. Abstr.* 20 (1981).

[27] P. R. Rosevear, H. N. Bramson, C. O'Brian, E. T. Kaiser, and A. S. Mildvan, *Biochemistry* (1983) in press.

[28] P. R. Rosevear, G. M. Smith, S. Meshitsuka, A. S. Mildvan, P. Desmeules, and G. L. Kenyon, in "Phosphorus Chemistry" (L. D. Quin and J. G. Verkade, eds.), A.C.S. Symposium 171, p. 125 (1981).

$$r = C\left[\left(\frac{qfT_{1p}T_{o.s.}}{T_{o.s.} - T_{1p}} - \tau_M\right)\left(\frac{3\tau_c}{1 + \omega_I^2\tau_c^2} + \frac{7\tau_c}{1 + \omega_s^2\tau_c^2}\right)\right]^{1/6} \quad (2)$$

The constant C is a product of known physical quantities, numerically equal to 812 Å/sec$^{1/3}$ for Mn^{2+}–proton interactions, 601 for Mn^{2+}–^{31}P interactions, and 705 for Cr^{3+}–proton interactions.[3,4,25] The normalization factor f is the concentration ratio of bound paramagnet to total substrate, ω_I is the nuclear precession frequency in radians/sec ($2\pi \times$ the NMR frequency, in Hz), and ω_s is the electron precession frequency in radians/sec ($2\pi \times 657 \times$ the proton NMR frequency). As pointed out elsewhere[4,25] the outer-sphere contribution to the relaxation rate ($1/T_{o.s.}$), which is usually less than 10% of $1/fT_{1p}$, is determined by measuring the residual value of $1/fT_{1p}$ of the substrate after displacing it from the active site by a competitive inhibitor. The lifetime (τ_M) of the complex is estimated by measuring the transverse relaxation rate, i.e., the paramagnetic NMR line broadening ($1/fT_{2p}$), as a function of temperature. If the outer-sphere contribution is small ($T_{o.s.} \gg T_{1p}$) and the relaxation time is not limited by the lifetime of the complex ($fT_{1p} \gg \tau_M$), then Eq. (2) simplifies to the more familiar form:

$$r = C\left[qfT_{1p}\left(\frac{3\tau_c}{1 + \omega_I^2\tau_c^2} + \frac{7\tau_c}{1 + \omega_s^2\tau_c^2}\right)\right]^{1/6} \quad (3)$$

The stoichiometry (q) is determined by metal and substrate binding studies as discussed above. The remaining unknown in Eq. (3) needed to calculate the distance (r), is the correlation time (τ_c), which is most appropriately determined by measuring $1/fT_{1p}$ at two or more frequencies. More than two frequencies are preferred because τ_c may itself have a frequency dependence.[4,5]

Conformation of Enzyme-Bound $Co(NH_3)_4ATP$.[7,26,27] The binding of nucleotides other than $Co(NH_3)_4ATP$ to the catalytic subunit of protein kinase induces two tight Mn^{2+} binding sites. To avoid the complexity of two paramagnetic reference points, distances from Mn^{2+} at the inhibitory site to the substrate β,γ-bidentate $Co(NH_3)_4ATP$ were determined[7] since this nucleotide induces only one Mn^{2+} binding site. The proton NMR studies were carried out at 100 and 360 MHz. To minimize strong solvent proton signals, stock solutions of salts (KCl), buffers, and $Co(NH_3)_4ATP$ are deuterated by cycles of lyophilizing and dissolving the residue in ^2H_2O at least three times. To exchange the coordinated ammine protons of $Co(NH_3)_4ATP$, after the first cycle of deuteration it has been found necessary to raise the pH* to a meter reading of 10 with KO^2H in ^2H_2O for 5 min followed by readjustment to pH* 6.8. To avoid strong buffer signals, Tris base deuterated at the carbon-bound protons is available from

Stohler Isotopes, Waltham, Mass. Enzyme solutions containing buffer are deuterated by at least two cycles of five-fold concentration and redilution in ^2H_2O. Concentration of the protein is carried out by one of three methods. Vacuum dialysis using the A. H. Thomas ultrafiltration cell and Schleicher and Schuell membrane filters is a safe but time consuming procedure. Ultrafiltration using Millipore immersible CX capsule filters (catalog No. PTGC 11K25) after having drawn at least 100 ml of distilled water through them in order to remove organic preservatives is a more rapid procedure with slightly greater loss of protein. Amicon filtration in a 10-ml Amicon cell at 15–20 psi maximum pressure using an Amicon PM10 membrane has been found to be a rapid and safe procedure for the catalytic subunit of protein kinase[27] and for creatine kinase[29] but may not be safe for all proteins. With this rapid method, six cycles of twofold concentrations and dilutions were carried out. To remove trace paramagnetic metal ions, small volumes of stock solutions are passed over microcolumns containing 0.1 ml of Chelex 100 which had been preadjusted to pH 7 and equilibrated with the stock solution. Enzyme solutions may also be treated in this way but at a more dilute stage to avoid significant loss of enzyme during the equilibration.

Samples (0.40 ml) were placed in 5-mm o.d. by 178-mm-long NMR tubes (Wilmad 507-PP) together with a clean, dry Teflon vortex plug (Wilmad WG-805). The solutions typically contained 50 μM catalytic subunit, 25 mM racemic $Co(NH_3)_4ATP$, 50 mM Tris–Cl buffer, pH 7.5 (measured in H_2O prior to deuteration), and 0.15 M KCl. The residual HDO signal was suppressed by direct irradiation using the proton decoupler. Longitudinal relaxation rates ($1/T_1$) were measured either by the inversion recovery or saturation recovery methods discussed in detail elsewhere[4] and transverse relaxation rates ($1/T_2$) were determined from linewidth measurements. The sample was titrated with $MnCl_2$ (in ^2H_2O) over a wide range of concentrations (10–140 μM) measuring $1/T_1$ and $1/T_2$ after each addition. For frequency dependence studies, the same diamagnetic sample was measured at several frequencies, a small number of appropriate concentrations of Mn^{2+} were added, and measurements were again made at the various frequencies.[7]

For the ^{31}P NMR studies, carried out at three frequencies, 40.5, 72.9, and 145.8 MHz, larger samples of 1.0–2.0 ml were made up in NMR tubes of 10 mm o.d. by 178 mm (Wilmad 513-1PP) using a clean, dry, Teflon vortex plug (Wilmad WG-805A). Deuteration of reagents was not necessary but ^2H_2O was added to 20% by volume for field frequency locking. Treatment of stock solutions or of the final sample with Chelex 100 (prior

[29] P. R. Rosevear, P. Desmeules, G. L. Kenyon, and A. S. Mildvan, *Biochemistry* **20**, 6155 (1981).

to titration with $MnCl_2$) is especially important in ^{31}P NMR. Other components present were as described above except that lower concentrations of catalytic subunit (10 μM), $Co(NH_3)_4ATP$ (7 mM), and $MnCl_2$ (2–6 μM) were sufficient because of the large paramagnetic effects of Mn^{2+}. Broadband proton irradiation was used to eliminate coupling of the α-P to the C-5 methylene protons. Because of the existence of a weak binary Mn^{2+} complex of $Co(NH_3)_4ATP$, relaxation experiments were also carried out in absence of the catalytic subunit, a necessary control.[7]

The observed paramagnetic effects of Mn^{2+} on the nuclei of $Co(NH_3)_4ATP$ were a weighted average of those due to the binary Mn^{2+} complex, $(1/fT_{1p})_{binary}$, and those due to the ternary complex with enzyme $(1/fT_{1p})_{ternary}$.

$$\frac{1}{T_{1p}} = \frac{[Mn^{2+}-Co(NH_3)ATP]}{[Co(NH_3)_4ATP]_{Total}} \left(\frac{1}{fT_{1p}}\right)_{binary} + \frac{[Mn^{2+}-enzyme-Co(NH_3)_4ATP]}{[Co(NH_3)_4ATP]_{Total}} \left(\frac{1}{fT_{1p}}\right)_{ternary} \quad (4)$$

Such mixed effects are analyzed either by titrations, selectively diluting the enzyme to determine $(1/fT_{1p})_{ternary}$ or, as in the present case, using the measured dissociation constants of Mn^{2+} from the binary complex (15 mM) and the ternary complex (130 μM) to calculate the distribution of Mn^{2+}.[7] Another experimental approach in the analysis of such systems is equilibration of the binary and ternary complexes by microdialysis and measurement of the relaxation rates on both sides of the dialysis membrane. The $(1/fT_{1p})_{binary}$ and $(1/fT_{1p})_{ternary}$ are used to calculate Mn^{2+}–nucleus distances in the respective complexes with Eqs. (2) and (3).[7]

In the absence of enzyme, the distance measurements showed Mn^{2+} to bind weakly to the adenine ring of $Co(NH_3)_4ATP$ (Fig. 1A and B).[7] On the catalytic subunit, the distance measurements showed that the inhibitory Mn^{2+} bridged the enzyme to the $Co(NH_3)_4ATP$ with coordination of the three phosphoryl groups (Fig. 1C–F)[7] in a structure of the following type

$$\begin{array}{c} E-ATP-M_{act} \\ \diagdown \diagup \\ M_{inh} \end{array}$$

Model building by hand and by computer[4,25] yielded two alternative conformations for the enzyme-bound $Co(NH_3)_4ATP$, which satisfied the distance from Mn^{2+} to the protons and phosphorus atoms of the bound substrate within van der Waals tolerance of the atoms (Fig. 1C–F).[7] In one conformation (Fig. 1C and D) the torsional angle (χ) at the glycosidic bond was anti (84 ± 10°) while in the other (Fig. 1E and F) a syn conformation (χ = 284 ± 10°) was found. Although the anti conformation is

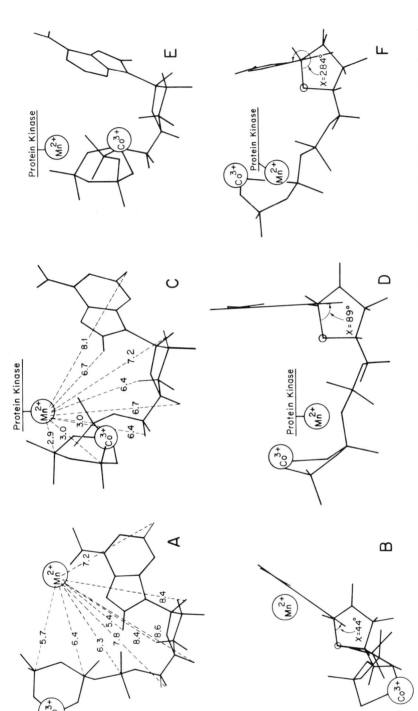

FIG. 1. Two views of the conformations of $Co(NH_3)_4ATP$ in the binary $Co(NH_3)_4ATP-Mn^{2+}$ complex (A,B) and in the ternary complex of protein kinase (anti solution, C,D; syn solution, E,F) consistent with the indicated distances (in Å) from Mn^{2+} measured by the paramagnetic probe-T_1 method.[7]

preferred on energetic grounds, it was not established by the paramagnetic probe-T_1 method because of the large distances to the Mn^{2+}. However, the NOE method, which measures interproton distances, showed that adenine H-8 is very near ribose H_2' but not near ribose H_1' ruling out the syn conformation.[26,27]

Principles of the Nuclear Overhauser Effect[30–32]

The nuclear Overhauser effect is defined as a change in the intensity or integrated area of a nuclear resonance when another resonance is irradiated with a radiofrequency (rf) field. This effect can be used qualitatively to detect the proximity of two protons which are less than ~5 Å apart, and, in certain cases, quantitatively to measure relative and absolute interproton distances over this range. Several comprehensive reviews of this phenomenon have been written.[30–32] The first use of an NOE in a macromolecular system, cytochrome c, was by Redfield and Gupta.[33] Only the essentials of the theory will be presented here. Absorption of energy from the selective application of rf irradiation to a nuclear spin resonance results in the perturbation of the energy levels of the nucleus from its equilibrium Boltzmann distribution in the magnetic field. Absorption of energy continues until the energy levels become equally populated. When this occurs the nuclear spin is said to be saturated and the resonance line in the spectrum disappears. Through natural relaxation processes, the saturated spin will dissipate this excess energy both as heat and by transfer to other nearest neighbor nuclei in the system which are dipolar coupled, ultimately, with continuous irradiation, setting up a steady state. The transfer of energy from the irradiated to an observed nucleus occurs by distance dependent dipolar interactions which increase the population of either the excited state or the ground state of the observed nucleus. The intensity of the observed nuclear resonance will therefore either decrease or increase depending on which of these effects predominate. Using the Bloch equations for dipolar coupled nuclear transitions in a system of two spin 1/2 nuclei and setting time derivatives equal to zero, Noggle and Schirmer[30] derived the steady-state equation for the fractional change in the intensity of resonance A on irradiation of resonance B [$f_A(B)$] which is a quantitative measure of the NOE:

[30] J. H. Noggle and R. E. Schirmer, "The Nuclear Overhauser Effect." Academic Press, New York, 1971.
[31] A. A. Bothner-By, *in* "Biological Applications of Magnetic Resonance" (R. G. Shulman, ed.), p. 177. Academic Press, New York, 1979.
[32] A. A. Bothner-By and R. Gassend, *Ann. N.Y. Acad. Sci.* **222**, 668 (1973).
[33] A. G. Redfield and R. K. Gupta, *Cold Spring Harbor Symp. Quant. Biol.* **36**, 405 (1971).

$$f_A(B) = \frac{\gamma_B}{\gamma_A} \cdot \frac{\sigma_{AB}}{\rho_A} \tag{5a}$$

In Eq. (5a) γ_B and γ_A are the gyromagnetic ratios of the nuclei. If both are protons the equation simplifies to

$$f_A(B) = \sigma_{AB}/\rho_A \tag{5b}$$

The parameter σ_{AB} is the cross-relaxation rate and ρ_A is the longitudinal relaxation rate of A or $1/T_{1A}$. These relaxation rates for pairs of protons are given by[30–32,34]

$$\sigma_{AB} = \frac{\gamma^4 \hbar^2}{10 r_{AB}^6} \left(\frac{6\tau_r}{1 + 4\omega_I^2 \tau_r^2} - \tau_r \right) \tag{6}$$

$$\rho_{AB} = \frac{\gamma^4 \hbar^2}{10 r_{AB}^6} \left(\tau_r + \frac{3\tau_r}{1 + \omega_I^2 \tau_r^2} + \frac{6\tau_r}{1 + 4\omega_I^2 \tau_r^2} \right) \tag{7}$$

where $\hbar$ is Planck's constant divided by 2π, r_{AB} is the interproton distance, τ_r is the correlation time or the time constant for reorientation of the interproton vector r_{AB}, and ω_I is the nuclear precession frequency in radians/sec. From Eqs. (5) through (7), the measured quantity $f_{AB}(B)$ is independent of distance. Although both σ_{AB} and ρ_A contain distance information, distances between individual pairs of protons are most appropriately calculated from the cross-relaxation rate σ_{AB}. The reason for this is that in a multiproton system with exchange it is difficult to isolate the effects of individual neighboring protons on ρ_A. Since $f_A(B)$ and ρ_A can be measured, σ_{AB}, in principle, can be calculated from the steady-state NOE according to Eq. (5b) in a two-proton system. However in the presence of multiple interacting protons, where secondary and tertiary NOEs can occur with time,[33,35,36] σ_{AB} is more accurately measured by the early time dependence of the Overhauser effect, $f_A(B)_t$, as the steady state is approached.[37,38] The relevant equation for a two spin system is

$$f_A(B)_t = \frac{\sigma_{AB}}{\rho_A}(1 - e^{-\rho_A t}) + \frac{\sigma_{AB}}{\rho_A - c}(e^{-\rho_A t} - e^{-ct}) \tag{8}$$

The approach to the steady-state NOE thus depends on σ_{AB}, the cross-relaxation rate constant for transfer of saturation from the irradiated spin B to the neighboring spin A, ρ_A the spin-lattice relaxation rate constant for spin A, t the duration of irradiation of spin B, and c a phenomenological

---

[34] I. Solomon, *Phys. Rev.* **99**, 559 (1955).
[35] A. Kalk and H. J. C. Berendsen, *J. Magn. Reson.* **24**, 343 (1976).
[36] P. J. Andree, *J. Magn. Reson.* **29**, 419 (1978).
[37] G. Wagner and K. Wuthrich, *J. Magn. Reson.* **33**, 675 (1979).
[38] A. Dubs, G. Wagner, and K. Wuthrich, *Biochim. Biophys. Acta* **577**, 177 (1979).

rate constant for saturation of the irradiated resonance (B). The constant c can be approximated by $1/2(1/T_{1B} + 1/T_{2B})$ where $1/T_{1B}$ and $1/T_{2B}$ are the longitudinal and transverse relaxation rates respectively for spin B.[37,39] When instantaneous saturation of the irradiated spin (B) is assumed, i.e., c is very large, the time evolution of the NOE from resonance B to resonance A[37,40] is given by

$$f_A(B)_t = \frac{\sigma_{AB}}{\rho_A} (1 - e^{-\rho_A t}) \qquad (9)$$

The experimentally measured values of $f_A(B)_t$ and ρ_A are fitted to Eq. (9) in order to evaluate σ_{AB}. An equation for a three-spin system of the form of Eq. (8) has been developed where the positioning of the third spin relative to the irradiated (B) and observed spin (A) can be varied.[37,38] In some cases this provides a better approximation of a multiproton environment. When σ_{AB} has been evaluated, the interproton distance r_{AB} can be calculated from

$$r_{AB} \text{ (in Å)} = (62.02) \left[\left(\frac{1}{\sigma_{AB}}\right) \left(\frac{6\tau_r}{1 + 4\omega_I^2 \tau_r^2} - \tau_r \right) \right]^{1/6} \qquad (10)$$

which was obtained by rearranging Eq. (6). The correlation time τ_r, or its upper limit is estimated from the diamagnetic T_{1A}/T_{2A} ratio using[34,41,42]

$$\frac{T_{1A}}{T_{2A}} = \frac{12\omega_I^4 \tau_r^4 + 37\omega_I^2 \tau_r^2 + 10}{16\omega_I^2 \tau_r^2 + 10} \qquad (11)$$

Calculation of the correlation time from σ_{AB}/ρ_A according to Eqs. (6) and (7) is usually inappropriate because multiple interproton interactions contribute to ρ_A, while only the cross-relaxation between protons A and B are contained in σ_{AB}.

An alternative method for determining τ_r which remains to be explored is a study of the frequency dependence of σ_{AB} according to Eq. (6).

Nuclear Overhauser effects may be used to study the conformation and interactions of small molecules with macromolecules, as will be discussed below for the conformation of $Co(NH_3)_4ATP$ binding to the catalytic subunit of protein kinase. Most of these studies have been performed in systems in which the small ligand is in fast exchange, i.e., is exchanging at a rate much greater than the relaxation rates, σ_{AB} and ρ_A. Under these

---

[39] N. C. M. Alma, B. J. M. Harmsen, W. E. Hull, G. van der Marel, J. H. van Boom, and C. W. Hilbers, *Biochemistry* **20**, 4419 (1981).
[40] J. Tropp and A. G. Redfield, *Biochemistry* **20**, 2133 (1981).
[41] T. J. Swift and R. E. Connick, *J. Chem. Phys.* **37**, 307 (1962).
[42] P. J. Stein and A. S. Mildvan, *Biochemistry* **17**, 2675 (1978).

conditions it has been found by James and Cohn[43] that at values of [free ligand]/[bound ligand] less than ~10, the observed NOE is independent of this ratio. Presumably this constancy of the NOE results from parallel changes of σ_{AB} and ρ_A over this range of relative concentrations. Under these conditions, the product of the measured NOE and ρ_A can directly yield σ_{AB} for the bound ligand, according to Eq. (5b). At higher ratios of free to bound ligand, the NOE was observed to decrease. Over this region, the σ_{AB} of the bound ligand is determined by measurements of the NOE and ρ_A at constant ligand concentration and at several concentrations of the macromolecule, and the data analyzed according to[32,44,45]

$$f_A(B)_{obs} = \frac{([L]_f/\sigma_{AB}) + ([L]_b/\rho_A)[f_A(B)]_{max}}{([L]_f/\sigma_{AB}) + ([L]_b/\rho_A)} \quad (12)$$

where $[L]_b$ and $[L]_f$ are the concentrations of bound and free ligand, respectively. Equation (12) is a special case of a more general equation derived by Bothner-By for a system which includes chemical exchange.[32]

Experimentally, in a mixture of free and bound ligands, one can easily distinguish NOEs from bound ligands by two criteria. First, because of the faster relaxation rates ($1/T_1 + 1/T_2$) on the macromolecule, the preirradiation time required to observe an NOE is generally <0.4 sec for the bound ligand, while the corresponding time for the free ligand is ~1 sec. Second, since $\omega_1\tau_r \gg 1.12$ for the bound ligand, the NOE is negative as may be seen from Eq. (6), while for the rapidly rotating free ligand, $\omega_1\tau_r < 1.12$ yielding a positive NOE.

Limitations of the NOE method for determining interproton distances result from the facts that the underlying theory assumes a fixed distance between the pairs of protons, and that the radius vector between them is rotating with a single time constant τ_r. As with the paramagnetic probe-T_1 method,[4,25] if several conformations coexisted with differing interproton distances, the NOE method would yield a nonlinear, or "root-mean-sixth" average conformation biased toward those conformers with shorter interproton distances. The combined use of both the NOE and paramagnetic probe-T_1 methods with several reference points provides a valuable test since the existence of a model consistent with both sets of data would indicate that a unique average conformation existed.[28] Motions more complex than simple rotation require the use of appropriate correlation

[43] T. L. James and M. Cohn, *J. Biol. Chem.* **249**, 2599 (1974).
[44] P. Balaram, A. A. Bothner-By, and E. Breslow, *J. Am. Chem. Soc.* **94**, 4017 (1972).
[45] A. A. Bothner-By, B. Lemarie, R. Walter, R. Tiao-Te Lo, L. D. Rabbani, and E. Breslow, *Int. J. Peptide Protein Res.* **16**, 450 (1980).

functions.[30,46] However, in practice, the numerical values of the frequency dependent terms [Eq. (6)] do not greatly differ, and uncertainties in these terms are diminished by $\sim\frac{1}{6}$ in distance calculations since a sixth root is taken.

Measurement of Nuclear Overhauser Effects

As mentioned above, interpretation of a steady state NOE in terms of an internuclear distance is straightforward. However, in macromolecular systems containing many protons, secondary and tertiary NOEs or "spin diffusion" may occur.[33,35,36] To take such effects into account, two types of transient measurements have been developed in order to measure the initial approach to the steady state NOE before the effects of spin diffusion dominate the system. Transient NOE measurements can be obtained by the following pulse sequence,[38]

$([RD-180°(\omega_B)-t_1\text{-observation pulse}]_{16}-$
$[RD-180°(\omega_{\text{off-res}})-t_1\text{-observation pulse}]_{16})_n$

where $180°$ (ω_B) is a selective $180°$ pulse at resonance frequency B, RD is a relaxation delay typically in the range of $5T_1$s, t_1 is a delay during which time the Overhauser effects are built up in the absence of irradiation, which is followed by an observation pulse. The entire process is repeated n times with n chosen to yield the desired signal/noise ($\geq 5:1$). The observation pulse width is chosen as $90°$ by separate measurements.[4] Typically, 16 transients are accumulated for the irradiation frequency (ω_B), and the pulse sequence is repeated with a selective $180°$ pulse ($\omega_{\text{off-res}}$) at a frequency in a region of the spectrum which has no resonances. The magnitude of the NOE is usually observed by subtraction of the free induction decay (FID) obtained following off-resonance irradiation from the FID obtained at ω_B, the on-resonance frequency.[38,40] Subtraction of the two free induction decays (FIDs) followed by Fourier transformation saves time compared to subtraction of the two Fourier transformed spectra, although both methods give equivalent results. The difference spectrum thus obtained will have positive peaks for positive NOEs ($\omega_I\tau_r \leq 1.12$) and negative peaks for negative NOEs ($\omega_I\tau_r > 1.12$).

In practice, however, the above pulse sequence has been shown to be of limited value because of the necessary compromise required between an RF power low enough for selective irradiation of a single resonance in crowded spectral regions, and high enough to yield usable signal to

[46] O. Jardetzky and G. C. K. Roberts, "NMR in Molecular Biology," p. 465. Academic Press, New York, 1981.

noise.[38] A preferable method to obtain NOE data with the necessary selectivity of the preirradiation pulse and optimum signal to noise was proposed by Redfield[40] and Wuthrich and co-workers[37,38] for which the phrase "driven NOEs" was coined. This method has been exclusively utilized in studying the conformation of the metal–nucleotide complex $Co(NH_3)_4ATP$ bound to the catalytic subunit of protein kinase.[26,27] According to Eqs. (8) and (9), the NOE is measured after varying the duration (t_1) of the selective preirradiation using the following pulse sequence

([RD-preirradiate (t_1,ω_B)-observation pulse]$_{16}$–

[RD-preirradiate ($t_1,\omega_{\text{off-res}}$)-observation pulse]$_{16}$)$_n$

where the relaxation delay RD is typically in the range of $5T_1$s. The proton decoupler is utilized to produce the selective preirradiation pulse of the desired duration. The preirradiation pulse (usually 32dB below 0.2 W) was typically varied between 50 msec and 1.5 sec. This range was sufficient to produce selective NOEs at short preirradiation times and also permitted the detection of secondary NOEs at longer times.

All NOE measurements on the conformation of $Co(NH_3)_4ATP$ bound to the catalytic subunit of protein kinase were carried out on a commercial (Bruker WM-250) NMR spectrometer, using 16 bit analog to digital conversion. The use of 16 bit analog to digital conversion provides better observable dynamic range which is important in observing small intensity changes of resonances in biological samples where relatively large amounts of residual HDO are present as a consequence of incomplete deuteration. To save time, three frequencies, ω_1, $\omega_{\text{off-res}}$, and ω_2, are each acquired in separate data blocks such that a single off resonance frequency inserted between the two test frequencies ω_1 and ω_2 serves as a control for both test frequencies.[40] Routinely 16 scans at each frequency are alternatively collected until the total number of scans necessary to obtain the desired signal to noise is obtained. The FID is then usually stored on a magnetic disk for processing at a future time. With commercially available software, a program can be used which will permit variation of not only t_1 but also ω_1, ω_2, and $\omega_{\text{off-res}}$, rendering the acquisition of spectra entirely under computer control.

The magnitude of the NOE is measured by subtraction of the FID of the $\omega_{\text{off-res}}$ control spectrum from the FID of either ω_1 or ω_2. The difference FID can then either be multiplied by an exponential filtering function to improve signal to noise at the expense of resolution and Fourier transformed or directly Fourier transformed. Importantly, the control spectrum ($\omega_{\text{off-res}}$) must be processed with the same parameters used to obtain the difference spectra. Calculation of the NOE to a given resonance is achieved by integrating the difference peak for that resonance in the difference spectrum and dividing it by the integrated intensity of the same resonance in the control spectrum which was obtained with $\omega_{\text{off-res}}$. If irradiation does not alter the line shape, then the difference in amplitude divided by the control amplitude also yields the NOE.

Determination of the cross-relaxation rate (σ_{AB}) using Eqs. (5b), (8), or (9) requires the measurement of not only the NOE but also the longitudinal relaxation rate ($\rho_A = 1/T_{1A}$). The selective-saturation recovery technique was utilized to obtain the relaxation rate ρ_A to avoid spin diffusion effects from other resonances.[40] The technique consists of applying a saturating preirradiation pulse to the resonance of interest and following the recovery of its magnetization as a function of time. The pulse sequence used for this is shown below,

$$[(t_1,\omega_A)\text{-}\tau\text{-observation pulse-RD}]_n$$

where t_1,ω_A is the time required for saturation of resonance A. For the resonances of Co(NH$_3$)$_4$ATP bound to protein kinase, $t = 0.1$ sec at 26 dB below 0.2 W. The recovery time τ is a variable delay between the saturating pulse and the observe pulse, the observation pulse is a 90° pulse, and RD is the relaxation delay, usually $5T_1$s. The recovery of magnetization over the time period $\sim 1/\rho_A$ is a first-order process and a plot of $\ln[I_0 - I(\tau)]$ vs τ yields ρ_A, where I_0 is the maximal intensity of the resonance and $I(\tau)$ is the observed intensity at time τ after the saturating pulse.

Transverse relaxation rates ($1/T_2$) were calculated from line width measurements at half-height ($\Delta\nu$) using the relation $1/T_2 = \pi\Delta\nu$. In the presence of complex spin–spin coupling which prevents the direct and accurate measurement of the line-width, the Carr–Purcell–Meiboom–Gill pulse sequence is preferable for measurement of T_2.[3] In exchange limited systems a study of T_2 as a function of temperature may be used to determine the contribution of exchange to the measured $1/T_2$ value.

In preparing samples, approximately 4.0 mg of the catalytic subunit of bovine heart protein kinase was used per experiment. The enzyme was deuterated by twofold concentration and dilution in a 10-ml Amicon cell (20 psi maximum pressure) using an Amicon PM10 membrane. Other components present were 10 mM deuterated Tris–Cl, pD 7.5, 150 mM KCl, and 0.1 mM DTT in ^{2}H$_2$O. The components of the mixture were passed through Chelex-100 (Bio-Rad) microcolumns to remove trace metal impurities. The catalytic subunit of protein kinase was concentrated to 0.35 ml (usually 0.22 mM in protein) and placed in a 5-mm-o.d. NMR tube (Wilmad 507-PP). A plug (Wilmad WG-805) was necessary to prevent vortex formation in the sample when spinning. The experiment was performed at 25°. Routinely, a ^{1}H NMR spectrum of the protein was acquired under the conditions of the NOE experiment before addition of

Co(NH$_3$)$_4$ATP. Deuterated Co(NH$_3$)$_4$ATP was then titrated into the protein sample (usually in three steps) to a final concentration in the range of 1.5–2.0 mM. Titration is useful to unambiguously identify the Co(NH$_3$)$_4$ATP resonances. It is particularly helpful in the identification of H3', H4', and H5' which are near the HDO resonance and have chemical shifts in the C$_\alpha$–H region of the protein spectrum. Usually, at each step in the titration, the protein spectrum alone is subtracted from the spectrum containing Co(NH$_3$)$_4$ATP and the difference spectrum of Co(NH$_3$)$_4$ATP analyzed. The final concentration of Co(NH$_3$)$_4$ATP is chosen to be high enough to obtain difference spectra with ≥5:1 signal to noise in 30–40 min with all resonances clearly observable and low enough to prevent dilution of the NOE due to a large pool of free Co(NH$_3$)$_4$ATP. The H2' resonance was always under the shoulder of the HDO line and could never be directly observed. It was found by decoupling experiments (observing a narrowing of H1'), that the chemical shift of H2' in the presence of the catalytic subunit was virtually unchanged from that of free Co(NH$_3$)$_4$ATP.

The initial NOE experiment usually consists of a survey of all the ^{1}H resonances of Co(NH$_3$)$_4$ATP at a constant preirradiation time (0.50 sec). Results from the initial survey are then analyzed to determine at which resonances a complete time dependence of the NOE will be measured. Choice of the initial preirradiation times is somewhat arbitrary, but generally first-order effects will be observed by 0.50 sec in protein–substrate systems when the substrate being observed is in fast exchange. Detailed measurement of the time dependence of the NOE requires decreasing preirradiation times ranging downward in 0.1 sec steps from 0.5 sec, until no effect is measured. The time dependence should also contain several time points longer than 0.50 sec (0.75, 1.00, and 1.50 sec).

An unexpected problem in the study of the catalytic subunit of protein kinase was the instability of the enzyme during the prolonged NMR experiment.[27] Enzyme assays should routinely be performed every 4–8 hr depending on the enzyme's stability. It was found that under the conditions of the NMR experiment the catalytic subunit of protein kinase lost activity with a half time of 10 hr. The loss of activity was coincident with a change in the pattern of observed NOEs of enzyme bound Co(NH$_3$)$_4$ATP. This effect will be further discussed below. For now, it is pertinent to state that the thoroughness of the NOE search was limited by the stability of the enzyme.

For enzyme-bound Co(NH$_3$)$_4$ATP a primary NOE of -5.1% was found from ribose H$_2'$ to adenine H$_8$.[27] No other NOEs were detected from any of the ribose protons to any of the adenine protons indicating that these were less than 1%. From the longitudinal relaxation rates of adenine H_8 (2.78 sec^{-1}) and adenine H_2 (1.10 sec^{-1}), the respective cross-relaxation rates from ribose H_2' to adenine H_8 and to adenine H_2 are 0.142 and ≤ 0.014 sec^{-1}, respectively. From the sixth root of the ratio of these values, it is concluded that ribose H_2' is at least 53% farther from adenine H_2 than from adenine H_8, establishing an anti conformation for the bound nucleotide as shown in Fig. 1C and D. Analogous arguments can be made that ribose H_3' and H_1' are at least 31% farther from adenine H-8 than is ribose H_2'. These relative distances are consistent with a 1'-endo-ribose pucker and a high anti-glycosidic torsional angle (Fig. 1C and D). The absolute distances from adenine H-8 to ribose H_2' (3.2 ± 0.6 Å), from adenine H-8 to ribose H-3' (≥ 4.2 Å), and from ribose H_1' to ribose H_2' (3.4 ± 0.7 Å), while of the correct order, appear from model building to be too great by ~17%. This is especially clear for the distance from ribose H_1' to ribose H_2' which is structurally constrained to the range 2.9 ± 0.1 Å. While the differences are within the experimental uncertainty of the distances determined by the NOE method (±20%), the mean values may be too large because the correlation time, calculated from the T_{1A}/T_{2A} ratio, is an upper limit, as has been seen in other systems.[42] The possibility also exists that the full σ_{AB} values were not measured at the available protein concentrations. Nevertheless, the mutual consistency of the interproton distances determined by NOE measurements[27] and the distances from Mn^{2+} determined by the paramagnetic probe-T_1 method[7] strongly indicate the existence of a unique conformation for $Co(NH_3)_4ATP$ bound to the active catalytic subunit of protein kinase.

Curiously, as the enzyme loses activity with time, a change occurs in the NOEs such that the primary effect between ribose H_2' and adenine H_8 disappears.[27] Hence the loss of activity appears to correlate with a perturbation of the enzyme structure at the ATP binding site, altering the conformation of the bound nucleotide substrate, although other interpretations are not excluded.

Conformation of Enzyme-Bound Heptapeptide Substrates and Analogs.[10] Since the presence of a bound nucleotide is needed to create divalent cation binding sites on protein kinase,[6] the conformation of peptides were studied by the paramagnetic probe-T_1 method in enzyme complexes which also contained nucleotides.[10] To avoid simultaneous occupancy of both the activating and inhibitory metal sites by paramagnetic metals, appropriate substitution-inert M(III)–nucleotide complexes were used. When Mn^{2+} at the inhibitory site was the paramagnetic reference point, complexes of the structure

$$\text{peptide--E--ATPCo(NH}_3)_4 \atop \diagdown \diagup \atop Mn^{2+}$$

were studied by titration of the ternary peptide–enzyme–nucleotide complex with $MnCl_2$. The Δ-isomer of β,γ-bidentate $Co(NH_3)_4ATP$ is itself a substrate in a slow phosphoryl transfer reaction to the serine-peptide (Leu-Arg-Arg-Ala-Ser-Leu-Gly). Hence $1/T_1$ and $1/T_2$ data on this ternary system were collected rapidly at only a small number of Mn^{2+} concentrations. Two alternative approaches were used to avoid such reactions. First, the serine-peptide was replaced by the alanine-peptide (Leu-Arg-Arg-Ala-Ala-Leu-Gly). Second, the active nucleotide complex was replaced by the $Co(NH_3)_4AMPPCP$ complex which does not transfer a phosphoryl group to the Ser-peptide.[9,10] To test for the presence of the β-pleated sheet structure, it was necessary to use another active peptide substrate, the Tyr-peptide (Leu-Arg-Arg-Tyr-Ser-Leu-Gly).

Deuterated samples (0.4 ml) for proton NMR typically contained 130 μM catalytic subunit, 6.0 mM peptide, 3.0 mM Co(III) nucleotide, 10 mM deuterated Tris–Cl buffer (pH 7.5), and 0.15 M KCl. These samples were titrated with $MnCl_2$ over the range 0 to 200 μM, measuring $1/T_1$ and $1/T_2$ as discussed above. Displacement experiments with 333 μM of the potent competitive inhibitor polyarginine removed the paramagnetic effects on $1/T_1$ and $1/T_2$ of the protons of the various peptides studied, establishing active site binding of the peptides, and a negligibly small value of the outer sphere ($1/T_{o.s.}$) contribution to the relaxation rate.[10]

To make use of a paramagnetic probe at the activating site, a complex of the following structure was studied:

$$\text{Ser-peptide–E–AMPPCP–}Cr^{3+}$$

Proton NMR samples containing 70 μM catalytic subunit and 8.0 mM Ser-peptide in 10 mM Bis-Tris or K^+ PIPES buffer (pD 6.0) and 0.15 M KCl were titrated with β,γ-bidentate Cr^{3+} AMPPCP, (0–75 μM) measuring $1/T_1$ and $1/T_2$ as discussed above. The lower pH was used to ensure stability of the Cr^{3+} complex. To study the effect on metal–substrate distances of a second metal at the inhibitory site, $MgCl_2$ was added to the above system, to form the following complex

$$\text{Ser-peptide–E–AMPPCP–}Cr^{3+}$$
$$\diagdown \ \diagup$$
$$Mg^{2+}$$

with no detectable change in relaxation rates, hence in the distances. Displacement experiments with the competitive inhibitor polyarginine (333 μM) carried out as mentioned above were used to establish active site binding of the peptides and a negligibly low value of $1/T_{o.s.}$.

Of the four major secondary structures found on the surface of proteins, the α-helix, the β-pleated sheet, the β-turn, and the coil, only the α-helix and the β-pleated sheet conformations for the enzyme-bound heptapeptide could be rigorously excluded by the paramagnetic probe-T_1 method. Kinetic studies from the literature with peptides of varying sequence provided some evidence against the obligatory requirement for any of the four possible β-turns within the heptapeptides studied.[10,16,17] Hence, if protein kinase has an absolute requirement for a specific secondary structure of the bound peptide or protein substrate then this structure is probably a coil (Fig. 2B). However a preference for a $β_{2-5}$ or a $β_{3-6}$ turn (Fig. 2A) cannot be excluded. Paramagnetic probe-T_1 studies with smaller peptides, with peptides enriched with ^{13}C in the backbone, and NOE measurements should define more precisely the conformation of the bound peptide substrate.

Kinetics of Substrate Exchange in Complexes of the Catalytic Subunit[7,10,16]

One of the major advantages of NMR for the study of the structure of enzyme complexes is that kinetic properties of the complexes are simultaneously measured. This provides a critical test of whether the complexes, the structures of which are being determined, are also forming and dissociating rapidly enough to participate in catalysis. The substrate complexes of protein kinase studied by NMR all survived this test.[7,10] As discussed in detail previously in this series,[3,4] the paramagnetic contribution to the transverse relaxation rate ($1/fT_{2p}$) sets a lower limit to $1/\tau_M$ the pseudo-first-order rate constant for the exchange of a substrate into a paramagnetic environment on an enzyme. In the case of the ternary complex of the catalytic subunit, Mn^{2+}, and $Co(NH_3)_4ATP$, a value of $1/fT_{2p}$ of the γ-P of 2.7×10^6 sec^{-1} was measured by the paramagnetic line broadening as described above.[7] This value sets a lower limit on the rate constants (k_{off}) for dissociation of Mn^{2+} out of the ternary complex and on the subsequent dissociation of $Co(NH_3)_4ATP$ from the binary complex. Since the equilibrium constant for dissociation (K_D) is equal to k_{off}/k_{on}, this limit together with the respective equilibrium constants for the dissociation of Mn^{2+} and for $Co(NH_3)_4ATP$ yielded lower limit k_{on} values for Mn^{2+} ($\geq 2 \times 10^{10}$ M^{-1} sec^{-1}) and for $Co(NH_3)_4ATP$ ($\geq 9 \times 10^9$ M^{-1} sec^{-1}) at the diffusion limit.[7]

Under certain conditions $1/fT_{2p}$ is equal to $1/\tau_M$. This is best determined by measuring $1/fT_{2p}$ as a function of temperature. A positive temperature dependence with high activation energy (>3 kcal/mol)[4,25,41] establishes such a mechanism. This was found to be the case for the $C_β$ methyl protons of the Ala-peptide in the presence of the quaternary complex of the catalytic subunit, Mn^{2+}, $Co(NH_3)_4ATP$, and the Ala-peptide (E_{act} = 7.3 ± 0.5 kcal/mol).[10] The simplest kinetic assumption, namely that $1/\tau_M = k_{off}$ of the Ala-peptide from this complex, directly yielded the value

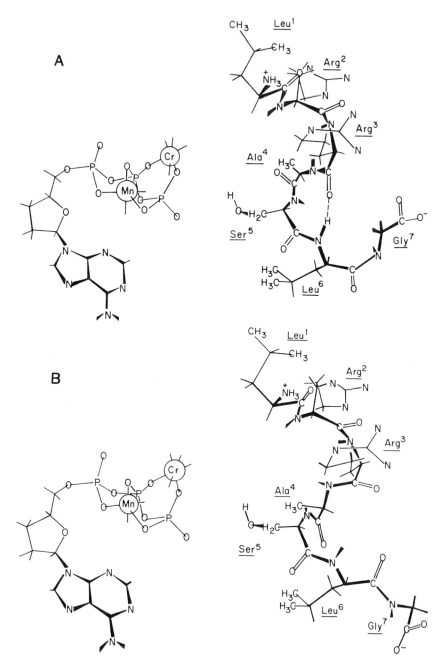

FIG. 2. Conformation and arrangement of the nucleotide and heptapeptide substrates at the active site of protein kinase consistent with distances obtained by NMR.[10] Cr occupies the activating site and Mn occupies the inhibitory site. A β_{3-6} turn (A) or a coil conformation (B) for the peptide substrate is shown, as discussed in the text.

of 5.1×10^3 sec^{-1} for k_{off} at 25°. From this value and the equilibrium constant, a k_{on} value was calculated (1.0×10^7 M^{-1} sec^{-1}) which was below the diffusion limit suggesting conformational selection or a conformational change of the peptide upon binding. Similar conclusions were reached for the Ser-peptide.[10] All of the k_{off} values greatly exceeded k_{cat} of the protein kinase reaction (30 sec^{-1}) showing that the complexes studied by NMR are kinetically competent to function in catalysis.[7,10,16]

Interactions of Regulatory and Catalytic Subunits

The regulatory dimer of protein kinase (R_2) inhibits the catalytic subunit by blocking the binding site of the peptide or protein substrate. This conclusion was based on NMR studies of the holoenzyme (R_2C_2).[11]

The binding of ADP to R_2C_2 (10.5 μM) in the presence of $MnCl_2$ (42 μM) was studied by PRR titrations over the range 0–40 μM ADP, as described above. The binding of Mn^{2+} to 40 μM catalytic subunit in the presence of 40 μM ADP and in the presence or absence of a threefold excess of regulatory subunit over catalytic subunit was studied by titration with $MnCl_2$ and monitored by EPR as described above.

One nucleotide binding site per catalytic subunit was found with the holoenzyme with a dissociation constant fivefold tighter than that found on the catalytic subunit alone. Two tight Mn^{2+} binding sites per enzyme–nucleotide complex were detected with dissociation constants indistinguishable from those found with the catalytic subunit alone. Hence inhibition of the catalytic subunit by the regulatory subunit does not result from occlusion of the metal nucleotide site nor from enhanced binding of the inhibitory metal.

Occlusion of the peptide site on the catalytic subunit by the regulatory subunit was detected by monitoring the paramagnetic effects of Mn^{2+}, at both the activating and inhibitory sites, on $1/T_2$ of the protons of the Ser-peptide and Ala-peptide as a phenomenological tool to detect active site binding of these peptides. In absence of R_2, complexes of the following structure were detected by paramagnetic broadening of the peptide protons:

$$\text{peptide–C–nucleotide–}Mn^{2+}$$
$$\phantom{\text{peptide–C–nucleotid}}\backslash/$$
$$\phantom{\text{peptide–C–nucleotide}}Mn^{2+}$$

The conditions for detecting such complexes are described above in the structural studies of the bound peptides. In the presence of the regulatory subunit, little paramagnetic line broadening of the peptide resonance was detected, indicating the formation of the following complex.

$$R \cdot \underset{\underset{Mn^{2+}}{\diagdown\diagup}}{C\text{-nucleotide}-Mn^{2+}}$$

Further evidence for such a complex, which was dissociated by cAMP, was obtained in the following system, in which $(R\text{-}P)_2C_2$ is replaced by $R\text{-}P \cdot C$ for simplicity.

$$\text{Ala-peptide} + R\text{-}P \cdot \underset{\underset{Mn^{2+}}{\diagdown\diagup}}{C\text{-}ATP\text{-}Mn^{2+}} \xrightarrow{\text{cAMP}} (\text{cAMP})_2 R\text{-}P$$

$$+ \text{Ala-peptide-}\underset{\underset{Mn^{2+}}{\diagdown\diagup}}{C\text{-}ATP\text{-}Mn^{2+}} \quad (13)$$

In a solution (0.4 ml) containing 22 μM holoenzyme (R_2C_2), 4.2 mM Ala-peptide, and 80 μM ATP, the addition of 195 μM MnCl$_2$ induced only a small paramagnetic broadening of the proton resonances of the Ala-peptide due to the weak binding of Mn^{2+} to the peptide. The addition of 600 μM cAMP induced a 5 Hz broadening of the resonances of the Ala-peptide as the regulatory subunit dissociated according to Eq. (13) permitting the active site binding of the Ala-peptide, and its enhanced relaxation by the enzyme-bound Mn^{2+}. Active site binding of the peptide was established by displacing it with the subsequent addition of 333 μM polyarginine which removed the 5 Hz line broadenings.[11] For these experiments, ATP was used as the nucleotide to insure phosphorylation of the regulatory subunit, and thereby to lower its affinity for C. This was necessary at the high level of R_2C_2 (22 μM) used for these experiments which exceeds by an order of magnitude the dissociation constant of R_2C_2 in the presence of cAMP. Thus when ADP was used as the nucleotide, under very similar conditions, no detectable broadening of the resonances of the Ser-peptide or Ala-peptide was induced by cAMP.

NMR Studies of the Regulatory Subunit

The phosphorus NMR of the type II regulatory dimer was examined in preliminary studies, in order to directly observe the covalently bound phosphate and the tightly bound cAMP molecules.[12] The sample of the regulatory subunit (0.267 mM monomer concentration) used for ^{31}P NMR studies contained 100 mM Tris–Cl buffer, pH 7.55, 5 mM 2-mercaptoethanol, and 1 mM EDTA in 25% [2]H$_2$O for field-frequency locking. ^{31}P NMR spectra were obtained on a Bruker WM-250 spectrometer operating at 101.27 MHz at 20° using a 10-mm sample tube with vortex plug. The minimal sample volume which did not sacrifice either signal to noise or resolution was found to be 2.0 ml.

Spectra were recorded with either gated broadband decoupling (10 dB below 10 W) or in the presence of ^{1}H coupling. Typically, NMR spectra were obtained with 512 to 4096 transients by averaging individual blocks of 512 transients each with 8192 data points. Since all of the phosphorus resonances were presumed bound to the regulatory subunit, a 90° observation pulse with a recycle time of 1.4 sec was typically used to allow full recovery of the signal. Chemical shifts were measured with respect to external 85% H_3PO_4 in a separate sample, similarly locked on an inner tube of ^2H_2O. Positive chemical shifts are defined as those upfield from 85% H_3PO_4 and negative chemical shifts are downfield.

The 101.27 MHz ^{31}P NMR spectrum of the type II regulatory subunit of cAMP-dependent protein kinase is shown in Fig. 3. Three phosphorus resonances in the region −4.60 to −4.25 ppm and a broad phosphorus resonance at +1.23 ppm are observed. The assignments, line widths, and integrated areas are given in the table. Integration of the resonances in this and other spectra, using methylphosphonate as an internal standard, gave phosphorus concentrations of 0.22, 0.20, 0.21, and 0.52 mM for peaks at −4.59, −4.44, −4.26, and +1.23 ppm, respectively. These concentrations are accurate to within 30%. When normalized to the concentration [R] of the regulatory subunit, the phosphorus species are present in the approximate ratio of 0.82, 0.75, 0.79, and 2.0 mol/mol R, respectively.

The addition of 3.4 μM catalytic subunit, MgADP, and an ADP regenerating system consisting of creatine kinase and creatine resulted in the rapid and complete loss of the resonance at −4.44 ppm. The phosphorus resonances at −4.59, −4.26, and +1.23 ppm were unaffected by the addition of the catalytic subunit. On the basis of the chemical shift of the phosphorus resonance at −4.44 ppm and the observation that the cata-

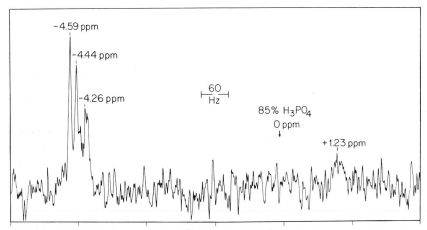

FIG. 3. ^{31}P NMR spectrum of the regulatory subunit of protein kinase at 101.3 MHz.[12] Conditions are given in the text and in the table.

³¹P RESONANCES IN THE TYPE II REGULATORY SUBUNIT OF
BOVINE HEART PROTEIN KINASE[a]

Chemical shift[b] (ppm)	Linewidth[c] (Hz)	Concentration of phosphorus[d] (mM)	mol phosphorus/mol R[e]	Tentative assignment[f]
−4.59	7	0.22	0.82	g
−4.44	7	0.20	0.75	P-serine
−4.26	12	0.21	0.79	g
1.23	32	0.52	2.0	3′,5′-cAMP

[a] P. R. Rosevear, A. S. Mildvan, C. O'Brian, and E. T. Kaiser, unpublished observations (1982).
[b] Chemical shifts from external 85% H_3PO_4; accurate to ±0.02 ppm.
[c] Linewidth measured in Hertz at half-maximal peak height.
[d] Concentration of phosphorus determined by addition of a known concentration of methylphosphonate. Concentrations are accurate to ±30%.
[e] Calculated from the phosphorus concentration using 0.267 mM as the monomer concentration for the regulatory subunit.
[f] NMR spectra of reference compounds at pH 7.55 yielded chemical shifts accurate to ±0.02 ppm for inorganic phosphate (−2.84 ppm), P-serine (−4.57 ppm), 3′,5′-cAMP (0.91 ppm), 3′-AMP (−4.61 ppm), and 5′-AMP (−4.33 ppm).
[g] See text.

lytic subunit of protein kinase catalyzes the specific loss of this resonance, it is proposed that the peak at −4.44 ppm is the exchangeable, autophosphorylated phosphoserine residue at position 95.[47] The small chemical shift difference (+0.13 ppm) of the phosphoserine resonance compared to standard phosphoserine suggests that it is slightly shifted due to the environment of the protein, but is apparently deprotonated.

The broad resonance at +1.23 ppm is assigned to tightly bound 3′,5′-cAMP based on its chemical shift, insensitivity to *E. coli* alkaline phosphatase (type III, Sigma), and presence in a stoichiometry of approximately 2 mol/mol R. This stoichiometry is in good agreement with that found by biochemical analysis (1.8 mol cAMP/mol R). The chemical shift of the bound cAMP is shifted upfield by 0.32 ppm compared with that of free cAMP at the identical pH. Such an upfield chemical shift could result from either electronic effects of the protein or a change in the O–P–O bond angle of 3′,5′-cAMP. It is clear that no ring-opening of cAMP has occurred upon binding to the regulatory subunit.

The addition of *E. coli* alkaline phosphatase (type III, Sigma) resulted in the rapid disappearance of resonances at −4.59 and −4.26 ppm indicat-

[47] K. Takio, S. Smith, E. Krebs, K. Walsh, and K. Titani, *Proc. Natl. Acad. Sci. U.S.A.* **79**, 2544 (1982).

ing that these phosphates were present as monoesters. Two alternative assignments for these signals are under consideration. The chemical shifts are consistent with the assignment of the resonances at −4.59 and −4.26 ppm as 3′-AMP and 5′-AMP, respectively (see the table). The reason why 3′-AMP and 5′-AMP would accompany the regulatory subunit during its extensive purification is unknown. They may result from breakdown of the tightly bound 3′,5′-cAMP. The fact that they are detected despite nearly full occupancy of the cAMP sites suggests that these nucleotides occupy different sites. Alternatively, these signals may represent phosphoserines which are not dischargeable by the catalytic subunit. The stoichiometry of approximately one per catalytic subunit is similar to that found by chemical analysis of the nonexchangeable phosphate content.[48] The difference in their chemical shifts could result from the asymmetry of the regulatory dimer. The widths of the resonances at −4.59, −4.44, and −4.26 ppm are relatively narrow indicating that the 3′-AMP, P-serine, and 5′-AMP phosphates are relatively mobile. The linewidth of the upfield resonance at +1.23 ppm assigned to cAMP is 32 Hz suggesting that the phosphorus is highly immobilized and tumbling at the rate of the entire protein.

Acknowledgments

This work was supported by National Institutes of Health Grants AM28616 and GM19037, National Science Foundation Grant PCM8121355, American Cancer Society Postdoctoral Fellowship PF2111 to Paul R. Rosevear, a Chaim Weizmann Fellowship to Joseph Granot, and an appropriation from the Commonwealth of Pennsylvania to the Institute for Cancer Research. Support for the computation was provided by National Institutes of Health Grant CA22780. The 360-MHz studies were done at the Middle Atlantic Regional NMR Facility which is supported by National Institutes of Health Grant RR-542.

[48] D. A. Flockhart, D. M. Watterson, and J. D. Corbin, *J. Biol. Chem.* **255,** 4435 (1980).

[13] Synthesis of Oligopeptides for the Study of Cyclic Nucleotide-Dependent Protein Kinases

By DAVID B. GLASS

The association of a protein kinase with its phosphate-accepting substrate is a complex protein–protein interaction. Investigations of such an interaction can be simplified by using peptides derived from protein substrates.[1] The first oligopeptides shown to be substrates for cAMP-dependent protein kinase were obtained from enzymic digests of myelin basic protein.[2] Daile et al.[3] prepared the first synthetic substrate for this enzyme, an octapeptide corresponding to a sequence contained in myelin basic protein. The synthetic peptides that have been used most frequently to study the substrate specificity of cAMP-dependent protein kinase are analogs of a sequence of amino acid residues at the site of phosphorylation in L-type pyruvate kinase.[4–7] Investigations using these peptides and peptides corresponding to sequences from other substrates have recently been reviewed.[8,9] Most studies of the substrate specificity of cGMP-dependent protein kinase have employed synthetic peptides and their analogs that correspond to sequences at phosphorylation sites in histone H2B.[9–11]

Because peptides used to study protein kinases are relatively short and small quantities of variations of one sequence are frequently required, the technique of stepwise solid-phase peptide synthesis[12,13] is most useful, particularly if automated. Classical methods of solution synthesis[14–17] provide alternative approaches to the synthesis of oligopeptides, but this chapter will be confined to the methods most commonly used in solid-phase synthesis of peptides as well as to the use of peptides to assay cyclic nucleotide-dependent protein kinases. I will stress those strategies for peptide synthesis that are applicable to commercially available re-

[1] C. Nolan, W. B. Novoa, E. G. Krebs, and E. H. Fischer, *Biochemistry* **3**, 542 (1964).
[2] P. Daile and P. R. Carnegie, *Biochem. Biophys. Res. Commun.* **61**, 852 (1974).
[3] P. Daile, P. R. Carnegie, and J. D. Young, *Nature (London)* **257**, 416 (1975).
[4] Ö. Zetterqvist, U. Ragnarsson, E. Humble, L. Berglund, and L. Engström, *Biochem. Biophys. Res. Commun.* **70**, 696 (1976).
[5] B. E. Kemp, D. J. Graves, E. Benjamini, and E. G. Krebs, *J. Biol. Chem.* **252**, 4888 (1977).
[6] J. R. Feramisco, D. B. Glass, and E. G. Krebs, *J. Biol. Chem.* **255**, 4240 (1980).
[7] J. Granot, A. S. Mildvan, H. N. Bramson, N. Thomas, and E. T. Kaiser, *Biochemistry* **20**, 602 (1981).
[8] E. G. Krebs and J. A. Beavo, *Annu. Rev. Biochem.* **48**, 923 (1979).
[9] D. B. Glass and E. G. Krebs, *Annu. Rev. Pharmacol. Toxicol.* **20**, 363 (1980).
[10] D. B. Glass and E. G. Krebs, *J. Biol. Chem.* **254**, 9728 (1979).
[11] D. B. Glass and E. G. Krebs, *J. Biol. Chem.* **257**, 1196 (1982).
[12] R. B. Merrifield, *J. Am. Chem. Soc.* **85**, 2149 (1963).
[13] R. B. Merrifield, *Biochemistry* **3**, 1385 (1964).
[14] F. M. Finn and K. Hofmann, in "The Proteins" (H. Neurath and R. L. Hill, eds.), 3rd Ed., Vol. 2, p. 105. Academic Press, New York, 1976.
[15] M. Bodanszky, Y. S. Klausner, and M. A. Ondetti, "Peptide Synthesis," 2nd ed. Wiley, New York, 1976.
[16] P. G. Katsoyannis and G. P. Schwartz, this series, Vol. 47, p. 501.
[17] E. Gross and J. Meienhofer, eds., "The Peptides: Analysis, Synthesis, Biology," Vol. 1. Academic Press, New York, 1979.

agents and equipment, specifically the use of chloromethylated polystyrene resin as solid support, the Boc[18] group for α-NH_2 protection, and DCC as a coupling reagent. The concern is certainly not to give a review of the field of solid-phase peptide synthesis, but to provide an overall guide to one approach for the synthesis of small peptides for use as phosphate-accepting substrates. It is hoped that this chapter will serve as a brief introduction to solid-phase peptide synthesis, particularly to individuals working in non-chemically oriented laboratories. For fuller details and coverage of synthetic methods not discussed here, the reader is referred to recent comprehensive reviews of solid-phase peptide synthesis appearing both in this series[19] and elsewhere.[20-24] Additional selected sources include several recent symposium proceedings.[25-27]

Synthesis of Protected Peptides

Principle. A schematic summary of the steps involved in solid-phase peptide synthesis is presented in Fig. 1. The N^{α}-Boc-protected[13] derivative of the COOH-terminal amino acid of the desired peptide is attached to the solid support, a cross-linked polystyrene resin. The deprotection step then removes the Boc moiety exposing the α-NH_2 group of this amino acid. Because deprotection employs acidolysis, the α-NH_2 group

---

[18] Abbreviations used: Boc, *tert*-butyloxycarbonyl; DCC, dicyclohexylcarbodiimide; TFA, trifluoroacetic acid; DIEA, diisopropylethylamine; CH_2Cl_2, dichloromethane; DMF, dimethylformamide; tosyl, toluene sulfonyl; PAM, phenylacetamidomethyl; HF, hydrogen fluoride; HBr, hydrogen bromide; HPLC, high-pressure liquid chromatography. Amino acids are of the L configuration unless otherwise indicated.

[19] M. S. Doscher, this series, Vol. 47, p. 578.

[20] J. M. Stewart and J. D. Young, "Solid Phase Peptide Synthesis." Freeman, San Francisco, California, 1969.

[21] R. B. Merrifield, in "The Chemistry of Polypeptides" (P. G. Katsoyannis, ed.), p. 335. Plenum, New York, 1973.

[22] J. Meienhofer, in "Hormonal Proteins and Peptides" (C. H. Li, ed.), Vol. 2, p. 45. Academic Press, New York, 1973.

[23] B. W. Erickson and R. B. Merrifield, in "The Proteins" (H. Neurath and R. L. Hill, eds.), 3rd Ed., Vol. 2, p. 255. Academic Press, New York, 1976.

[24] G. Barany and R. B. Merrifield, in "The Peptides: Analysis, Synthesis, Biology" (E. Gross and J. Meienhofer, eds.), Vol. 2, p. 1. Academic Press, New York, 1980.

[25] R. B. Merrifield, G. Barany, W. L. Cosand, M. Engelhard, and S. Mojsov, in "Peptides: Proceedings of the Fifth American Peptide Symposium" (M. Goodman and J. Meienhofer, eds.), p. 488. Wiley, New York, 1977.

[26] E. Gross and J. Meienhofer, eds., "Peptides, Structure and Biological Function." Pierce Chemical Company, Rockford, Illinois, 1979.

[27] R. B. Merrifield, in "Chemical Synthesis and Sequencing of Peptides and Proteins" (T.-Y. Liu, A. N. Schechter, R. L. Heinrikson, and P. G. Condliffe, eds.), p. 41. Elsevier, Amsterdam, 1981.

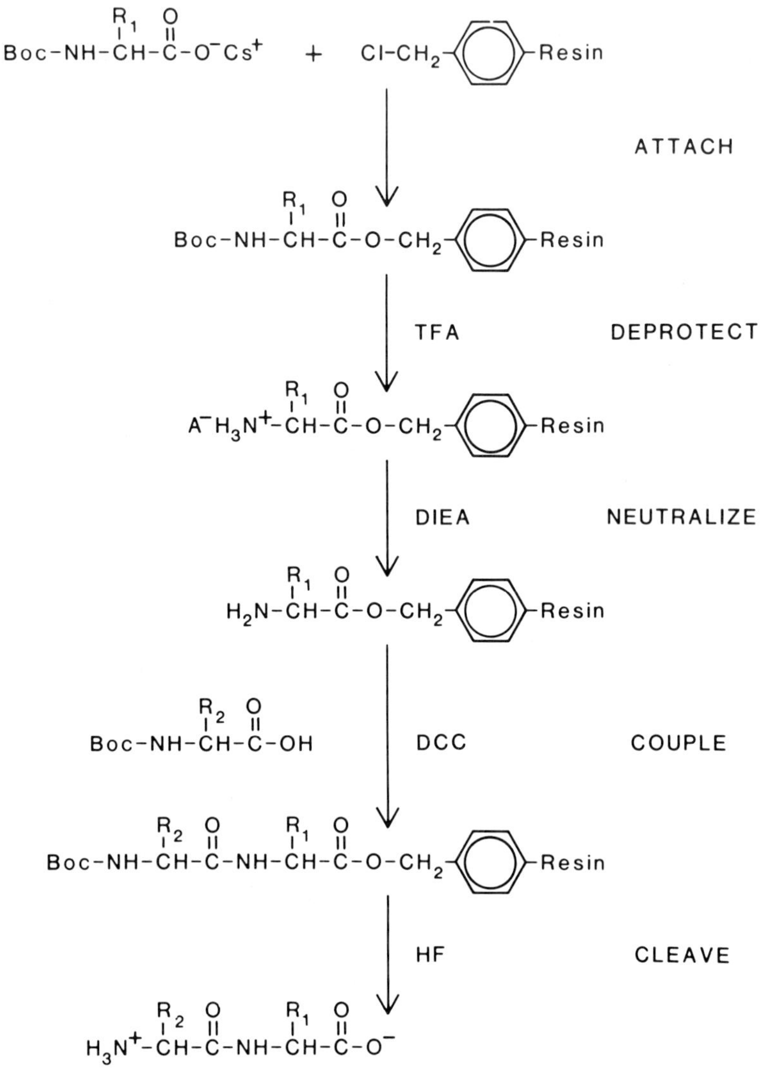

$$\text{Boc-NH-CH(R}_1\text{)-C(=O)-O}^-\text{Cs}^+ \quad + \quad \text{Cl-CH}_2\text{-C}_6\text{H}_4\text{-Resin}$$

ATTACH

$$\text{Boc-NH-CH(R}_1\text{)-C(=O)-O-CH}_2\text{-C}_6\text{H}_4\text{-Resin}$$

TFA   DEPROTECT

$$\text{A}^-\text{H}_3\text{N}^+\text{-CH(R}_1\text{)-C(=O)-O-CH}_2\text{-C}_6\text{H}_4\text{-Resin}$$

DIEA   NEUTRALIZE

$$\text{H}_2\text{N-CH(R}_1\text{)-C(=O)-O-CH}_2\text{-C}_6\text{H}_4\text{-Resin}$$

$$\text{Boc-NH-CH(R}_2\text{)-C(=O)-OH} \quad \text{DCC} \quad \text{COUPLE}$$

$$\text{Boc-NH-CH(R}_2\text{)-C(=O)-NH-CH(R}_1\text{)-C(=O)-O-CH}_2\text{-C}_6\text{H}_4\text{-Resin}$$

HF   CLEAVE

$$\text{H}_3\text{N}^+\text{-CH(R}_2\text{)-C(=O)-NH-CH(R}_1\text{)-C(=O)-O}^-$$

FIG. 1. The basic reactions involved in solid-phase peptide synthesis. Adapted from Merrifield et al.[25]

must be deprotonated in a neutralization step before the next residue can be coupled. An excess of the next N^α-Boc-amino acid is then added and peptide bond formation is affected by DCC. The steps of deprotection, neutralization, and coupling are repeated for each additional amino acid (a synthetic cycle), as the peptide grows stepwise from the COOH-terminus. In between each step in a synthetic cycle, excess reagents are conveniently removed from the peptidyl-resin by merely washing the insoluble resin with an appropriate solvent, usually dichloromethane. When the NH_2-terminal residue of the peptide has been coupled, the completed peptide is cleaved from the solid support by acidolysis with anhydrous HF or HBr in TFA (see Cleavage and Deprotection).

The success of solid-phase peptide synthesis depends on each reaction going to essentially total completion. If such high yields are not obtained at each step, products of incomplete reactions will accumulate resulting in a large number of heterogeneous side products.[19,24] The larger the synthesis attempted, the more difficult it will be to purify the desired peptide from closely related side products.

Reagents[28]

Chloromethylated polystyrene-*co*-divinylbenzene (1%) resin
Benzhydrylamine or *p*-methylbenzhydrylamine polystyrene-*co*-divinylbenzene (1%) resin
N^α-Boc-aminoacylpolystyrene-*co*-divinylbenzene (1%) resins
N^α-Boc-amino acids with specifically protected side chains (see listing under Side Chain Protecting Groups)
Dichloromethane
50% TFA in dichloromethane
5% DIEA in dichloromethane, prepared fresh daily
25% DCC in dichloromethane
Isopropanol

Equipment. Instruments for the fully automated synthesis of peptides have been described[23] and models are commercially available from Beckman Instruments, Inc. (Palo Alto, CA) and Vega Biochemicals (Tuscon, AZ). A manual peptide synthesizer is marketed by Peninsula Laboratories, Inc. (San Carlos, CA). The most simple and inexpensive apparatus needed for manual solid-phase synthesis consists of a glass reaction vessel containing a fritted disk and a mechanical shaker to rock the reaction vessel and thus agitate the resin within it.[20]

Procedure

Choice of Resin and Attachment of the COOH-Terminal Amino Acid. Although the recently developed PAM-resin[29] is the most suitable solid

---

[28] Commercial sources of reagents for solid-phase peptide synthesis include Bachem, Inc. (Marina Del Rey, CA), Beckman Instruments, Inc. (Palo Alto, CA), Calbiochem (La Jolla, CA), Lab Systems, Inc. (San Mateo, CA), Peninsula Laboratories, Inc. (San Carlos, CA), Pierce Chemical Co. (Rockford, IL), Research Plus, Inc. (Bayonne, NJ), Serva Biochemicals (Garden City Park, NY), and Vega Biochemicals (Tuscon, AZ), among others.

[29] A. R. Mitchell, B. W. Erickson, M. N. Ryabtsev, R. S. Hodges, and R. B. Merrifield, *J. Am. Chem. Soc.* **98**, 7357 (1976).

support for the synthesis of large peptides, its extra expense is not necessarily warranted for the shorter peptides of interest here. The standard 1% crosslinked chloromethylated polystyrene resin is commercially available with a choice of Boc-aminoacyl substitutions in the range of 0.3 to 0.6 mEq/g. Alternatively, the COOH-terminal amino acid, as the cesium salt, can be esterified to the chloromethylated resin as described by Gisin.[30] Although chloromethylated resin has been used successfully in the synthesis of many peptides, it has several important disadvantages that should be kept in mind. The benzyl ester bond that couples the peptide to the resin is relatively labile under the conditions of the deprotection step (33–50% TFA in CH_2Cl_2), resulting in the gradual loss of peptide chains from the resin during each synthetic cycle.[31] Peptide chains synthesized on this resin can undergo premature, irreversible termination due to trifluoroacetylation by resin-bound trifluoroacetoxymethyl groups.[32] In addition, quaternary ammonium sites can be formed by reaction of residual chloromethyl groups with the tertiary amine used in the neutralization step, yielding a resin with undesirable ion exchange properties.[24,30]

For the synthesis of peptides containing a COOH-terminal amide group, the benzhydrylamine resin[33,34] or the p-methylbenzhydrylamine resin[35] is recommended. Protected amino acids are attached to these resins with DCC.[34] Functional amine groups of these resins that are not fully acylated during the attachment step should be acetylated or otherwise blocked.[12,36] Cleavage of the completed peptide from these resins with anhydrous HF yields the peptide as a COOH-terminal amide.

Once the scale of the peptide synthesis is chosen, the appropriate amount of Boc-aminoacyl resin is placed in the reaction vessel and washed with dichloromethane to prepare it for the synthetic cycle.

Deprotection, Neutralization, and Coupling Steps. A typical sequence of operations for a synthetic cycle is shown in Table I. When the Boc group is used for α-NH_2 protection[13] in conjugation with syntheses on chloromethylated resins, it is removed during the deprotection steps by exposure to 50% TFA in CH_2Cl_2 for 30 min. A short prewash with TFA serves to prevent dilution of the TFA in the subsequent treatment by the

---

[30] B. F. Gisin, *Helv. Chim. Acta* **56**, 1476 (1973).
[31] B. Gutte and R. B. Merrifield, *J. Biol. Chem.* **246**, 1922 (1971).
[32] S. B. H. Kent, A. R. Mitchell, M. Engelhard, and R. B. Merrifield, *Proc. Natl. Acad. Sci. U.S.A.* **76**, 2180 (1979).
[33] P. G. Pietta and G. R. Marshall, *J. Chem. Soc., Chem. Commun.* **1970**, 650 (1970).
[34] P. G. Pietta, P. F. Cavallo, G. Marshall, and M. Pace, *Gazz. Chim. Ital.* **103**, 483 (1973).
[35] K. Channabasavaiah and J. M. Stewart, *Biochem. Biophys. Res. Commun.* **86**, 1266 (1979).
[36] L. D. Markley and L. C. Dorman, *Tetrahedron Lett.* **21**, 1787 (1970).

TABLE I
TYPICAL SCHEDULE OF OPERATIONS FOR A SYNTHETIC CYCLE IN A
DOUBLE-COUPLING PROTOCOL[a]

Operation	Step	Reagent[b]	Time (min)	Number of times
Deprotection	1	50% TFA/CH_2Cl_2	1	1
	2	50% TFA/CH_2Cl_2	25	1
Wash	1	CH_2Cl_2	1	5
	2	Isopropanol	1	3
	3	CH_2Cl_2	1	5
Neutralization	1	5% DIEA/CH_2Cl_2	5	2
Wash	1	CH_2Cl_2	1	5
	2	Isopropanol	1	3
	3	CH_2Cl_2	1	5
Coupling	1	Boc-amino acid	5	1
	2	DCC	60	1
Wash	1	CH_2Cl_2	1	5
	2	Isopropanol	1	3
	3	CH_2Cl_2	1	5
Neutralization	1	5% DIEA/CH_2Cl_2	5	2
Wash	1	CH_2Cl_2	1	5
	2	Isopropanol	1	3
	3	CH_2Cl_2	1	5
Recouple	1	Boc-amino acid	5	1
	2	DCC	60	1
Wash	1	CH_2Cl_2	1	5
	2	Isopropanol	1	3
	3	CH_2Cl_2	1	5

[a] Adapted from Hodges and Merrifield[44] and Feinberg and Merrifield[69]; variations are possible.

[b] Typical volumes of reagent for deprotection, neutralization, and wash steps are 10–20 ml/g of resin. Coupling reactions are usually conducted in volumes approximately two-thirds of those above.

CH_2Cl_2 that swells the resin. This procedure leaves the α-NH_2 group of the peptidyl resin in the form of the trifluoroacetate salt. Brief treatment with 5% DIEA[37] in CH_2Cl_2 is used to neutralize the protonated α-NH_2 group. DIEA, rather than the originally introduced 10% triethylamine, is used because it minimizes formation of quaternary ammonium sites on the resin that result from alkylation of the tertiary amine by unsubstituted chloromethyl groups remaining on the resin.[24,30,38]

The standard wash protocol includes a step with isopropanol which

[37] S.-S. Wang and R. B. Merrifield, *J. Am. Chem. Soc.* **91**, 6488 (1969).
[38] P. Fankhauser, P. Fries, and M. Brenner, *Helv. Chim. Acta* **56**, 2516 (1973).

shrinks the resin. The shrinking of the resin and the reswelling in CH_2Cl_2 promote the complete removal of excess reagents before the coupling step.[39]

After deprotection and neutralization, coupling of the next amino acid is usually performed twice (Table I). Use of such a double-coupling protocol is somewhat expensive in terms of Boc-amino acids and time, but if automated synthesis is used without routine monitoring of the coupling step, double-coupling helps to ensure complete peptide bond formation before the next synthetic cycle. For most Boc-amino acids, a threefold molar excess with respect to the initial substitution of the resin is used. Boc-amino acid dissolved in CH_2Cl_2 is added to the deprotected and neutralized peptidyl resin and the two are mixed for approximately 5 min. At this time, an appropriate volume of 25% DCC in CH_2Cl_2 calculated to deliver an equimolar amount of DCC to Boc-amino acid is added to the reaction vessel. The coupling reaction is allowed to proceed for 1 to 2 hr. The volume of solvent in the coupling step should be kept to a minimum so as to increase the concentration of the reactants and drive the reaction to completion. The reactions involved in the activation of the α-COOH group of the incoming amino acid by DCC, the nucleophilic attack on the α-NH_2 amino group of the peptidyl resin, and the generation of side products have been well described.[19,23,24,40] After washing, neutralization is repeated and the second coupling is performed. A subsequent wash completes one synthetic cycle. Several of the conditions described above are modified for the coupling of certain amino acids, particularly asparagine and glutamine (see below).

It is wise to routinely monitor completeness of the coupling reaction during each synthetic cycle. Several techniques are available to do this.[23,24] Reaction of either ninhydrin[41] or fluorescamine[42] with residual free amino groups on small samples of the peptidyl resin can be done rapidly, but these tests give only semiquantitative results. The use of picric acid[43] or imidazolium picrate[44] to monitor coupling reactions is quantitative and nondestructive and has been automated.[44] If monitoring indicates complete (>99+%) peptide bond formation after a first coupling, double coupling is not necessary.

[39] L. Corley, D. H. Sachs, and C. B. Anfinsen, *Biochem. Biophys. Res. Commun.* **47**, 1353 (1972).
[40] D. H. Rich and J. Singh, in "The Peptides: Analysis, Synthesis, Biology" (E. Gross and J. Meienhofer, eds.), Vol. 1, p. 241. Academic Press, New York, 1979.
[41] E. Kaiser, R. L. Colescott, C. D. Bossinger, and P. I. Cook, *Anal. Biochem.* **34**, 595 (1970).
[42] A. M. Felix and M. H. Jimenez, *Anal. Biochem.* **52**, 377 (1973).
[43] B. F. Gisin, *Anal. Chim. Acta* **58**, 248 (1972).
[44] R. S. Hodges and R. B. Merrifield, *Anal. Biochem.* **65**, 241 (1975).

At the end of the synthesis, the protected peptidyl resin is washed extensively, dried *in vacuo,* and weighed. The increase in weight of the resin during synthesis is an approximate measure of the extent of peptide chain growth that has occurred. The dried peptidyl resin can be stored at room temperature until the cleavage-deprotection step.

Side Chain Protecting Groups. The side chains of alanine, glycine, isoleucine, leucine, phenylalanine, proline, and valine do not contain reactive functional groups that must be blocked during peptide synthesis. The side chains of the other, common amino acids are reactive under some or all of the conditions of peptide synthesis and must be blocked by protecting groups. These protecting groups must be stable to all the reaction steps in a synthetic cycle but be susceptible to selective removal at the end of the synthesis. The choices of side chain protecting groups compatible with the use of the Boc moiety for α-NH_2 group protection are briefly summarized below. The reader should refer to more extensive discussions for details, possible side reactions, and alternative strategies.[23,24,25] All of the N^α-Boc-protected amino acids mentioned below are commercially available. It is suggested that their purity be checked by thin-layer chromatography prior to use.[20]

One or more arginine or lysine residues are present in almost all peptides of interest as substrates of cyclic nucleotide-dependent protein kinases.[9] The basic guanidino moiety of arginine can be protected with either the nitro or tosyl groups. The N^G-nitro or N^G-tosyl derivatives of N^α-Boc-arginine are sparingly soluble in CH_2Cl_2, so they are dissolved in a small volume of DMF and then CH_2Cl_2 is added so that the coupling reaction is conducted in a mixed solvent of DMF:CH_2Cl_2 (1:3).[44] The N^G-tosyl derivative is probably the best choice if cleavage and deprotection will be by anhydrous HF (see Cleavage and Deprotection), as HF cleavage of the N^G-nitro derivative can yield ornithine in some sequences.[46] The N^G-nitro derivative is suitable for peptides to be cleaved by HBr-TFA, but the nitro group is stable to this treatment and must be removed by a subsequent catalytic hydrogenation step.[20,24]

N^α-Boc-lysine in which the ε-NH_2 group is protected with the benzyloxycarbonyl moiety can be successfully used in very short peptides or in instances where lysine is one of the last residues to be coupled. However, the benzyloxycarbonyl protecting group is relatively acid labile,[47,48] and repetitive exposure to TFA during N^α-Boc deprotection will cause a

---

[45] E. Gross and J. Meienhofer, eds., "The Peptides: Analysis, Synthesis, Biology," Vol. 3. Academic Press, New York, 1980.
[46] D. Yamashiro, J. Blake, and C. H. Li, *J. Am. Chem. Soc.* **94**, 2855 (1972).
[47] D. Yamashiro and C. H. Li, *Int. J. Pept. Protein Res.* **4**, 181 (1972).
[48] B. W. Erickson and R. B. Merrifield, *J. Am. Chem. Soc.* **95**, 3750 (1973).

significant degree of deblocking. Any free ε-NH$_2$ groups present during coupling of subsequent amino acids will produce branched-chain products. A halogenated benzyl derivative such as the 2-chlorobenzyloxycarbonyl group[49] has greater acid stability and is the ε-NH$_2$ protecting group of choice of those commercially available.

The β-imidazole ring of histidine can be protected as the N^{im}-tosyl or N^{im}-dinitrophenyl derivatives. The tosyl blocking group is removed by anhydrous HF. The N^{im}-dinitrophenyl group is totally stable to these acidolytic cleavage procedures and must be separately removed in a deblocking step employing thiolysis[50] or ammonolysis.[51] Both of these derivatives of N^α-Boc-histidine are also best dissolved in DMF : CH$_2$Cl$_2$ (1 : 3).

N^α-Boc-N^ε-2-chlorobenzyloxycarbonyl-lysine and N^α-Boc-N^{im}-tosylhistidine are most frequently (and inexpensively) available as their *tert*-butylamine and dicyclohexylamine salts, respectively, and must be converted to their corresponding free acids before use.[19,24]

Serine and threonine are the residues phosphorylated by cyclic nucleotide-dependent protein kinases. The β-hydroxyl groups of the side chains of these amino acids are blocked during peptide synthesis as their O-benzyl ethers. The β- and γ-carboxyl groups of aspartic acid and glutamic acid, respectively, are protected as their O-benzyl esters. The benzyl group on all four of these amino acids is readily removed by either HF or HBr-TFA.[22-24] Benzyl protection of the side chain of tyrosine is only moderately stable under conditions removing the N^α-Boc group.[47,48] The phenolic hydroxy group of tyrosine is best protected as the O-2,6-dichlorobenzyl derivative.[48] This blocking group has sufficient stability to Boc deprotection conditions and also minimizes the intramolecular alkylation of the 3-position of the tyrosine ring observed during deprotection of O-benzyltyrosine with HF.[48]

N^α-Boc-tryptophan can be used without protection of its β-indole function, but it is susceptible to oxidative decomposition under acidic conditions. It can be protected by running the synthesis under an atmosphere of nitrogen and by including reducing agents in the TFA solutions used for deprotection.[52] Tryptophan residues in peptides appear to be suitably stable under the conditions of HF cleavage.[53] The β-indole ring of

---

[49] B. W. Erickson and R. B. Merrifield, *J. Am. Chem. Soc.* **95**, 3757 (1973).
[50] S. Shaltiel and M. Fridkin, *Biochemistry* **9**, 5122 (1970).
[51] J. M. Stewart, M. Knight, A. C. M. Paiva, and T. Paiva, in "Progress in Peptide Research" (S. Lande, ed.), p. 59. Gordon & Breach, New York, 1972.
[52] J. Blake, K.-T. Wang, and C. H. Li, *Biochemistry* **11**, 438 (1972).
[53] S. Sakakibara, Y. Shimonishi, Y. Kishida, M. Okada, and H. Sugihara, *Bull. Chem. Soc. Jpn.* **40**, 2164 (1967).

tryptophan is best protected as its N^n-formyl derivative.[54–56] The formyl group is stable to HF treatment,[24] but is removed in weakly alkaline media.[55,56]

Cysteine and methionine are infrequently found in the sequence of amino acids around the phosphorylation sites of substrates of cyclic nucleotide-dependent protein kinases and therefore have not been used as residues in synthetic peptide substrates. The β-thiol function of cysteine can be protected by either the S-4-methylbenzyl[57] or S-3,4-dimethylbenzyl[58] group. These blocking groups are removed by anhydrous HF. Methionine can be used unprotected, but there is a danger of S-alkylation during removal of Boc groups with TFA.[23] A scavenger such as methyl ethyl sulfide should be used during cleavage and deprotection of methionine-containing peptides.[20] Alternatively, sulfoxide derivatives of N^α-Boc-methionine can be used during synthesis, and methionine can be regenerated by reduction of methionine sulfoxide with mercaptoethanol subsequent to cleavage and deprotection.[31] The use of norleucine which is isosteric with methionine avoids these problems.[24] Norleucine may be a suitable substitute for methionine in peptides for the study of protein kinases.

Incorporation of N^α-Boc-derivatives of asparagine or glutamine into peptides by a DCC-catalyzed coupling step results in the dehydration of the neutral carboxamide groups of these amino acids to their corresponding ω-nitriles.[59] Therefore, in many cases DCC is not used and these amino acids are incorporated as their active esters.[60,61] The p-nitrophenyl esters of these Boc-amino acids are dissolved in DMF for the coupling reaction, since they do not react sufficiently well in CH_2Cl_2. Since peptide bond formation by this route is rather slow, a sixfold molar excess of active ester is added to the peptidyl resin and is allowed to react for from 4 to 24 hr. Because the reaction is run in DMF, this solvent should be used for the washes immediately before and after the coupling reaction. The use of these active esters is compatible with most synthetic strategies

---

[54] A. Previero, M. A. Coletti-Previero, and J.-C. Cavadore, *Biochim. Biophys. Acta* **147**, 453 (1967).
[55] D. Yamashiro and C. H. Li, *J. Org. Chem.* **38**, 2594 (1973).
[56] M. Ohno, S. Tsukamoto, and N. Izumiya, *J. Chem. Soc., Chem. Commun.* **1972**, 663 (1972).
[57] B. W. Erickson and R. B. Merrifield, *Isr. J. Chem.* **12**, 79 (1974).
[58] D. Yamashiro, R. L. Noble, and C. H. Li, *J. Org. Chem.* **38**, 3561 (1973).
[59] D. V. Kashelikar and C. Ressler, *J. Am. Chem. Soc.* **86**, 2467 (1964).
[60] M. Bodanszky, *in* "The Peptides: Analysis, Synthesis, Biology" (E. Gross and J. Meienhofer, eds.), Vol. 1, p. 105. Academic Press, New York, 1979.
[61] M. Bodanszky and V. du Vigneaud, *J. Am. Chem. Soc.* **81**, 5688 (1959).

since the free carboxamido groups of asparaginyl and glutamyl residues already in the peptide chain are not dehydrated by DCC in subsequent coupling reactions.[23,61] An alternative strategy to that of active ester coupling involves derivatives of N^α-Boc-asparagine or -glutamine in which their ω-carboxamido functions are blocked by the xanthenyl group.[62] These derivatives are commercially available and can be incorporated by the direct action of DCC, but because of the poor solubility of their acyl urea by-products formed during DCC coupling slight changes are needed in the wash protocol.[62]

Cleavage and Deprotection

The benzyl ester bond linking the protected peptide to the resin is relatively stable to TFA used during the repetitive deprotection steps but is readily cleaved by strong anhydrous acids such as HBr in TFA[13] or liquid HF.[53,63] HBr-TFA requires less expensive equipment than does HF and is less hazardous.[64] However, HBr-TFA does not deblock several of the commonly used side chain protecting groups, particularly those of arginine, histidine, and cysteine.[20] Thus an extra deprotection step must be included in the protocol. In addition, HBr-TFA causes destruction of tryptophan residues and is not suitable for cleavage of peptide amides from benzhydrylamine resins. Treatment with anhydrous HF is now the method of choice for most peptides because removal of side chain protecting groups occurs concurrently with cleavage of the peptide from the solid support. The need for a subsequent deprotection step is thus eliminated. The two-step cleavage and deprotection reactions can, however, be of use in the synthesis of peptides as substrates of protein kinases. Synthesis of protected peptides containing N^G-nitroarginine followed by cleavage in HBr-TFA can yield free peptides that are deprotected at all residues except arginine. Such analogs can be useful in subsequent enzymic studies,[11] because arginine residues are important determinants of specificity of cyclic nucleotide-dependent protein kinases.[4,5]

Cleavage-Deprotection by Anhydrous HF. The use of anhydrous HF is best conducted in a vacuum apparatus constructed of a polyfluoroethylene material such as Kel-F.[64] Details of such an apparatus have been published,[20,53,65] and an HF cleavage apparatus is commercially available (Peninsula Laboratories, San Carlos, CA). Anhydrous HF is one of the

[62] L. C. Dorman, D. A. Nelson, and R. C. L. Chow, in "Progress in Peptide Research" (S. Lande, ed.), p. 65. Gordon & Breach, New York, 1972.
[63] J. Lenard and A. B. Robinson, *J. Am. Chem. Soc.* **89**, 181 (1967).
[64] S. Sakakibara, in "Chemistry and Biochemistry of Amino Acids, Peptides and Proteins" (B. Weinstein, ed.), p. 51. Dekker, New York, 1971.
[65] J. Lenard, *Chem. Rev.* **69**, 625 (1969).

strongest inorganic acids known and is highly toxic. Suitable equipment and precautions should attend its use.[66]

The application of HF to cleavage and deprotection in solid phase peptide synthesis has been well documented.[53,63-65] The use of the HF cleavage apparatus has been described in detail.[20,64] Briefly, the protected peptidyl resin along with anisole and a Teflon-coated stirring bar are placed in a reaction vessel attached to the vacuum line of the apparatus. The anisole is used as a carbonium and nitronium ion scavenger[67] to prevent damage to the peptide by electrophiles. HF is dried over cobaltic trifluoride in a reservoir vessel and is then transferred by vacuum distillation to the reaction vessel which is cooled in liquid nitrogen. The peptidyl resin is usually stirred in liquid HF containing 10% (v/v) anisole for 30 to 60 min at 0° using an ice-water bath to jacket the reaction vessel. The HF and anisole are then removed under vacuum to a trap vessel cooled in liquid nitrogen. The dried resin is washed with ethyl acetate to remove residual anisole, and the cleaved peptide is extracted from the resin with dilute acetic acid. Lyophilization of this extract yields the crude, deprotected peptide.

Proper reaction conditions to minimize anisole-dependent side reactions involving glutamic acid residues have been suggested.[68,69] The N- to O-acyl shift in serine or threonine-containing peptides does not appear to be a serious problem.[63] A commercial service of custom HF cleavage is available.[70]

Cleavage by HBr-TFA. Cleavage by HBr-TFA can be carried out in a standard glass-fritted filter funnel with associated adapters and tubing as described by Stewart and Young.[20] The operation should be conducted in a good hood. The peptidyl resin is suspended in neat TFA and HBr is bubbled through it for 60 to 90 min at room temperature.[13] Although acetic acid has been used as a solvent for the reaction, it is not suitable for the peptides of interest here because serine or threonine residues can be acetylated under these conditions.[71] To prevent bromination of tyrosine residues, the HBr should be scrubbed free of bromine by bubbling it through a trap containing anisole in TFA.[20] If the peptide contains tyrosine, cysteine, or methionine residues, an excess of anisole should be used in the cleavage vessel as a scavenger to prevent the formation of benzylated products.[20,67] After the reaction is terminated, the TFA is collected and residual peptide is extracted from the resin on the fritted

[66] C. M. Sharts and W. A. Sheppard, *Org. React.* **21**, 125 (1974).
[67] F. Weygand and W. Steglich, *Z. Naturforsch. Abt. B* **14**, 472 (1959).
[68] S. Sano and S. Kawanishi, *J. Am. Chem. Soc.* **97**, 3480 (1975).
[69] R. S. Feinberg and R. B. Merrifield, *J. Am. Chem. Soc.* **97**, 3485 (1975).
[70] Custom HF cleavage is offered by several of the companies listed under footnote 28.
[71] E. D. Nicolaides and H. A. DeWald, *J. Org. Chem.* **28**, 1926 (1963).

filter with additional TFA. The TFA is removed from the pooled extracts by rotoevaporation without heating. The residue is dissolved in 90% acetic acid or other suitable solvent which is further evaporated to remove residual HBr and TFA. Extraction of anisole, if present, with ether or ethyl acetate is followed by lyophilization yielding the cleaved peptide.

At this stage, the peptide will require catalytic hydrogenation to deprotect N^G-nitroarginine or N^{im}-tosylhistidine if these residues are present. Catalytic hydrogenation can be carried out in an apparatus constructed of standard laboratory glassware.[20] The crude peptide is dissolved in 90% acetic acid, and palladium oxide (5%) on barium sulfate is added as catalyst. Hydrogen gas is slowly bubbled through the mixture with stirring for 12 to 24 hr. The extent of hydrogenolysis of the N^G-nitro group can be conveniently monitored by the disappearance of absorbance at 271 nm due to N^G-nitroarginine.[72] The deprotection of the imidazole ring of histidine can be followed by the appearance of a positive Pauly test.[20] The catalyst is removed by centrifugation and the deprotected peptide is recovered from the supernatant solution. Sulfur-containing amino acids tend to poison the catalyst used in the hydrogenation reaction. Therefore, peptides containing cysteine derivatives or methione are best cleaved and deprotected with anhydrous HF.

The procedures described above for the synthesis, cleavage, and deprotection of peptides have been highly selective. It should be emphasized that numerous chemical approaches and synthetic strategies are available, and the procedures outlined here are not necessarily the best current choices for all, or even most, synthetic problems. If used carefully, they are applicable to the synthesis of small (usually hexa- to dodeca-) peptides useful in the study of cyclic nucleotide-dependent and possibly other[73] protein kinases.

Purification and Analysis of Peptides

Crude peptides from the cleavage-deprotection step(s) can be purified by any number of techniques generally applicable to peptides and proteins. These approaches include crystallization, countercurrent distribution, partition chromatography,[74] gel filtration, ion-exchange chromatography, and affinity chromatography, if applicable. The reader is referred to previous volumes in this series[75] for detailed accounts of these methods. Ion-exchange chromatography on SP-Sephadex using pyridine-acetate buffers[75] is particularly useful for the basic peptides synthesized as

[72] B. Riniker and R. Schwyzer, *Helv. Chim. Acta* **44**, 674 (1961).
[73] T. Hunter, *J. Biol. Chem.* **257**, 4843 (1982).
[74] D. Yamashiro, *in* "Hormonal Proteins and Peptides" (C. H. Li, ed.), Vol. 9, p. 25. Academic Press, New York, 1980.
[75] This series, Vols. 11, 25, 47.

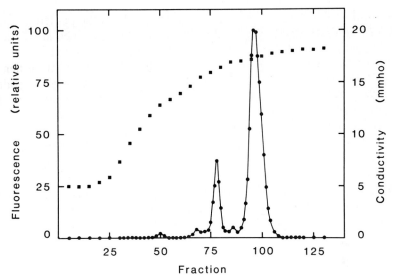

FIG. 2. Chromatography of crude cleavage-deprotection product from the synthesis of Arg-Lys-Ala-Ser-Arg-Lys-Glu on SP-Sephadex developed with a convex gradient of 0.2 M pyridine acetate, pH 3.1 to 2.0 M pyridine-acetate, pH 5.0 [W. A. Schroeder, this series, Vol. 25, p. 203 (1972)]. Fractions were assayed for fluorescence (●) after reaction with fluorescamine and were monitored for conductivity (■). This peptide was used in a study of cyclic GMP-dependent protein kinase [D. B. Glass and E. G. Krebs, *J. Biol. Chem.* **257**, 1196 (1982)]. See text for details.

substrates of protein kinases (Fig. 2). In this example, the major peak was the desired product. The minor peak had the same amino acid composition on acid hydrolysis as did the major peak and was tentatively identified as a mixture of peptides in which one of the two nitroarginine residues was incompletely deprotected. Recently, preparative scale HPLC on either ion-exchange or reverse-phase columns has been perfected for peptides.[76–81] HPLC is now probably the method of choice for final purifica-

[76] J. E. Rivier, *J. Liq. Chromatogr.* **1**, 343 (1978).
[77] H. P. J. Bennett, C. A. Browne, D. Goltzman, and S. Solomom, in "Peptides, Structure and Biological Function" (E. Gross and J. Meienhofer, eds.), p. 121. Pierce Chemical Comp., Rockford, Illinois, 1979.
[78] J. Rivier, J. Desmond, J. Spiess, M. Perrin, W. Vale, R. Eksteen, and B. Karger, in "Peptides, Structure and Biological Function" (E. Gross and J. Meienhofer, eds.), p. 125. Pierce Chemical Comp., Rockford, Illinois, 1979.
[79] F. E. Regnier and K. M. Gooding, *Anal. Biochem.* **103**, 1 (1980).
[80] L. E. Henderson, R. Sowder, and S. Oroszlan, in "Chemical Synthesis and Sequencing of Peptides and Proteins" (T.-Y. Liu, A. N. Schechter, R. L. Heinrikson, and P. G. Condliffe, eds.), p. 251. Elsevier, Amsterdam, 1981.
[81] T. F. Gabriel, J. Michalewsky, and J. Meienhofer, in "Perspectives in Peptide Chemistry" (A. Eberle, R. Geiger, and T. Wieland, eds.), p. 195. Karger, Basel, 1981.

tion of synthetic peptides; however, crude cleavage products should at least be purified by gel filtration prior to application to HPLC columns. Fluorescamine[82] is a convenient and sensitive detection agent for peptides in all chromatographic systems where they cannot be analyzed by continuous measurement of UV absorbance. The Sakaguchi test for arginine[20] is useful, especially for rapid screening of fractions, since so many peptide substrates of cyclic nucleotide-dependent kinases contain this amino acid.

With the potential for termination, deletion, and addition peptides[19] in the synthetic product and side reactions possible during deprotection, it cannot be overemphasized that the final product must be rigorously analyzed for purity and fidelity to the desired peptide. A peptide may be a poor substrate for a protein kinase because of its intended structure or as a result of some unrecognized problem in its synthesis. Only careful analysis of peptides will allow valid conclusions to be drawn in subsequent enzymic studies. Typical analytical techniques include high-voltage electrophoresis, polyacrylamide electrophoresis, thin layer chromatography in several solvents on different supports including reverse-phase media, amino acid analysis after both acid and enzymic hydrolysis, analysis for optical isomers,[83] and possibly end-group and sequence analysis. Again, previous volumes in this series should be consulted.[75] Radiolabeled peptides (see below) should also be analyzed for radiochemical purity.

Assay of Protein Kinases Using Synthetic Peptide Substrates

Principle. Oligopeptides can be phosphorylated by cyclic nucleotide-dependent protein kinases under the same reaction conditions as for protein substrates,[84] but the [^{32}P]phosphopeptide products cannot be isolated by the standard method employing precipitation by trichloroacetic acid onto filter papers.[85] Electrophoresis of reaction mixtures has been used to separate labeled phosphopeptide products from reactants,[2-4] but this technique is cumbersome for processing large numbers of samples. [^{32}P]Phosphopeptides can be separated from excess [γ-^{32}P]ATP and inorganic [^{32}P]phosphate by ion-exchange chromatography on anion-exchange[86] or

---

[82] S. Udenfriend, S. Stein, P. Böhlen, W. Dairman, W. Leimgruber, and M. Weigele, *Science* **178**, 871 (1972).

[83] D. S. Kemp, *in* "The Peptides: Analysis, Synthesis, Biology" (E. Gross and J. Meienhofer, eds.), Vol. 1, p. 315. Academic Press, New York, 1979.

[84] R. Roskoski, Jr., this Volume [1].

[85] G. M. Carlson and D. J. Graves, *J. Biol. Chem.* **251**, 7480 (1976).

[86] B. E. Kemp, E. Benjamini, and E. G. Krebs, *Proc. Natl. Acad. Sci. U.S.A.* **73**, 1038 (1976).

cation-exchange columns,[87] but the former procedure is more convenient. Alternatively, the method based on ion-exchange binding of products to squares of phosphocellulose paper first described for phosphoproteins[88] has been adapted for isolation of phosphopeptides.[84,89,90] The phosphocellulose paper (Method 1) and anion-exchange resin (Method 2) techniques are described below. Both of these methods are applicable to isolation of peptides or proteins, and are therefore useful for the direct comparison of peptide and parent protein as substrates in a simultaneous assay.

Reagents

Cellulose phosphate paper, grade P81 (Whatman)
Acetic acid, 30 and 15%
Acetone
Toluene scintillation fluid: 6 g of 2,5-diphenyloxazole; 0.1 g of 1,4-bis[2-(5-phenyloxazolyl)]benzene; 1 liter of toluene
Anion-exchange resin AG 1-X8, acetate form, 200–400 mesh (Bio-Rad)

Procedure

Method 1. After incubation for the appropriate period of time, the reaction is terminated by spotting 50 μl of the reaction mixture onto a cellulose phosphate paper square (2 × 2 cm) previously numbered with a lead pencil. The squares are immediately immersed in 30% acetic acid in a beaker lined with a wire mesh basket[91] to facilitate handling of the papers and to protect them from a magnetic stirring bar. The beaker should contain 6–10 ml of acetic acid per cellulose phosphate paper square. The papers are washed at room temperature for 20 min with slow rotation of the stirring bar. The wire mesh basket containing the papers is then removed from the beaker and the acetic acid is discarded into an appropriate radioactive waste container. The papers are then washed three times in 15% acetic acid for 15 min each time. The papers are washed for 5 min in acetone and then dried using a heat lamp or a stream of warm air from a hair dryer. After being placed in vials with the toluene-based scintillant, the papers are counted.

---

[87] G. Tessmer and D. J. Graves, *Biochem. Biophys. Res. Commun.* **50**, 1 (1973).
[88] J. J. Witt and R. Roskoski, Jr., *Anal. Biochem.* **66**, 253 (1975).
[89] G. W. Tessmer, J. R. Skuster, L. B. Tabatabai, and D. J. Graves, *J. Biol. Chem.* **252**, 5666 (1977).
[90] D. B. Glass, R. A. Masaracchia, J. R. Feramisco, and B. E. Kemp, *Anal. Biochem.* **87**, 566 (1978).
[91] J. D. Corbin and E. M. Reimann, this series, Vol. 38, p. 287.

Blank (no enzyme) values usually correspond to less than 0.5% of the total amount of [γ-^{32}P]ATP that is spotted on a paper. Acceptable blanks are also obtained if less concentrated acetic acid is used in all the washes.

The capacity of a 2 × 2 cm cellulose phosphate square is at least 375 nmol of peptide (Leu-Arg-Arg-Ala-Ser-Leu-Gly) and recovery after all washes is usually 96–98%. However, the method is accurate only for peptides containing two or more basic residues in addition to a free NH_2-terminus. Peptides having a net positive charge of two or less (prior to phosphorylation) are not quantitatively retained on the ion-exchange papers under the wash conditions. The presence of high ionic strength in the reaction mixture during spotting exacerbates the loss of these peptides from the papers. The recovery of peptides having a net positive charge of three or greater is not decreased by salt concentrations up to 150 mM in the reaction mixture.

Method 2. Reactions are terminated in individual assay tubes by the addition of acetic acid to a final concentration of 30%. The final volume is not critical, but the range of 0.2 to 0.5 ml is convenient. The terminated reaction mixture is applied to a small column (polypropylene Econocolumn; Bio-Rad) containing 2 ml of AG 1-X8 anion-exchange resin equilibrated with 30% acetic acid. The eluant is collected directly into a scintillation vial. A plastic rack can be fashioned to conveniently position 100 columns over a tray of scintillation vials. The reaction tube is rinsed out with 1 ml of 30% acetic acid, which is allowed to completely run into the column. The [^{32}P]phosphopeptide (as well as all dephosphopeptide) is then quantitatively eluted into the scintillation vial with an additional 3 ml of 30% acetic acid. The radioactivity is conveniently measured by counting Cerenkov radiation.[92] This has the disadvantage of lower counting efficiency, but does not require the expense of scintillation fluid and allows ready recovery of phosphopeptide from individual samples for further purification and analysis. The AG 1-X8 can be stored to allow the bound radioactivity to decay before the resin is regenerated.

Each column has an essentially unlimited capacity for phosphopeptide which passes directly through, while [γ-^{32}P]ATP is retained. Thus, the limitation is the capacity of the anion exchange resin for ATP, which is very high (1.4 mEq/ml) and is not a factor in typical protein kinase assays even if saturating concentrations of ATP are used. The blanks depend to a large degree on the amount and nature of peptide used since many of the highly positively charged peptide substrates bind small amounts of [γ-^{32}P]ATP noncovalently under the separation conditions.

The anion-exchange column procedure is useful for all but the most

---

[92] H. H. Ross, *Anal. Chem.* **41,** 1260 (1969).

acidic peptides, which can be retarded on the resin. However, most all the peptide substrates phosphorylated by cyclic nucleotide-dependent protein kinases are basic, not acidic.[5,8,9] Care must be taken to prevent contamination of the columns or the AG 1-X8 resin by even trace quantities of cation exchanger.

General Comments and Applications

The stoichiometric phosphorylation of synthetic peptide substrates should be demonstrated to rule out the possibility of trace contaminants in the peptide preparation being the major phosphate acceptors. If more than one hydroxy amino acid is present in a particular peptide substrate,[10] the phosphorylated residue(s) should be clearly identified if the peptide is to serve as a valid model for the phosphorylation of a specific site in a parent protein substrate. Determination of stoichiometry of phosphorylation of the peptide is a helpful first step in this process. As with all substrates of protein kinases, the nature of the phosphate bond in the product can be determined by assessing acid and base stability[93] and identifying O-phosphate esters of serine or threonine after partial acid hydrolysis.[5,94]

Although several innovative uses have been made of synthetic peptides in the study of cyclic nucleotide-dependent protein kinases, their most common application has been as substrates to study the determinants of specificity of these enzymes.[8,9] The importance of specific residues within the primary sequence of the peptide is usually assessed by comparing the K_m and V_{max} values of a series of peptide analogs in which selective substitutions of amino acids have been made.[5] In addition, the importance of the position of certain amino acids relative to the phosphorylatable residue can be determined by such a kinetic analysis. This is demonstrated for the cyclic AMP-dependent protein kinase in Table II. It can be observed that the position of the two arginine residues relative to the serine in the model peptide sequence from pyruvate kinase is critical for facile phosphorylation by the enzyme. Analyses of the kinetic parameters for peptides as substrates does not, however, clearly delineate whether particular amino acids are important because they are recognized and bound by the kinase or because they produce conformations of the peptide that are more favorable for phosphorylation. Studies using physicochemical techniques such as nuclear magnetic resonance are being conducted to determine the conformation of a peptide substrate when bound

[93] G. Taborsky, *Adv. Protein Chem.* **28**, 1 (1974).
[94] N. K. Schaffer, this series, Vol. 11, p. 702.

TABLE II
KINETIC CONSTANTS FOR CYCLIC AMP-DEPENDENT PROTEIN KINASE WITH SYNTHETIC PEPTIDES
CONTAINING ARGININE RESIDUES AT VARYING DISTANCES FROM THE
PHOSPHORYLATABLE SERINE[a]

Peptide sequence	Apparent K_m (μM)	V_{max} (μmol/min/ mg)	Ratio V_{max}/K_m
Gly-Gly-Gly-Gly-Gly-Gly-Gly-*Arg*-*Arg*-Ser-Leu-Gly	1080 ± 47	14.7 ± 0.3	0.013
Gly-Gly-Gly-Gly-Gly-Gly-*Arg*-*Arg*-Ala-Ser-Leu-Gly	2.4 ± 0.3	18.2 ± 0.5	7.58
Gly-Gly-Gly-Gly-Gly-*Arg*-*Arg*-Gly-Ala-Ser-Leu-Gly	400 ± 38	8.4 ± 0.4	0.021
Gly-Gly-Gly-Gly-*Arg*-*Arg*-Gly-Gly-Ala-Ser-Leu-Gly	4070 ± 413	2.7 ± 0.3	0.0007
Gly-Gly-Gly-*Arg*-*Arg*-Gly-Gly-Gly-Ala-Ser-Leu-Gly	290 ± 30	0.05 ± 0.01	0.0002
Gly-*Arg*-*Arg*-Gly-Gly-Gly-Gly-Gly-Ala-Ser-Leu-Gly	960 ± 61	0.26 ± 0.01	0.0003

[a] Reproduced from Feramisco et al.[6] with permission.

at the active site of cyclic AMP-dependent protein kinase.[7] Thus, synthetic peptides may be useful in determining the role of secondary or higher orders of structure in dictating substrate specificity.

In addition to studies of substrate specificity, peptide substrates modeled after sequences in intact proteins are useful in delineating substrate-directed, rather than enzyme-directed, influences of effector molecules.[8] Peptides have helped indicate that the apparent stimulation of cyclic GMP-dependent protein kinase activity toward histone substrates by high ionic strength[10] or by the stimulatory modulator protein[95] is due to effects on the histones to make them better substrates and not on the kinase itself. In another application, peptide substrates synthesized to contain specific reporter groups have been used to develop continuous recording spectrophotometric[96] or fluorometric[97] assays of cyclic AMP-dependent protein kinase.

Analogs of peptide substrates in which an alanine residue has been substituted for the phosphorylatable serine residue act as competitive inhibitors of either cyclic AMP-dependent or cyclic GMP-dependent protein kinases.[86,98,99] Such synthetic inhibitor and substrate peptides have

[95] M. Shoji, N. L. Brackett, J. Tse, R. Shapira, and J. F. Kuo, *J. Biol. Chem.* **253**, 3427 (1978).
[96] H. N. Bramson, N. Thomas, W. F. DeGrado, and E. T. Kaiser, *J. Am. Chem. Soc.* **102**, 7156 (1980).
[97] D. E. Wright, E. S. Noiman, P. B. Chock, and V. Chau, *Proc. Natl. Acad. Sci. U.S.A.* **78**, 6048 (1981).
[98] J. R. Feramisco and E. G. Krebs, *J. Biol. Chem.* **253**, 8968 (1978).
[99] D. B. Glass, L. J. McFann, M. D. Miller, and C. E. Zeilig, in "Protein Phosphorylation" (O. M. Rosen and E. G. Krebs, eds.), p. 267. Cold Spring Harbor Laboratory, Cold Spring Harbor, New York, 1981.

certain advantages over intact proteins as well-defined substrates, dead-end inhibitors, and product inhibitors for studies of the kinetic mechanism of the bireactant reactions catalyzed by the protein kinases.[99–101] In addition, the potential exists for the development of modified synthetic peptides which would serve as irreversible inhibitors or would be suitable for affinity or photoaffinity labeling of the active sites of these enzymes.

For all of these applications, chemical modification of peptides can be useful in blocking or modifying determinants recognized by the protein kinases. Readers are referred to previous volumes of this series for suitable methods of chemical modification.[75] For incorporation into synthetic peptides of unusual amino acids that are not commercially available as their Boc-protected derivatives, the Boc derivatives can conveniently be synthesized using 2-*tert*-butyloxycarbonyloxyimino-2-phenylacetonitrile.[102] The dissociation constants and other binding parameters of protein kinases for synthetic peptides can be easily measured by equilibrium dialysis. However, radiolabeled peptides are needed for these studies. The NH_2-terminus of a peptide can be labeled to form the N^α-[^{3}H]acetyl peptide with either [^{3}H]acetic acid and DCC or [^{3}H]acetic anhydride while the peptide is still on the solid-phase resin or with [^{3}H]acetic anhydride when the purified peptide is in solution.[103] Tritium-labeled amino acids can be converted to their N^α-Boc derivatives as referenced above and used to synthesize radiolabeled peptides. Obviously, tyrosine-containing peptides can be synthesized and iodinated by standard techniques.

The heptapeptide corresponding to the pyruvate kinase sequence has even been microinjected into *Xenopus* oocytes where it is phosphorylated *in vivo*.[104] If potent, irreversible inhibitor peptides of protein kinases were developed, they would be interesting candidates for microinjection into various cell types.

---

[100] A. H. Pomerantz, V. G. Allfrey, R. B. Merrifield, and E. M. Johnson, *Proc. Natl. Acad. Sci. U.S.A.* **74,** 4261 (1977).

[101] D. W. Bolen, J. Stingelin, H. N. Bramson, and E. T. Kaiser, *Biochemistry* **19,** 1176 (1980).

[102] M. Itoh, D. Hagiwara, and T. Kamiya, *Tetrahedron Lett.* **49,** 4393 (1975).

[103] J. F. Riordan and B. L. Vallee, this series, Vol. 25, p. 494.

[104] J. L. Maller, B. E. Kemp, and E. G. Krebs, *Proc. Natl. Acad. Sci. U.S.A.* **75,** 248 (1978).

[14] Affinity Labeling of cAMP-Dependent Protein Kinases

By SUSAN S. TAYLOR, ANTHONY R. KERLAVAGE, and MARK J. ZOLLER

Affinity labeling, that is, the use of specific reactive substrate and/or cofactor analogs, has proven to be highly informative in terms of providing insight into specific binding sites on a variety of proteins. The usefulness of such an approach can be at many levels. At a more general level, affinity labeling can be used as a tool for the selective covalent modification of a single protein in a mixture of other proteins. Frequently, affinity labeling can be utilized to correlate or distinguish the role of a given site in the overall functioning of a particular protein. Affinity labeling can also be used as a tool for identifying functional domains in proteins or for general localization of a particular site in the polypeptide chain. Ultimately, affinity labeling can be used to identify specific amino acids that contribute to or are in close proximity to a given binding site.

In utilizing affinity labeling there are several important criteria to consider. First is specificity. It is crucial that the label be directed to a specific site and that the parent compound of the label be able to specifically protect against covalent incorporation. In general, analogs with a high affinity are advantageous, particularly when it is desired to label one protein selectively in a cell extract. Second, when the affinity analog is used as an inhibitor of enzymatic activity, it is important to establish the kinetics of inhibition showing that pseudo-first-order kinetics are followed. Although specific sites can be identified when random diffusion-mediated labeling is followed, to identify specific sites related to function is difficult to achieve. Third, the stoichiometry of incorporation should be established, and that stoichiometry should be consistent with the stoichiometry of binding of the parent compound. Finally, the site should be localized in the polypeptide chain by (1) identifying the specific amino acid that is modified; (2) identifying the labeled site within a large peptide fragment or domain generated from the native protein; and (3) ultimately sequencing the peptide that contains the site of modification.

We have utilized affinity labeling to probe the structure of cAMP-dependent protein kinase. This kinase is an excellent model for the application of affinity labeling since it contains a variety of nucleotide binding sites as well as a peptide substrate binding site and exists in multiple aggregation states. The native enzyme is an inactive tetramer containing two regulatory (R) subunits and two catalytic (C) subunits. cAMP promotes dissociation into a regulatory subunit dimer and two active catalytic subunits[1]:

$$R_2C_2 \underset{}{\overset{4cAMP}{\rightleftharpoons}} R_2\text{-}(cAMP)_4 + 2C$$

The dissociated catalytic subunit is an ATP:protein phosphotransferase that phosphorylates a variety of proteins but recognizes the following general sequence: Arg-Arg-X-Ser(P)-X, where the serine or threonine to be phosphorylated is preceded by two basic residues, usually arginines.[1] Consequently, there are binding sites for both cAMP and ATP, as well as for peptides, and both nucleotide and peptide analogs have been used to modify the isolated regulatory and catalytic subunits as well as the holoenzyme.[2-6] Only two nucleotide analogs will be described here, one an example of an alkylating affinity reagent and the other an example of a photoactivatable affinity reagent.

Affinity Labeling of the ATP-Binding Site

One of the most successful ATP analogs that has been used is the alkylating reagent, fluorosulfonylbenzoyl 5'-adenosine (FSO_2BzAdo), first synthesized and utilized by Colman[7] and shown in Fig. 1. This analog has been utilized to covalently modify a number of proteins and has been shown to modify a variety of nucleotide-binding sites including ATP, NAD, and even cAMP sites.[7-11] It is not therefore absolutely specific for ATP sites, and can be more accurately described as a probe for adenine nucleotide binding sites where the alkylating fluorosulfonyl group lies in a position that closely approximates the γ-phosphate of ATP even though FSO_2BzAdo lacks the negative charges of ATP. In the case of the catalytic subunit of cAMP-dependent protein kinases, FSO_2BzAdo has been shown to be specific for the ATP-binding site of the catalytic subunit.[2,12]

---

[1] E. G. Krebs and J. A. Beavo, *Annu. Rev. Biochem.* **48**, 923 (1979).
[2] M. J. Zoller and S. S. Taylor, *J. Biol. Chem.* **254**, 8363 (1979).
[3] A. R. Kerlavage and S. S. Taylor, *J. Biol. Chem.* **255**, 8483 (1980).
[4] J. Hoppe and W. Friest, *Eur. J. Biochem.* **93**, 141 (1979).
[5] A. Kupfer, V. Gani, J. S. Jiménez, and S. Shaltiel, *Proc. Natl. Acad. Sci. U.S.A.* **76**, 3073 (1979).
[6] H. N. Bramson, N. Thomas, R. Matsueda, N. C. Nelson, S. S. Taylor, and E. T. Kaiser, *J. Biol. Chem.* **257**, 10575 (1982).
[7] R. F. Colman, P. K. Pal, and J. L. Wyatt, this series, Vol. 46, p. 240.
[8] P. K. Pal, W. J. Wechter, and R. F. Colman, *J. Biol. Chem.* **250**, 8140 (1975).
[9] F. S. Esch and W. S. Allison, *J. Biol. Chem.* **253**, 6100 (1978).
[10] J. L. Wyatt and R. F. Colman, *Biochemistry* **16**, 1333 (1977).
[11] L. Weng, R. L. Heinrikson, and T. E. Mansour, *J. Biol. Chem.* **255**, 1492 (1980).
[12] C. S. Hixon and E. G. Krebs, *J. Biol. Chem.* **254**, 7509 (1979).

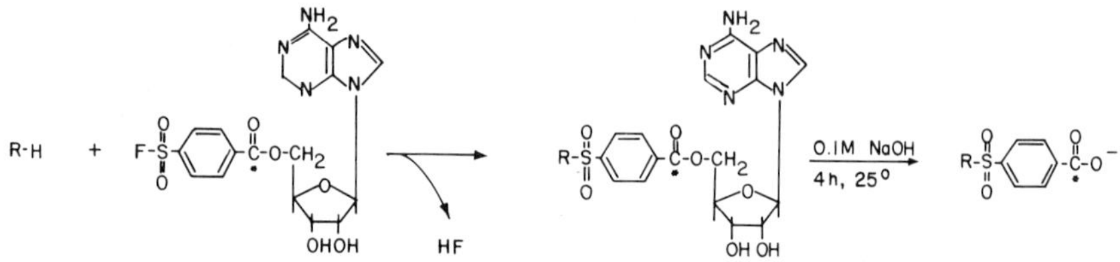

FIG. 1. The structure and reactivity of fluorosulfonylbenzoyl-5'-adenosine. Covalent modification by FSO_2BzAdo is indicated where R represents a reactive group on a protein. Following covalent modification, the sulfonylbenzoyl 5'-adenosine can be hydrolyzed as indicated by treatment with mild alkali. The asterisk (*) indicates the location of ^{14}C in the radioactive analog.

At low concentrations (≤1 mM) alkylating reagents such as FSO_2BzAdo tend to have the advantage of reacting primarily when they are bound in close proximity to a nucleophilic group on the surface of a protein. Nonspecific interaction with nucleophiles on the surface of the molecule is usually minimal as is competition with solvent when the analog is bound at a specific site on the surface of the protein. Another advantage of alkylating reagents is that if the analog is bound in an environment that contains an appropriately positioned nucleophile, stoichiometry usually can be readily achieved.

p-Fluorosulfonylbenzoyl 5'-adenosine. Nonradioactive FSO_2BzAdo was synthesized according to Colman.[7] $FSO_2[^{14}C]BzAdo$ was synthesized according to Esch and Allison,[13] as were carboxybenzoylsulfonyl-lysine (CBS-lys) and CBS-tyrosine. The original Colman procedure[7] can also be used as an alternative procedure for synthesizing radioactive FSO_2BzAdo. This latter procedure, although much simpler, has the disadvantage that the isotopic label is in the adenine ring, and, therefore, is lost if hydrolysis occurs at the alkali-labile ester bond. For instance, radioactivity is usually lost when the radioactively labeled protein is subjected to standard sodium dodecyl sulfate-polyacrylamide gel electrophoresis run under standard conditions at pH >8. The reagent here was labeled in the benzoic acid moiety so that the labile ester bond could be taken advantage of to selectively purify the modified peptide with the radioactivity always remaining with the peptide. The disadvantage of the benzoic acid-labeled FSO_2BzAdo is that the specific activity is less than what can be achieved with adenosine-labeled FSBzAdo. The latter is also now commercially available (New England Nuclear).

Kinetics of Inhibition. In order to establish the kinetics of inhibition, the activity of the catalytic subunit, purified from porcine heart,[14] was

[13] F. S. Esch and W. S. Allison, *Anal. Biochem.* **84**, 642 (1978).
[14] M. J. Zoller, N. C. Nelson, and S. S. Taylor, *J. Biol. Chem.* **256**, 10837 (1981).

measured as a function of FSO_2BzAdo concentration where the maximal concentration of FSO_2BzAdo is limited by its solubility in aqueous solution (~1 mM). Enzyme concentrations of 0.1 to 2.0 mg/ml were utilized at pH 6.8–7.2 in 50 mM Tris–HCl, phosphate, or HEPES containing 10% glycerol. Caution must be taken to remove any residual 2-mercaptoethanol or dithiothreitol from the reaction mixture. The reaction (70 µl) was initiated by the addition of 1 µl FSO_2BzAdo (70 mM) dissolved in dimethyl sulfoxide (DMSO) and was incubated at 37°. Aliquots were removed at designated times for direct assay. When the reaction was carried out on a preparative scale, the reaction was terminated by adding 2-mercaptoethanol to a final concentration of 50 mM. Controls in the presence of equal concentrations of DMSO but minus FSBzAdo are also essential. When the log of v/v_0 was plotted vs time, where v and v_0 represent the enzymatic activity at any given time and at zero time, the linearity indicated a pseudo-first-order reaction, and at equivalent concentrations ATP-Mg^{2+} afforded complete protection against inhibition. In addition, saturation kinetics were observed (Fig. 2) indicating that FSO_2BzAdo was bound reversibly to the enzyme (E), prior to irreversible

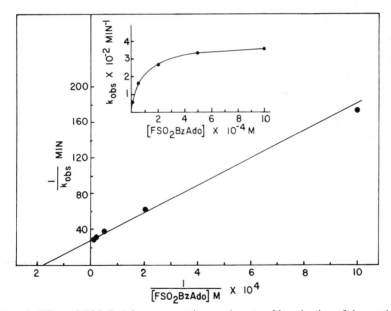

FIG. 2. Effect of FSO_2BzAdo concentration on the rate of inactivation of the catalytic subunit of cAMP-dependent protein kinase. The C-subunit (0.15 mg/ml) was incubated at varying concentrations of FSO_2BzAdo. At each concentration the k_{obs} was determined from the slope of ln v/v_0 versus time. Reproduced from Zoller and Taylor.[2]

inactivation

$$E + I \underset{k_{-1}}{\overset{k_1}{\rightleftharpoons}} EI \overset{k_2}{\rightarrow} E_i \qquad (1)$$

where I = FSO_2BzAdo and E_i = irreversibly inactivated enzyme. The observed rate of inhibition (k_{obs}) can therefore be expressed as

$$k_{obs} = \frac{k_2}{(K_i/[I]) + 1} \qquad (2)$$

where $K_i = (k_{-1} + k_2)/k_1$. As indicated in Fig. 2, the K_i (FSO_2BzAdo) was calculated to be 5.7×10^{-5} M in contrast to a K_m (ATP) of 1.2×10^{-5} M.[2]

Stoichiometry. Having established first order kinetics, the stoichiometry of the reaction was determined. To achieve this, $FSO_2[^{14}C]BzAdo$ was utilized. The C-subunit (15 mg; 0.5–2.0 mg/ml) was treated with $FSO_2[^{14}C]BzAdo$ as described above until >90% of the enzymatic activity was inhibited. After terminating the reaction with 50 mM 2-mercaptoethanol, the protein was precipitated with ammonium sulfate and resuspended in buffer containing 6 M guanidine-HCl. Excess reagent was removed by gel filtration in the presence of 6 M guanidine-HCl. Following dialysis, the inhibited protein was shown to contain 0.9 mol FSO_2BzAdo/mol catalytic subunit, consistent with the binding of 1 mol ATP/mol catalytic subunit.[2]

Modified Amino Acid. Although cysteine and histidine residues could be alkylated by this reagent, the most stable amino acid derivatives are tyrosine and lysine. To establish which amino acid was modified in the C-subunit, the ^{14}C-covalently labeled protein was hydrolyzed with 6 N HCl for 24 hr at 105° and then subjected to paper electrophoresis at pH 8.9 for 1 hr at 2 kV. Electrophoresis carried out in conjunction with standard CBS-Lys and CBS-Tyr indicated that all of the radioactivity was associated with CBS-lysine.[2]

Isolation of Modified Peptide. Isolation of the peptide that was covalently modified was achieved by again using $FSO_2[^{14}C]BzAdo$. In this case, it was particularly important to have the label in the benzoic acid portion of the molecule, since the alkaline lability of the ester bond was taken advantage of in the purification. After isolating the modified protein as described above, treatment with trypsin was carried out (1 : 200 w/w trypsin : C-subunit; 50 mM NH_4HCO_3, pH 8.3; 4 hr; 37°). The tryptic peptides were then diluted 1 : 20 with 1% pyridine : 1% collidine (pH 8.0) and eluted from DEAE-Sephadex A-25 (12 ml) with a 500-ml linear gradient decreasing in pH from 8.0 to 5.0, where the pH of the second buffer was adjusted with glacial HAc (Fig. 3, top). The major peak of radioactivity was pooled, lyophilized, and then incubated with 1 ml of 0.1 M NaOH for 4 hr at room temperature which is sufficient to cleave the ester bond

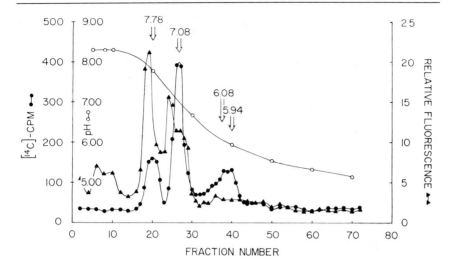

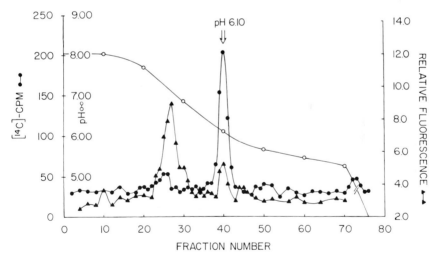

Fig. 3. DEAE-Sephadex A-25 chromatography of the tryptic peptides of $FSO_2[^{14}C]BzAdo$-inactivated catalytic subunit. The upper graph is the elution of the complete mixture of tryptic peptides immediately following proteolysis. The major peak of radioactivity (tubes 24–29) was pooled, lyophilized, and subjected to mild alkaline hydrolysis as described. Following alkaline hydrolysis the peptide solution was reapplied to the same DEAE-Sephadex A-25 column and eluted under identical conditions as indicated in the lower graph.

between the benzoyl and adenosine moiety but insufficient to cleave most peptide bonds. After desalting by gel filtration using Sephadex G-25 in 50 mM NH$_4$OH, the peptide was diluted 1:2 with 1% pyridine:1% collidine and reapplied to an identical DEAE column (Fig. 3, bottom). The modified peptide was specifically retained and eluted at pH 6.1, whereas the other peptides in the pooled fraction eluted at pH 7.08 as previously.[2,14] An alternative procedure for isolating FSBzAdo-modified peptides using a Sepharose-borate column that specifically recognizes the vicinyl hydroxyl groups on the ribose ring has also been described.[15] The modified peptide, using as little as 0.1 mg total protein for a tryptic digestion, can also be isolated by HPLC using procedures similar to those described in the next section.

After determining the amino acid composition, the sequence of the modified peptide indicated below was determined by both manual dansyl Edman degradation and solid phase automated sequencing:

Glu-Thr-Gly-Asn-His-Phe-Ala-Met-Lys(CBS)-Ile-Leu-Asp-Lys

A smaller peak of radioactivity (at pH 7.78) was also observed in the initial DEAE-column (Fig. 3, top). This fraction was treated in an identical manner and was shown to contain the same peptide plus an additional two residues that resulted from incomplete tryptic digestion. The modified cyanogen bromide peptide from this region has also been isolated directly by HPLC without relying on alkaline hydrolysis and has provided an extended sequence for this region of the molecule.[14]

General Application of FSO$_2$BzAdo Labeling to Protein Kinases. It might be emphasized that a homologous lysine residue has been subsequently modified in cGMP-dependent protein kinase.[16] In addition, it is also interesting that p60^{src}, a tyrosine-specific kinase, appears to be homologous to the C-subunit based on a comparison of the C-subunit sequence and the DNA-derived p60^{src} sequence.[17] Furthermore, one of the regions that is most highly conserved in this comparison is the region surrounding the FSO$_2$BzAdo-modified lysine in the C-subunit. FSO$_2$BzAdo has also been shown to inhibit and covalently modify the epidermal growth factor receptor[18] which is thought to be a tyrosine kinase. This particular analog may, therefore, have rather general applicability in the comparison of various protein kinases.

[15] A. E. Annamalai, P. K. Pal, and R. F. Colman, *Anal. Biochem.* **99**, 85 (1979).
[16] E. Hashimoto, K. Takio, and E. G. Krebs, *J. Biol. Chem.* **257**, 727 (1982).
[17] W. C. Barker and M. O. Dayhoff, *Proc. Natl. Acad. Sci. U.S.A.* **79**, 2836 (1982).
[18] S. A. Buhrow, S. Cohen, and J. V. Staros, *J. Biol. Chem.* **257**, 4019 (1982).

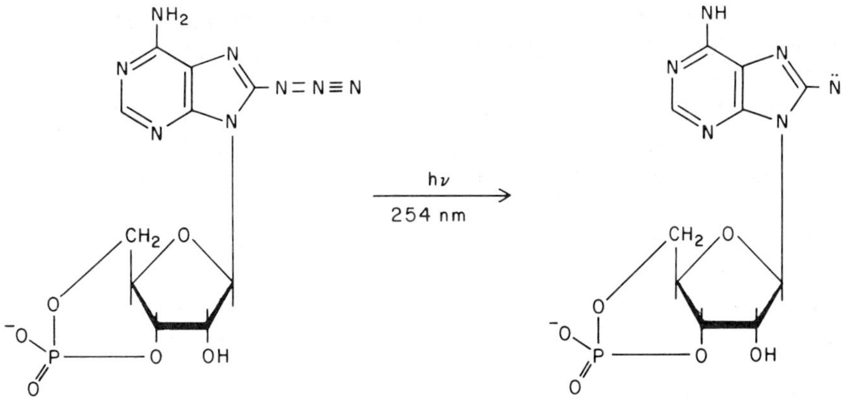

FIG. 4. Structure of 8-azido-cAMP and of the activated nitrene generated by photolysis. Reproduced from Zoller and Taylor.[2]

Affinity Labeling of cAMP-Binding Sites

In order to label the cAMP-binding site(s) of the regulatory subunit, a photoaffinity analog, 8-azido-cAMP (N_3-cAMP), was utilized (Fig. 4). N_3-cAMP was first synthesized by Haley et al.[19,20] who demonstrated it to be a specific probe for covalently modifying cAMP-binding sites. A major advantage of this particular photoaffinity analog is its high affinity for the regulatory subunit ($K_d \sim 10^{-7}$ M) which makes it feasible to specifically label the regulatory subunit even in whole cell extracts where R accounts for only a minor fraction of the total protein. It is only at higher concentrations of the analog that significant nonspecific labeling of other proteins is encountered. In contrast to alkylating reagents, a major disadvantage with photoaffinity probes in general is that the solvent often competes very effectively for the activated group, in this case a nitrene, that is generated by photolysis, thereby making stoichiometric incorporation difficult to achieve. Furthermore, the nitrene can potentially insert into any amino acid side chain in close proximity thereby lowering the chance of specifically labeling a single site. These inherent difficulties were not apparent in the labeling studies of the regulatory subunit of cAMP-dependent protein kinase II, presumably due in part to the high affinity of R for

[19] A. H. Pomerantz, S. A. Rudolph, B. E. Haley, and P. Greengard, *Biochemistry* **14**, 3858 (1975).
[20] B. E. Haley, this series, Vol. 46, p. 339.

N_3-cAMP and the specific environment of the cAMP-binding site. The general utilization of photoaffinity labeling has recently been reviewed.[21]

Photoaffinity Labeling. N_3-[^{32}P]cAMP and N_3-[^{3}H]cAMP were synthesized according to Haley[20] and can also be purchased from New England Nuclear or Schwarz-Mann. The type II regulatory subunit (R^{II}) of cAMP-dependent protein kinase was purified from porcine heart.[22] R^{II} (10–50 μM) was incubated with 8-N_3-cAMP for 20 min at 30° in 50 mM potassium phosphate free of 2-mercaptoethanol, transferred to a 1 ml quartz cuvette containing a spinbar, and irradiated at room temperature with a UVS-11 lamp (254 nm) with gentle stirring for 4 min at a distance of 3 cm. For gels, incubation and irradiation were carried out in glass tubes (0.6 × 5 cm) without stirring, using 4 μg R^{II}, 0.5 μM potassium phosphate or MES, pH 6.8 in a total volume of 0.05 ml. Samples were applied immediately to SDS–polyacrylamide gels and electrophoresed according to Laemmli.[23]

Specificity of 8-N_3-cAMP Binding. The specificity of N_3-cAMP labeling was demonstrated initially using a homogenous preparation of R^{II}. As indicated in Fig. 5, N_3-[^{3}H]cAMP was specifically and covalently incorporated into both the phosphorylated and dephosphorylated forms of R^{II}. The incorporation of radioactivity was unaffected by a 20-fold molar excess of ATP or AMP over 8-N_3-cAMP but was nearly abolished by a 20-fold excess of cAMP. The incorporation of radioactivity also was abolished by omission of irradiation of the protein: N_3-cAMP mixture, and greater than 90% was abolished by preirradiation of N_3-cAMP, demonstrating the photodependence of the incorporation. Preirradiation of R^{II} had no effect upon covalent incorporation.[3]

Correlation of Structural Domains with Function. Since the R-subunit of cAMP-dependent protein kinase has been shown to be labile to very specific limited proteolysis,[24–26] N_3-cAMP also was utilized to localize the cAMP-binding sites with respect to the proteolytic fragments. Limited proteolysis of R^{II} generates two major domains: a smaller dimeric fragment corresponding to the N-terminal region of the molecule and associated with maintaining the dimeric structure of native R^{II} and a larger C-terminal domain that is monomeric. Photoincorporation of N_3-[^{3}H]cAMP was utilized to confirm that the cAMP-binding properties of R were retained exclusively with the C-terminal domain. Photoincorporation was

---

[21] V. Chowdhry and F. H. Westheimer, *Annu. Rev. Biochem.* **48**, 293 (1979).
[22] M. J. Zoller, A. R. Kerlavage, and S. S. Taylor, *J. Biol. Chem.* **254**, 2408 (1979).
[23] U. K. Laemmli, *Nature (London)* **227**, 680 (1970).
[24] R. L. Potter and S. S. Taylor, *J. Biol. Chem.* **254**, 2413 (1979).
[25] R. L. Potter and S. S. Taylor, *J. Biol. Chem.* **254**, 9000 (1979).
[26] S. R. Rannels and J. D. Corbin, *J. Cyclic Nucleotide Res.* **6**, 201 (1980).

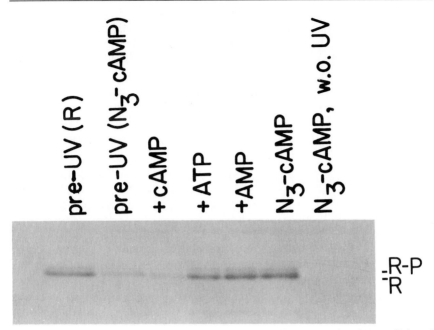

FIG. 5. Autoradiograph of an SDS–polyacrylamide gel demonstrating the specificity of incorporation of 8-N$_3$cAMP into R$_u^{II}$. R$_u^{II}$ refers to urea-treated R^{II} prepared as described previously.[3] R$_u^{II}$ (4 μg) was incubated with 0.5 μM 8-N$_3$-[^{3}H]cAMP in 10 mM Mes (pH 6.8) and 12 μM concentrations of various nonradioactive nucleotides as indicated. Channels 1 (far left) and 2: 20-min preirradiation of R$_u^{II}$ and 8-N$_3$-[^{3}H]cAMP, respectively. Channel 6: 8-N$_3$-[^{3}H]cAMP alone. Channel 7: the incubation mixture was not photolyzed. Bands were cut from the gel and counted in a toluene scintillation mixture containing 3% Protosol. The amounts (in femtomoles) of 8-N$_3$-[^{3}H]cAMP incorporated into R$_u^{II}$ for each channel were pre-UV(R), 32; pre-UV(8-N$_3$cAMP), 3; cAMP, 3; ATP, 30; AMP, 32; 8-N$_3$-cAMP, 34; no-UV, 0. R and R-P, dephosphorylated and phosphorylated R-subunit. Reproduced from Kerlavage and Taylor.[3]

readily achieved whether photolysis was carried out prior to or after proteolysis and in both cases, photoincorporation of radioactivity was associated only with the C-terminal domain. In addition, the purified C-terminal domain was photolabeled to the same extent as native R^{II}, indicating that the isolated domain retained its cAMP-binding properties.[25]

Isolation of Modified Peptide. In order to further localize the site(s) of covalent modification, R^{II} (27 μM; 3–10 mg total protein) was covalently modified with N$_3$-[^{3}H]cAMP (82 μM). Excess reagent was removed by gel filtration on Sephadex G-25 in 0.2 M Tris, pH 7.0 containing 6 M guanidine-HCl, and a stoichiometry of approximately 0.5 mol N$_3$-cAMP/mol R monomer was determined. Following dialysis against water and lyophili-

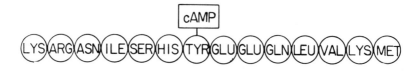

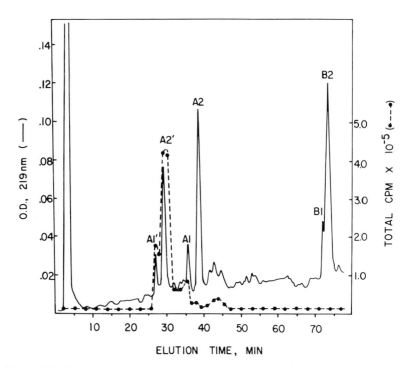

FIG. 6. HPLC separation of the Sephadex G-50 fraction containing the N_3-[^{3}H]cAMP-labeled CNBr peptide from R^{II}. Separation was carried out on a C_{18} μBondapak column with a flow rate of 1.0 ml/min, using a 50-min gradient of 0 to 50% where buffer A = 0.1% TFA and B = acetonitrile. Reproduced from Kerlavage and Taylor.[3]

zation, N_3-[^{3}H]cAMP-labeled R^{II} was carboxymethylated,[27] cleaved with cyanogen bromide,[28] and applied to a column of Sephadex G-50 (22 × 197 cm) equilibrated in 1% acetic acid. The elution profile showed a single major radioactive peak which was pooled and further resolved by HPLC (Fig. 6). Three sets of doublets (A1'–A2'; A1–A2, B1–B2) were observed

[27] A. M. Crestfield, S. Moore, and W. H. Stein, *J. Biol. Chem.* **238**, 622 (1963).
[28] E. Steers, G. R. Craven, C. B. Anfinsen, and J. L. Bethione, *J. Biol. Chem.* **240**, 2478 (1965).

and greater than 90% of the eluted radioactivity was associated with the first of these doublets (A1'–A2'). The peaks within each doublet were shown to be the homoserine and homoserine lactone forms of the same peptide.[3] Amino acid composition data and sequence analysis showed peptide A' to be the N_3-cAMP-modified form of peptide A. When the two peptides were mixed and sequenced, radioactivity was associated with the seventh residue of the peptide identifying the tyrosine residue at this position as the specific site of N_3-cAMP modification (Fig. 6).

The importance of HPLC in the isolation and characterization of small quantities of modified peptides should be strongly emphasized. Although an intervening step of gel filtration was utilized here, it is often quite feasible to apply very small amounts (as little as 0.1 mg) of a total cyanogen bromide cleavage mixture or of a total tryptic digest and resolve most of the components. A C_{18} μBondapak column, using a 0–60% gradient of 0.1% TFA vs acetonitrile, has proved to be generally effective in separating most mixtures of peptides. For example, a 2-hr gradient is sufficient to resolve most of the tryptic peptides derived from 0.1–0.2 mg of both R- and C-subunits with yields sufficient to obtain at least an amino acid composition and to determine the N-terminal amino acid residue by the dansylation procedure of Hartley.[29] For larger peptides, it is sometimes necessary to use a larger pore size (300 Å) column (RP-P-Synchrompak, Synchrom Inc., Linden, Indiana) and/or isopropanol rather than acetonitrile as the second buffer.

Correlation of 8-N_3-cAMP Labeling with the cAMP-Binding Sites of R^{II}. One of the primary objectives of affinity labeling is to be able to probe as specifically as possible various sites on proteins and to determine what role those sites play in the overall functioning of the molecule. The labeling of R^{II} with N_3-cAMP posed some particularly interesting questions since, as was indicated earlier, each R^{II} monomer is known to have two high-affinity binding sites for cAMP,[1,30,31] referred to often as site 1 and 2. Nevertheless, photoaffinity labeling with N_3-cAMP indicated that only one major site was modified. For this reason, the stoichiometry of both binding and photoincorporation was determined rigorously using holoenzyme (R_2C_2). In addition, the effect of N_3-cAMP on activation and dissociation of holoenzyme was established and compared to the parent compound, cAMP.[32]

---

[29] B. S. Hartley, *Biochem. J.* **119**, 805 (1971).
[30] J. D. Corbin, P. H. Sugden, L. West, D. A. Flockhart, T. M. Lincoln, and D. McCarthy, *J. Biol. Chem.* **253**, 3997 (1978).
[31] P. H. Sugden and J. D. Corbin, *Biochem. J.* **159**, 423 (1976).
[32] A. R. Kerlavage and S. S. Taylor, *J. Biol. Chem.* **257**, 1749 (1982).

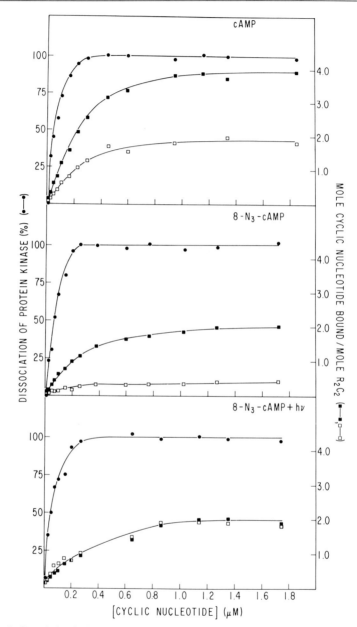

FIG. 7. Correlation between dissociation of protein kinase and cyclic nucleotide binding. Purified dephosphoholoenzyme (20.7 nM) was incubated for 30 min at 4° with various concentrations of either [^{3}H]cAMP (15,000 cpm/pmol) or 8-N$_3$-[^{3}H]cAMP (3448 cpm/pmol) in a final volume of 180 μl of buffer B (50 mM potassium phosphate, 5 mM 2-mercapto-

Since R^{II} has no known catalytic activity, the stoichiometry of N_3-cAMP binding and incorporation was determined by Millipore filtration. Two assays were utilized: Assay A,[33] which measures binding to only one class of cAMP binding sites, and Assay B, which measures total cAMP binding to both sites.[31] The results indicated in Fig. 7 showed that N_3-cAMP bound preferentially with a high affinity to only one class of cAMP binding sites. If possible, it is also advisable to determine binding constants by an alternative and independent method such as equilibrium dialysis. When photoactivation was carried out using holoenzyme, a stoichiometry of 1 mol N_3-cAMP/mol R monomer following photoincorporation was readily achieved using 1 μM N_3-cAMP and 20 nM R^{II} monomer. Cyanogen bromide cleavage of the N_3-[^{3}H]cAMP-labeled R^{II} followed by HPLC indicated that only a single peptide was labeled corresponding to the tyrosine residue identified previously.[32] N_3-cAMP was also shown to be equivalent to cAMP in the activation of the holoenzyme (Fig. 7) and in the dissociation of holoenzyme,[33] thus indicating that occupation of one site alone may be sufficient to activate and dissociate the holoenzyme.

In summary, 8-N_3-cAMP has provided an unusual example of stoichiometric photolabeling of a single amino acid residue. In addition, it has helped to resolve the specific contribution of one class of cAMP binding sites in the overall functioning of the cAMP-dependent enzyme.

Acknowledgment

This work was supported in part by U.S. Public Health Service Grant GM-19301.

---

[33] G. N. Gill and G. M. Walton, this series, Vol. 38, p. 376.

---

ethanol, 1 mg/ml of bovine serum albumin). Duplicate 20-μl aliquots were removed for phosphotransferase activity plus and minus cAMP. Millipore filtration Assay A (□—□), Millipore filtration Assay B (■—■). Percentage dissociation was calculated as the ratio of phosphotransferase activity minus cAMP to that plus cAMP. Top, holoenzyme incubated with [^{3}H]cAMP. Center, holoenzyme incubated with 8-N_3-[^{3}H]cAMP. Bottom, holoenzyme incubated with 8-N_3-[^{3}H]cAMP followed by photolysis. Reproduced from Kerlavage and Taylor.[32]

[15] Photoaffinity Labeling of the Regulatory Subunit of cAMP-Dependent Protein Kinase

By ULRICH WALTER and PAUL GREENGARD

Photoaffinity labeling as a special type of affinity labeling has been extensively used to study binding proteins for the cyclic nucleotides cAMP and cGMP. Initially, cyclic nucleotide binding proteins were analyzed by direct photoaffinity labeling with cAMP and cGMP.[1] N^6-(Ethyl-2-diazomalonyl)-cAMP has been shown to label specifically the regulatory subunit of cAMP-dependent protein kinase of erythrocyte membranes.[2] The most extensively used photoaffinity analog of cAMP has been 8-N_3-[^{32}P]cAMP, first synthesized and used by B. E. Haley.[3] This chapter describes the use of 8-N_3-[^{32}P]cAMP as a photoaffinity label for the regulatory subunit of cAMP-dependent protein kinase.

Synthesis and Characterization of 8-N_3-[^{32}P]cAMP

8-N_3-[^{32}P]cAMP is synthesized from [^{32}P]cAMP via a two-step synthetic procedure originally described by Haley.[3] The original procedure has been somewhat modified, as described in detail elsewhere.[4,5] 8-N_3-[^{32}P]cAMP and 8-N_3-[^{3}H]cAMP are now commercially available from ICN and NEN, respectively. Preparations of 8-N_3-[^{32}P]cAMP with specific radioactivities up to 160 Ci/mmol can be obtained. However, for most experimental conditions 8-N_3-[^{32}P]cAMP with a specific radioactivity of about 10 Ci/mmol is quite sufficient. The purity and photolability of each batch of 8-N_3-[^{32}P]cAMP should always be checked. This can be done by thin-layer chromatography using Eastman TLC cellulose sheets with a solvent system of 1-butanol/acetic acid/H_2O (5:2:3, v:v:v) as has been described in detail.[4,5] 8-N_3-[^{32}P]cAMP when stored in the dark at $-20°$ in absolute methanol is stable for at least 4 weeks.

Photoaffinity Labeling Procedure

Photoaffinity labeling experiments are performed by placing solutions into individual wells of glass or ceramic serology plates, i.e., Pyrex spot plates. An individual well may hold up to 400 µl. Except for special

---

[1] R. S. Antonoff and J. J. Ferguson, *J. Biol. Chem.* **249**, 3319 (1974).
[2] C. S. Rubin, *J. Biol. Chem.* **250**, 9044 (1975).
[3] B. E. Haley, *Biochemistry* **14**, 3852 (1975).
[4] U. Walter, I. Uno, A. Y.-C. Liu, and P. Greengard, *J. Biol. Chem.* **252**, 6494 (1977).
[5] B. E. Haley, this series, Vol. 46, p. 339.

experimental conditions, the serology plates are placed on ice and kept at 4° throughout the experiment. It is also recommended that the samples be kept in the dark during incubations with 8-N$_3$-[^{32}P]cAMP by covering the serology plates with aluminum foil. However, short exposures to room light do not cause significant breakdown of 8-N$_3$-[^{32}P]cAMP. Typical reaction mixtures (final volume, 0.1 ml) contain a buffer such as 50 mM sodium morpholinoethane sulfonate (pH 6.2), 10 mM MgCl$_2$, 1 mM 1-isobutylmethylxanthine (IBMX), 0.5 mM 2-mercaptoethanol, 1 μM 8-N$_3$-[^{32}P]cAMP (5–10 Ci/mmol) and various amounts of protein up to 200 μg. The presence of Ca^{2+}-chelating agents and protease inhibitors may be helpful with crude tissue extracts. This photoaffinity labeling procedure can be used with various buffer systems over a relatively broad pH range. Incubations are usually carried out in the absence or presence of an excess of nonradioactive cAMP, i.e., 20–50 μM, to detect specific labeling.

Before photolysis, preincubations are carried out for 30–60 min to allow reversible 8-N$_3$-[^{32}P]cAMP binding to reach maximum. The samples are then irradiated for 10 min at 254 nm. This is achieved by placing a UV hand lamp over the entire serology plate at a distance of about 4–8 cm. Useful lamps were found to be a UVS-11 Mineralite (UV-Products, Inc., USA) or Fluotest Type 5303 (Original Hanau Quarzlampen GMBH, F.R.G.). Under most conditions, maximal covalent labeling is observed within 5 min of the onset of UV light exposure. Longer exposure may cause some crosslinking of protein. Covalent labeling of proteins is abolished when 8-N$_3$-[^{32}P]cAMP is exposed to UV light before the incubation procedure.

Total covalent incorporation of 8-N$_3$-[^{32}P]cAMP can be measured by counting acid precipitates. However, for complex protein mixtures it is more useful to analyze the sample by sodium dodecyl sulfate (SDS)–polyacrylamide gel electrophoresis and autoradiography. After photolysis, the entire reaction mixture is mixed with 50 μl of a SDS-containing "stop solution," boiled for 2 min, and then subjected to SDS–polyacrylamide gel electrophoresis. Gels are then stained for protein, destained, dried, and examined by autoradiography as described elsewhere.[4] Radioactive bands localized by autoradiography may be cut out of the dried gel and counted by liquid scintillation spectrometry to determine the absolute amount of radioactivity incorporated. Alternatively, densitometric tracings of the autoradiograph may be obtained.

Photoaffinity Labeling of the Type II cAMP-Dependent Protein Kinase

A typical photoaffinity labeling experiment with a Type II cAMP-dependent protein kinase purified 800-fold from bovine heart according to the procedure of Rubin et al.[6] is shown in Fig. 1. The regulatory subunit (M_r 54,000–57,000) of the holoenzyme was maximally phosphorylated both in the absence and presence of cAMP since the catalytic subunit is able to phosphorylate the regulatory subunit both in the dissociated and undissociated state of the holoenzyme.[7] Both in the absence and presence of ATP the enzyme covalently incorporated 8-N$_3$-[^{32}P]cAMP into the regulatory subunit, which could be completely blocked by an excess of unlabeled cAMP. A small decrease in the electrophoretic mobility of the regulatory subunit was observed when the photoaffinity labeling procedure was carried out in the presence of ATP resulting in a shift of the apparent molecular weight from 54,000 to 57,000 as estimated from SDS–polyacrylamide gel electrophoresis (Fig. 1). This effect, due to the phosphorylation of the regulatory subunit,[8] explains the observation that the Type II regulatory subunit (R-II) in crude extracts under certain conditions appears as a doublet on SDS–polyacrylamide gels since this regulatory subunit may exist in phospho- and dephospho-forms.

Reversible binding of [^{3}H]cAMP or 8-N$_3$-[^{32}P]cAMP to the regulatory subunit (R-II) of Type II cAMP-dependent protein kinase and photoactivated 8-N$_3$-[^{32}P]cAMP incorporation occur with similar concentration dependency (Fig. 2). Half-maximal binding, analyzed by the Millipore filtration method,[8,9] occurred at 0.09 μM [^{3}H]cAMP or 0.33 μM 8-N$_3$-[^{32}P]cAMP, as calculated from double reciprocal plots (Fig. 2). This small difference in affinity between cAMP and 8-N$_3$-[^{32}P]cAMP is also reflected in a small difference in the ability of cAMP and 8-N$_3$-cAMP to stimulate the histone kinase activity of the Type II cAMP-dependent protein kinase: half-maximal stimulation was obtained with 0.06 μM cAMP and with 0.25 μM 8-N$_3$-cAMP.[10]

Photoaffinity labeling of the regulatory subunit R-II with 8-N$_3$-[^{32}P]cAMP displayed a similar concentration dependency with half-maximal incorporation at 0.14 μM 8-N$_3$-[^{32}P]cAMP (Fig. 2). Under saturating conditions, 8-N$_3$-[^{32}P]cAMP was able to covalently label the identical number of cAMP-binding sites as that determined by the Millipore filtration method. In addition, the amount of regulatory subunit R-II measured by autophosphorylation was in close agreement with the amount of regulatory subunit determined by the photoactivated incorporation of 8-N$_3$-[^{32}P]cAMP.[10] With the purest preparation of 8-N$_3$-[^{32}P]cAMP and homogeneous Type II cAMP-dependent protein kinase reconstituted from

[6] C. S. Rubin, J. Erlichman, and O. M. Rosen, this series, Vol. 38, p. 308.
[7] R. Rangel-Aldao and O. M. Rosen, *J. Biol. Chem.* **251**, 7526 (1976).
[8] F. Hofmann, J. A. Beavo, P. J. Bechtel, and E. G. Krebs, *J. Biol. Chem.* **250**, 7795 (1975).
[9] G. N. Gill and G. M. Walton, this series, Vol. 38, p. 376.
[10] U. Walter and P. Greengard, *J. Cyclic Nucleotide Res.* **4**, 437 (1978).

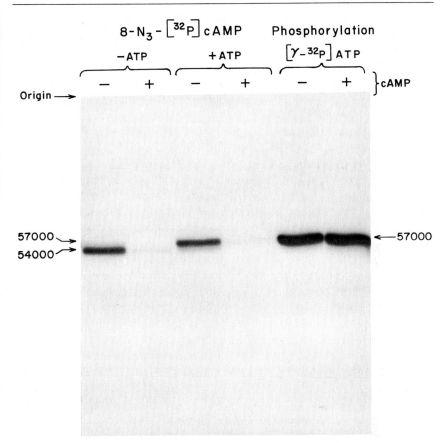

FIG. 1. Autoradiograph showing autophosphorylation of, and 8-N$_3$-[^{32}P]cAMP incorporation into, the regulatory subunit of type II cAMP-dependent protein kinase. The enzyme, 800-fold purified from bovine heart, was about 30% pure according to densitometric tracing of Coomassie blue-stained SDS–polyacrylamide gels. Photoaffinity labeling was carried out under standard conditions using 1 μM 8-N$_3$-[^{32}P]cAMP, in the absence or presence of 1 mM ATP, with and without 20 μM cAMP. Phosphorylation (autophosphorylation) was carried out in the absence or presence of 10 μM cAMP. All samples contained 0.5 μg of protein and were analyzed by SDS–polyacrylamide gel electrophoresis and autoradiography. The enzyme preparation used, which was completely in the dephospho form at the start of the experiment, maximally incorporated 1.51 pmol of 8-N$_3$-[^{32}P]cAMP and 1.56 pmol of [^{32}P]phosphate from [γ-^{32}P]ATP per 0.5 μg per protein. From U. Walter and P. Greengard, *J. Cyclic Nucleotide Res.* **4**, 437 (1978), reprinted with permission.

purified regulatory and catalytic subunit,[11] a maximal covalent incorporation of 0.95 pmol 8-N$_3$-[^{32}P]cAMP/pmol of regulatory subunit monomer could be obtained (data not shown).

[11] S. M. Lohmann, U. Walter, and P. Greengard, *J. Biol. Chem.* **255**, 9985 (1980).

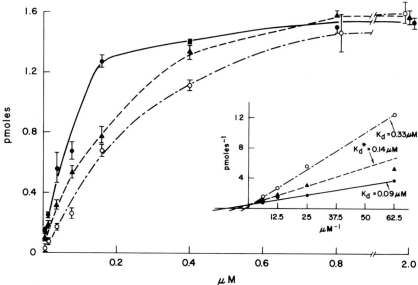

FIG. 2. Concentration-dependent photoactivated incorporation of 8-N_3-[^{32}P]cAMP (▲---▲) into, and reversible binding of [^{3}H]cAMP (●—●) or 8-N_3-[^{32}P]cAMP (○---○) to, Type II cAMP-dependent protein kinase from bovine heart. The protein kinase (0.5 μg of protein, preparation identical to that used in Fig. 1) was analyzed for reversible binding under standard conditions using the indicated concentrations of [^{3}H]cAMP or 8-N_3-[^{32}P]cAMP. The same amount of enzyme was analyzed for the photoactivated incorporation of 8-N_3-[^{32}P]cAMP under standard conditions using the indicated concentrations of 8-N_3-[^{32}P]cAMP. All results are expressed as picomoles of [^{3}H]cAMP or 8-N_3-[^{32}P]cAMP per 0.5 μg of enzyme protein. The mean value and standard deviation of triplicate determinations are shown. Dissociation constants (K_d) were calculated from double-reciprocal plots of the results (inset). The K_d value (*) given for 8-N_3-[^{32}P]cAMP incorporation represents the concentration of 8-N_3-[^{32}P]cAMP producing half-maximal incorporation, and is not a true dissociation constant. From U. Walter and P. Greengard, *J. Cyclic Nucleotide Res.* **4**, 437 (1978), reprinted with permission.

Recently, however, it has been demonstrated that the regulatory subunits of Type I and Type II cAMP-dependent protein kinase bind 2 mol of cAMP/mol of regulatory monomer[12,13] and that half of the cAMP bound is lost during the process of Millipore filtration.[14,15] Other studies have shown that two different intrachain cAMP-binding sites of the regulatory subunits can be distinguished by their cAMP-dissociation rate and cyclic

[12] J. D. Corbin, P. H. Sugden, L. West, D. A. Flockhart, T. M. Lincoln, and D. McCarthy, *J. Biol. Chem.* **253**, 3997 (1978).
[13] W. Weber and H. Hilz, *Biochem. Biophys. Res. Commun.* **90**, 1073 (1979).
[14] S. E. Builder, J. A. Beavo, and E. G. Krebs, *J. Biol. Chem.* **255**, 2350 (1980).
[15] D. Øgreid and S. O. Døskeland, *FEBS Lett.* **121**, 340 (1980).

nucleotide specificity.[16] 8-Substituted cAMP-analogs such as 8-N$_3$-cAMP appear to have a marked selectivity for one of the two cAMP-binding sites.[16] All these results considered together suggest that 8-N$_3$-[^{32}P]cAMP selectively and quantitatively labels one of the two distinct cAMP-binding sites of the regulatory subunit of cAMP-dependent protein kinase. This conclusion is in agreement with two recent studies which reported that the regulatory subunit R-II covalently labeled with 8-N$_3$-[^{32}P]cAMP contained only a single modified peptide and tyrosine residue.[17,18] The latter study[18] also confirmed our results[10] that the Type II cAMP-dependent protein kinase maximally incorporates 1 mol of 8-N$_3$-cAMP/mol of regulatory subunit monomer.

Photoaffinity Labeling of the Regulatory Subunits of cAMP-Dependent Protein Kinase in Crude Tissue Extracts

Of the many proteins present in crude extracts such as bovine heart cytosol, only a few proteins are labeled to a significant extent when the photoaffinity labeling procedure is carried out with 8-N$_3$-[^{32}P]cAMP concentrations up to 1 μM (Fig. 3). Two proteins with molecular weight of about 55,000 (R-II) and 47,000 (R-I) specifically incorporated 8-N$_3$-[^{32}P]cAMP since an excess of nonradioactive cAMP prevented their labeling. R-I and R-II were identified as the regulatory subunits of Type I and Type II cAMP-dependent protein kinase, respectively, by a variety of methods[4] including experiments in which R-I and R-II were specifically precipitated by an immunological procedure[11] using antisera directed against the Type I and Type II regulatory subunits (Fig. 3).

The regulatory subunits R-I and R-II, as analyzed by photoaffinity labeling with 8-N$_3$-[^{32}P]cAMP, have been detected in a wide variety of tissue extracts examined. R-I and R-II exist as soluble and membrane-bound forms and specific labeling of both forms by 8-N$_3$-[^{32}P]cAMP has been demonstrated in experiments using homogenates, membrane preparations, and soluble extracts.[4,19]

The concentration dependency of photoactivated incorporation of 8-N$_3$-[^{32}P]cAMP into R-I and R-II present in crude extracts is similar to that observed with purified preparations. When using crude tissue extracts and an 8-N$_3$-[^{32}P]cAMP concentration (i.e., 1 μM) saturating for R-I and R-II, the amount of specific photoactivated incorporation of 8-N$_3$-[^{32}P]cAMP into R-I and R-II is proportional to the amount of extract

[16] S. R. Rannels and J. D. Corbin, *J. Biol. Chem.* **255,** 7085 (1980).
[17] A. R. Kerlavage and S. S. Taylor, *J. Biol. Chem.* **255,** 8483 (1980).
[18] A. R. Kerlavage and S. S. Taylor, *J. Biol. Chem.* **257,** 1749 (1982).
[19] U. Walter, P. Kanof, H. Schulman, and P. Greengard, *J. Biol. Chem.* **253,** 6275 (1978).

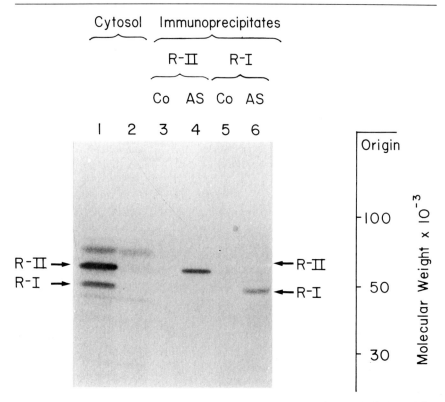

FIG. 3. Autoradiograph showing the photoaffinity labeling and immunological precipitation of the regulatory subunits R-I and R-II of cAMP-dependent protein kinase from bovine heart cytosol. Bovine heart cytosol (170 μg of protein) was labeled with 1 μM 8-N_3-[^{32}P]cAMP in the presence of 1 mM ATP under standard conditions in the absence (lane 1) or presence (lane 2) of 50 μM cAMP. Cytosol samples labeled with 8-N_3-[^{32}P]cAMP in the absence of cAMP were also analyzed by immunological precipitation using a preimmune serum (Co, lane 3) and an antiserum against the Type II regulatory subunit R-II of bovine heart (AS, lane 4) or using a preimmune serum (Co, lane 5) and antiserum against the Type I regulatory subunit R-I of bovine lung (AS, lane 6). Photoaffinity-labeled cytosol samples (lanes 1 and 2) and the immunoprecipitates (lanes 3–6) were analyzed by SDS–polyacrylamide gel electrophoresis and autoradiography.

protein used (up to 200 μg) and agrees reasonably well with the total amount of specific cAMP-binding sites determined by the Millipore filtration method.[4,19–22] The tissue distribution of R-I and R-II as measured by the photoaffinity labeling procedure also agrees well with the tissue distri-

[20] S. M. Lohmann, U. Walter, and P. Greengard, *J. Cyclic Nucleotide Res.* **4**, 445 (1978).
[21] R. Rangel-Aldao, J. W. Kupiec, and O. M. Rosen, *J. Biol. Chem.* **254**, 2499 (1979).
[22] C. S. Rubin, R. Rangel-Aldao, D. Sarkar, J. Erlichman, and N. Fleischer, *J. Biol. Chem.* **254**, 3797 (1979).

bution of Type I and Type II cAMP-dependent protein kinase holoenzymes determined by DEAE-cellulose chromatography.[4,19] However, in special cases regulatory subunits may also exist as "free cAMP-binding proteins"[23] not associated with the catalytic subunit.

Several problems are often encountered when applying the photoaffinity labeling procedure to crude tissue extracts. Nonspecific labeling is often observed with 8-N_3-[^{32}P]cAMP concentrations found necessary to saturate both the R-I and R-II regulatory subunit. Low concentrations of 2-mercaptoethanol (i.e., 0.5 mM) significantly reduce such nonspecific labeling without affecting the specific labeling of R-I and R-II. However, high concentrations of 2-mercaptoethanol (i.e., 10 mM) also reduce the specific labeling of R-I and R-II. The presence of free cyclic nucleotides or cyclic nucleotides bound to R-I and R-II may also interfere with the photoaffinity labeling of crude tissue extracts. Conditions which allow the exchange of bound and free cyclic nucleotides,[23] or pretreatment of the tissue extracts with urea,[17,24] may be required to achieve maximal labeling of R-I and R-II. The interference by phosphodiesterases present in crude tissue extracts is minimized by the inclusion of a phosphodiesterase inhibitor such as 1-isobutylmethylxanthine (IBMX) in the incubation mixture for the photoaffinity labeling. The regulatory subunits R-I and R-II are also subject to significant proteolysis in crude tissue extracts which affects the pattern of specific cAMP-binding proteins. In fact, photoaffinity labeling with 8-N_3-[^{32}P]cAMP is a valuable tool to study proteolytic fragments of the regulatory subunits R-I and R-II. The presence of Ca^{2+}-chelating agents and protease inhibitors such as Aprotinin (Trasylol) during preparation and incubation of crude tissue extracts reduces proteolysis of R-I and R-II.

Concluding Remarks

Photoaffinity labeling with 8-N_3-[^{32}P]cAMP in combination with SDS–polyacrylamide gel electrophoreses and autoradiography can be used to identify specific cAMP-binding proteins in crude tissue extracts and in purified preparations. The procedure has been successfully applied to the two types of regulatory subunits of cAMP-dependent protein kinase as outlined in this chapter and to the cAMP-receptor protein of *E. coli*.[25] The procedure can be used to study qualitatively and, with proper control

---

[23] U. Walter, M. R. C. Costa, X. Breakefield, and P. Greengard, *Proc. Natl. Acad. Sci. U.S.A.* **76**, 3251 (1979).

[24] K. Schwechheimer and F. Hofmann, *J. Biol. Chem.* **252**, 7690 (1977).

[25] H. Aiba and J. S. Krakow, *Biochemistry* **19**, 1857 (1980).

experiments, quantitatively the tissue and subcellular distributions of the R-I and R-II regulatory subunits.

In combination with immunological methods, this technique may allow the detection and examination of microheterogeneity within these two classes of regulatory subunit.[11,21,22] Photoaffinity labeling with 8-N_3-[^{32}P]cAMP is an important tool to study proteolytic fragments of cAMP-binding proteins and the chemical nature of their cAMP-binding site. A similar procedure using 8-N_3-[^{32}P]cIMP[26,27] or 8-N_3-[^{32}P]cGMP[28] has been developed to investigate the cGMP-binding site of cGMP-dependent protein kinase.

Acknowledgments

The authors were supported by grants from the Deutsche Forschungsgemeinschaft (Wa 366/2 and Wa 366/3) to U. W. and by Grants MH-17387 and NS-08440 from the U.S. Public Health Service to P. G.

[26] J. E. Casnellie, D. J. Schlichter, U. Walter, and P. Greengard, *J. Biol. Chem.* **253**, 4771 (1978).
[27] U. Walter, *Eur. J. Biochem.* **118**, 339 (1981).
[28] R. L. Geahlen, B. E. Haley, and E. G. Krebs, *Proc. Natl. Acad. Sci. U.S.A.* **76**, 2213 (1979).

[16] Use of 1,N^6-Etheno-cAMP as a Fluorescent Probe to Study cAMP-Dependent Protein Kinase

By HILLARY D. WHITE,[1] STEPHEN B. SMITH, and EDWIN G. KREBS

The fluorescent analog of cAMP, 1,N^6-etheno-cAMP (εcAMP), was originally synthesized and used to activate cAMP-dependent protein kinase by Secrist *et al.*[2,3] We have used this analog in a variety of assays to study various aspects of the activation–deactivation mechanism of the cAMP-dependent protein kinases:

$$R_2C_2 + 4 \text{ cyclic nucleotide} \underset{\text{deactivation}}{\overset{\text{activation}}{\rightleftharpoons}} R_2(\text{cyclic nucleotide})_4 + 2C \qquad (1)$$

The inactive holoenzyme is denoted here by R_2C_2 and the regulatory

[1] American Cancer Society Postdoctoral Fellow to HDW (PF1774). HDW, SBS, and EGK were supported by the Howard Hughes Medical Institute.
[2] J. A. Secrist, J. R. Barrio, N. J. Leonard, and G. Weber, *Biochemistry* **11**, 3499 (1972).
[3] J. A. Secrist, J. R. Barrio, N. J. Leonard, C. Villar-Palasi, and A. G. Gilman, *Science* **177**, 279 (1972).

dimer and catalytic subunit by R_2 and C, respectively. Most of the assays using εcAMP are based on the fact that when εcAMP is bound to the regulatory dimer (R_2εcAMP$_4$) its fluorescence is quenched approximately 90% with the Type I R and approximately 60% with the Type II R. We shall briefly mention three assays using this cAMP analog and then describe in detail the protocol for a fourth εcAMP-based assay that is especially useful because of its sensitivity to cooperativity phenomena.

Activation Kinetic Assay. The activation kinetic assay does not rely on the fluorescent properties of εcAMP, but is followed instead by phosphoryltransferase activity: R_2C_2 holoenzyme is titrated with εcAMP to effect the release of active catalytic subunit, which is then free to phosphorylate a peptide substrate in the presence of [γ-^{32}P]ATP:

$$R_2C_2 + 4 \text{ εcAMP} \rightarrow R_2\text{εcAMP}_4 + 2C \quad (2)$$
$$\text{inactive} \qquad\qquad\qquad\qquad \text{active}$$

$$\text{peptide} + [\gamma\text{-}^{32}\text{P}]\text{ATP} \xrightarrow[\text{Mg}^{2+}]{C} [^{32}\text{P}]\text{phosphorylpeptide} + \text{ADP} \quad (3)$$

At any specific total concentration of εcAMP, the occupancy of R_2 sites for εcAMP is assumed to be proportional to the phosphoryltransferase activity at that same concentration of εcAMP relative to the activity at saturating concentrations of εcAMP. The concentration of free εcAMP can then be calculated and plotted against the corresponding relative fluorescence to give a saturation curve that can be quantitatively analyzed.

Cyclic Nucleotide Exchange Kinetic Assay. This assay and those following all use R_2 that has been fully saturated with εcAMP (R_2εcAMP$_4$) as described below. A large excess of cAMP (1000 times the concentration of cyclic nucleotide sites) is added to the R_2εcAMP$_4$ complex at time zero and the increase in fluorescence due to the appearance of free εcAMP is monitored with time:

$$R_2\text{εcAMP}_4 + 4\text{cAMP} \xrightarrow{\text{excess cAMP}} R_2\text{cAMP}_4 + 4 \text{ εcAMP} \quad (4)$$
$$\text{fluorescent}$$
$$\text{signal}$$

The reaction can be followed in the absence of C or at various ratios of C relative to R_2.

Deactivation Kinetic Assay. The deactivation kinetic assay is followed by stopped-flow fluorometry. An excess of C, sufficient to induce all the εcAMP to be released from R_2εcAMP$_4$, is rapidly mixed with the R_2εcAMP$_4$ complex:

$$R_2\text{εcAMP}_4 + 2C \xrightarrow{\text{excess C}} R_2C_2 + 4 \text{ εcAMP} \quad (5)$$
$$\text{fluorescent}$$
$$\text{signal}$$

The increase in fluorescence due to the release of εcAMP is monitored on an oscilloscope. If the Type I enzyme is used, MgATP is included to facilitate the recombination of R_2 and 2C.

Deactivation Equilibrium (C Titration) Assay. The $R_2\varepsilon cAMP_4$ complex can be titrated with C causing the release of εcAMP from R_2 and the concomitant reappearance of its fluorescent signal:

$$R_2\varepsilon cAMP_4 + 2C \xrightleftharpoons{\text{excess C}} R_2C_2 + 4 \underset{\substack{\text{fluorescent} \\ \text{signal}}}{\varepsilon cAMP} \tag{6}$$

We have found this C titration or deactivation equilibrium assay (as opposed to the activation kinetic or phosphoryltransferase assay) especially useful in monitoring the cooperativity involved in the activation–deactivation mechanisms of the cAMP-dependent protein kinases. To illustrate its usefulness we present below the details for preparing the Type I $R_2\varepsilon cAMP_4$ complex and the C titration (deactivation equilibrium) assay using this complex.

C Titration (Deactivation Equilibrium) Assay

Materials

Etheno-cAMP. The fluorescent analog of cAMP (εcAMP) was obtained from Sigma Chemical Co. The characteristic absorption spectrum and the extinction coefficient ($5.6 \times 10^3 \, M^{-1} \, cm^{-1}$) of εcAMP are given by Secrist et al.[2,3] Care should be taken to keep stock solutions of this compound in vessels shielded from light to prevent photodecomposition.

MEK Buffer. MEK buffer was used throughout these studies and consisted of 5 mM morpholinopropanesulfonic acid, 1 mM ethylenediaminetetraacetic acid, and 100 mM KCl, adjusted to pH 7.0, and to which 1 mM dithiothreitol was added before use.

Purification and Concentration Determination of the cAMP-Dependent Protein Kinase Subunits. Type I R_2 and C subunits were purified from bovine skeletal muscle by established procedures: the holoenzyme was partially purified by DEAE-cellulose column chromatography and further treated by cAMP-Sepharose affinity chromatography (cAMP elution) to obtain R_2[4,5] and by CM-Sephadex differential batch chromatography (+/− cAMP) followed by Sephacryl S-200 column chromatography to obtain C.[6] Care was taken to determine final protein concentrations

---

[4] W. L. Dills, C. D. Goodwin, T. M. Lincoln, J. A. Beavo, P. J. Bechtel, and J. D. Corbin, *Adv. Cyclic Nucleotide Res.* **10**, 199 (1979).
[5] S. B. Smith, H. D. White, B. M. Flug, and E. G. Krebs, unpublished data.
[6] P. J. Bechtel, J. A. Beavo, and E. G. Krebs, *J. Biol. Chem.* **252**, 2691 (1977).

accurately (for example, $R_2\varepsilon cAMP_4$, see below) so that the occupancy of the cyclic nucleotide sites on R_2 with $\varepsilon cAMP$ could be determined. Amino acid analyses of the protein stock solutions were therefore used together with spectral data to obtain the 0.1% solution extinction coefficient of $R_2\varepsilon cAMP_4$ at 280 nm: 0.98 liter g^{-1} cm^{-1}.

Procedures

Preparation of $R_2\varepsilon cAMP_4$. Purified $R_2 cAMP_4$ (23 mg), which had been exhaustively dialyzed to remove any unbound cAMP, was vacuum concentrated (in a dialysis bag) to a volume of 1–1.5 ml. To this was added an 800-fold excess of $\varepsilon cAMP$ (132 mg) relative to the concentration of cyclic nucleotide sites present. The solution was immediately readjusted to a pH of 7.0. The large excess of $\varepsilon cAMP$ was required to ensure quantitative exchange with cAMP, due to the lower binding affinity of R_2 for $\varepsilon cAMP$ relative to cAMP (approximately 8×).[7] After equilibration overnight at 4° and then for 1 hr at 30°, $R_2\varepsilon cAMP_4$ was separated from unbound cyclic nucleotide and from any R_2 proteolysis products (when present) by Sephadex G-200 (fine) column chromatography. One milliliter fractions were collected from the 1.4 × 100 cm column at a flow rate of approximately 5 ml/hr. The $R_2\varepsilon cAMP_4$, R_2-derived fragments (if present), and unbound $\varepsilon cAMP$ peak fractions were eluted from the column at 60, 90, and 150 ml, respectively. Fractions containing $R_2\varepsilon cAMP_4$ were identified by the characteristic spectrum shown in Fig. 1. The wavelength maximum can be seen at 277 nm with shoulders at 260, 270, and 282 nm, due to the bound $\varepsilon cAMP$. $R_2\varepsilon cAMP_4$-derived fragment(s) show similar spectral characteristics, indicating at least partial saturation with $\varepsilon cAMP$. Aliquots of representative fractions across the $R_2\varepsilon cAMP_4$ peak were then analyzed by SDS–polyacrylamide electrophoresis to determine if the fractions to be pooled contained any detectable trace of the R_2 proteolytic fragments. It is important that no R_2 proteolysis products are present because they bind cyclic nucleotide with affinities different from those of the unproteolyzed R_2.[7] The unproteolyzed $R_2\varepsilon cAMP_4$ fractions were pooled and an aliquot was assayed spectrophotometrically for protein concentration (see above). Another aliquot was diluted into a cuvette to 150 nM $R_2\varepsilon cAMP_4$ (the approximately average physiological concentration) and a 1000-fold excess of cAMP relative to cyclic nucleotide binding sites was added to exchange with the bound fluorescent cyclic nucleotide. The resultant increase in fluorescence was recorded using a Turner Model 430 fluorometer at an excitation wavelength of 315 nm and an emission wavelength of 420 nm. This was compared with the fluorescence of an

---

[7] Observations by H. D. White, S. B. Smith, and E. G. Krebs (1981).

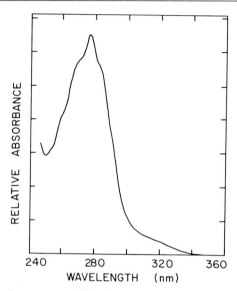

FIG. 1. Absorption spectrum of $R_2\varepsilon cAMP_4$, prepared as described in the text.

εcAMP standard solution to determine the concentration of εcAMP released from the protein and thereby verify the concentration of cyclic nucleotide binding sites present. The concentration of sites on R_2 for cyclic nucleotide was and should be, of course, four times that of the regulatory dimer.[8]

C Titration Assay. The following procedure for the titration of $R_2\varepsilon cAMP_4$ with C provides a sensitive assay for monitoring the cooperative interactions involved in the activation–deactivation mechanism of the cAMP-dependent protein kinase: The $R_2\varepsilon cAMP_4$ complex was diluted in a cuvette with MEK buffer to 150 nM in a 2.2-ml volume in the presence of 5 mM Mg(OAc)$_2$ and 100 μM ATP. At an excitation wavelength of 315 nm and an emission wavelength of 420 nm, the fluorometer was adjusted to a 0–80% scale by using a buffer blank and a 600 nM εcAMP standard solution, respectively. (A value of 80% instead of 100% was chosen because of the small C-induced enhancement of εcAMP fluorescence, see below.) Small aliquots of a fresh, concentrated[9] C stock solution (10–12 mg/ml in MEK buffer) were then added to the $R_2\varepsilon cAMP_4$ solution and after equilibration, the relative fluorescence values were recorded. Fluorescence values were corrected for volume changes and for a C-induced enhancement of εcAMP fluorescence. The latter correc-

[8] S. E. Builder, J. A. Beavo, and E. G. Krebs, *J. Biol. Chem.* **255**, 2350 (1980).
[9] The catalytic subunit can be conveniently vacuum concentrated overnight in a dialysis bag.

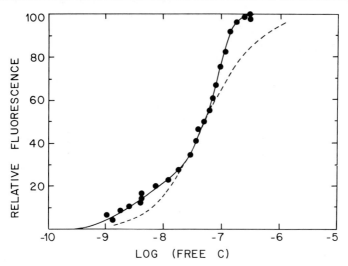

FIG. 2. C titration (deactivation equilibrium) assay data. The release of εcAMP from $R_2\varepsilon cAMP_4$ was followed fluorometrically by titrating $R_2\varepsilon cAMP_4$ with C as described in the text. For comparison, a theoretical curve indicative of a noncooperative process is shown by the dashed line.

tion was quantitated by titrating as above but in the absence of the $R_2\varepsilon cAMP_4$ complex and in the presence of 600 nM εcAMP. The C-induced increase in εcAMP fluorescence is assumed to be due to media effects since the effect was linear with increasing C concentrations even at a 100-fold excess of C relative to εcAMP. The end point for each titration, when additional aliquots of C produced no increase in corrected fluorescence, was confirmed by adding a large excess of cAMP in a negligible volume to the cuvette at the end of the titration. This ensured that any remaining bound εcAMP was released from the regulatory subunits. It was assumed that at any point during the titration, the percentage of the maximum corrected fluorescence was equal to the percentage of εcAMP released from R_2 and to the percentage of C bound to R_2. Figure 2 shows the resulting titration plot for the Type I kinase. For comparison a theoretical curve expected for a noncooperative process is also shown in the same figure (dashed line). The lower portion of the curve is indicative of negative cooperativity while the upper portion is indicative of positive cooperativity. The C titration (deactivation) assay therefore provides a technique by which one can monitor the cooperative processes that are a part of the activation–deactivation mechanism of cAMP-dependent protein kinases.[7] In contrast such information has not been available via the phosphoryltransferase (activation kinetic) assay procedure.[7]

[17] Using Analogs to Study Selectivity and Cooperativity of Cyclic Nucleotide Binding Sites

By STEPHEN R. RANNELS and JACKIE D. CORBIN

Cyclic AMP-dependent protein kinase holoenzyme is activated when cAMP binds to a dimeric regulatory subunit (R), and the active catalytic subunit (C) dissociates from the holoenzyme complex according to the following stoichiometry:[1]

$$R_2C_2 + 4 \, cAMP \rightleftharpoons R_2 \cdot cAMP_4 + 2C$$

It is now known that the two intrachain cyclic nucleotide binding sites of each R monomer are different, by virtue of their unique rates of nucleotide dissociation and site-selectivity of cAMP analogs.[2] A technique was developed to study the relative selectivity of various cyclic nucleotide analogs for protein kinase binding sites. We have also applied the results obtained in site-selectivity studies to show that the binding of one class of cAMP analogs which prefers a single intrachain site can greatly stimulate the binding of another class of site-selective analogs to the other intrachain site.[3] This cooperativity in binding of cyclic nucleotides to both sites could be an important mechanism for amplifying the magnitude and rate of response to an *in vivo* hormonal elevation of intracellular cAMP levels. The methodology used to demonstrate site selectivity and cooperativity of cyclic nucleotides in binding to protein kinase could easily be applied to studies of ligand interactions with multiple binding sites of any receptor.

Materials and Methods

The R subunits of both isozymes I and II, which contain bound cAMP, are purified from various tissues as previously described.[1-4] Two different holoenzymes are used for the experiments. Partially purified holoenzymes are obtained by following procedures up to the cAMP-affinity column step. The resuspended $(NH_4)_2SO_4$ pellet is dialyzed against 10 mM potassium phosphate, 1 mM EDTA (pH 6.8) (KP/EDTA), and stored at -20 or $4°$. Reconstituted holoenzyme is prepared by first dialyzing cAMP-saturated homogeneous R subunit against KP/EDTA for 4 days at

[1] J. D. Corbin, P. H. Sugden, L. West, D. A. Flockhart, T. M. Lincoln, and D. McCarthy, *J. Biol. Chem.* **253**, 3997 (1978).
[2] S. R. Rannels and J. D. Corbin, *J. Biol. Chem.* **255**, 7085 (1980).
[3] S. R. Rannels and J. D. Corbin, *J. Biol. Chem.* **256**, 7871 (1981).
[4] S. R. Rannels and J. D. Corbin, *J. Cyclic Nucleotide Res.* **6**, (3), 201 (1980).

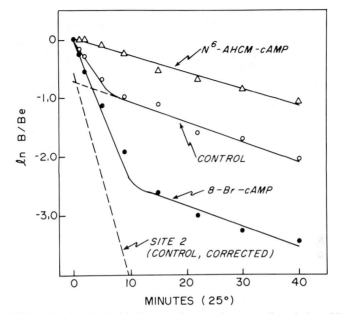

FIG. 1. Effect of analogs in the binding reaction on subsequent dissociation of [^{3}H]cAMP during a chase with excess unlabeled cAMP.

4°. Homogeneous C subunit, prepared as described before,[5] is added to the R at twice its molar ratio and the mixture is dialyzed against the same buffer for an additional 5 days. [^{3}H]cAMP is from New England Nuclear, [^{3}H]cIMP and 8-N$_3$-[^{32}P]cAMP are from ICN. All other nucleotides are purchased from Sigma or prepared as previously described.[6]

Selectivity of Analogs for cAMP-Binding Sites

Regulatory subunits saturated with cAMP, holoenzyme prepared from purified subunits, or partially purified holoenzymes are incubated for 45 min at 25° with a binding reaction mixture containing 2 M NaCl, 0.5 mg/ml histone mixture (Sigma Type IIA), and 1 μM [^{3}H]cAMP in 50 mM potassium phosphate, 1 mM EDTA, pH 6.8. We have previously shown that incubation and Millipore filtration using these conditions results in full stoichiometric binding of cAMP to the R subunit (2 mol cAMP/mol R monomer).[1] However, high salt and histone are not required to observe the selective analog binding described below. As shown in Fig. 1, 100 μl

[5] P. H. Sugden, L. A. Holladay, E. M. Reimann, and J. D. Corbin, *Biochem. J.* **159**, 409 (1976).
[6] J. D. Corbin, S. R. Rannels, D. A. Flockhart, A. M. Robinson-Steiner, M. C. Tigani, S. O. Døskeland, R. Suva, and J. P. Miller, *Eur. J. Biochem.* **125**, 259 (1982).

of 0.2 μM R subunit is incubated with 400 μl of [^{3}H]cAMP binding mixture and 50 μl H$_2$O or cAMP analog. At 45 min, when equilibrium exchange is complete, a 40-μl aliquot is withdrawn and filtered through Millipore membranes.[1] For filtration, each sample is immediately cooled to 4° by adding it to 1 ml of KP/EDTA. The sample mixture is then passed over the Millipore filter (0.45 μm), the reaction tube washed with an additional 1 ml of cold buffer, and the filter is washed with 4 ml of buffer. To the remaining volume, 5 μl of 10 mM cAMP is added (approximately 100-fold excess) and 40-μl aliquots are subsequently filtered at the times indicated. In the control (no cAMP analog added), dissociation of [^{3}H]cAMP occurs from two distinct sites which are present in approximately equal amounts, the percentage of each obtained by extrapolation of the lesser slope to the y axis (dashed line) and calculating the antilog. Analogs of cAMP are added at concentrations which inhibit [^{3}H]cAMP equilibrium binding by approximately 75%. If an analog being tested could compete selectively with [^{3}H]cAMP for one of the two binding sites (a selectivity relative to cAMP), then an inhibition of [^{3}H]cAMP binding of greater than 50% assures that the remaining bound [^{3}H]cAMP is mainly in the other site. Some analogs are sufficiently site selective that an inhibition greater than 50% is difficult to achieve.[2] Combinations of two such analogs of opposite selectivity inhibit [^{3}H]cAMP binding more efficiently than cAMP itself.[6] For analogs with high site selectivity, a near 50% inhibition of [^{3}H]cAMP binding is adequate. However, to see good selective effects of other analogs, they may have to be added in concentrations which inhibit [^{3}H]cAMP binding by as much as 90%. Thus, in Fig. 1, 50 μl of 18 μM N^6-aminohexylcarbamoylmethyl-cAMP (N^6-AHCM-cAMP) inhibits [^{3}H]cAMP binding by 62%. The remaining [^{3}H]cAMP occupies only the site from which cAMP dissociates slowly (Site 1) following a chase with nonradioactive cAMP. Alternatively, inhibition of [^{3}H]cAMP binding with 8-Br-cAMP shifts the dissociation curve downward, indicating that the remaining bound [^{3}H]cAMP occupies mainly the site from which cAMP dissociates more rapidly (Site 2). N^6-AHCM-cAMP, cIMP, or other 6-substituted analogs show a selectivity relative to cAMP for Site 2, whereas 8-substituted analogs prefer Site 1.[2,6] In addition, alteration at C-2 tends to render these nucleotides Site 1-selective. Because cAMP-binding sites are fully saturated at all times, specific effects of analogs to alter [^{3}H]cAMP dissociation rates are unlikely. In fact, the dissociation rates observed for Site 2 after correction for the slow Site 1 component are similar for all experimental protocols. This correction is accomplished in the following manner. A continuation of the Site 1 slope is drawn through the ordinate, as indicated by the dashed line in Fig. 1 for the control curve. For any time point along this line, the fraction of

[³H]cAMP which remains bound at that time (B/Be) is subtracted from the value of the uncorrected curve for Site 2. The negative logarithm of these values can be replotted as a corrected slope for Site 2 (shown in Fig. 1). The corrected dissociation rate constant is 0.35 min^{-1} compared to the uncorrected value of 0.15 min^{-1}, indicating that the observed rate of dissociation for Site 2 (when occupied by half of the total bound [³H]cAMP) is in error by a factor of greater than 2. The dissociation rate constant for Site 1 is approximately unaltered (0.03 min^{-1}). The time ($t_{1/2}$) required for dissociation of 50% of the bound [³H]cAMP from a particular site can be calculated from the dissociation rate constant (K) using the equation: $t_{1/2} = \ln 0.5/-K$. Using the values above the $t_{1/2}$ would be 23.0 and 2.0 min, respectively, for Sites 1 and 2. The technique described above can be used to demonstrate different intrachain binding sites in crude extracts, partially purified and purified holoenzymes, or homogeneous R subunits.

As discussed, the amount of [³H]cAMP bound to Site 1 or 2 of R can readily be determined following an equilibration period during which complete nucleotide ([³H]cAMP and analog) exchange and full saturation of both binding sites occur. This is done by extrapolation of the slow component of the dissociation curve and calculating the amount of [³H]cAMP bound to Site 1 by multiplying the antilog of this extrapolated value times the total [³H]cAMP bound at 0 time. This number is subtracted from the total [³H]cAMP binding (Sites 1 + 2 at zero time) to obtain Site 2 binding. Thus, the degree of inhibition by analogs of [³H]cAMP binding to Site 1 or 2 can be determined. A quantitative determination of the affinity of an analog relative to cAMP for either Site 1 or 2 has been reported by Øgreid et al.[7] They determined the ratio K_IcAMP/K_I analog from plots of R_0/C_x versus i, where R_0 is the total concentration of each binding site (1 or 2), C_x is the amount of [³H]cAMP bound, and i is the variable free concentration of each analog tested. This plot yielded a straight line for all analogs tested, the K_IcAMP/K_I analog being the ratio of the slopes obtained for unlabeled cAMP and the analog. The inhibition of [³H]cAMP binding by analogs was reversible and competitive.[7]

Cooperativity in Binding of Site-Selective Analogs

A direct application of analog site selectivity was realized by observing that one site-selective analog could stimulate the binding of the class of site-selective analogs which bind to the other intrachain binding site.[3] This stimulation of binding is referred to as "cooperative" binding, and is

[7] D. Øgreid, S. O. Døskeland, and J. P. Miller, *J. Biol. Chem.* **258**, 1041 (1983).

obtained using this more direct method which is in most cases preferable over the usual kinetic approaches of determination of cooperativity. Stimulation of the binding of one site-selective analog to one intrachain site by another site-selective analog is really a specialized case of the site-specific binding described in Fig. 1. Selectivity is thus relative to another analog, rather than cAMP itself. Cyclic AMP itself, however, does stimulate cAMP and analog binding, but the magnitude of these effects is diminished by virtue of its reduced selectivity. A typical experiment to test the effect of a C-8 analog on [^{3}H]cIMP binding is shown in Fig. 2.

To 320 μl of 14.7 mM potassium phosphate (pH 6.8) containing 13 nM [^{3}H]cIMP is added 112 μl 1 mg/ml histone (Type IIA, Sigma), 80 μl of H$_2$O or 100 nM 8-N$_3$-cAMP, and 80 μl of reconstituted Type II holoenzyme (1.7 nM R subunit) diluted with KP/EDTA containing 1 mg/ml bovine serum albumin as described under Methods. Following incubation at 4° for the times indicated, 74 μl is removed and filtered. It can be seen that equimolar concentrations of 8-N$_3$-cAMP markedly stimulate [^{3}H]cIMP binding. The increased [^{3}H]cIMP binding has been shown in other experiments to be at Site 2.[3] It is also apparent that this cooperative stimulation by 8-N$_3$-cAMP results from an increase in the rate of [^{3}H]cIMP binding. The stimulatory effect decreases with time because the binding in the presence of 8-N$_3$-cAMP levels off and the binding in the absence of 8-N$_3$-cAMP continues to increase. The concentration of the C-8 analog required for maximum stimulation is usually, but not always, approximately equimolar or less compared with the C-6 analog [^{3}H]cIMP.[3,6] Higher concentrations tend to lower [^{3}H]cIMP binding by

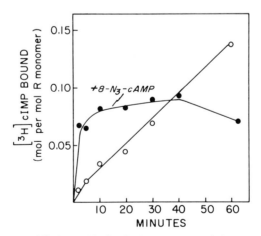

FIG. 2. Time course of [^{3}H]cIMP binding in the presence and absence of a Site 1-selective analog.

competing at Site 2. Such competition might also explain the lack of stimulation observed at the longer times as seen in Fig. 2.

The R subunit of the reconstituted holoenzyme used for some of these studies potentially retains low amounts of cAMP, even after extensive dialysis. Thus, the binding of two site-selective analogs could be complicated by the presence of nonradioactive cAMP. All binding studies are thus done at 4° and for short times (5–10 min) to minimize exchange reactions. The first observation of cooperative binding of cyclic nucleotide analogs (at 30°) was for a partially purified holoenzyme.[3] Histones or other polycationic molecules were not required to see the effect, presumably because other natural factors present in the preparation could function in their place. The effects reported previously were also of lesser magnitude. The higher temperature used in those experiments may have caused the C-8 analogs to also bind to Site 2, as seen in Fig. 2 after longer incubations. Optimal analog stimulation of Site 2 binding of [^{3}H]cIMP is obtained at short times, low temperature, low concentrations of the cyclic nucleotides, and when the Site 1-directed C-8 analog is approximately equimolar with [^{3}H]cIMP.

Stimulation of binding of Site 1 directed analogs by Site 2 analogs, or cooperativity in the reverse direction, is also observed with the reconstituted holoenzyme used for Fig. 2 (not shown), but only slight effects of this type were observed previously in the partially purified holoenzyme. Cooperative binding requires the presence of C subunit, suggesting a function in the physiological activation of protein kinase. Because cooperativity appears to operate in both directions, it is not yet understood whether binding to the two different sites occurs in an ordered or random sequence. It is possible that interchain binding cooperativity is also being observed. Several tables which correlate analog site selectivity and the ability of various analogs to stimulate [^{3}H]cIMP binding to purified Type II R subunit and holoenzyme have been reported.[3,6] Those analogs which select Site 1 relative to cAMP also result in a greater stimulation of [^{3}H]cIMP binding in cooperative studies. The same trend is observed for Type I holoenzyme; however, nucleotide binding specificity is different from the Type II isozyme and MgATP causes an enhanced stimulation, possibly by simply slowing down the binding reaction.[2,6] Because none of the analogs are absolutely site-specific, some "cross-site" binding must occur, and the magnitude of the cooperative binding effect probably represents a minimum. It is not clear if cooperative binding is solely an intrachain mechanism or if both chains of the R dimer are involved. For example, occupancy of both intrachain sites during cooperative binding could subsequently result in nucleotide binding to the other chain by a similar process. Thus, cooperativity in this case may reflect interaction between domains in a single peptide chain and/or subunit interaction. The use of site-selective analogs allows one to more readily observe what we believe is a physiological mechanism which operates *in vivo*.

"Blocking" Nucleotide Binding Sites with Analogs

Another application for the use of cyclic nucleotide analogs in studies of protein kinases is illustrated in Fig. 3, in which photoaffinity labeling of cGMP-dependent protein kinase using 8-N_3-[^{32}P]cAMP in the presence of competing cyclic nucleotides is shown. In preparation A, beef lung en-

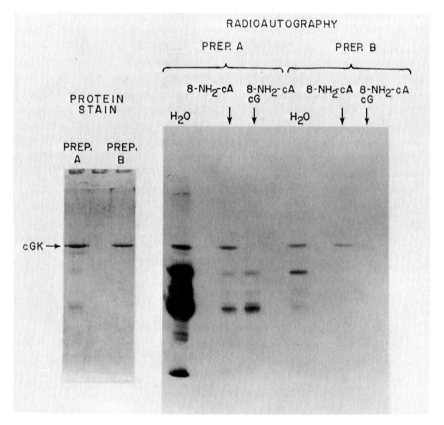

FIG. 3. Photoaffinity labeling of cGMP-dependent protein kinase with 8-N_3-[^{32}P]cAMP in the presence of cyclic nucleotide analogs.

zyme is purified through the affinity column step.[8] For preparation B, the enzyme is purified by the affinity column step followed by sucrose density gradient centrifugation.[8] First, 50 µl of enzyme (~4 µg) is added to the incubation tube (12 × 75 mm) followed by the addition of 5 µl 1 mM 8-NH$_2$-cAMP, 10 mM cGMP, or H$_2$O where indicated. After 30 min at 4°, 4 µl 110 µM 8-N$_3$-[^{32}P]cAMP (ICN, specific activity ~50 Ci/mmol) is added and the mixture is exposed to UV light (254 nm) immediately for 1 hr. Following the addition of 5 µl 10% SDS, gel electrophoresis is carried out.[9] The gels are dried and autoradiographs prepared by 16 hr exposure to Kodak X-Omat RP (XRP-1) film in a Dupont Quanta II cassette with intensifying screen. Preparation A contains numerous visible contaminating lower molecular weight proteins when assessed by Coomassie blue staining. These cAMP-binding proteins, which are mainly R subunits and R subunit fragments, are readily labeled with 8-N$_3$-[^{32}P]cAMP. Addition of excess 8-NH$_2$-cAMP, however, blocks most of this labeling, but has a lesser effect on labeling of the cGMP-dependent protein kinase. In the more highly purified enzyme (preparation B), 8-NH$_2$-cAMP completely eliminates labeling of the cAMP-dependent protein kinase R subunits (and fragments), so that only cGMP-dependent protein kinase is specifically tagged by the photoaffinity label. This is indicated in the last lane where added cGMP completely eliminates 8-N$_3$-[^{32}P]cAMP binding. Thus, 8-NH$_2$-cAMP, which binds more readily to cAMP-binding proteins than to the cGMP-dependent kinase, can be used to "block" the cAMP-binding sites during 8-N$_3$-cAMP photoaffinity labeling.

The use of specific analogs of ligands which bind to receptor proteins has been an important tool in studies of cAMP-dependent protein kinases. We are applying these techniques of analog selectivity and binding cooperativity to studies of kinase activation in intact fat cells, extracts of rat hearts, and in studies of phosphodiesterase. Differences in analog selectivity for Type I and Type II binding sites suggest a potential for selective activation of a single protein kinase isozyme in an intact tissue (or *in vitro*) where both enzymes are present. Also, as shown in Fig. 3, the labeling of cGMP-dependent protein kinase with 8-N$_3$-cAMP is achieved in a mixture of kinases by specifically "blocking" cyclic nucleotide binding sites with site-selective analogs. This approach could be useful for any mixture of cyclic nucleotide binding proteins.

---

[8] T. M. Lincoln, W. L. Dills, Jr., and J. D. Corbin, *J. Biol. Chem.* **252**, 4269 (1977).
[9] U. K. Laemmli, *Nature (London)* **227**, 680 (1970).

[18] Reversible Autophosphorylation of Type II cAMP-Dependent Protein Kinase: Distinction between Intramolecular and Intermolecular Reactions

By JACK ERLICHMAN, RAPHAEL RANGEL-ALDAO, and ORA M. ROSEN

Many proteins are subject to reversible covalent modification by phosphorylation and it is widely recognized that this modification is a critical component of the physiological response of mammalian cells to specific stimuli. The cAMP-dependent protein kinases (PK) catalyze the initial phosphorylation reaction in cAMP-mediated pathways. In most tissues and cells there are two classes of cAMP-dependent protein kinases, designated types I and II by their order of elution from DEAE-cellulose. Both classes are composed of four polypeptide chains, consisting of a regulatory (R) dimer and two catalytic (C) subunits. The R components of the two types of kinase are different from each other whereas the C components appear to be both identical or very similar.[1] Both types I and II PKs are themselves subject to covalent modification by phosphorylation.[2,3] Each R monomer may contain up to 2 mol of phosphate; however, the mechanism by which R subunits are phosphorylated is different for each class of kinase[4,5] and may also be different for each site present in a R monomer.[6] Phosphorylation of the single, readily exchangeable site on type II R enhances binding of cAMP to R[7] and slows the rate of reassociation of R and C subunits when cAMP levels are lowered.[8] In this chapter we describe techniques to measure the reversible phosphorylation of the exchangeable site on type II R and distinguish between intramolecular and intermolecular phosphorylation of this site.

Preparation of Protein Kinase

Type II cAMP-dependent protein kinase, the predominant form of cAMP-dependent PK in bovine cardiac muscle, was purified from this tissue by the method of Rubin *et al.* and had a specific activity of 800 nmol

---

[1] D. B. Glass and E. G. Krebs, *Annu. Rev. Pharmacol. Toxicol.* **20**, 363 (1980).
[2] J. Erlichman, R. Rosenfeld, and O. M. Rosen, *J. Biol. Chem.* **249**, 5000 (1974).
[3] R. L. Geahlen and E. G. Krebs, *J. Biol. Chem.* **255**, 1164 (1980).
[4] R. Rangel-Aldao and O. M. Rosen, *J. Biol. Chem.* **251**, 7526 (1976).
[5] R. L. Geahlen and E. G. Krebs, *J. Biol. Chem.* **255**, 9375 (1980).
[6] D. A. Flockhart, D. M. Watterson, and J. D. Corbin, *J. Biol. Chem.* **255**, 4435 (1980).
[7] F. Hofmann, J. A. Beavo, P. J. Bechtel, and E. G. Krebs, *J. Biol. Chem.* **250**, 7795 (1975).
[8] R. Rangel-Aldao and O. M. Rosen, *J. Biol. Chem.* **251**, 3375 (1976).

^{32}P transferred to protamine /min/mg protein.[9] R and C subunits were prepared by dissociation of the holoenzyme in the presence of cGMP followed by chromatography on DEAE-cellulose[9] or ω-aminohexyl-agarose.[8] Purified R and C subunits from heart can also be prepared by a combination of chromatography on DEAE-cellulose and affinity chromatography on 8-(6-aminohexyl)amino-cAMP-Sepharose 4B.[10] C subunits are then isolated from the material that does not adsorb to the affinity resin by chromatography on CM-Sephadex.[11] The kinase holoenzyme and its subunits prepared by these methods are greater than 90% pure when assessed by sodium dodecyl sulfate–polyacrylamide gel electrophoresis.

Preparation of Protein Kinase Inhibitor

Protein kinase inhibitor was purified according to the method described by Walsh et al.[12] Its specific activity was 650 units/mg of protein. One unit is defined as the amount needed to inhibit the incorporation of 1 pmol of ^{32}P into histone when 0.08 μg of purified protein kinase was incubated for 15 min at 37° in 200 μl containing 50 mM potassium phosphate buffer pH 7.1, 10 mM MgSO$_4$, 50 μg of histone IIA, 0.5 mg of bovine serum albumin, and 2 μM cAMP.

Phosphorylation of Protein Kinase Holoenzyme

$$R_2 \cdot C_2 + 2ATP \rightarrow [R\text{-}^P]_2 \cdot C_2 + 2ADP$$

Purified protein kinase (10–100 μg) is incubated for 15 min at 23° in 100 μl of 50 mM potassium phosphate buffer, pH 7.0, containing 50 μM [γ-^{32}P]ATP (>1000 cpm/pmol), 10 mM MgSO$_4$, and 1 mM dithiothreitol. Upon completion of the reaction the amount of ^{32}P incorporated into the enzyme is determined by one of two methods: (1) An aliquot of the enzyme is precipitated with cold 10% trichloroacetic acid, the precipitate is collected on glass fiber filters, and assayed for radioactivity by liquid scintillation spectrometry. (2) An aliquot is transferred to another tube containing 200 μl of 50 mM potassium phosphate buffer (pH 7.1), containing 0.1 M EDTA and 2 μM [^{3}H]cAMP. The contents are poured onto Millipore filters, washed with 10 ml of the same buffer, dried, and assayed for ^{32}P and ^{3}H. Phosphorylation is expressed as moles ^{32}P incorporated per mole [^{3}H]cAMP bound at saturation. Unincorporated ^{32}P is removed

[9] C. S. Rubin, J. Erlichman, and O. M. Rosen, J. Biol. Chem. **247**, 36 (1972).
[10] W. Weber and H. Hilz, Eur. J. Biochem. **83**, 215 (1978).
[11] P. J. Bechtel, J. A. Beavo, and E. G. Krebs, J. Biol. Chem. **252**, 2691 (1977).
[12] D. A. Walsh, C. D. Ashby, C. Gonzalez, D. Calkins, E. H. Fischer, and E. G. Krebs, J. Biol. Chem. **246**, 1977 (1971).

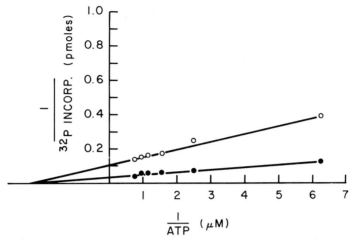

FIG. 1. Autophosphorylation at different concentrations of ATP. The concentrations of protein kinase were 71.2 pmol/ml (O—O) and 213.7 pmol/ml (●—●). The specific activity of the ATP was 2200 cpm/pmol. Reactions were carried out at 4° for 2 min in the absence of cAMP. From Rosen and Erlichman.[14]

from samples by dialysis or filtration on Sephadex G-25. Since only one of the two cAMP binding sites present on the R monomer is measured using this assay[13] the ratio of $^{32}P/^{3}H$ approaches one when unphosphorylated enzyme is fully labeled.[8]

With most preparations of enzyme, phosphorylation is 80% complete in less than 1 min and 100% complete after 10 min of incubation in the presence or absence of cAMP. The enzyme incorporates 1.0–1.9 mol of phosphate per mol of holoenzyme. It is likely that this variation in ^{32}P incorporation reflects endogenous phosphate present in the purified enzyme.[14] The rate of ^{32}P incorporation can be slowed by either lowering the temperature to 4° or decreasing the concentration of [γ-^{32}P]ATP.[14] The apparent K_m for ATP for autophosphorylation is 0.4 μM (Fig. 1) compared to 12 μM ATP when exogenous proteins are used as substrates.[9]

Phosphorylation of Isolated R Subunits

$$R_2 + 2ATP \xrightarrow[\text{cAMP}]{\text{C}} [R\text{-}P]_2 + 2ADP$$

The procedure for phosphorylating isolated regulatory subunits is identical to autophosphorylation of holoenzyme except that catalytic

[13] J. D. Corbin, P. H. Sugden, L. West, D. A. Flockhart, T. M. Lincoln, and D. McCarthy, *J. Biol. Chem.* **253**, 3997 (1978).
[14] O. M. Rosen and J. Erlichman, *J. Biol. Chem.* **250**, 7788 (1975).

amounts of C can be used to phosphorylate excess R subunits. The reaction requires cAMP to prevent reassociation of R and C. Heat-inactivated holoenzyme may be used in place of purified R subunits as the substrate.[14] Since heat-inactivated R is no longer capable of interacting with C,[6] cAMP is not needed if purified C is used instead of holoenzyme.

Distinction between Intramolecular and Intermolecular Autophosphorylation

There are a number of ways by which the phosphorylated form of the protein kinase holoenzyme could be generated: (1) intermolecular phosphorylation of R by the catalytic subunit in the presence of cAMP followed by reassociation of the subunits; and (2) intermolecular or intramolecular phosphorylation catalyzed by the intact holoenzyme. If phosphorylation occurs by an intermolecular reaction involving free C then conditions that inhibit catalytic subunit activity (e.g., addition of protein kinase inhibitor protein) should prevent phosphorylation of R. The initial velocity of the phosphorylation reaction should be proportional to the enzyme concentration if an intermolecular mechanism is involved and independent of enzyme concentration if the reaction is intramolecular. Finally, addition of catalytic subunit to the holoenzyme should affect the rate of phosphorylation of R if intermolecular phosphorylation is occurring.

Effect of Protein Kinase Inhibitors on Protein Kinase Activity

Holoenzyme is incubated in 50 μl of 50 mM potassium phosphate buffer, pH 7.1 at 0° with various concentrations of either protein kinase inhibitor protein, R subunits, or nonsubstrate peptides. The mixtures are then assayed at 37° for 15 min in the presence or absence of cAMP for (1) protein kinase activity using histone as substrate or (2) autophosphorylation. Typical results are shown in Figs. 2 and 3 and Table I. Addition of protein kinase inhibitor blocks histone phosphorylation in the presence or absence of cAMP but has no effect on autophosphorylation in the absence of cAMP (Fig. 2). Similarly, the addition of an inhibitor peptide such as Lys-Tyr-Thr has no effect on autophosphorylation in the absence of cAMP but blocks phosphorylation of R in the presence of cAMP (Fig. 3). R subunits in excess of C are phosphorylated only when cAMP is added to dissociate the holoenzyme (Table I).

Effect of Enzyme Concentration on the Rate of Autophosphorylation

Initial velocity of autophosphorylation is measured at several enzyme dilutions. Incubations are carried out in 50 μl of 50 mM potassium phos-

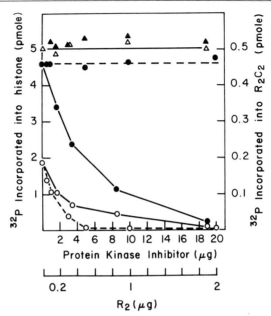

FIG. 2. Effect of cAMP-binding protein and protein kinase inhibitor on protein kinase activity. Holoenzyme (0.08 μg) was incubated in 20 μl of 50 mM potassium phosphate buffer, pH 7.1, for 5 min at 0° with either protein kinase inhibitor or cAMP-binding protein at the concentrations depicted on the abscissa. The mixtures were then assayed at 37° for 15 min for either (a) protein kinase activity using histone as substrate or (b) self-phosphorylation in the presence of 5 μM [γ-^{32}P]ATP (9000 cpm/pmol). For assays using histone as substrate [see (a), above] the symbols are (●—●) R_2C_2 plus protein kinase inhibitor plus 2 μM cAMP; (○—○) R_2C_2 plus protein kinase inhibitor; (●---●) R_2C_2 plus R_2 plus 2 μM cAMP; (○---○) R_2C_2 plus R_2. For self-phosphorylation assays [see (b), above] the symbols used are (△—△) R_2C_2 plus protein kinase inhibitor, (▲—▲) R_2C_2 plus R_2. From Rangel-Aldao and Rosen.[4]

phate buffer pH 7.1 at 4°. Lowering the temperature to 4° slows the reaction rate sufficiently to measure initial velocities when short incubation times are used (see Fig. 4). The velocity of autophosphorylation in the absence of cAMP is not affected by a 100-fold dilution of the enzyme (Fig. 4). Furthermore, when C subunits are added to excess R in the absence of cAMP, phosphorylation of R is stoichiometric with the amount of holoenzyme formed. In the presence of cAMP, maximal phosphorylation of R is independent of the concentration of C and all of the R subunits can be phosphorylated at each concentration of C (Fig. 5). When holoenzyme is used as a substrate for C the rate of autophosphorylation is unaffected by the addition of C subunits.[4] These studies indicate that the C subunit catalyzes the intramolecular autophosphorylation of its own R

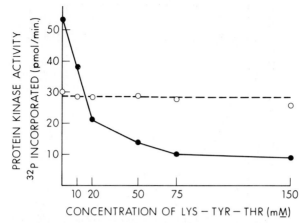

FIG. 3. Effect of Lys-Tyr-Thr on self-phosphorylation of protein kinase. Holoenzyme (2.85 pmol) was incubated at 4° for 1 min in 50 μl of 50 mM potassium phosphate buffer, pH 7.1, containing 10 mM MgSO$_4$, 0.5 μM [γ-^{32}P]ATP (13,180 cpm/pmol), 50 μg of bovine serum albumin, with (●—●) or without (○---○) 1 μM cAMP. The reaction was stopped by the addition of 200 μl of 0.1 M EDTA, pH 4.0 and the mixture filtered on Millipore filters. The filters were washed twice with 10 ml of the EDTA solution, dried, and assayed for ^{32}P in a solution of Omnifluor/toluene, 4 g/liter. The total amount of ^{32}P incorporated into protein kinase was 3.7 pmol. From Rosen et al.[16]

TABLE I
PHOSPHORYLATION OF cAMP-BINDING PROTEIN BY PROTEIN KINASE IN THE PRESENCE AND ABSENCE OF cAMP[a,b]

R$_2$C$_2$ (pmol)	R$_2$ (pmol)	cAMP	[^{32}P]Phosphate incorporated (pmol)
5.70	—	—	6.40
5.70	—	+	6.40
5.70	15	—	7.03
5.70	15	+	20.17

[a] Holoenzyme was incubated in 25 μl of 50 mM potassium phosphate buffer, pH 7.1, with or without cAMP-binding protein. The mixtures were then assayed for self-phosphorylation in the presence of 0.2 mM [γ-^{32}P]ATP (1000 cpm/pmol) and, where indicated, 8 μM cAMP. Incubations were for 10 min at 37° in a final volume of 50 μl. R$_2$C$_2$, holoenzyme; R$_2$, cAMP-binding protein.
[b] From Rangel-Aldao and Rosen.[4]

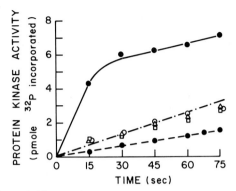

FIG. 4. Effect of protein kinase concentration on self-phosphorylation. Holoenzyme was phosphorylated at three different concentrations at 4° in 50 μl of 50 mM potassium phosphate, pH 7.1, containing 10 mM $MgSO_4$, 50 μg of bovine serum albumin, 5 μM [γ-^{32}P]ATP (30,000 cpm/pmol), and, where indicated, 1.0 μM cAMP. The ^{32}P incorporated into the holoenzyme at each concentration was multiplied by the dilution factor in order to obtain the values depicted on the ordinate. The symbols used are (O–·–O), 22.8 pmol of R_2C_2; (△–·–△), 1.14 pmol of R_2C_2; (□–·–□), 0.228 pmol of R_2C_2; (●—●), 22.8 pmol of R_2C_2 + cAMP; (●---●), 0.228 pmol of R_2C_2 + cAMP. From Rangel-Aldao and Rosen.[4]

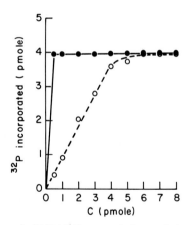

FIG. 5. Phosphorylation of cAMP-binding protein by catalytic subunit in the presence or absence of cAMP. cAMP-binding protein (4 pmol) was incubated at 4° for 1 min with the indicated amounts of catalytic subunit in 40 μl of 50 mM potassium phosphate buffer, pH 7.1, containing 10 mM $MgSO_4$, with (●—●) or without (O---O) 1.0 μM cAMP. The reaction was initiated by the addition of 100 pmol of [γ-^{32}P]ATP (21,000 cpm/pmol) and incubations were carried out at 4° for 20 min. From Rangel and Rosen.[4]

subunit when present in the intact holoenzyme but can also catalyze the intermolecular phosphorylation of R when the holoenzyme is dissociated by cAMP. The kinetic parameters of the intramolecular autophosphorylation are distinguishable from those characteristic of the intermolecular phosphorylation. The K_m for ATP is 30-fold lower. The velocity is significantly greater for the intramolecular phosphorylation. In the absence of cAMP, only those R subunits associated with C can serve as a substrate. In contrast, the intermolecular phosphorylation of R is analogous to phosphorylation of exogenous protein substrates such as histones or protamine. Phosphorylation is blocked by inhibitors of catalytic activity such as the protein kinase inhibitor. The rate of phosphorylation of excess R is stimulated by cAMP and proportional to the concentration of C. Since the total concentration of ATP in mammalian cells is greater than 1 mM and the kinase exists principally as a holoenzyme in the unstimulated cell, it is likely that the intramolecular reaction is responsible for maintaining the holoenzyme in the phosphoform.

Reversal of Autophosphorylation and Dephosphorylation

$$[R\text{-}^P]_2 \cdot C_2 + 2ADP \rightarrow R_2C_2 + 2ATP$$

$$[R\text{-}^P]_2 \cdot C_2 \xrightarrow{\text{phosphoprotein phosphatase}} R_2 \cdot C_2 + 2P_i$$

There are two mechanisms for dephosphorylating phospho-R. The first is reversal of the autophosphorylation reaction. This reaction is monitored by measuring the disappearance of radioactivity from ^{32}P-labeled protein kinase concomitant with the formation of charcoal-adsorbable radioactivity which is identified chromatographically as [^{32}P]ATP.[14]

The standard reaction mixture (0.1 ml) contains 50 mM Tris–maleate buffer, pH 5.5, 5 mM MgSO$_4$, 20 μM cAMP (where indicated), and ^{32}P-labeled protein kinase containing approximately 1 pmol of ^{32}P/pmol of enzyme. Reactions are generally incubated for 1 min at 23° or for 30 min in ice and are terminated by the rapid addition of 2.0 ml of cold, 0.2 mM Tris–HCl buffer, pH 7.5, followed by filtration on Millipore filters. The filters quantitatively retain the protein kinase. The filtrates (including a 3-ml wash with the Tris–HCl buffer) are collected into tubes containing 10 μl of 1.0 N HCl. Norit (0.3 ml of a 1/10 suspension in water) is added with shaking and then collected on Whatman glass fiber (GF/C) filters. Controls in which either MgSO$_4$ or ADP are omitted from the reaction mixture are always less than 10% of the experimentally determined values.

Millipore and glass fiber filters are counted by liquid scintillation spectrometry. The reaction is dependent upon the presence of active protein

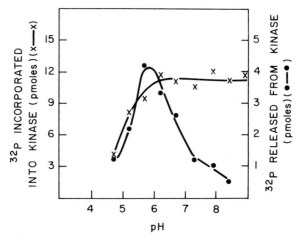

Fig. 6. Activity of the forward and reverse reactions at different pH values. The forward reaction (X—X) was carried out using 28 pmol of protein kinase and 4 μM ATP (1300 cpm/pmol). Incubations were for 1 min at 0°. The reverse reaction (●—●) was performed with 0.02 mM ADP, 5 mM Mg^{2+}, and 14 pmol of enzyme containing 0.74 pmol of ^{32}P/pmol of enzyme, 341 cpm/pmol of ^{32}P. Incubations were for 30 min at 0°. Both reactions were carried out in the absence of cAMP. Buffers were 0.05 M Tris–maleate, adjusted to give the indicated pH at 0°. From Rosen and Erlichman.[14]

kinase, Mg^{2+}, and ADP. Other nucleotides including GDP cannot substitute for ADP. Although the forward autophosphorylation reaction is optimal between pH 6 and 8.5, the reverse reaction has a sharp pH optimum at pH 5.5 (Fig. 6).

The rate of the reverse reaction, like the forward reaction, is extremely rapid and only linear (in the presence of cAMP) for 2 to 3 min at 23°. The rate is dramatically enhanced over a wide range of ADP concentrations by the addition of cAMP (Fig. 7).

The concentration of ADP required for half-maximal activation of the reverse reaction (in the presence of cAMP) is about 15 μM, only 35- to 40-fold greater than the K_m for ATP in the forward reaction. The concentration of ADP required to reverse the autophosphorylation reaction at more physiological pH values may, however, be substantially greater than that needed at pH 5.5.

The second mechanism, a phosphoprotein phosphatase reaction, is in keeping with the precedent established for other proteins whose biological activity is regulated by reversible phosphorylation. A phosphoprotein phosphatase was purified from bovine cardiac muscle by the procedure of Chou et al.[15] and used to study dephosphorylation of phospho-PK. Reac-

[15] C. K. Chou, J. Alfano, and O. M. Rosen, J. Biol. Chem. **252**, 2855 (1977).

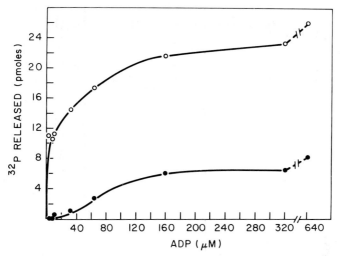

FIG. 7. Effect of ADP on the reverse reaction. Phosphorylated protein kinase (54 pmol containing 0.74 pmol of ^{32}P/pmol of enzyme, 85.2 cpm/pmol of ^{32}P) was incubated with 5 mM Mg^{2+} and the indicated concentration of ADP for 1 min at 23° in the presence (O—O) or absence (●—●) of 20 μM cAMP. From Rosen and Erlichman.[14]

tions are carried out in 50 μl of 50 mM Tris–HCl buffer pH 7.4 containing 5 mM dithiothreitol, 90 ng purified phosphoprotein phosphatase, and either ^{32}P-R_2 or ^{32}P-R_2C_2. As shown in Table II, the phosphatase is far more effective in removing ^{32}P from R_2 than from holoenzyme. Reassociation of PK with C appears to block accessibility of the phosphate group to the phosphatase. Other phosphoprotein phosphatases such as potato acid phosphatase[16] also prefer phospho-R to phospho R_2C_2 as a substrate for dephosphorylation.

Application of Autophosphorylation to the Study of Protein Kinase Structure and Regulation

Studies from several laboratories have used autophosphorylation as a probe to elucidate the structure and regulation of type II cAMP-dependent PKs. Early studies from this laboratory showed that the phosphoform of PK was more susceptible to net dissociation by cAMP than dephospho-PK.[2] Rangel and Rosen[8,17] extended those studies to show that the degree of phosphorylation of R markedly affects its ability to recombine with C to regenerate holoenzyme. The sequence of the autophos-

[16] O. M. Rosen, R. Rangel-Aldao, and J. Erlichman, *Curr. Top. Cell. Regul.* **12**, 39 (1977).
[17] R. Rangel-Aldao and O. M. Rosen, *J. Biol. Chem.* **252**, 7140 (1977).

TABLE II
EFFECT OF cAMP ON PHOSPHOPROTEIN PHOSPHATASE ACTIVITY ON THE
PHOSPHORYLATED FORMS OF PROTEIN KINASE AND cAMP-BINDING PROTEIN[a,b]

Substrate	cAMP (0.1 mM)	Phosphatase activity (pmol ^{32}P released/5 min)
1. Phosphoprotein kinase (3 μM)	−	2.7
	+	13.7
2. Phosphoprotein kinase (3 μM) + catalytic subunit (6.6 μM)	−	2.2
	+	12.7
3. Phospho-cAMP-binding protein (5.7 μM)	−	13.5
	+	14.0
4. Phospho-cAMP-binding protein (5.7 μM) + catalytic subunit (0.67 μM)	−	12.9
	+	16.6
5. Phospho-cAMP-binding protein (5.7 μM) + catalytic subunit (3.3 μM)	−	3.2
	+	15.5
6. Phospho-cAMP-binding protein (5.7 μM) + catalytic subunit (6.6 μM)	−	1.8
	+	13.8

[a] The reactions were performed with 90 ng of purified ethanol-treated phosphatase in a final volume of 25 μl. Incubations were for 5 min at 37°. Protein kinase and cAMP-binding protein contained 7.8 nmol ^{32}P/mg protein and 12.1 nmol ^{32}P/mg protein, respectively. The specific activity of the ^{32}P-labeled substrates was 965 cpm/pmol ^{32}P.
[b] From Chou et al.[15]

phorylation site in bovine heart PK has been determined by Krebs and Beavo[18] and is homologous to the sequence present in the cAMP binding fragment in porcine skeletal muscle.[19] Studies by Flockhart et al.[6] have shown that this autophosphorylation site is located within the catalytic subunit binding domain. Many investigators have used autophosphorylation of type II PKs to covalently radioactively label the enzyme so that the subunits can be easily identified during various procedures such as immunoprecipitation and polyacrylamide gel electrophoresis.[20]

[18] E. G. Krebs and J. A. Beavo, Annu. Rev. Biochem. **48**, 923 (1979).
[19] R. L. Potter and S. S. Taylor, J. Biol. Chem. **254**, 9000 (1979).
[20] C. S. Rubin, R. Rangel-Aldao, D. Sarkar, J. Erlichman, and N. Fleischer, J. Biol. Chem. **254**, 3797 (1979).

[19] Use of Immunological Approaches to Identify a Brain Protein Kinase Isozyme

By DWIJEN SARKAR, JACK ERLICHMAN, NORMAN FLEISCHER, and CHARLES S. RUBIN

Type II cAMP-dependent protein kinases (PKs) from many mammalian tissues appear to be very closely related on the basis of their physicochemical and kinetic properties, subunit compositions, and mode of activation by cAMP.[1] For example, homogeneous preparations of the prototypic heart and skeletal muscle type II PKs are virtually identical in size, shape, ion-exchange properties, and isoelectric point. Furthermore, both enzymes contain regulatory (R) subunits with an apparent M_r of 55,000 that are subject to intramolecular phosphorylation.[1-3] Although type II PKs from other tissues have not been studied in similar detail, the available data suggest that their properties parallel those of the prototypic enzymes.

In the absence of comparative amino acid sequence data and detailed information on the three-dimensional structures of various type II PKs, the possible occurrence of multiple PK II species, exhibiting subtle but significant differences in structure and function, could not be excluded. The immunological techniques of indirect immunoprecipitation and competitive binding immunoassay can be combined with functional assays (e.g., cAMP binding measurements, autophosphorylation of R II, photoaffinity labeling of R II) to address the question of whether there are subclasses of type II PKs. Antibodies raised against bovine cerebral cortex R II have been employed as highly specific and sensitive probes for identifying and characterizing two tissue-specific subclasses of type II PKs.

Principles

Specific, high-titer anti-R II serum can be used in conjunction with experimental samples containing known concentrations of R II subunits to construct immunoprecipitation and competitive binding assays that directly assess similarities and differences among PKs.[4] The content of R II subunits in experimental samples can be directly determined by per-

---

[1] H. G. Nimmo and P. Cohen, *Adv. Cyclic Nucleotide Res.* **8**, 145 (1977).
[2] O. M. Rosen, J. Erlichman, and C. S. Rubin, *Adv. Cyclic Nucleotide Res.* **5**, 253 (1975).
[3] F. Hofmann, J. A. Beavo, P. J. Bechtel, and E. G. Krebs, *J. Biol. Chem.* **250**, 7795 (1975).
[4] J. Erlichman, D. Sarkar, N. Fleischer, and C. S. Rubin, *J. Biol. Chem.* **255**, 8179 (1980).

forming equilibrium cAMP binding assays. To ensure accurate measurements (1) high concentrations of [^{3}H]cAMP (5 to 10 μM) are used and (2) the results obtained in standard cyclic nucleotide binding determinations[5] are compared to the data obtained in the presence of excess exogenous catalytic subunit (C) and MgATP. The latter conditions facilitate the displacement of tightly bound cAMP from R II and permit the estimation of total R II content.[6] Measurements performed with C and MgATP are typically required for samples of purified R II eluted from cAMP-agarose affinity columns with high concentrations of cAMP and samples obtained from cells or tissues treated with hormones, cAMP analogs, and/or cAMP phosphodiesterase inhibitors. Normal tissue samples generally contain negligible amounts of free R II-cAMP complexes.[6] It should be noted that the standard cAMP binding assay only measures one of the two cAMP-binding sites on R II[7,8] but this has no effect on the analyses and interpretations described below. If the quantification of the absolute number of cAMP binding sites on R II is desired the assay described by Corbin et al.[7] or equilibrium analysis[7,8] may be employed.

Once the concentration of R II subunits has been established several types of immunochemical analyses can be employed to evaluate whether two or more protein kinases are very closely related (i.e., are immunologically indistinguishable or distinct). (1) Equal amounts of holoenzymes are subjected to indirect immunoprecipitation by varying amounts of anti-R II serum, and phosphotransferase and cAMP-binding activities are determined in the supernatant fractions. (2) Equal concentrations of protein kinases are first dissociated with a saturating concentration of [^{3}H]cAMP and then immunoprecipitated with varying amounts of anti-R II serum. [^{3}H]cAMP-binding activity is measured in both supernatant and pellet fractions. (3) Competitive binding radioimmunoassays are carried out using homogeneous ^{125}I-labeled brain R II and anti-brain R II serum. In this version of radioimmunoassay the relative potencies of specific type II protein kinases in displacing ^{125}I-labeled R II from binding sites on the antibodies are determined. Graphical representations of the data obtained by these methods allow for a qualitative assessment of relatedness of PKs by inspection. The concentrations of antibody required to precipitate 50% of a fixed amount of PK in (1) and (2) and the concentration of PK needed to displace 50% of ^{125}I-R II in (3) provide quantitative parameters proportional to the average affinity of the antibodies for PK II. Large differences

[5] C. S. Rubin, J. Erlichman, and O. M. Rosen, this series, Vol. 38, p. 308.
[6] F. Hofmann, P. J. Bechtel, and E. G. Krebs, *J. Biol. Chem.* **252**, 1441 (1977).
[7] J. D. Corbin, P. H. Sugden, L. West, D. A. Flockhart, T. M. Lincoln, and D. McCarthy, *J. Biol. Chem.* **253**, 3997 (1978).
[8] S. E. Builder, J. A. Beavo, and E. G. Krebs, *J. Biol. Chem.* **255**, 2350 (1980).

in affinities imply significant differences in primary and/or higher order structure between different PKs.

Materials

Fresh bovine tissues were obtained from a local slaughterhouse and transported to the laboratory on ice.

Goat anti-rabbit IgG was purchased from Research Products International. Carrier-free Na^{125}I (17 Ci/mg) was obtained from Amersham/Searle. [^{3}H]cAMP (34 Ci/mmol) and [γ-^{32}P]ATP (25 Ci/mmol) were obtained from New England Nuclear. DEAE-cellulose (DE-52) was obtained from Whatman; Sephadex G-150 and CNBr-activated Sepharose 4B were purchased from Pharmacia; bovine serum albumin (Fraction V) was from Pentex Inc.; cAMP was obtained from Sigma.

Preparation of Partially Purified Type II Protein Kinases

Cytosolic type II PKs were partially purified by ion-exchange chromatography on DEAE-cellulose as previously described[4] to eliminate as much as possible the type I protein kinases and proteins that interfered with the accurate determination of catalytic and cAMP binding activities. It could not be ruled out, however, that slight contamination by type I free R could occur, particularly in tissues, such as rat heart, which contain mainly this form.

Preparation of R Subunits from Brain Protein Kinase II

R subunits were purified from cerebral cortex protein kinase II by a combination of ion-exchange chromatography on DEAE-cellulose and affinity chromatography on 8-(6-aminohexyl)amino-cAMP-Sepharose 4B using the procedure of Weber and Hilz.[9] The subunit preparation appeared to be nearly homogeneous as judged by the coincidence of Coomassie blue-stained and ^{32}P-labeled bands of M_r 55,000 after photoaffinity labeling with 8-azido-[^{32}P]cAMP, electrophoresis on sodium dodecyl sulfate–polyacrylamide gels, and autoradiography.[10] Aliquots of R II preparations [1.5 mg of protein/ml in 10 mM potassium phosphate buffer, pH 6.8, containing 4 mM EDTA, 1 mM 2-mercaptoethanol, and 25 mM benzamidine (buffer A)] were stored in liquid N$_2$ without loss of binding or antigenic activities.

[9] W. Weber and H. Hilz, *Eur. J. Biochem.* **83**, 215 (1978).
[10] C. S. Rubin, R. Rangel-Aldao, D. Sarkar, J. Erlichman, and N. Fleischer, *J. Biol. Chem.* **254**, 3797 (1979).

Protein Kinase and cAMP-Binding Assays

cAMP-dependent protein kinase activities were determined according to the method of Rubin et al.[5] cAMP-binding assays were performed by a modification[5] of the method of Gilman.[11]

Preparation of Antisera

Antisera to highly purified brain R II were prepared in New Zealand white rabbits. R II (1 mg in 0.67 ml of buffer A) was emulsified with an equal volume of Freund's complete adjuvant. Two aliquots (0.1 ml) of the emulsion were injected into two footpads, and the remainder was injected intradermally at multiple sites. The rabbits received three additional intradermal injections of R II (1 mg) in incomplete adjuvant at 10-day intervals. Seven days after the final injection anti-R II titers reached maximal levels which were maintained for 3 weeks. Serum collected 7 to 20 days after the last injection was used in all the experiments described below.

Indirect Immunoprecipitation of Protein Kinase

Partially purified type II protein kinases were incubated with varying amounts of anti-brain R II serum in 100 μl of buffer A. Preimmune rabbit serum (0 to 10 μl) was added to maintain a constant volume of serum in all assays. Incubations were performed for 2 to 15 hr at 4°. Binding was maximal at 2 hr and unaltered for a period of 24 hr. At the completion of the first incubation goat anti-rabbit IgG (5 μl/μl of rabbit serum) was added, and the incubation was continued for 2 hr. Immunoprecipitates were pelleted by sedimentation at 8000 g, and the supernatant solutions were assayed for cAMP-binding and protein kinase activities.

Preparation of ^{125}I-R II

Purified bovine brain R II was iodinated according to the procedure of Fraker and Speck.[12] R II (10 μg) was mixed with 0.5 mCi of Na^{125}I (carrier free) in 40 μl of 0.25 M potassium phosphate buffer, pH 7.4 and then transferred to a glass tube which had been coated with 4 μg of the water-insoluble oxidizing agent 1,3,4,6-tetrachloro-3α,6α-diphenyl glycoluril (Pierce Chemical Co.) After 10 min at 25° the reaction was terminated by removing the contents from the reaction tube. ^{125}I-R II was separated from unreacted ^{125}I by gel filtration on a column (1.2 × 10 cm) of Sephadex G-150 which was equilibrated and eluted with 50 mM potassium

[11] A. G. Gilman, *Proc. Natl. Acad. Sci. U.S.A.* **67**, 305 (1970).
[12] F. J. Fraker and J. C. Speck, *Biochem. Biophys. Res. Commun.* **80**, 849 (1978).

phosphate buffer, pH 7.4, containing 0.15 M NaCl (buffer B). Fractions containing ^{125}I-R II were pooled, supplemented with 1% albumin, and stored frozen at $-20°$ for up to 4 weeks. The specific activity of ^{125}I-R II was 17 μCi/μg of protein.

Competitive Displacement Radioimmunoassays

Competitive binding radioimmunoassays for cAMP-dependent protein kinases were carried out by a modification of the procedure of Fleischer *et al.*[13] Incubations were performed in polystyrene tubes (12 × 75 mm, Sarstedt). Anti-brain R II serum was diluted 1:100,000 in buffer B containing 1% bovine serum albumin, and aliquots (0.1 ml) corresponding to 0.1 nl of serum were added to 200 μl of buffer B containing 3 mM EDTA, 1% normal rabbit serum, 0.44 ng of ^{125}I-R II (15,000 dpm), and varying amounts of type II cAMP-dependent protein kinases. After incubating the samples for 16 hr at 4° goat anti-rabbit IgG (10 μl) was added, and the incubations were continued for 2 hr at 4°. Samples were then mixed with 1 ml of buffer B and sedimented for 10 min at 4000 g at 4°. The supernatant solutions were removed by aspiration and radioactivity in the immunoprecipitates was measured in a Packard autogamma spectrometer. Results are expressed as percentage of added ^{125}I-R II bound to anti-R II IgG. Under these conditions, 35% of the tracer was bound by the antibodies in the absence of competing, nonradioactive ligand.

Indirect Immunoprecipitation of Brain and Heart Type II Protein Kinase

The relative reactivities of type II protein kinases from brain and heart with anti-brain R II immunoglobulins were examined in an indirect immunoprecipitation assay using identical enzyme concentrations (Fig. 1). Although the serum was capable of binding both enzymes, the antibodies clearly recognized the neural kinase with higher affinity. One microliter of serum complexed more than 60% of the cerebral cortex enzyme but could precipitate only 10% of the heart kinase. The same results were obtained whether cAMP-binding activity (Fig. 1A) or phosphotransferase activity (Fig. 1B) was measured.

Since the antibodies were prepared against dissociated brain R II it was possible that the differences observed were related to interactions between catalytic and R subunits in the holoenzyme. To address this point, samples containing 0.8 pmol of either brain or heart protein kinase II were dissociated with a saturating concentration of [^{3}H]cAMP and then

---

[13] N. Fleischer, O. M. Rosen, and M. Reichlin, *Proc. Natl. Acad. Sci. U.S.A.* **73**, 54 (1976).

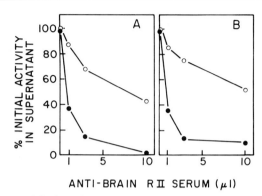

FIG. 1. Immunoprecipitation of cAMP-binding and catalytic activities of type II protein kinases by anti-brain R II serum. Brain protein kinase (50 pmol) and heart protein kinase (52 pmol) were incubated with the indicated amounts of rabbit anti-brain R II serum for 16 hr and subsequently precipitated by the addition of anti-rabbit IgG serum. Immunoprecipitates were isolated by centrifugation, and aliquots of the supernatant were assayed for cAMP-binding activity (A) and phosphotransferase activity (B). Data are presented as a percentage of original activity remaining in the supernatant fraction. (●) Cerebral cortex cytosolic kinase; (○) heart kinase. (From Erlichman et al.[4] reprinted with permission.)

immunoprecipitated with anti-brain R II serum (see the table). Fifty percent of the cAMP-binding activity from brain PK II was complexed by 25 nl of antiserum, whereas 20-fold more serum was required to precipitate a similar amount of heart R II. Catalytic activity was not precipitated in these assays.

Thus, indirect immunoprecipitation experiments readily detect antigenic differences between type II PKs from cerebral cortex and heart, thereby establishing the utility and sensitivity of immunological analysis in the characterization of subclasses of PKs.

Identification of Two Subclasses of Type II Protein Kinases via Radioimmunoassay Analysis

When a specific, high-titer antiserum is available, competitive-displacement radioimmunoassay analysis provides a powerful and sensitive technique for distinguishing structural differences among related proteins.[14] In competitive-binding immunoassays, type II protein kinases from liver, skeletal muscle, heart, and kidney exhibited similar displacement profiles and were less than 5% as potent as the cerebral cortex

---

[14] A. Nisonoff, M. Reichlin, and E. Margoliash, *J. Biol. Chem.* **245**, 940 (1970).

INDIRECT IMMUNOPRECIPITATION OF cAMP-BINDING ACTIVITY
BY ANTI-BRAIN R II SERUM[a]

Anti-brain R II serum (μl)	cAMP-binding activity precipitated (%)	
	Brain kinase II	Heart kinase II
0	0	0
0.01	14	1
0.025	50	5
0.05	66	15
0.10	73	23
0.25	84	37
0.50	88	58
1.00	97	79
2.50	99	90

[a] Type II protein kinases (0.8 pmol) from brain and heart were incubated with a saturating concentration of [^{3}H]cAMP. The indicated amounts of anti-brain R II serum were then added, and indirect immunoprecipitations were carried out. The amounts of [^{3}H]cAMP bound to R II in aliquots of the supernatant solutions were determined as described above. The binding of [^{3}H]cAMP to immunoprecipitated R II was quantitated by resuspending the precipitate in 0.5 ml of buffer A (0°), collecting and washing (10 ml of buffer A) the precipitate on glass-fiber filters (Whatman GF/C), and determining the radioactivity on the dried filters. The two methods of determining the proportion of cAMP-binding activity precipitated yielded virtually identical results.

enzyme (Fig. 2). The nonneural kinases apparently constitute a highly related subset of type II enzymes that are immunologically distinct from cerebral cortex protein kinase II.

To determine whether the cerebral cortex enzyme was unique or closely related to other kinases, type II protein kinases from several regions of the brain and from neuroendocrine tissues were compared with the cerebral cortex enzyme in additional radioimmunoassays. Type II protein kinases from bovine cerebellum, midbrain, medulla, and brainstem as well as from pineal gland, posterior pituitary, anterior pituitary, and adrenal medulla generated families of coincident competitive-displacement curves that were indistinguishable from the curve obtained with cerebral cortex protein kinase II. Thus, neural and neuroendocrine cells contain a second, distinct subset of neural-specific type II protein kinases (see ref. 4).

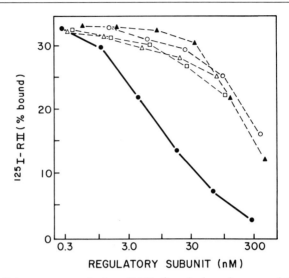

FIG. 2. Radioimmunoassays for type II PKs from brain and nonneural tissues. Type II enzymes were prepared from brain (●), muscle (○), heart (△), kidney (□), and liver (▲). (From Erlichman et al.[4] reprinted with permission.)

Further Applications of Immunological Analyses in the Characterization of PK II Isozymes

Competitive binding and immunoprecipitation assays have also been adapted to (1) study the conservation of type II PK subclasses in several mammalian species, (2) identify multiple forms of R II in cerebral cortex, and (3) monitor the expression of neural-specific and nonneural PKs in human tumors.[15] Representative experiments are summarized below.

Conservation of Subclasses in Several Species

To determine whether the preceding findings might have more general implications for the structural properties of neural protein kinases in other mammals, type II protein kinases were isolated from the hearts and brains of rats and hamsters and then compared with bovine neural and nonneural type II kinases in a competitive-binding assay (Fig. 3). The neural type II enzymes from rat and hamster were considerably more immunoreactive in the assay than any of the nonneural enzymes, including the homologous bovine muscle protein kinase II, and approached to within 25% of the potency of bovine brain R II. These preliminary data suggest that the

[15] C. S. Rubin, N. Fleischer, D. Sarkar, and J. Erlichman, Cold Spring Harbor Conf. Cell Proliferation **8**, 1333 (1981).

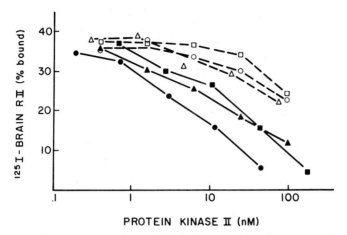

FIG. 3. Radioimmunoassays of type II PKs from neural and nonneural tissues from several species. Competitive binding assays were performed using type II kinases from bovine brain (●), and skeletal muscle (○), rat brain (▲), and heart (△), and hamster brain (■) and heart (□). (From Rubin et al.[15] reprinted with permission.)

disparate antigenic properties of the two type II subclasses may be conserved in a number of species.

Preliminary Evidence for Multiple Species of R II in Cerebral Cortex

Type II protein kinases from bovine cerebral cortex and skeletal muscle were preincubated in 25 mM 2-(N-morpholino)ethanesulfonic acid buffer (pH 6.8) in the presence or absence of 10 mM MgCl$_2$ and 0.5 mM ATP for 15 min at 0°. Subsequently, the samples were photoaffinity-labeled with 8-N$_3$-[^{32}P]cAMP and immunoprecipitated with specific anti-R II serum as described by Rubin et al.[10] Immunoprecipitates were then analyzed by electrophoresis on 0.1% SDS–10% polyacrylamide gels and autoradiography as described previously.[10] The autophosphorylated nonneural PK migrated as a monodisperse species with a molecular weight of 57,000 on SDS–polyacrylamide gels. Dephosphorylated R II subunits from both neural-specific and nonneural kinases exhibited molecular weights of 55,000. However, complete conversion from the 55,000 form to the 57,000 form upon autophosphorylation was only a characteristic of the nonneural enzymes.

In contrast, autophosphorylation of R subunits in neural-specific protein kinases consistently resulted in the appearance of two R II polypeptides with molecular weights of 57,000 and 55,000.[10,15] Both components of the 57,000/55,000 doublet are cAMP-binding proteins by virtue of their

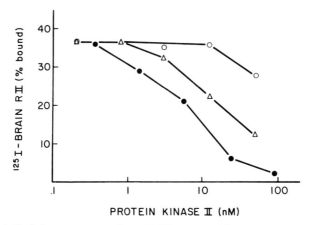

FIG. 4. Radioimmunoassays of type II PKs prepared from a human pheochromocytoma and a human adrenal cortical adenoma. Competitive-displacement radioimmunoassays were performed using type II kinases from bovine cerebral cortex (●), human pheochromocytoma (△), and human adrenal cortical adenoma (○). (From Rubin et al.[15] reprinted with permission.)

labeling with 8-N_3-[^{32}P]cAMP. Although the structural basis for the autophosphorylation-dependent alteration in apparent molecular weight is not known, the results raise the possibility of multiple R II subunits in neural-specific type II kinases.

Expression of PK II Isozymes in Human Tumors

The potential applicability of correlating the expression of neural-specific protein kinase II with the biological and pathological characteristics of human neuroendocrine tumors has been examined. Cytosolic protein kinase II derived from an adrenal-medullary tumor (pheochromocytoma) displayed a high level of immunoreactivity in a competitive-binding immunoassay constructed with anti-brain R II serum and using bovine brain protein kinase II as a standard (Fig. 4). The properties of the kinase are therefore consistent with the neuroendocrine origin of the malignancy. The immunoassay also correctly identified the type II protein kinase from a tumor of the adrenal cortex as a weakly reactive, nonneural enzyme.

[20] Using Mutants to Study cAMP-Dependent Protein Kinase

By MICHAEL GOTTESMAN

The genetic analysis of cultured cells which respond to cyclic AMP makes it possible to define the steps involved in cAMP action in intact cells. Several cell lines are available in which cAMP effects are both profound and easily quantified, and many of these have proved to be appropriate for genetic analysis. The genetic analysis of cAMP effects in cultured cells was pioneered by Tomkins and Coffino and their co-workers, who showed that viable mutants of mouse S49 lymphoma cells with altered cAMP-dependent protein kinases could be isolated and analyzed biochemically.[1] Examples of other cell lines studied genetically include the mouse macrophage line J774.2 which responds to cAMP by increased phagocytosis and growth inhibition[2]; the mouse Cloudman melanoma line in which cAMP stimulates melanin production[3]; neuroblastoma lines which respond to cAMP by shape change and neurite outgrowth[4]; and mouse adrenal Y-1 cells in which steroidogenesis is under cAMP control.[5]

We have been studying the response to cAMP of the Chinese hamster ovary (CHO) fibroblast. The advantages of this cell line for genetic analysis include the ease with which stable mutants can be obtained,[6] its rapid doubling time ($t_{1/2}$ = 12 hr) and ease of cloning in monolayer culture or agar and its relatively simple and stable karyotype.[7] CHO cells respond to cAMP within minutes by cessation of cell surface blebbing activity and reduction in ruffling activity[8,9]; within 1 to 6 hr by increased cell adhesion, elongation, and alignment; and by stimulation of the activity of several enzymes including ornithine decarboxylase[10] and transglutaminase,[11] and

[1] V. Daniel, G. Litwack, and G. M. Tomkins, *Proc. Natl. Acad. Sci. U.S.A.* **70,** 76 (1973).
[2] N. Rosen, J. Piscitello, J. Schneck, R. J. Muschel, B. R. Bloom, and O. M. Rosen, *J. Cell. Physiol.* **98,** 125 (1979).
[3] J. M. Pawelek, *J. Cell. Physiol.* **98,** 619 (1979).
[4] R. Simantov and L. Sachs, *J. Biol. Chem.* **250,** 3236 (1975).
[5] N. S. Gutmann, P. A. Rae, and B. P. Schimmer, *J. Cell. Physiol.* **97,** 451 (1978).
[6] L. Siminovitch, *Cell* **7,** 1 (1976).
[7] L. L. Deaven and D. F. Peterson, *Chromosoma* **41,** 129 (1973).
[8] M. M. Gottesman, A. LeCam, M. Bukowski, and I. Pastan, *Somat. Cell Genet.* **6,** 45 (1980).
[9] A. W. Hsie and T. T. Puck, *Proc. Natl. Acad. Sci. U.S.A.* **68,** 358 (1971).
[10] U. Lichti and M. M. Gottesman, *J. Cell. Physiol.* **113,** 433 (1982).
[11] P. G. Milhaud, P. J. A. Davies, I. Pastan, and M. M. Gottesman, *Biochim. Biophys. Acta* **630,** 476 (1980).

within 12 hr by a reduction in growth rate[8,9] and decreased transport of glucose and amino acids (system A).[12] The precise mechanism of none of these phenomena is known, and their time courses are very different, suggesting that both direct and indirect effects of cAMP might be involved. We began to isolate cAMP-resistant mutants of CHO cells with the aim of genetically defining all of the steps involved in cAMP action and clarifying the role of cAMP-dependent protein kinase in each of these cAMP effects. To date, we have characterized eight mutants resistant to the growth inhibitory effects of cAMP and all of them have defects in one or both cAMP-dependent protein kinases. Despite a variety of different defects in cAMP-dependent protein kinase activity, these mutants are resistant to all of the cAMP effects outlined above, indicating that in CHO cells activation of cAMP-dependent protein kinase is an obligatory step for all known cAMP effects.

Growth and Handling of Chinese Hamster Ovary Cells

All of the CHO lines in use in our laboratory are derived from a CHO Pro⁻5 line of cells sent to us by L. Siminovitch.[13] This cell line was originally established in culture from the ovary of a Chinese hamster by Puck et al.[14] There are at least two major derivatives of this original cell line in current use: the CHO-K1 line (ATCC CCL 61), known as the "Denver" subline, and the CHO Pro⁻5 line we use in our studies, known as the "Toronto" subline. After receipt in our laboratory, the CHO Pro⁻5 cells were subcloned. One subclone, which we called 10001, was chosen because of its fibroblastic morphology and its sensitivity to cAMP-induced morphological change and growth inhibition. This subclone is the parental clone for all of the mutants isolated in this laboratory. Cells are routinely grown in monolayer culture in the alpha modification of minimal essential medium with Earle's salts (Flow Laboratories) containing 10% fetal calf serum (Associated Biomedical Systems, Inc.), 2 mM glutamine, penicillin (50 units/ml), and streptomycin (50 μg/ml). Cells should divide every 12 hr; if growth is slower inadequate medium or serum or contamination with a pathogen should be suspected. Sera are screened for ability to support clonal growth at high efficiency ($\geq$80%) at a concentration of 1%, lack of toxicity at 10% concentration, ability to support cAMP-induced morphological change, and ability to support growth of CHO cells in suspension. Suspension cultures when needed are maintained at 37° at from 1×10^4 to 1×10^6 cells/ml in a Gyrotory Water Bath Shaker (New

[12] A. LeCam, M. M. Gottesman, and I. Pastan, *J. Biol. Chem.* **255**, 8103 (1980).
[13] P. Stanley, V. Caillibot, and L. Siminovitch, *Cell* **6**, 121 (1975).
[14] T. T. Puck, S. J. Ciecuira, and A. Robinson, *J. Exp. Med.* **108**, 945 (1958).

Brunswick Scientific Co., Inc.) at 160 rpm in tightly capped 125-ml glass bottles pregassed with CO_2. Cells grown in inadequate serum will clump after 24 hr in suspension and growth of cells will not be logarithmic. Approximately one out of three tested sera is adequate by all of the above criteria.

The parental subclone 10001 was grown to a population of several million cells and cells were frozen in complete medium containing 10% glycerol and stored in the vapor phase of a liquid nitrogen tank in individual 1 ml Nunc cryostatic tubes (5 × 10^5 cells/ml). The parental line is routinely discarded and new cells defrosted after cells have been in culture for approximately 100 generations (2 months). This reduces the likelihood of genetic drift which would make the parental line less suitable as a control for the studies we perform on mutant response to cAMP. Individual mutant lines are frozen promptly after isolation, either in 10% glycerol or in 7% DMSO. Cells are frozen by wrapping them in insulating foam and freezing at $-70°$. We have found that survival of CHO cells frozen in 7% DMSO approaches 100%, but that if cells are allowed to defrost and the DMSO is not immediately diluted (as might happen if a freezer failed), all of the cells die, while some survival is possible for up to 24 hr at room temperature in 10% glycerol after defrosting.

Isolation of cAMP Resistant Mutants

Cyclic AMP analogs do not kill CHO cells or completely inhibit their growth. If 10^5 or more CHO cells are plated in the presence of a cAMP analog or an agent which increases intracellular cAMP levels on a 100-mm tissue culture dish the cells will grow to confluence. Under these conditions it is not possible to isolate clones which are relatively resistant to the growth inhibitory properties of the cAMP. We have found, however, that CHO cell growth in suspension or in semisolid medium is more inhibited by cAMP than growth in monolayer culture.

Three selective agents have been used to isolate cAMP resistant CHO cells: cholera toxin (Schwarz-Mann), which stimulates adenylate cyclase activity and raises intracellular cAMP levels; methylisobutylxanthine (Aldrich Chemical Co.), which inhibits cyclic nucleotide phosphodiesterase activity and also raises intracellular cAMP levels; and the poorly hydrolyzable cAMP derivative 8-Br-cAMP (Sigma). We have not used dibutyryl cAMP in these studies because our initial experiments indicated that our cAMP resistant mutants showed a morphological and growth inhibitory response to dibutyryl-cAMP which was not distinguishable from their response to sodium butyrate, indicating that CHO cells were able to generate biologically active butyrate from this cAMP analog.[8] CHO cells do respond by elevating cAMP levels when treated with prostaglandin E_1, but the elevation is too transient to affect growth of the cells. Isoproterenol is not an appropriate selective agent since CHO cells have few β-adrenergic receptors.[15] Our initial studies indicated that the same kinds of mutants affecting cAMP-dependent protein kinase were obtained regardless of the selective agent or agents, or whether cells were selected in semisolid medium or suspension. Therefore, the following protocol has been designed to reproducibly yield cAMP resistant CHO cells.

1. CHO cells (5×10^5) are plated on each of four 100-mm tissue culture dishes in the presence of 15 ml of complete medium. After growth overnight at 37°, medium is removed from three dishes and cells are mutagenized with ethylmethanesulfonate (EMS) or ultraviolet light as previously described.[8]

2. After mutagenesis, cell survival is determined by plating cells in nonselective liquid medium. Plate 100 and 1000 cells per 100-mm dish and count clones after 7 days. Colonies are scored after staining for 20 min with a 0.5% aqueous solution of methylene blue. The unmutagenized dish of cells serves as the control; survival after mutagenesis should be approximately 10–20% relative to untreated cells. Since it takes 1 week to determine cell survival by this technique, it will be necessary to determine precise mutagenesis conditions (concentration of agent and duration of treatment) before the actual selection is carried out.

3. After mutagenesis, cells are grown for 4 to 7 days to allow expression of cAMP resistant phenotypes before selection. Cells should be replated at lower density during this time period to allow logarithmic growth. At the end of this period, cells are placed in selective medium (see below) and are also tested for mutagenesis by quantitation of the appearance of ouabain-resistant clones. Cells (1×10^6) from control and mutagenized populations are plated in 100-mm tissue culture dishes containing 2 mM ouabain (Sigma) prepared by dilution from a 0.4 M stock solution of ouabain in DMSO. After 7–10 days growth in ouabain-containing medium, the mutagenized population of cells should show at least 5- to 10-fold more resistant clones than the control population.[16]

4. After the 4- to 7-day expression period, 5×10^5 cells are plated in agar containing 8-Br-cAMP as follows: A 5% stock solution of agar (Difco) in water is prepared by autoclaving 5 g of agar in 100 ml of water. Agar lots must be prescreened for toxicity to CHO cells; one out of two

[15] A. Levitzky, personal communication.
[16] M. S. Rabin and M. M. Gottesman, *Somat. Cell Genet.* **5**, 571 (1979).

lots we tested was found to be nontoxic. If nontoxic agar is not available, highly purified agarose (Indubiose, Fisher Scientific) may be used. Agar is allowed to cool no lower than 45° and 10 ml of 5% agar is mixed with 90 ml of complete medium at 45° (final agar concentration, 0.5%). 8-Br-cAMP (Sigma) is added to a final concentration of 1 mM from a stock solution of 0.1 M 8-Br-cAMP in Dulbecco's phosphate-buffered saline without Ca^{2+} or Mg^{2+}. Ten milliliters of this 0.5% agar is pipetted into a 100-mm tissue culture dish and allowed to harden for 10 min at room temperature. A suspension containing 5 × 10^5 cells in 1.5 ml of complete medium at 37% is mixed with 3 ml of 0.5% agar (45°), 8-Br-cAMP is added to a concentration of 1 mM, and this suspension of cells in agar is quickly pipetted on top of the 0.5% hardened agar layer.

5. After incubation for 10 days at 37°, cAMP resistant CHO clones will be apparent as large colonies seen against a background of very small, cAMP-sensitive colonies (frequency of appearance, 1 out of 10^5 to 10^6 plated cells). These colonies are picked by pipetting them from the soft agar and subcloned in monolayer culture in the presence of 1 mM 8-Br-cAMP at least twice to free them of contamination with cAMP-sensitive colonies. These are cloned again in nonselective medium and tested for resistance to 8-Br-cAMP. We pick colonies from monolayer culture by removing all culture medium and scraping the colony off the dish with a pipet and then aspirating the cells in a drop of medium.

Characterization of cAMP Resistant CHO Cells

To determine whether our cAMP resistant CHO cells had altered cAMP-dependent protein kinases or were defective in some other step in the mechanism of action of cAMP, we first characterized the activity of the cAMP-dependent protein kinases in crude extracts. Using histone as substrate, we found several mutants which appeared to have abnormal protein kinase activity and a few with normal cAMP-stimulated protein kinase activity. Our further analysis of these mutants, which is outlined below, revealed that of eight mutants analyzed biochemically, all seemed to have some alteration in cAMP-dependent protein kinase despite the fact that two of these had apparently normal activity in crude extracts. The steps used in this analysis, the results of which are summarized in Table I, were as follows:

1. Prepare crude extracts of CHO cells by Dounce homogenization of approximately 1 × 10^7 cells suspended in hypotonic buffer.[8] The 15,000 g supernatants of these extracts are analyzed for cAMP-dependent protein

TABLE I
Cyclic AMP Resistant CHO Mutants

Strain number	cAMP-dependent activity in extracts	Type I protein kinase	Type II protein kinase	8-Azido-[^{32}P]cAMP binding	Phosphorylation of 52K protein	Putative defect	Reference
10001	Normal	Normal	Normal	Normal	Present	Wild-type	a, b
10215(D)[h]	Shift in dose–response	Present	Absent	Normal	Absent	Altered catalytic subunit	c
10248(D)	Shift in dose–response	Present	Absent	Altered binding to RI	Absent	Altered Type I regulatory subunit	d, e
10987(D)	Normal	Present	Absent	Normal	Absent	Unknown	f
11348(D)	Normal	Present	Absent	Normal	Absent	Unknown	f
10226(D)	Shift in dose–response	Present	Absent	Normal	Not tested	Unknown	f
10223 (R)[h]	Decreased activity	Absent	Present	Normal (PKI peak missing)	Present	Unknown	b
10260(R)	Little or no activity	Absent	Very reduced	Normal (PKI peak missing)	Present	Reduced levels of catalytic subunit	b, g
10265(R)	Decreased activity	Absent	Reduced	Normal (PKI peak missing)	Present	Reduced levels of catalytic subunit	b

[a] M. M. Gottesman, A. LeCam, M. Bukowski, and I. Pastan, *Somat. Cell Genet.* **6**, 45 (1980).
[b] T. J. Singh, C. Roth, M. M. Gottesman, and I. Pastan, *J. Biol. Chem.* **256**, 926 (1981).
[c] D. Evain, M. M. Gottesman, I. Pastan, and W. B. Anderson, *J. Biol. Chem.* **254**, 6931 (1979).
[d] T. J. Singh, I. Pastan, and M. M. Gottesman, in preparation.
[e] M. M. Gottesman, T. Singh, A. LeCam, C. Roth, J. C. Nicolas, F. Cabral, and I. Pastan, *Cold Spring Harbor Conf. Cell Proliferation* **8**, 195 (1981).
[f] M. Leitschuh, C. Roth, and M. M. Gottesman, unpublished data.
[g] M. Murtaugh, A. Steiner, and P. Davies, *J. Cell Biol.* **95**, 64 (1982).
[h] The designations D or R refer to whether the mutants are recessive (R) or dominant (D) in somatic cell hybrids.

kinase activity by the method of Corbin and Reimann[17] using calf thymus histone as substrate. More than 95% of the total protein kinase activity is present in this supernatant fraction. Extracts are assayed at a variety of cAMP concentrations from 10^{-8} to 10^{-4} M. Three classes of mutants were obtained by this analysis: (1) those with apparently normal amounts of active enzyme and normal response to cAMP; (2) mutants with a shifted dose–response curve for cAMP but significant amounts of activity at high cAMP levels; and (3) mutants with little or no cAMP stimulated protein kinase activity at all cAMP concentrations tested.

2. Genetic analysis of these mutants involved fusion of mutant cells with a genetically marked derivative of the wild-type parent and determination of the cAMP resistance of the resulting hybrid.[8] Standard techniques were used for polyethylene glycol-mediated cell fusion and hybrid selection.[8] Both dominant and recessive mutants were obtained, as summarized in Table I. Of the three recessive mutants we have so far characterized, two (10265 and 10260) fall into the same complementation group.[18]

3. CHO cells contain the two usual cAMP-dependent protein kinases. In order to analyze each enzyme independently, it was necessary to separate Type I and Type II cAMP-dependent protein kinases by chromatography on DEAE cellulose. These enzyme activities are present in approximately equal amounts. Extracts for these columns were prepared from a minimum of 10^8 cells (approximately 20–30 mg of total cell protein). Column fractions eluted with a linear sodium chloride gradient were analyzed for the presence of cAMP stimulated histone kinase activity[18] and were photoaffinity-labeled with 8-azido-[^{32}P]cAMP to detect cAMP binding proteins.[19] Pooled peak fractions of Type I and Type II cAMP-dependent protein kinases were analyzed for activity as a function of cAMP dose in order to detect mutants in which one enzyme had shifted dose–response to cAMP. The two peaks of enzyme activity and a peak of binding activity which eluted between the two enzyme peaks on the DEAE columns were analyzed for 8-azido-[^{32}P]cAMP binding activity by polyacrylamide gel electrophoresis of the labeled samples.[18] These analyses, also summarized in Table I, indicated that some of our mutants lacked either/or both Type I and Type II cAMP-dependent protein kinase activity, and two had altered cAMP stimulation of the Type I enzyme. In one case (Mutant 10215), this defective activation by cAMP was attributed to an abnormal catalytic subunit as indicated by altered substrate

---

[17] J. D. Corbin and E. M. Reimann, in "Methods in Enzymology" (J. G. Hardman and B. W. O'Malley, eds.), p. 287. Academic Press, New York, 1976.
[18] T. J. Singh, C. Roth, M. M. Gottesman, and I. Pastan, *J. Biol. Chem.* **256,** 926 (1981).
[19] U. Walter, I. Uno, A. Y.-C. Liu, and P. Greengard, *J. Biol. Chem.* **252,** 6494 (1977).

specificity of this subunit,[20] and, in the second case, the problem appeared to reside in reduced affinity for cAMP binding of the Type I regulatory subunit.[21]

4. Because all of these analyses were based on activity of cAMP-dependent protein kinases using histone as substrate, it still seemed possible that some of our mutant protein kinases might retain histone kinase activity, but lose activity against specific cellular substrates. Accordingly, we analyzed cAMP-dependent phosphorylations of endogenous substrates both in crude homogenates and in intact cells.[22] Nonoverlapping sets of substrates were obtained by these two different techniques: in crude extracts we found prominent cAMP-dependent phosphorylation of proteins with molecular weights of 300,000, 34,000, 26,000, and 17,000; in intact cells the cAMP-dependent phosphorylation of a 52,000 molecular weight family of proteins was seen.[22] The phosphorylations observed in cell homogenates were reduced in all of the mutants tested, while the 52,000-dalton protein whose phosphorylation was observed in intact cells was not phosphorylated only in those mutants lacking Type II protein kinase (see Table I). This analysis allows us to conclude with some confidence that the two protein kinases serve different functions in intact CHO cells since the Type II kinase only seems to be required for phosphorylation of the 52,000 dalton protein. Furthermore, homogenization of cells seems to result in altered substrate specificity of the kinases, either because of changes in enzyme localization, substrate localization or structure, or the strikingly different conditions of assay. Recent studies have allowed us to identify other substrates for cAMP-dependent protein kinase in CHO cells including the cytoskeletal protein vimentin[23] and the *src* gene product of Rous sarcoma virus.[24]

Use of cAMP-Dependent Protein Kinase Mutants

The availability of multiple CHO mutants with altered cAMP-dependent protein kinases allows us to begin to dissect the complex biochemistry and substrate specificity of this multisubunit enzyme as outlined in the previous section. In addition, these mutants serve as a tool with which to

---

[20] D. Evain, M. M. Gottesman, I. Pastan, and W. B. Anderson, *J. Biol. Chem.* **254**, 6931 (1979).

[21] T. J. Singh, I. Pastan, and M. M. Gottesman, in preparation.

[22] A. LeCam, J. C. Nicolas, T. J. Singh, F. Cabral, I. Pastan, and M. M. Gottesman, *J. Biol. Chem.* **256**, 933 (1981).

[23] M. M. Gottesman, T. Singh, A. LeCam, C. Roth, J. C. Nicolas, F. Cabral, and I. Pastan, *Cold Spring Harbor Conf. Cell Proliferation* **8**, 195 (1981).

[24] C. Roth, N. Richert, I. Pastan, and M. M. Gottesman, in preparation.

test individual hypotheses concerning the possible role of cAMP-dependent protein kinases in the response of cells to cAMP and other drugs thought to act through a cAMP-dependent mechanism.

Table II summarizes our current knowledge concerning the responsiveness of wild-type CHO cells and various mutants to a variety of agents. All of the tested mutants appear to be resistant to every one of the cAMP stimulated responses we have tested. In contrast, the mutants are quite sensitive to other agents which inhibit cell growth such as butyrate and 8-Br-AMP, which have been postulated to act through cAMP-dependent mechanisms. Other drugs, such as the tumor promoter 12-O-tetradecanoylphorbol 13-acetate (TPA) and interferon, give somewhat reduced

TABLE II
RESPONSE OF CHO CYCLIC AMP-DEPENDENT PROTEIN KINASE MUTANTS TO VARIOUS AGENTS[a]

Treatment	Response of wild-type parental CHO cells	Response of mutants	References
1 mM 8-Br-cAMP, 100 ng/ml cholera toxin, or 0.5 mM methylisobutylxanthine	Loss of blebs	No response	b
	Elongation	No response	b
	Growth inhibition	No response	b
	Decreased V_{max} for glucose, amino acid transport	No response	c
	Stimulation of ornithine decarboxylase	No response	d
	Stimulation of transglutaminase	No response	e
1 mM sodium butyrate	Cell flattening	Normal	b
	Growth inhibition	Normal	b
1 mM 8-Br-AMP	Growth inhibition	Slight reduction	f
	Cell rounding	Normal	f
100 ng/ml TPA	Stimulation of ornithine decarboxylase activity	Some reduction	d
100 units/ml human β-interferon	Antiviral activity against VSV, EMC virus	Some reduction	g
2000 units/ml human β-interferon	Growth inhibition	Normal	g

[a] In general, at least mutants 10215 (no PKII), 10248 (no PKII), and 10260 (No PKI, very little PKII) have been tested.
[b] M. M. Gottesman, A. LeCam, M. Bukowski, and I. Pastan, *Somat. Cell Genet.* **6**, 45 (1980).
[c] A. LeCam, M. M. Gottesman, and I. Pastan, *J. Biol. Chem.* **255**, 8103 (1980).
[d] U. Lichti, and M. M. Gottesman, *J. Cell. Physiol.* **113**, 433 (1982).
[e] P. G. Milhaud, P. J. A. Davies, I. Pastan, and M. M. Gottesman, *Biochim. Biophys. Acta* **630**, 476 (1980).
[f] M. M. Gottesman, in preparation.
[g] D. Banerjee, K. Baksi, and M. M. Gottesman, in preparation.

responses in mutant cells, indicating that cAMP-dependent protein kinase may have a facilitating, but not obligatory role, in these responses.

Acknowledgments

I would like to thank my many scientific colleagues in the Laboratory of Molecular Biology for discussions and ideas which have contributed to this work, with special thanks to my collaborators Charles Roth, Fernando Cabral, Toolsee Singh, Jean-Claude Nicolas, Alphonse LeCam, Pierre Milhaud, and Ira Pastan, who provided unswerving support and encouragement. I am also grateful to Norma Pollekoff, who typed the manuscript.

[21] Protein Modulation of Cyclic Nucleotide-Dependent Protein Kinases

By GORDON M. WALTON and GORDON N. GILL

Principle

A poly(L-arginine) binding site has been identified on cGMP-dependent protein kinase and on the regulatory subunits of type I and type II cAMP-dependent protein kinases.[1,2] Interaction of poly(L-arginine) with this site results in a time-, temperature-, and concentration-dependent inactivation of cyclic nucleotide binding activity. The presence of cyclic nucleotide at saturation concentrations protects cyclic nucleotide binding sites against inactivation. Protection of binding activity by cyclic nucleotide against inactivation by various concentrations of poly(L-arginine) demonstrates noncompetitive kinetics, indicating the existence of separate binding sites for both ligands.

Poly(L-arginine) binding to cyclic nucleotide-dependent protein kinases affects kinase activity as well. The interaction of poly(L-arginine) with cGMP-dependent protein kinase results in a rapid activation of basal catalytic activity (assayed in the absence of cGMP) and a significant enhancement in the rate of autophosphorylation. Prolonged interaction demonstrates a subsequent irreversible loss of total kinase activity as assayed with exogenous substrates, while autophosphorylation of the enzyme remains fully functional. The interaction of poly(L-arginine) with type II cAMP-dependent protein kinase stimulates autophosphorylation

[1] G. M. Walton and G. N. Gill, *J. Biol. Chem.* **255**, 1603 (1980).
[2] G. M. Walton and G. N. Gill, *J. Biol. Chem.* **256**, 1681 (1981).

of the receptor subunit in the absence but not in the presence of cAMP.[3] In the presence of cAMP, dissociation of the catalytic subunit occurs, and is not further affected by interactions occurring on the regulatory subunit. Competitive inhibition of the catalytic subunit by poly(L-arginine) when assayed with exogenous substrates or by autophosphorylation has been observed.[4-6] However, the concentrations of poly(L-arginine) required for competitive inhibition of catalytic activity (0.76 μM) are an order of magnitude higher than the concentrations required for noncompetitive inhibition of cyclic nucleotide binding activity (8.8 nM).[2,4]

Histones are effective substrates for the cyclic nucleotide-dependent protein kinases. In addition to the phosphorylation sites, core histones (H2A, H2B, H3, and H4), but not histone H1, contain arginine-rich domains which compete with poly(L-arginine) for the poly(L-arginine) binding site on the enzyme. Interaction of core histones with the poly(L-arginine) binding site of cGMP-dependent protein kinase in the absence of cGMP results in activation of catalytic activity and inactivation of cGMP binding. Subsequent loss of total kinase activity assayed with or without cGMP occurs. The effects of core histones are time-, temperature-, and concentration-dependent and apparently irreversible.[1,2] Core histones inactivate cAMP binding to receptor subunits of type I and type II cAMP-dependent protein kinases.[2] Histones also induce subunit dissociation and activation of type I cAMP-dependent protein kinase which dissociates more readily than the type II isoenzyme.[7,8]

Modulator proteins containing acidic domains [poly(L-arginine) binding sites] prevent the interaction of poly(L-arginine) and core histones with cyclic nucleotide-dependent protein kinases.[1] Modulator proteins have not been observed to reverse the activation/inactivation process once completed. The interaction of modulator proteins with poly(L-arginine)-rich domains does not prevent the phosphorylation of histones, though very acidic polymers (e.g., poly(L-aspartate) and poly(L-glutamate)), which effectively prevent inactivation of cyclic nucleotide binding, are capable of inhibition of catalytic activity.[9] 2,3-Butanedione specifically modifies arginine residues in poly(L-arginine) and core histones to

---

[3] O. M. Rosen and J. Erlichman, *J. Biol. Chem.* **250**, 7788 (1975).
[4] J. G. Demaille, K. A. Peters, and E. H. Fischer, *Biochemistry* **16**, 3080 (1977).
[5] Y. S. Chiu and M. Tao, *J. Biol. Chem.* **253**, 7145 (1978).
[6] J. L. Bear, J. G. Zalitis, and A. G. MacKinlay, *Biochem. Biophys. Res. Commun.* **84**, 450 (1978).
[7] J. D. Corbin, S. L. Keely, and C. R. Park, *J. Biol. Chem.* **250**, 218 (1975).
[8] E. Miyamoto, G. L. Petzold, J. S. Harris, and P. Greengard, *Biochem. Biophys. Res. Commun.* **44**, 305 (1971).
[9] G. N. Gill, C. E. Monken, and G. M. Walton, *Cold Spring Harbor Conf. Cell Proliferation* **8**, 251 (1981).

prevent their interaction with the poly(L-arginine) binding site of the enzymes. However, modification of arginine residues, which are required for substrate site recognition, also inhibits phosphate incorporation.[1,10]

Materials for Assays

Potassium phosphate, monobasic and dibasic; dithiothreitol; magnesium chloride; bovine serum albumin (crystalline); poly(L-arginine) hydrochloride, type II-B, $M_r = 60,000$ (Sigma Chemical Co.); histone, type II-A (Sigma Chemical Co.); cAMP and cGMP; [^{3}H]cAMP and [^{3}H]cGMP (20–40 Ci/mmol); [γ-^{32}P]ATP (25 Ci/mmol); trichloroacetic acid; 2,5-diphenyloxazole; *p*-bis-2-(5-phenyloxazolyl)benzene; toluene; filter paper discs 3MM (Whatman Ltd.); and Millipore filters, 0.45 µm (Millipore Corp.).

Procedures

Purification of Proteins. cGMP-dependent protein kinase is purified from bovine lung by affinity chromatography as described elsewhere and in this series.[1,11] Preparation of cAMP-dependent protein kinase and/or receptor subunits of type I or type II have been described.[12-14] Individual histones are prepared as described by Oliver *et al.*,[15] modulator protein as described by Shoji *et al.*,[16] and troponin as described by Stull and Buss.[17]

Interaction of Poly(L-arginine) or Histone with Cyclic Nucleotide-Dependent Protein Kinase. Incubations are carried out in 0.5-ml conical polyethylene or polypropylene test tubes on an ice bath by combining 15 µl of H$_2$O, 5 µl of 0.3 *M* potassium phosphate, pH 6.8, containing 10 m*M* dithiothreitol, 5 µl of 50 m*M* magnesium chloride, 10 µl of 10 mg/ml of bovine serum albumin, and 5 µl of enzyme containing 2–80 ng protein. Interactions are initiated with the addition of 5 µl of 10–1000 µg/ml of poly(L-arginine) or histone and the reactions incubated in a 30° water bath. Interactions are terminated after 5–60 min with the addition of

---

[10] T. M. Lincoln and J. D. Corbin, *Proc. Natl. Acad. Sci. U.S.A.* **74**, 3239 (1977).
[11] P. J. Bechtel, J. A. Beavo, and E. G. Krebs, *J. Biol. Chem.* **252**, 2691 (1977).
[12] M. J. Zoller, A. R. Kerlavage, and S. S. Taylor, *J. Biol. Chem.* **254**, 2408 (1979).
[13] W. L. Dills, Jr., J. A. Beavo, P. J. Bechtel, and E. G. Krebs, *Biochem. Biophys. Res. Commun.* **62**, 70 (1975).
[14] F. Hofmann, J. A. Beavo, P. J. Bechtel, and E. G. Krebs, *J. Biol. Chem.* **250**, 7795 (1975).
[15] D. Oliver, K. R. Sommer, S. Panyim, S. Spiker, and R. Chalkley, *Biochem. J.* **129**, 349 (1972).
[16] M. Shoji, N. L. Brackett, J. Tse, R. Shapira, and J. F. Kuo, *J. Biol. Chem.* **253**, 3427 (1978).
[17] J. T. Stull and J. E. Buss, *J. Biol. Chem.* **252**, 851 (1977).

cAMP or cGMP or modulator protein and returned immediately to an ice bath; subsequently binding, kinase, or autophosphorylation activities are determined.

Cyclic Nucleotide Binding Activity. The initial interaction period is terminated with the addition of 10 µl of 2 µM [³H]cAMP or [³H]cGMP to reaction mixtures containing 40–80 ng of enzyme and H_2O to a total volume of 50 µl. Incubations are carried out for 30 min on an ice bath. Quantitation of binding sites is performed using Millipore filters to separate bound from free cyclic nucleotide.[18,19] Reaction mixtures are transferred to a filter reservoir containing 5 ml of ice-cold wash buffer (10 mM potassium phosphate, pH 6.8, and 5 mM magnesium chloride). Protein–ligand complex binds to the filter and free ligand is removed by filtering the solution under vacuum, followed by a thorough washing with cold wash buffer. Filters are dried under a heat lamp for 10 min and counted in 1–2 ml of scintillation fluid (4 g of 2,5-diphenyloxazole and 50 mg of *p*-bis-2-(5-phenyloxazolyl)benzene per liter of toluene. Background is determined in assays without enzyme. Binding of cyclic nucleotide to the kinases in plastic reaction tubes is enhanced 5–20% by poly(L-arginine), all five histones, troponin, and modulator protein. This nonspecific effect is independent of temperature and greater in glass tubes, where stimulation of binding was observed to be as great as 100%. To correct for the stimulation of binding when such proteins are present during the interaction and assay periods, controls are performed by adding the same concentrations of protein subsequent to the initial interaction period. Because the nonspecific enhancement is less in plastic tubes, these are used for all experiments.

Cyclic Nucleotide-Dependent Protein Kinase Activity. The initial interaction period is terminated with the addition of 5 µl of 2 mg/ml of troponin to the reaction mixture containing 2–20 ng of enzyme and immediately cooled in an ice bath. Five microliters of 10 µM cAMP or cGMP are added when required and water is added to a total volume of 45 µl. To initiate the assay, 5 µl of 1 mM [γ-³²P]ATP (300–600 cpm/pmol) is added and reactions are incubated at 30° for 5–10 min. The reactions are terminated by transferring a 40-µl aliquot to a paper disc and immersing in ice-cold 15% (w/v) trichloroacetic acid. After 10–15 min, three additional washes of 10–15 min are performed in 5% trichloroacetic acid. The first and third washes are at 0–5° while the second wash is heated to 90°. The filters are rinsed with acetone, dried, and counted in scintillation fluid as described above for Millipore filters. Background is determined with reactions performed in the absence of troponin.

[18] G. N. Gill and G. M. Walton, this series, Vol. 38, p. 376.
[19] G. N. Gill and G. M. Walton, *Adv. Cyclic Nucleotide Res.* **10**, 93 (1979).

Self-Phosphorylation of Cyclic Nucleotide-Dependent Protein Kinases. The initial interaction period is terminated by quickly transferring the reaction mixtures containing 40–80 ng of enzyme to an ice bath. Five microliters of 10 μM cAMP or cGMP are added when required and water is added to a total volume of 45 μl. The assay is initiated with 5 μl of 5 μM ATP (~20,000 cpm/pmol) and incubated for 5–10 min at 30°. The reaction is terminated by transferring a 40-μl aliquot to a paper disc and immersing in ice-cold 15% (w/v) trichloroacetic acid. Filters are processed and radioactivity is quantitated as described for the kinase assay. Background is determined in reactions performed in the absence of enzyme.

Comments

Determination of the fraction of total binding sites remaining (B/B_0) as a function of time of interaction results in a linear plot of log B/B_0 versus time where B and B_0 are binding sites observed at indicated time and zero time, respectively (Fig. 1). Pseudo-first-order rate constants for the inactivation of cyclic nucleotide binding by various concentrations of poly(L-arginine) and core histones are obtained. When saturation kinetics are observed, inhibitor dissociation constants (K_I) are determined. A K_I of 8.8 and 14.9 nM was calculated for poly(L-arginine) inactivation of binding to cGMP- and cAMP-dependent protein kinase, respectively.[2] Condi-

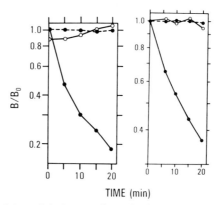

FIG. 1. Effect of poly(L-arginine) on cyclic nucleotide binding activity versus time. *Left:* cGMP-dependent protein kinase with 40 μg/ml of poly(L-arginine). (●—●), plus poly(L-arginine) at 30°; (●---●), plus poly(L-arginine) at 0°; (○—○), minus poly(L-arginine) during 30° interaction period, plus poly(L-arginine) during 0° binding assay. *Right:* Receptor subunit from type II cAMP-dependent protein kinase with 100 μg/ml of poly(L-arginine). (●—●), plus poly(L-arginine); (○—○), plus poly(L-arginine), plus 0.4 μM [³H]cAMP; (●---●), minus poly(L-arginine) during 30° interaction period, plus poly(L-arginine) during 0° binding assay. From G. M. Walton and G. N. Gill, *J. Biol. Chem.* **256**, 1681 (1981).

tions that favor enzyme instability, e.g., pH, absence of sulfhydryl reagent, low magnesium ion, potassium ion, or low protein concentrations can enhance the rate of the activation/inactivation process during the interaction period. A minimum concentration of 1–2 mg/ml of bovine serum albumin is required for stability of cGMP-dependent protein kinase under conditions described above. Higher concentrations of bovine serum albumin free of modulator proteins will not have a significant effect on the interaction with poly(L-arginine).

Because histones undergo substantial changes in secondary and tertiary structure in response to changes in ionic strength, significant changes occur in the rate and the kinetics of the activation/inactivation process. Low ionic strength conditions demonstrated saturation kinetics for inactivation of binding activity of cGMP-dependent protein kinase with each of the core histones and K_I values ranged from 0.65 μM for H2B to 2.6 μM for H4.[2] High ionic strength conditions demonstrated nonsaturation kinetics and the inactivation of cGMP binding increased linearly with increases in histone concentration. Ionic strength also affects the activation of cGMP-dependent protein kinase activity at low histone concentrations. Low ionic strength has been shown to favor activation of catalytic activity in the absence of cGMP. High ionic strength conditions inhibited activation of low histone levels and cGMP-dependent catalytic activity was maintained. High substrate levels of histone inhibit total kinase activity, even at high ionic strength conditions as is the case with poly(L-arginine) (see the table). Inhibition of total kinase activity by high concentrations (10–50 μg/ml) of poly(L-arginine) is not likely due to competitive inhibition at the substrate site because reversibility of inacti-

EFFECT OF POLY(L-ARGININE) ON cGMP-DEPENDENT PROTEIN KINASE ACTIVITY

		Troponin phosphorylation	
Poly(L-arginine)[a] (μg/ml)	Modulator protein	−cGMP (pmol/10 min)	+cGMP (pmol/10 min)
—	−	18.8	72.1
—	+	10.1	83.3
10	−	69.6	85.2
10	+	11.9	80.2
50	−	31.1	38.4
50	+	45.4	59.8

[a] Interaction period was 30 min in the absence or presence of 1 mg/ml of modulator protein.[16]

vation was not observed and self-phosphorylation is enhanced and functional.

In studying the effects of poly(L-arginine) and core histones on kinase activity, the use of a substrate devoid of arginine-rich domains is advantageous. Troponin is such a substrate with a site phosphorylated by both cAMP- and cGMP-dependent protein kinases and the additional advantage of containing modulator protein activity as well.[1] Histone H1 and the synthetic peptides representing the phosphorylation sites of histone H2B are also devoid of arginine-rich domains and effective substrates for both enzymes.[20,21] Interaction of poly(L-arginine) or core histones can be effectively prevented by modulator proteins capable of binding arginine domains. Several effective modulator proteins have been identified. In addition to the modulator protein described by Shoji et al.,[16] soybean and lima bean trypsin inhibitors,[1] and calmodulin[22] are effective. Studies involving modulator proteins and kinase activity require purified proteins free of endogenous phosphate incorporation. While the poly(L-arginine) binding site on the enzyme is specific for arginine-rich domains in such proteins as core histones and protamine, the anionic regions of the enzymes and the modulator proteins involved have not been identified.

Acknowledgment

Studies from the authors' laboratory were supported by NIH Research Grant AM 13149 from the National Institute of Arthritis, Diabetes, Digestive and Kidney Diseases.

[20] D. B. Glass and E. G. Krebs, *J. Biol. Chem.* **254**, 9728 (1979).
[21] C. E. Zeilig, T. A. Langan, and D. B. Glass, *J. Biol. Chem.* **256**, 994 (1981).
[22] T. Yamaki and H. Hidaka, *Biochem. Biophys. Res. Commun.* **94**, 727 (1980).

[22] Substrate-Directed Regulation of cAMP-Dependent Phosphorylation

By M. R. EL-MAGHRABI, T. H. CLAUS, and S. J. PILKIS

Multifunctional phosphoprotein phosphatase has been shown to be regulated by the specific interaction of ligands with the phosphoprotein substrates of the enzyme, i.e., substrate-directed effects.[1] However, only a few enzymes have been identified whose cAMP-dependent phosphory-

[1] E. G. Krebs and J. A. Beavo, *Annu. Rev. Biochem.* **48**, 923 (1979).

lation is affected by their interaction with allosteric effectors. These include for example the type L isozyme of pyruvate kinase, 6-phosphofructo-1-kinase and 6-phosphofructo-2-kinase/fructose-2,6-bisphosphatase. We will describe specific conditions under which substrate-directed regulation of phosphorylation is observed for the above enzymes as well as some general principles involved in such regulation.

Phosphorylation of pyruvate kinase, 6-phosphofructo-1-kinase, or 6-phosphofructo-2-kinase/fructose-2,6-bisphosphatase is catalyzed by the catalytic subunit of cAMP-dependent protein kinase, which can be purified from rat liver by cAMP elution from DEAE-Sephacel, hydroxylapatite chromatography, and gel filtration on Sephadex G-100 superfine.[2] About 3 μg of catalytic subunit (M_r 40,000) can be purified from 1 g of rat liver with an activity of 2 to 3 μmol P_i incorporated into histone/min/mg protein. The incorporation of ^{32}P into the various substrates was measured by the method of Corbin and Reimann.[3] In none of the examples cited below did the effectors tested influence the phosphorylation of histone catalyzed by cAMP-dependent protein kinase.

Pyruvate Kinase (ATP:pyruvate 2-O-phosphotransferase). Cyclic AMP-dependent protein kinase catalyzed phosphorylation caused a decrease in the affinity of pyruvate kinase for phosphoenolpyruvate and for the allosteric activator fructose 1,6-bisphosphate and an increase in its affinity for the allosteric inhibitors alanine and ATP.[4-7] Pyruvate kinase can be purified to homogeneity from rat liver by $(NH_4)_2SO_4$ fractionation, DEAE-cellulose chromatography, and substrate elution from Blue Sepharose.[8] All buffers should contain a proteolytic inhibitor such as 0.1 mM phenylmethylsulfonylfluoride since the amino-terminal peptide containing the phosphorylation site is very susceptible to proteolysis. Using this procedure, more than 100 μg of homogeneous enzyme can be purified from 1 g of rat liver. One mole of P_i/mole of subunit can be incorporated into serine residues of this tetrameric enzyme (subunit M_r = 58,000). Pyruvate kinase is a very good substrate for cAMP-dependent phosphorylation (K_m = 17 μM) and the initial rates of its phosphorylation are linear only during the incorporation of the first mole of phosphate per tetramer.

[2] M. R. El-Maghrabi, W. S. Haston, D. A. Flockhart, T. H. Claus, and S. J. Pilkis, *J. Biol. Chem.* **255**, 668 (1980).
[3] J. D. Corbin and E. M. Reimann, this series, Vol. 38, p. 287.
[4] L. Engstrom, *Curr. Top. Cell. Regul.* **13**, 29 (1978).
[5] T. H. Claus and S. J. Pilkis, in "Biochemical Actions of Hormones" (G. Litwack, ed.), Vol. 8, p. 209. Academic Press, New York, 1981.
[6] J. P. Riou, T. H. Claus, and S. J. Pilkis, *Biochem. Biophys. Res. Commun.* **73**, 591 (1976).
[7] M. R. El-Maghrabi, T. H. Claus, M. M. McGrane, and S. J. Pilkis, *J. Biol. Chem.* **257**, 233 (1982).
[8] J. P. Riou, T. H. Claus, and S. J. Pilkis, *J. Biol. Chem.* **253**, 656 (1978).

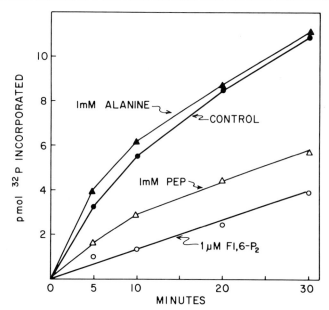

FIG. 1. The effect of fructose 1,6-bisphosphate, phosphoenolpyruvate, and alanine on the rate of phosphorylation of pyruvate kinase. Dephosphorylated pyruvate kinase (30 μg) was incubated with 0.3 mM [γ-^{32}P]ATP (300 cpm/pmol), 5 mM MgCl$_2$, and 50 mM Tris–HCl, pH 7.4, at 30° with no additions (●), with 1 μM fructose 1,6-bisphosphate (F1,6-P$_2$) (○), 1 mM phosphoenolpyruvate (PEP) (△), or with 1 mM alanine (▲) and the reaction started by the addition of 5 μU of catalytic subunit.

The molar ratio of pyruvate kinase to catalytic subunit should be kept at 200:1 or greater. In order to study the influence of effectors on the initial rates of pyruvate kinase phosphorylation, the 2 to 3 mol of P$_i$ already present in cAMP-dependent sites of the isolated enzyme should be removed by incubation with phosphoprotein phosphatase or with ADP and catalytic subunit in the reverse reaction.[2] The effects of fructose 1,6-bisphosphate, phosphoenolpyruvate, and alanine on the initial rate of phosphorylation of dephosphorylated pyruvate kinase are shown in Fig. 1. The allosteric inhibitor alanine caused a slight increase in initial rate of phosphorylation while the allosteric activator fructose 1,6-bisphosphate inhibited the initial rate by more than 80%. Phosphoenolpyruvate also inhibited the phosphorylation of pyruvate kinase but was not as effective as fructose 1,6-bisphosphate. These results suggest that pyruvate kinase is a "better" substrate for the cAMP-dependent protein kinase when the enzyme is allosterically inhibited and a "poorer" substrate when it is allosterically activated. Consistent with this notion is the influence of pH on the effect of fructose 1,6-bisphosphate and alanine on pyruvate kinase

TABLE I
INFLUENCE OF pH ON THE EFFECTS OF FRUCTOSE
1,6-BISPHOSPHATE AND ALANINE ON THE RATE OF
PHOSPHORYLATION OF PYRUVATE KINASE[a]

	pmol/min		
pH	Control	+F1,6-P$_2$	+Alanine
6	3.99	3.94 (− 1.3)	5.41 (+35.6)
7	5.09	3.74 (−26.5)	6.34 (+24.6)
8	10.12	6.62 (−34.6)	10.21 (+ 0.9)

[a] Dephosphorylated pyruvate kinase (30 μg) was incubated with 0.3 mM [γ-^{32}P]ATP (300 cpm/pmol), 5 mM MgCl$_2$, and 50 mM Tris–HCl at the specified pH and where indicated with 100 μM fructose 1,6-bisphosphate (F1,6-P$_2$) or with 1 mM alanine and the reaction started by the addition of 5 μU of catalytic subunit. The percentage change from control values is given in parentheses.

phosphorylation (Table I). At low pH values, the enzyme is in an active form and the rate of phosphorylation is low and unaffected by fructose 1,6-bisphosphate but more sensitive to stimulation by alanine. Conversely, at high pH values pyruvate kinase is in an inactive form and the rate of its phosphorylation is high and more sensitive to inhibition by fructose 1,6-bisphosphate. Similar findings have been reported by Feliu et al. for the rat liver enzyme,[9] by Eigenbrodt and Schoner on the chicken liver enzyme,[10] and by Berglund et al. for pyruvate kinase from pig liver.[11]

6-Phosphofructo-1-kinase (ATP : D-Fructose 6-phosphate 1-phosphotransferase). Although the physiological significance of the cAMP-dependent phosphorylation of this enzyme is uncertain, its rate of phosphorylation is also modulated by allosteric effectors. 6-Phosphofructo-1-kinase can be purified from rat liver by (NH$_4$)$_2$SO$_4$ fractionation, DEAE-Sephadex (A50) chromatography, and gel filtration on Ultrogel 2A.[12] The procedure makes use of the aggregation of the enzyme in the presence of 100 mM (NH$_4$)$_2$SO$_4$ (MW ca. 15 × 10^6) to achieve a separation of the enzyme from contaminating proteins during gel filtration. Approximately 4 μg of homogeneous 6-phosphofructo-1-kinase can be purified from 1 g of rat

[9] J. E. Feliu, L. Hue, and H.-G. Hers, *Eur. J. Biochem.* **81**, 609 (1977).
[10] E. Eigenbrodt and W. Schoner, *Hoppe-Seylers Z. Physiol. Chem.* **358**, 1057 (1977).
[11] L. Berglund, O. Ljungstrom, and L. Engstrom, *J. Biol. Chem.* **252**, 613 (1977).
[12] S. J. Pilkis, M. R. El-Maghrabi, and T. H. Claus, *Arch. Biochem. Biophys.* **215**, 379 (1982).

TABLE II
INFLUENCE OF ALLOSTERIC EFFECTORS ON THE
RATE OF PHOSPHORYLATION OF
6-PHOSPHOFRUCTO-1-KINASE[a]

Addition	pmol/min	
	pH 7.0	pH 8.5
None	0.19	0.59
AMP 20 μM	0.45	0.59
AMP 500 μM	0.55	0.6
F2,6-P$_2$ 2 μM	0.22	0.6
F2,6-P$_2$ 10 μM	0.44	0.61
ATP 1 mM	0.12	0.54
ATP 2 mM	0.08	0.44

[a] 6-Phosphofructo-1-kinase (16 μg) was incubated at 30° with 0.3 mM [γ-^{32}P]ATP (800 cpm/pmol), 6 mM MgCl$_2$, 10 mM mercaptoethanol, 35 mM potassium phosphate, and 20 mM Tris–HCl, pH 7.0 or pH 8.5 with or without the indicated concentrations of effectors and the reaction started by the addition of 0.8 mU of catalytic subunit.

liver using this procedure. One mole of P_i/mole of subunit can be incorporated into serine residues of this tetrameric enzyme (subunit M_r = 83,000). As in the case of pyruvate kinase, protease inhibitors should be included during the purification to prevent proteolytic cleavage of the carboxyl-terminal phosphopeptide. The K_m for 6-phosphofructo-1-kinase phosphorylation is high (120 μM) relative to that of pyruvate kinase and a molar ratio of enzyme to catalytic subunit of 10:1 or less should be used for measurement of initial rates of phosphorylation. Table II shows that phosphorylation of rat hepatic 6-phosphofructo-1-kinase by the catalytic subunit of cAMP-dependent protein kinase can be modulated by AMP, fructose 2,6-bisphosphate, ATP, and H$^+$. In contrast to pyruvate kinase, conditions that favor the active form of the enzyme (AMP, fructose 2,6-bisphosphate, elevated pH) enhance phosphorylation while conditions that favor the inactive form (low pH, ATP) decrease the rate of phosphorylation. Fructose 1,6-bisphosphate, a product of the 6-phosphofructo-1-kinase-catalyzed reaction and a weak allosteric activator of the enzyme, has no measurable effect on the rate of its phosphorylation. Kemp et al. have reported similar findings for the rabbit muscle enzyme.[13]

[13] R. G. Kemp, L. G. Foe, S. P. Latshaw, R. Poorman, and R. L. Heinrikson, J. Biol. Chem. 256, 7282 (1981).

6-Phosphofructo-2-kinase/Fructose 2,6-bisphosphatase (ATP:D-fructose 6-phosphate 2-phosphotransferase/fructose 2,6-bisphosphate 2-0-phosphohydrolase). The activities of the newly discovered 6-phosphofructo-2-kinase/fructose-2,6-bisphosphatase have been found to be subject to cAMP-dependent regulation. Phosphorylation resulted in the inhibition of the phosphotransferase activity[14-16] and activation of the phosphohydrolase activity.[17-19] We have presented evidence suggesting that the two activities are present in the same enzyme, i.e., the enzyme is bifunctional, and that phosphorylation of the protein was accompanied by reciprocal effects on the enzyme activities.[18,19] Moreover, preliminary studies indicated that the rate of phosphorylation of the enzyme was also modulated by substrate-directed effects. Homogeneous preparations of 6-phosphofructo-2-kinase/fructose-2,6-bisphosphatase can be obtained from rat liver by polyethylene glycol fractionation, DEAE-Sephadex (A50) chromatography, and substrate elution from phosphocellulose.[18,19] Approximately 3 µg of protein can be purified from 1 g of liver. This dimeric protein (subunit M_r = 55,000) can be phosphorylated by the catalytic subunit of cAMP-dependent protein kinase to the extent of 1 mol of P_i/mol of subunit. The K_m for phosphorylation appears to be very low and a molar ratio of enzyme to catalytic subunit of approximately 400 : 1 or more is satisfactory for measurement of initial rates of phosphorylation.

The influence of fructose 2,6-bisphosphate and fructose 6-phosphate on the rate of phosphorylation of 6-phosphofructo-2-kinase/fructose-2,6-bisphosphatase is shown is Fig. 2. Fructose 2,6-bisphosphate at 80 µM resulted in a 60% inhibition of phosphorylation, while at 8 µM a 12% inhibition was observed (data not shown). Fructose 6-phosphate on the other hand had no effect up to concentrations of 200 µM. Elucidation of the physiological significance of substrate-directed regulation of 6-phosphofructo-2-kinase/fructose-2,6-bisphosphatase phosphorylation awaits further investigation.

Other Enzymes. Atmar and Kuehn[20] have reported substrate-directed

---

[14] E. Van Schaftingen, D. R. Davies, and H.-G. Hers, *Biochem. Biophys. Res. Commun.* **103,** 362 (1981).
[15] M. R. El-Maghrabi, T. H. Claus, J. Pilkis, and S. J. Pilkis, *Proc. Natl. Acad. Sci. U.S.A.* **79,** 315 (1982).
[16] E. Furuya, M. Yokoyama, and K. Uyeda, *Proc. Natl. Acad. Sci. U.S.A.* **79,** 325 (1982).
[17] E. Furuya, M. Yokoyama, and K. Uyeda, *Biochem. Biophys. Res. Commun.* **105,** 264 (1982).
[18] M. R. El-Maghrabi, T. H. Claus, J. Pilkis, E. Fox, and S. J. Pilkis, *J. Biol. Chem.* **257,** 7603 (1982).
[19] M. R. El-Maghrabi, E. Fox, J. Pilkis, and S. J. Pilkis, *Biochem. Biophys. Res. Commun.* **106,** 794 (1982).
[20] V. Atmar and G. Kuehn, this volume [39].

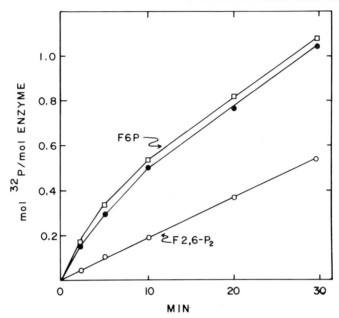

FIG. 2. The effect of fructose 2,6-bisphosphate and fructose 6-phosphate on the rate of phosphorylation of 6-phosphofructo-2-kinase/fructose-2,6-bisphosphatase. Ten micrograms of the kinase/bisphosphatase were incubated with 0.3 mM [γ-^{32}P]ATP (1000 cpm/pmol), 6 mM MgCl$_2$, and 20 mM TES, pH 7.5 at 30°, with no additions (●), with 80 μM fructose 2,6-bisphosphate (F2,6-P$_2$) (○), or with 200 μM fructose 6-phosphate (F6P) (□) and the reaction started by the addition of 10 μU of catalytic subunit.

regulation of ornithine decarboxylase phosphorylation catalyzed by polyamine-stimulated protein kinase. Also, Yeh et al.[21] reported that the rate of phosphorylation and inactivation of partially purified acetyl-CoA carboxylase was affected by adenine nucleotides, and it has been observed that phenylalanine hydroxylase phosphorylation is modulated by effectors.[22]

Conclusion. Although there are only a limited number of reports on effector-directed modulation of cAMP-dependent phosphorylation, a number of general observations can be made. Many of the proteins which are subject to protein kinase-catalyzed phosphorylation exhibit allosteric properties, i.e., they interact with low-molecular-weight ligands and undergo conformational changes resulting in transitions between active and inactive forms. It is reasonable to assume that the phosphorylation of many if not all enzymes showing effector-induced conformational

[21] L.-A. Yeh, K.-H. Lee, and K.-H. Kim, *J. Biol. Chem.* **255**, 2308 (1980).
[22] A. Doskeland, D. Ogrid, S. O. Doskeland, and T. Flatmark, personal communication.

changes may be regulated by substrate-directed effects though very few enzymes have been studied in this regard. When studying substrate-directed phosphorylation it is important to remove all endogenous phosphate from the enzyme preparation. For example, preparations of pyruvate kinase which contain significant amounts of phosphate in cAMP-dependent sites are phosphorylated at much slower rates than preparations which have been completely dephosphorylated.[2] The high phosphate containing form (>2 mol P_i/mol enzyme) also does not exhibit ligand-dependent changes in its rate of phosphorylation.[2] Also, the concentrations of the effectors tested should be within the known physiological range.

Whether effector-modulated phosphorylation of enzymes is important in the physiological modulation of enzyme activity is uncertain, but such modulation has been shown to occur in intact cells. For example, the phosphorylation states of pyruvate kinase and 6-phosphofructo-1-kinase have been shown to be regulated by allosteric effectors in isolated rat hepatocytes.[23,24] It is likely that examination of additional proteins will reveal that substrate-directed regulation of phosphorylation is widespread and at least in some cases important in the *in vivo* regulation of enzyme activity.

[23] T. H. Claus, M. R. El-Maghrabi, and S. J. Pilkis, *J. Biol. Chem.* **254**, 7855 (1979).
[24] T. H. Claus, J. R. Schlumpf, M. R. El-Maghrabi, and S. J. Pilkis, *J. Biol. Chem.* **257**, 7541 (1982).

[23] Use of Microinjection Techniques to Study Protein Kinases and Protein Phosphorylation in Amphibian Oocytes

By JAMES L. MALLER

It has become fundamentally clear in recent years that one of the most basic mechanisms of hormone action involves changes in the phosphorylation state of key regulatory enzymes. Phosphorylation can involve either activation or inhibition of activity and be regulated by activity changes in protein kinase and/or protein phosphatase. Changes in the activity of these enzymes may depend on alterations in cyclic nucleotide levels, increases in free calcium, or unknown factors. In spite of the acknowledged importance of protein phosphorylation, with the exception of the enzymes of glycogen metabolism, little is known about the specific sub-

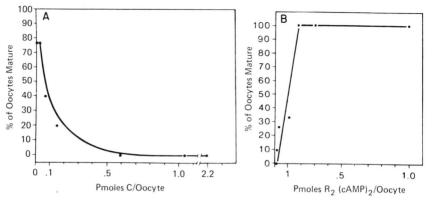

FIG. 1. Dose-dependent effect of microinjected protein kinase subunits on oocyte maturation. (A) Homogeneous catalytic subunit of cAMP-dependent protein kinase was injected to the indicated amounts in a group of 25–30 oocytes and the effect on subsequent progesterone-induced maturation assessed. The results show that an increase in cAMP-dependent protein kinase is sufficient to block maturation; 0.1 pmol represents an internal concentration of 1×10^{-7} M. (B) Homogeneous Type II regulatory subunit of cAMP-dependent protein kinase was injected into oocytes, and after further incubation, oocytes underwent oocyte maturation in the absence of progesterone. These results show that a decrease in cAMP-dependent protein kinase activity is sufficient to induce oocyte maturation. [Taken from J. L. Maller and E. G. Krebs, *J. Biol. Chem.* **252**, 1712 (1977).]

strates whose phosphorylation state is changed by hormones. This lack of knowledge is due both to the relative lack of specificity of many protein kinases for phosphorylation *in vitro* and to the inability to assess the importance of phosphorylation changes seen *in vivo*. An approach that shows great promise for addressing these problems is microinjection of cells with protein kinases, specific substances that affect protein kinase activity, or specific substrates and their antibodies that may alter cell function. This approach is most feasible in amphibian oocytes, where microinjection is sufficiently quantitative and reproducible to permit dose–response curves and biochemical analysis to be performed.

An example of the power of this technique is shown in Fig. 1, where the injection of the catalytic and regulatory subunits of cAMP-dependent protein kinase demonstrated that an observed decrease in cAMP was both necessary and sufficient to control the hormonal response.[1] In Fig. 2, the effect of protein phosphatase inhibitor-2 was determined for the dephosphorylation of injected phosphorylase *a* in the intact oocyte. The results show that only about 40% of the dephosphorylation can be attributed to phosphatase 1, the class of enzyme inhibited by phosphatase inhibitor-2 and subject to regulation by cAMP.

[1] J. L. Maller and E. G. Krebs, *J. Biol. Chem.* **252**, 1712 (1977).

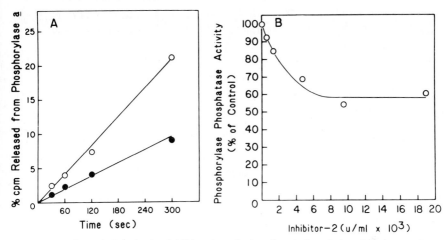

FIG. 2. Effect of phosphatase inhibitor-2 on dephosphorylation of phosphorylase a after microinjection into *Xenopus* oocytes. (A) ^{32}P-labeled phosphorylase a was injected into 5–10 oocytes and the rate of dephosphorylation determined as a function of time in the absence (○—○) and presence (●—●) of coinjected phosphatase inhibitor-2. Reactions were terminated by dropping oocytes in ice-cold 10% trichloroacetic acid and the extent of dephosphorylation was measured by liquid scintillation counting of soluble and insoluble fractions. (B) Dose–response curve for inhibition of phosphorylase a dephosphorylation by phosphatase inhibitor-2. The same experiment as in A was carried out with increasing concentrations (in a constant volume) of Inh-2 coinjected with phosphorylase a. The abscissa is the final intracellular concentration of inhibitor-2. Half-maximal inhibition seen at 1500 units/ml corresponds to about 0.6 μM Inh-2. Phosphatase inhibitor-2 was prepared as described by J. G. Foulkes and P. Cohen, *Eur. J. Biochem.*, **105**, 195 (1980). [Data from J. G. Foulkes and J. L. Maller, *FEBS Lett.* **150**, 155 (1982).]

Microinjection of somatic cells is also possible with current technology but it is difficult if not impossible to reproducibly inject the same volume into each cell, identify or recover only injected cells alive after injections, or do a sufficient number for subsequent biochemical analysis. The volume of the oocyte is 500,000× that of a typical somatic cell. Thus, about 500,000 somatic cells would need to be injected to monitor the amount of material (50 nl) that can be injected into one oocyte. For this reason, this article is limited to procedures for *Xenopus* oocytes.

Equipment

A dissecting microscope is required both for isolation of oocytes and the injection itself. The most commonly used final magnification is about 25×. The microscope of choice for most investigators is the Wild M7A Stereozoom scope because of superior optics and a wider field of view, which increases the number of oocytes that can be injected with a single needleful of solution. However, any inexpensive fixed objective or zoom dissecting microscope (e.g., American Optical Stereostar) can be used. A suitable micromanipulator is the Narishige MM3 (Labtron Scientific, Farmingdale, NY) with movement in all three axes, available in left- or right-handed models. The micromanipulator can be purchased with a base or attached to a small ring stand.

A micropipet puller is necessary for the construction of micropipets. A suitable instrument is the Model 700C vertical pipet-puller from David A. Kopf Instruments (Tujunga, CA). Also needed is a 0.2-ml syringe microburet (Gilmont Instruments, Great Neck, NY) that is connected to the micropipet to provide the force for delivery of solutions.

Solutions

Two solutions, medium OR2[2] and modified Barth,[3] are in general use for the culture of *Xenopus* oocytes. No serum or macromolecular precursors are required for culture of these cells; they obtain all needed precursors from breakdown of stored components. For incubation labeling, the medium is supplemented with 200 μg/ml gentamicin sulfate (Sigma) and for incubations over 8 hr 1 mM sodium pyruvate or oxaloacetate is present in the medium.

The compositions of the two solutions are given in the table. The final pH is 7.6 in either case. For incubation studies with $^{32}P_i$ and medium OR2, the phosphate is left out of the medium to increase intracellular specific activity. It is also left out when $^{32}P_i$ is injected to prevent loss of label from

COMPOSITION OF OOCYTE CULTURE SOLUTIONS

Salt	Concentration (mM)	
	OR2	Modified Barth
NaCl	83	88
KCl	2.5	1.0
MgSO$_4$	—	0.82
MgCl$_2$	1.0	—
CaCl$_2$	1.0	0.41
Ca(NO$_3$)$_2$	—	0.33
NaHCO$_3$	—	2.4
HEPES	5	10
NaH$_2$PO$_4$	1.0	—

[2] R. A. Wallace, D. W. Jared, J. N. Dumont, and M. W. Sega, *J. Exp. Zool.* **184**, 321 (1973).
[3] C. C. Ford and J. B. Gurdon, *J. Embryol. Exp. Morphol.* **37**, 203 (1977).

the cell. There is no compelling reason for choosing one medium over the other; studies in the author's laboratory use medium OR2.

Preparation of Micropipets

Fifty-microliter capillary pipets (VWR), are pulled by hand over a small flame to an external diameter of 200–350 μm. The constricted region should be centered in the length of the pipet and extend for a distance of 4–5 cm. The pilot light on a standard Bunsen burner is suitable for this purpose. The pipet is carefully placed in the vertical pipet-puller with the heating coil centered in the partially constricted region. The length of the constricted region to be calibrated later will be that portion below the heating coil, and the desired length of this region will vary depending on the microscope and optics employed. The standard coil supplied with the pipet-puller is used in my laboratory at a heater value of 37 and solenoid setting of 10. The heating element is turned on, and after automatic shut off, the pipet is removed and cut at the most tapered point. Only the lower portion is generally used. In some cases, this procedure can be shortened by manually pulling the pipet downward while heating in the coil. The tapered point of the pipet is severed under the dissecting scope with watchmakers forceps about 2 cm past the region of the initial hand-pulled constriction. The tip diameter should be in the range of 10–30 μm. The other end of the micropipet is pulled to a taper over a small flame to facilitate leak-proof insertion into the tubing connecting the syringe microburet to the micromanipulator. The pipet is calibrated by marking with a fine-tipped water-insoluble marker the 200–350 μm diameter section of the pipet into 10–15 intervals of approximately 0.5–1 mm by alignment of the needle under the dissecting scope with a micrometer vernier. Needles are pooled in groups differing by 50 μm (internal diameter) in the 200–350 μm total diameter section. Needles within a given size group and calibration setting generally deliver the same volume between marks ±10%. A range of internal diameters from 100 to 250 μm covers with suitable calibration marks a range of injected volumes of 10–100 nl. With practice, over a hundred micropipets can be prepared in a day, and they can be stored indefinitely.

Isolation of Oocytes

Female *Xenopus laevis* can be obtained from either the South African Snake Farm, Fish Hoek, S.A. or NASCO Scientific, Ft. Atkinson, WI. Animals from South Africa, ordered by telegram, arrive in about 10 days and are in the author's experience of higher quality and less expensive. Animals are decapitated and the ovary removed into a 100 mm petri dish of medium OR2. Alternatively, animals can be anesthetized by hypothermia or immersion in neutral 0.12% MS222 (Tricaine) for 30 min followed by surgical removal of part of the ovary through a dorsal incision. *Xenopus* appears to be able to regenerate ovarian tissue, with young animals much more efficient in this regard.

The ovary is cut into small pieces so that oocytes are exposed to the medium, rinsed through two changes of OR2, and incubated in OR2 at room temperature. Two methods are in general use for the isolation of oocytes. The method of choice is manual dissection from the ovarian follicle with watchmaker's forceps (Dumont, #5) although this method requires practice and manual dexterity. Only the very largest, unblemished, Stage VI[4] oocytes should be chosen. The oocytes should be at least 1.3 mm in diameter. The quality of the oocytes is a very important variable in any experiment. One way to assess quality is to verify that at least 90% of an oocyte sample matures in response to progesterone, as evidenced by a white spot in the animal pole 3–6 hr after steroid (1 μM). The oocyte is grasped with watchmaker's forceps at the stalk connecting it to the ovarian mesentery. Another pair of forceps grasps the stalk above this point and peels it off the oocyte, often "popping" the oocyte out of the ovarian follicle bag enclosing the oocyte. In the ovary, oocytes are surrounded by three cell layers: squamous epithelium, thecal cells, and follicle cells. Manual dissection removes all but the follicle cells. Following dissection, oocytes are transferred in groups of 10 to fresh medium with a cut-off pasteur pipet; with practice, over 300 oocytes per hour can be manually dissected from the ovary. If it is desirable to also remove follicle cells, as for example before incubation labeling with $^{32}P_i$, Pronase treatment is employed. Manually dissected oocytes (25–50) are transferred to a vial containing 2 ml of 50 μg/ml Pronase in OR2 and incubated at room temperature for 7 min. The reaction is terminated by addition of 1 ml of OR2 containing 10 mg/ml bovine serum albumin (BSA) and five changes of medium containing 1 mg/ml BSA. The BSA should be highly purified and freed of insulin and insulin-like growth factors, since these can induce oocyte maturation. A suitable product is available from Reheis Chemicals (Phoenix, Arizona). The effectiveness of follicle cell removal is assessed by the failure of an oocyte maturation response to human chorionic gonadotropin (100 IU/ml).

Collagenase Method. An alternative method in use in some laboratories involves collagenase treatment. Washed ovarian pieces are blotted dry on Kimwipes and placed in vials containing 5 mg/ml collagenase (Sigma, Type II) in OR2. The ratio of collagenase solution to tissue should

[4] J. N. Dumont, *J. Morphol.* **136**, 153 (1972).

be at least 2 : 1. The vials are gently shaken horizontally for 2-3 hr at room temperature. Different lot numbers of collagenase vary markedly in effectiveness. The contents of the vial are dispersed into a large (100 mm) petri dish containing OR2 and the ovarian fragments shaken vigorously in the dish with watchmaker's forceps. This procedure selectively releases many of the large Stage VI oocytes. It is also possible to do a complete digestion of the ovary by a longer digestion (8 hr or overnight at 1-4 mg/ml) but the viability of the cells is often reduced. The released oocytes are taken up in a cut-off pasteur pipet and washed through three additional petri dishes containing OR2. The oocytes are examined under the dissecting scope and large, unblemished Stage VI oocytes are selected and transferred to another dish. This selection procedure is tedious and affords no real savings in time over the manual dissection method. However, it is possible to isolate larger numbers of oocytes with this method. Collagenase-treated oocytes still retain the follicle cell layer after short incubation; with longer incubations in collagenase, follicle cells may also be removed.[5]

Labeling of Oocytes. $^{32}P_i$ labeling can be achieved either by incubation in $^{32}P_i$ in OR2 or by injection of $^{32}P_i$. Higher intracellular specific activities can be achieved by microinjection (10^4 cpm/pmol), although this is not usually necessary. Available evidence indicates that injected $^{32}P_i$ results in a constant intracellular specific activity within 6 min, labeling essentially all γ and β positions of all four nucleoside triphosphates.[6] Incubation labeling is generally at 100-200 μCi/ml in phosphate-free medium for 2-4 hr to achieve significant labeling of the pool. Oocytes injected or incubated with $^{32}P_i$ should be examined before further use for damage, and any morphologically altered oocytes removed.

Injection Procedure

The syringe microburet and 20-30 cm of tubing are filled with a light oil (e.g., Beckman Diffusion Pump Oil) and a calibrated needle is inserted into the open end of the tubing and fixed in place on the micromanipulator. The needle is filled with oil up to about 2 cm before the beginning of the initial constricted region. The apparatus is adjusted so that the needle can be moved in the field of view of the microscope.

Filling of Micropipets. The solution to be injected is taken up in a 5 μl Lang-levy pipet and kept (if necessary) on ice in an ice bucket next to the microscope. The solution is dispensed in a single drop onto a siliconized

---

[5] T. J. Mohun, C. D. Lane, A. Colman, and C. C. Wylie, *J. Embryol. Exp. Morphol.* **61**, 367 (1981).
[6] J. L. Maller, M. Wu, and J. C. Gerhart, *Dev. Biol.* **58**, 295 (1977).

depression slide which is placed on the microscope stage so that the drop is at the extreme left of the field. The micropipet is lowered so the tip is also at the left of the field and the calibrated region is visible to the right. The tip is lowered in the drop and the syringe microburet used to suck solution into the needle. Filling of the calibrated region should not proceed beyond the field of view. The needle is raised from the drop and the remaining solution withdrawn into the Lang-levy pipet. Oocytes in OR2 in a 60 mm petri dish cover are placed on the stage and oriented to the left side of the field. The micropipet is lowered into the solution and placed next to the first oocyte. By focusing up and down, it should be possible to see in the same field the oocyte, the tip of the needle, and the meniscus of the solution to be injected in the calibrated region of the pipet. The oocyte is anchored with watchmaker's forceps, the tip inserted into the animal pole just above the equator, inserted about one-third of the way through the oocyte and the syringe microburet turned to deliver the region between two calibration marks into the oocyte. The needle is withdrawn from the oocyte, inserted in the next oocyte, and the operation repeated. Generally, four to six oocytes can be injected per minute. Ten or fifteen oocytes can usually be injected with a single needleful of solution. After a needleful has been delivered, injected oocytes are transferred to a separate dish and the micropipet refilled. Usually more oocytes than are needed should be injected so that removal of damaged oocytes does not limit the experimental analysis. A typical experiment would utilize 100–500 oocytes and less than 10% of the oocytes develop abnormal morphology over an 8-hr period. For longer incubations, attrition increases and the medium should be changed at least daily.

Extraction of Phosphorylated Proteins from Oocytes. Extraction is carried out by homogenization on ice in a 2 ml Dounce homogenizer. The yolk protein and pigment is removed by centrifugation at 7000 g, and the supernatant removed free of lipid with a Pasteur pipet. Reproducibility is improved if multiple protease inhibitors are present along with 100 mM fluoride and 10 mM EGTA, the latter to inhibit the abundant Ca^{2+}-activated protease present in yolk platelets.

[24] Determination of the cAMP-Dependent Protein Kinase Activity Ratio in Intact Tissues

By JACKIE D. CORBIN

There are many circumstances in endocrinological studies in which it is useful to know the level of activation of cAMP-dependent protein kinase. In a typical example, after hormonal treatment of a particular tissue, the protein kinase activity ratio is often seen to change significantly with little or no measurable change in the cAMP level. This lack of correlation may be due in some instances to the limitations of the cAMP assay in detecting small changes that occur in the physiological range of concentration rather than to a real lack of parallel changes in the two parameters. On the other hand, because of the strong cooperativity in binding of cAMP to the two different cAMP binding sites of the protein kinase, one might expect small increases in cAMP to cause large increases in bound cAMP and protein kinase activation.[1] Thus, measurement of the activity ratio in such a situation is most informative. The techniques for determination of the cAMP-dependent protein kinase activity ratio in intact adipose tissue have been outlined in an earlier volume of this series.[2] Several investigations have since revealed that the same methods are inadequate for some other tissues. The existence of two main isozymic forms of the enzyme, referred to as type I and type II, and the widely varying distribution of these forms in tissues and species is the principal underlying cause.[3] It is necessary, therefore, to reexamine the original method and to outline procedures and assays which will, as closely as possible, estimate the cellular protein kinase activity ratio in tissues containing mainly type I, mainly type II, or a mixture of the two. When hormones or other agents cause changes in cAMP levels in tissues the protein kinase is altered according to the following equation[4]:

$$R_2C_2(\text{inactive}) + 4\ \text{cAMP} \rightleftharpoons R_2\text{cAMP}_4 + 2C(\text{active}) \tag{1}$$

R represents the regulatory subunit and C represents the catalytic subunit of the enzyme. From examination of this equation, assay of protein kinase activity in tissue extracts will detect the active enzyme form only. Mea-

---

[1] S. R. Rannels and J. D. Corbin, *J. Biol. Chem.* **256**, 7871 (1981).
[2] T. R. Soderling, J. D. Corbin, and C. R. Park, this series, Vol. 38, p. 358.
[3] J. D. Corbin and S. L. Keely, *J. Biol. Chem.* **252**, 910 (1977).
[4] J. D. Corbin, P. H. Sugden, L. West, D. A. Flockhart, T. M. Lincoln, and D. McCarthy, *J. Biol. Chem.* **253**, 3997 (1978).

surement of the total enzyme activity can be achieved by addition of high cAMP to the assay, which shifts the equilibrium essentially completely to the right. This is the basis for the determination of the protein kinase activity ratio (activity in absence of cAMP divided by activity in presence of cAMP), which represents the fraction of the total enzyme (R_2C_2 + C) in the active C form.

Preparation of Tissue Extracts

The choice of homogenizer depends on the nature of the tissue examined. It may be necessary to perform preliminary studies of the effects of time and intensity of homogenization, and in some cases the effect of freezing–thawing of the tissue. We have found that a too vigorous homogenization can shift Eq. (1) to the right, and that the free C is more labile than R_2C_2 to denaturation during homogenization. The choice of buffer also depends on the tissue to be studied. This is due primarily to the isozyme pattern of the tissue and to the presence of certain variable inhibitors such as ATPases and phosphodiesterases. The homogenization procedures that we have found optimal for rat heart, epididymal fat pad, and hepatocytes are outlined in Table I. These procedures are presented in more detail in previous publications.[5-7]

Some tissues, such as rat heart, can be stored frozen before homogenization. For other tissues, such as rat hepatocytes, freezing–thawing was found to increase the enzyme activity ratio. We have also found that freezing–thawing of homogenates or supernatant fractions may increase the activity ratio. It is recommended that the protein kinase assay be performed as soon after homogenization as possible, preferably within 30 min.

Protein Kinase Assay

The filter paper assay procedure which is recommended for determination of cAMP-dependent protein kinase activity has been presented in detail in an earlier volume of this series.[8] For convenience, the assay is described again with only slight modification of the original method. In principle, the assay is based on incorporation of ^{32}P from [γ-^{32}P]ATP into a suitable protein substrate. Although many proteins or peptides may be used as substrate, a histone mixture is the one of choice in our laboratory because of its low-cost availability, high level of ^{32}P incorporation, lack of endogenous protein kinases, and stability and precipitability. We have

[5] S. L. Keely, J. D. Corbin, and C. R. Park, *J. Biol. Chem.* **250**, 4832 (1975).
[6] T. R. Soderling, J. C. Corbin, and C. R. Park, *J. Biol. Chem.* **248**, 1822 (1973).
[7] A. D. Cherrington, F. D. Assimacopoulos, S. C. Harper, J. D. Corbin, C. R. Park, and J. H. Exton, *J. Biol. Chem.* **251**, 5209 (1976).
[8] J. D. Corbin and E. M. Reimann, this series, Vol. 38, p. 287.

TABLE I
PREPARATION OF TISSUE EXTRACTS

Rat tissue	Approximate isozyme distribution (% type I)	Homogenization buffer	Homogenization procedure
Heart	80	10 mM potassium phosphate (pH 6.8), 10 mM EDTA, 0.5 mM 1-methyl-3-isobutylxanthine	Use 15 ml of buffer per gram of frozen and powdered perfused hearts. Homogenize using three up-and-down strokes of a drill motor-driven Teflon pestle in a glass homogenizer tube. Centrifuge at 27,000 g for 20 min at 4°. Assay supernatant
Epididymal fat pad	10	10 mM potassium phosphate (pH 6.8), 10 mM EDTA, 0.5 mM 1-methyl-3-isobutylxanthine, 0.5 M NaCl	Use 1 ml of buffer per gram of incubated fat pad. Homogenize by procedure above. Centrifuge at 12,000 g for 5 min at 4°. Assay infranatant
Hepatocytes	50	Centrifuge 1 ml of cell suspension (40 mg of tissue) at 50 g for 15 sec. Remove supernatant and resuspend cells in 2 ml of cold homogenizing buffer [10 mM potassium phosphate (pH 6.8), 10 mM EDTA, 0.5 mM 1-methyl-3-isobutylxanthine, and 150 mM KCl]	Homogenize for 10 sec with Ultraturrax homogenizer. Assay homogenate

also used with at least equal success the artificial peptide substrate as described by Kemp and Clark,[9] which utilizes a different procedure from that described below for isolating ^{32}P-labeled substrate for counting. This peptide, which is commercially available, may be particularly advantageous for tissue extracts containing high levels of cAMP-independent histone kinases.

Reagents for Assay

> Potassium phosphate, 50 mM, pH 6.8, with or without 6 μM cAMP
> Histone mixture (type II-A, Sigma Chemical Company), 30 mg/ml in H$_2$O

[9] B. E. Kemp and M. G. Clark, *J. Biol. Chem.* **253**, 5147 (1978).

Mg[γ-^{32}P]ATP (18 mM magnesium acetate 1 mM [γ-^{32}]ATP) in H$_2$O. Specific activity ~100 cpm/pmol. The [γ-^{32}P]ATP can be obtained from commercial sources or prepared by the method of Walseth and Johnson.[10]
Enzyme solution, 0.25–10 × 10^{-6} units/ml
Trichloroacetic acid, 10%
Ethanol, 95%
Ethyl ether

Procedure for Assay

When using adipose tissue or hepatocytes, 150 or 50 mM NaF, respectively, is included in the phosphate reagent. Equal volumes of the potassium phosphate, histone, and Mg[^{32}P]ATP are combined. Two mixtures, one with and one without cAMP, are prepared. The mixtures are stable for several weeks with repeated freezing and thawing. To disposable glass test tubes (1.2 × 7.5 cm) are added 50 μl of the mixtures. The reactions are initiated by pipetting 20 μl (adipose tissue or hepatocytes) or 10 μl (heart) of the appropriate buffer (blank) or enzyme solution into the mixtures, mixing, and placing the reaction tubes in a water bath at 30°. After incubating 5 min, 50 μl of the reaction mixtures is spotted on filter paper squares (Whatman 31 ET, 2 × 2 cm) numbered with pencil lead. The squares are dropped immediately into a beaker containing 10% trichloroacetic acid at room temperature. The beaker should contain an inverted stainless-steel screen dish lodged at the bottom with sufficient space for a magnetic stirring bar to be placed underneath. This set-up allows for stirring the trichloroacetic acid without contact between stirring bar and paper squares. The tearing of bits of paper from the squares during washing is thus avoided. The beaker should contain approximately 5 ml of trichloroacetic acid per filter square. The filter squares are washed at room temperature for 15 min at the slowest possible rotating speed of the stirring bar. The trichloroacetic acid is poured into an appropriate radioactive waste container and replaced with fresh 10% trichloroacetic acid. This wash procedure is repeated once for 15 min and then two more times at 5 min each (total = four washes). The filter squares are then washed in 95% ethanol and then ethyl ether for 1 min each. The squares are dried for approximately 5 min with a hair dryer, and then placed into 10 ml of scintillant (either Amersham ACS, which contains surfactant, or Beckman Ready-Solv EP, which contains emulsifier, can be used) for counting. Since the radioactive protein remains adsorbed to the filter squares during counting, the squares can be removed and the scintillant reused

[10] T. F. Walseth and R. A. Johnson, *Biochim. Biophys. Acta* **562**, 11 (1979).

TABLE II
HORMONAL ELEVATION OF PROTEIN KINASE ACTIVITY
RATIO IN TISSUES[a]

Rat tissue	Protein kinase activity ratio	
	Basal	Hormone-stimulated
Heart	0.15	0.35
Epididymal fat pad	0.25	0.59
Hepatocytes	0.35	0.52

[a] The hormones used were epinephrine (0.1 μM, 5 min) for heart, epinephrine (1 μM, 10 min) for epididymal fat pad, and glucagon (0.2 mM, 1.3 min) for hepatocytes.

following counting. If scintillant not containing surfactant or emulsifier is used, a yellow color appears after reuse a few times. The ethanol and ether washes can also be saved and reused. Stirring is normally included to ensure efficient washing, although we have found that low blanks can be obtained even without stirring. The blank is normally equivalent to 10 pmol or less.

A unit of activity is defined as incorporation of 1 μmol of ^{32}P per minute. The protein kinase activity ratio is the enzyme activity in the absence of cAMP divided by that in the presence of cAMP. Typical basal and hormone-elevated activity ratios[5-7] are shown in Table II.

Pitfalls

There are numerous potential pitfalls (Table III), several of which are present for many other enzymes in crude extracts, which one must be aware of in measurements of the protein kinase activity ratio. These problems have been discussed in detail and procedures to circumvent them have been presented as shown in the references of Table III. As pointed out in these references and in the earlier volume of this series,[2] because of these pitfalls the measurements of activity ratio by this procedure may be semiquantitative. The artifacts are minimized by the suggested additions to the homogenization medium for rat heart, adipose tissue, and liver. However, for tissues not studied before the potential pitfalls should be carefully considered. Of particular concern is the question of whether or not the ratio measures an event which occurs intracellularly or during preparation and assay of enzyme activity. One method to approach the answer to this question involves the use of charcoal in the

TABLE III
POTENTIAL PITFALLS IN DETERMINATION OF THE
PROTEIN KINASE ACTIVITY RATIO

	Reference
ATPases	13
Phosphodiesterases	7, 13
Phosphoprotein phosphatases	13
Release of sequestered cAMP	11
Artificial RC dissociation (histone, salt)	7, 12, 14
C binds to particles	7, 18
Salt inhibits kinase	7, 13, 15
Artificial R and C reassociation	7, 12, 13
ATP effects on reassociation	14
Endogenous or added cAMP too high	8, 16
Isozyme distribution (tissue and species)	3, 7, 14
cAMP-independent protein kinases	7, 13
Presence of endogenous substrates	8
Heterogeneity of cells in a tissue	6, 7
Proteases	4, 16

homogenizing medium.[11] Palmer et al.[12] have recommended the inclusion of exogenous protein kinase in the medium in order to correct for activation–inactivation artifacts.

[11] J. D. Corbin, S. L. Keely, T. R. Soderling, and C. R. Park, in "Advances in Cyclic Nucleotide Research" (P. Greengard and A. Robinson, eds.), p. 265. Raven, New York, 1975.
[12] W. K. Palmer, J. M. McPherson, and D. A. Walsh, J. Biol. Chem. 255, 2663 (1980).
[13] J. D. Corbin, T. R. Soderling, and C. R. Park, J. Biol. Chem. 248, 1813 (1973).
[14] J. D. Corbin, S. L. Keely, and C. R. Park, J. Biol. Chem. 250, 218 (1975).
[15] J. D. Corbin, C. O. Brostrom, R. L. Alexander, and E. G. Krebs, J. Biol. Chem. 247, 3736 (1972).
[16] H. Iwai, M. Inamasu, and S. Takeyama, Biochem. Biophys. Res. Commun. 46, 824 (1972).
[17] J. D. Corbin, C. O. Brostrom, C. A. King, and E. G. Krebs, J. Biol. Chem. 247, 7790 (1972).
[18] S. L. Keely, J. D. Corbin, and C. R. Park, Proc. Natl. Acad. Sci. U.S.A. 72, 1501 (1975).

[25] Radiolabeling and Detection Methods for Studying Metabolism of Regulatory Subunit of cAMP-Dependent Protein Kinase I in Intact Cultured Cells

By ROBERT A. STEINBERG

Alterations in expression of cAMP-dependent protein kinase subunits accompanying kinase activation, tissue differentiation, cellular transformation, cell cycle traverse, or somatic mutations have generated interest in kinase subunit metabolism and its intracellular regulation.[1-3] The most direct approach for studying specific protein metabolism in intact cells involves incorporation of radiolabeled amino acids and subsequent quantitation of radioactivity associated with the protein of interest. Synthesis rates are estimated by the relative labeling of the protein in a short pulse, and turnover rates are determined by the loss of label from the protein during a chase period after labeling. Protein modifications can be followed using radioisotopes specific for the modifying groups or, for some modifications, by alterations in electrophoretic mobility of the protein.

Implementation of an isotopic labeling approach to study specific protein metabolism requires a means of resolving radioactivity in the protein of interest from that in other cellular constituents. Such purification imposes serious obstacles for study of proteins occurring in cells at low abundances; a high degree of purification is necessary, and it is critical that recoveries of the protein of interest be uniform (or calculable) between samples to be compared. Cyclic AMP-dependent protein kinase presents additional difficulties since subunit dissociation accompanying intracellular activation modifies the protein's physical properties.

We have had considerable success using high-resolution two-dimensional polyacrylamide gel electrophoresis[4] of crude cell extracts to resolve type I regulatory subunit (R_I) of protein kinase for studies of its synthesis and degradation in cultured S49 mouse lymphoma cells.[2,3] Radiolabeled cells are extracted in a highly denaturing buffer to provide both reproducible patterns of extracted proteins and essentially quantitative solubilization of R_I. Since the two-dimensional gel procedure resolves polypeptides differing by less than a single unit of charge, it has also proven useful for studying the modification of R_I by phosphorylation and

---

[1] R. A. Jungmann and D. H. Russell, *Life Sci.* **20**, 1787 (1977).
[2] R. A. Steinberg and D. A. Agard, *J. Biol. Chem.* **256**, 10731 (1981).
[3] R. A. Steinberg and D. A. Agard, *J. Biol. Chem.* **256**, 11356 (1981).
[4] P. H. O'Farrell, *J. Biol. Chem.* **250**, 4007 (1975).

mutation,[3,5] and for studying the effector-dependent phosphorylation of intracellular kinase substrates.[6] This chapter discusses procedures for combining metabolic labeling with two-dimensional gel analysis to study the regulation of R_I metabolism. For detailed and explicit description of the two-dimensional gel procedure, readers should consult articles by O'Farrell[4] and by O'Farrell and O'Farrell.[7]

Metabolic Labeling

General Considerations

Good resolution in two-dimensional gel electrophoresis depends upon low protein loading, so cellular proteins must be labeled to high specific radioactivity for detection of R_I. [^{35}S]Methionine is the isotope of choice for most protein labeling studies since it is available at high specific activity, has sufficient energy for detection by direct autoradiography of dried gels, and is relatively inexpensive. For detection of R_I in whole cell extracts, gels should be loaded with $0.2-2 \times 10^6$ acid-precipitable ^{35}S cpm in 10–20 µg of protein. The higher numbers of counts are advisable for label-chase experiments where the proportion of counts in R_I falls sharply as a function of time of chase.

Optimal labeling conditions vary with cell type, so appropriate conditions should be established for the cells under study. For short-term labeling experiments the concentration of methionine and the ratio of medium to cells can be reduced substantially to facilitate efficient utilization of isotope. A compromise must be struck, however, between considerations of metabolic balance and isotope utilization, since excessive cell concentration leads to rapid depletion of nutrients and/or generation of toxic waste products, and methionine concentrations below about 2.5 µM result in depression of cellular protein synthesis. Since cellular metabolism may be affected by changes of media and/or the physical manipulations associated with these changes, cells should be allowed recovery periods after such manipulations where experimentally feasible. Experimental protocols should be designed to avoid or to control for differences in physical treatment of cell cultures.

R_I from a variety of sources consists of two forms differing in isoelectric points by about 0.1 pH unit. In several cases this electrophoretic

[5] R. A. Steinberg, P. H. O'Farrell, U. Friedrich, and P. Coffino, *Cell* **10**, 381 (1977).
[6] R. A. Steinberg and P. Coffino, *Cell* **18**, 719 (1979).
[7] P. H. O'Farrell and P. Z. O'Farrell, *in* "Methods in Cell Biology," Vol. 14, p. 407. Academic Press, New York, 1977.

heterogeneity has been shown to result from phosphorylation.[3,5,8] Regulation of R_I phosphorylation can be studied using the electrophoretic difference in [^{35}S]methionine-labeled material or by differences in labeling with [^{32}P]- or [^{33}P]phosphate. Electrophoretic separation of [^{35}S]methionine-labeled forms of R_I provides the more versatile approach for studying phosphorylation, since visualization of both phosphorylated and nonphosphorylated forms of the protein allows direct determination of the extent of phosphorylation without concern about changes in rate of phosphate turnover or in amount of R_I. Phosphate labeling is essential only to demonstrate that phosphorylation causes the electrophoretic alteration and to analyze the peptide sites phosphorylated.

Medium for [^{35}S]Methionine Labeling

Methionine-free medium is prepared as a 10× concentrate of normal growth medium (Dulbecco's modified Eagle's Medium for S49 lymphoma cells) lacking methionine, sodium bicarbonate, glutamine and serum, and acidified to pH 1–2 with hydrochloric acid to prevent precipitation. The solution is sterilized by filtration and stored at 4° for up to 6 months. Serum is dialyzed against four changes of 0.15 M sodium chloride and portions stored frozen; small portions of 200 mM L-glutamine are also stored frozen. Sterile solutions of 1 M HEPES (N-2-hydroxyethylpiperazine-N'-2-ethanesulfonic acid), pH 7.4, 0.5 mM L-methionine, and 5% w/v sodium bicarbonate are stored at 4°. Low-methionine labeling medium (50 ml) is prepared by adding 5 ml 10× concentrate and 0.5 ml 5% sodium bicarbonate to about 30 ml twice-distilled water, then adjusting to pH 7.0–7.4 (based on phenol red indicator dye) with 10 N sodium hydroxide while bubbling with 5% carbon dioxide (or whatever concentration is appropriate for the incubators to be used). Glutamine is then added to 4 mM, methionine to 5 μM, and HEPES to 10 mM. Bubbling is stopped, 5 ml dialyzed serum is added, volume is adjusted to 50 ml with water, and the medium is filtered to sterilize. It is convenient to prepare low-methionine medium in advance of use and incubate overnight to ensure pH equilibration. Stocks of 10× medium are tested by preparing 1× medium with normal methionine content and testing its ability to support cell growth. High specific activity [^{35}S]methionine is stored in small portions at −70°.

Labeling Suspension Cells

The following is a protocol we have developed for S49 mouse lymphoma cells; it will require modification for cells of different sizes and

[8] R. L. Geahlen and E. G. Krebs, *J. Biol. Chem.* **255**, 9375 (1980).

metabolic rates. Exponentially growing cells are harvested by centrifugation, washed once by resuspension and centrifugation in low methionine medium, and resuspended at a density of 2.5×10^6 cells per ml (0.25% volume/volume suspension). After at least 2 hr of preincubation at 37° (during which effectors may be added), [^{35}S]methionine is added and incubation continued. The total preincubation plus labeling period should not exceed 6.5 hr. Incorporation is linear between 2 and 6.5 hr at a rate of about 10% of exogenous methionine per hour. Labeling 0.25 ml of such a culture at 200 μCi/ml [^{35}S]methionine for 30 min yields about 3.5×10^6 cpm of incorporated radioactivity in 20–30 μg of protein (sufficient material for two to three gels). Where metabolic stability is not a major concern, more efficient isotope utilization in short pulses can be achieved by increasing cell concentration to 2.5×10^7 per ml; at this density S49 cells will incorporate about 40–50% of exogenous methionine in an hour.

Incorporations are stopped by centrifuging cells in a microcentrifuge for 5 sec at ca. 10,000 g, aspirating media, and lysing with gel sample buffer [9.5 M urea, 1.6% Ampholines, pH 5–7 (LKB), 0.4% Ampholines, pH 3.5–10, 2% w/v Nonidet P-40, 5% 2-mercaptoethanol][4] to which an additional 25 mg of urea has been added per ml to bring the aqueous cell volume up to about 9.5 M in urea. This extraction buffer is prewarmed at 37° until all urea is dissolved, and 100 μl is used to extract 2.5×10^6 cells. The concentrations of contaminating serum proteins can be reduced by washing cells once with ice-cold physiological saline (containing 2 mM L-methionine) before lysing, but this is generally unnecessary.

Labeling Cells Attached to Plastic

The procedure outlined above for suspension cells is modified as follows. Cells are grown in 17-mm-diameter wells of multiwell dishes (available from Falcon or Costar) to greater than 80% confluence. Medium is aspirated, cells are washed with 0.5–1 ml low-methionine medium, then cells are preincubated for 2 or more hr with 0.5 ml low-methionine medium. For pulse-labeling, all but 100 μl of medium is removed from cells and 10–50 μCi [^{35}S]methionine is added. For labeling durations longer than about 0.5 hr, a greater volume of medium should be left to avoid dehydration or excessive depletion of methionine. Incorporation is stopped by aspirating labeling medium, rinsing quickly with 0.2–0.5 ml ice-cold physiological saline (containing 2 mM L-methionine), and then extracting with 50–100 μl high urea gel sample buffer (as described above).

Label-Chase Procedures

Chase protocols are appropriate for studies of protein turnover and modification. To avoid unnecessary metabolic perturbation, "conditioned medium" (prepared by filter sterilization of a culture supernatant) is used in washes and chase, media are prewarmed to 37°, and all operations are performed at 25–37°. When possible, cells are allowed at least an hour recovery time before sampling. For chase times longer than about 8 hr, antibiotics (200 units/ml penicillin and 200 µg/ml streptomycin) and fresh glutamine (to 4 mM) are included in chase media.

For attached cells a chase is initiated by aspirating label, rinsing with 0.2–0.5 ml conditioned medium, then replacing with 0.5–1.0 ml conditioned medium. For suspension cultures, samples are layered carefully onto cushions of whole serum (using a volume of serum at least twice the volume of sample), and the cells are washed by centrifugation through this layer; medium and serum are aspirated and cells resuspended in conditioned medium. These procedures remove more than 99% of unincorporated radioactivity, dilute residual isotope by increasing culture volumes from those used for labeling, and reduce methionine specific radioactivity by an additional 40-fold through restoration of normal methionine concentration; it is unnecessary therefore to add additional methionine to chase media.

Sample Handling and Determination of Incorporated Radioactivity

Since the two-dimensional gel system uses protein charge as one parameter for separation, it is essential to avoid the introduction of artifactual charge heterogeneity. Urea-denatured samples are kept frozen at −70° in small sealed tubes except for short periods of time required for assaying incorporated radioactivity or loading portions onto gels; heating is avoided. Cells labeled for purification of R_I (below) are frozen as aspirated pellets or monolayers and stored at −70° for subsequent extraction.

To quantify incorporated radioactivity in extracts, 1-µl samples are added to 99 µl water, then 20 µl of these 100-fold dilutions is coprecipitated with 25 µl 2.5% (w/v) bovine serum albumin using 4 ml ice-cold 5% trichloroacetic acid. Precipitates are collected onto glass fiber filters (Whatman GF/C) by filtration, tubes and filters rinsed with 5% trichloroacetic acid, and filters counted in a suitable scintillation cocktail (e.g., ACS, Amersham). Total cellular incorporation of [^{35}S]methionine (e.g., in experiments to assess labeling conditions) is monitored by dissolving culture samples in 0.1 ml 1 N sodium hydroxide, then precipitating and filtering as above; the sodium hydroxide solubilization reduces background radioactivity to negligible levels.

Two-Dimensional Gel Analysis

The procedure and apparatus for running high-resolution two-dimensional polyacrylamide gels are described elsewhere in great detail.[4,7] For resolution of R_I in crude cell extracts, first dimension isoelectric focusing gel composition is modified by increasing Ampholine concentration to 6% and increasing the amounts of TEMED and ammonium persulfate to ensure polymerization at this higher Ampholine concentration. Ten milliters of first dimension gel solution contains 5.5 g urea, 0.957 ml of water, 1.33 ml of acrylamide solution (28.38%, w/v, acrylamide, 1.62% N,N'-methylenebisacrylamide), 2 ml 10%, w/v, Nonidet P-40, 1.2 ml 40% Ampholines, pH 5–7, 0.3 ml 40% Ampholines, pH 3.5–10, 15 μl TEMED, and 15 μl 10%, w/v, ammonium persulfate. First dimension gels are run for 7000 V-hr (including a 400 V-hr prerun and a final hour at 800 V). Second dimension sodium dodecyl sulfate gels are 7.5% in polyacrylamide.

To ensure comparable gel patterns first dimension gels for all samples of an experiment are run together. Equal acid-precipitable cpm of each sample are loaded so that radioactivity in R_I represents the proportion of total incorporated radioactivity in R_I. It is not necessary to run all second dimension gels at the same time so long as the same reagents are used and tracking dye is allowed to migrate a standard distance. Gels from the same experiment are subjected to autoradiography and the autoradiograms developed in a single batch. Standardization wedges containing graded amounts of ^{14}C-labeled protein[4] are exposed with each set of gels to determine the linear response range of the film and to aid in quantitation.

Identification of R_I in Two-Dimensional Gel Patterns

R_Is from a variety of mammalian species have apparent subunit molecular weights of about 50,000 and isoelectric points of 6.0–6.3 in the O'Farrell two-dimensional gel system; they generally resolve into two species differing by about a unit in charge that correspond, in at least some cases, to nonphosphorylated and phosphorylated forms of the molecule.[3,5,8] Since a number of unrelated protein species have gel mobilities similar to those of one or the other form of R_I, it is essential to confirm the identities of putative R_I spots. Mobilities of bonafide R_I species are determined using samples purified on a cAMP affinity column; material mock-purified on a control column is used to distinguish R_I from contaminating proteins.

Columns containing 15–30 µl bed volumes of resin are prepared in 1 ml disposable syringe barrels using glass fiber filter discs (Whatman GF/C) as supports and disposable three-way stopcocks to control flow rates (extra outflow ports are sealed with the rubber gaskets from 1 ml disposable syringes). N^6-(2-Aminoethyl)-cAMP-Sepharose[9] is used for specific adsorption of R_I (a variety of other cAMP affinity resins should work equally well); control Sepharose is prepared by reacting cyanogen bromide activated Sepharose 4B (Pharmacia) with ethanolamine.[5] Washed, frozen, [^{35}S]methionine-labeled cells (above) are lysed with buffer containing 10 mM Tris, pH 7.5, 2 mM dithiothreitol, 2 mM ethylenediaminetetraacetic acid, 2 mM L-methionine, and 0.5% w/v Nonidet P-40 to give about 5 mg/ml protein in the extract. The extract is centrifuged 5 min at about 10,000 g in a micro centrifuge and portions of the supernatant fraction are loaded onto affinity or control columns that have been washed extensively with SB [10 mM Mes (4-morpholinoethanesulfonic acid), pH 6.6, 1 mM dithiothreitol, 2 mM ethylenediaminetetraacetate, 1 mM L-methionine, 0.3 mg/ml egg albumin (5× crystallized), 0.5% Nonidet P-40]. Samples containing 5–20 × 10^6 acid-precipitable cpm are allowed to enter column beds in portions equal to 2/3 of the bed volume, and each portion is allowed at least 20 min to bind. Columns may be loaded and run at 4° or at room temperature; R_I binding is favored at the higher temperature, but proteolysis might cause problems in some preparations. After loading is complete stopcocks are opened and columns are washed three times with SB, two times with SB containing 1 M sodium chloride, two times more with SB, and two times with SB lacking methionine and egg albumin. Each wash is 10 column volumes and two to three drops of additional buffer are used to initiate each buffer change. To elute columns, stopcocks are removed, and liquid below column beds is carefully withdrawn with tissue paper wicks. Gel sample buffer (75–150 µl) is applied to columns at room temperature, and the eluates are collected into tubes containing crystalline urea in the proportion of 10 mg urea per 15 µl column bed volume. (Most of the bound radioactivity is removed in the first two column volumes of eluate). When the urea is fully dissolved, 2–5 µl samples of the eluates are taken into scintillation fluid for counting. This affinity-purification procedure results in about 50- to 200-fold enrichment of R_I. Figure 1A shows the R_I region of a gel pattern from S49 mouse lymphoma cell proteins purified on N^6-(2-aminoethyl)-cAMP-Sepharose; Fig. 1B shows the same region from a gel of material

[9] W. L. Dills, Jr., J. A. Beavo, P. J. Bechtel, and E. G. Krebs, *Biochem. Biophys. Res. Commun.* **62,** 70 (1975).

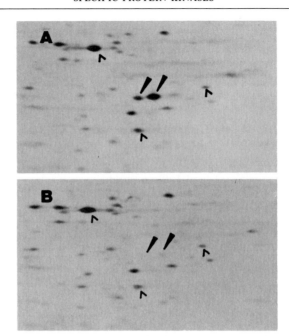

FIG. 1. Localization of R_I in two-dimensional gel patterns. An extract from S49 cells labeled for 2.5 hr with [^{35}S]methionine at 100 μCi/ml was affinity-purified on N^6-(2-aminoethyl)-cAMP-Sepharose (A) or mock-purified on control Sepharose (B); 10^5 cpm of column eluates were subjected to two-dimensional gel electrophoresis. Patterns shown are from 5-day autoradiographic exposures; the acidic ends of isoelectric focusing dimensions are at the right, and the high-molecular-weight regions of sodium dodecyl sulfate dimensions are at the top. Arrowheads indicate the positions of R_I species with the nonphosphorylated form on the left and the phosphorylated form on the right. Carets in this figure and in Fig. 2 indicate the positions of three major cellular proteins that can be used to aid in orientation of the patterns; the designated protein below R_I is nonmuscle actin.

eluted from control Sepharose. Positions of the two forms of R_I are indicated by arrowheads. The positions of three contaminating species that are abundant proteins in crude cell extracts are indicated by carets; these can be used to align the purified protein patterns with patterns from crude cell extracts (see Fig. 2).

A problem more difficult than the localization of R_I species is that of determining whether or not all radioactivity in the position of R_I species corresponds to R_I protein. In principle it should be possible to remove R_I from an extract by passage through an affinity column or by immune precipitation, but, in our hands, neither procedure completely removes R_I. Our studies on the metabolism of R_I in S49 cells suggest another approach to this problem, but its applicability to other systems has yet to be tested. In untreated S49 cells or in cells treated with $N^6,O^{2'}$-dibutyryl-cAMP R_I has a half-life in excess of 8 hr, while in cells treated with 8-bromo-cAMP R_I has a half-life of less than 1 hr.[2] Therefore, a chase of 8 hr in the presence of 8-bromo-cAMP will reduce labeling of R_I to less than 1% the level in cells chased without drug or with dibutyryl-cAMP. Figure 2 shows gel patterns from cells labeled for 1 hr with [^{35}S]methionine then

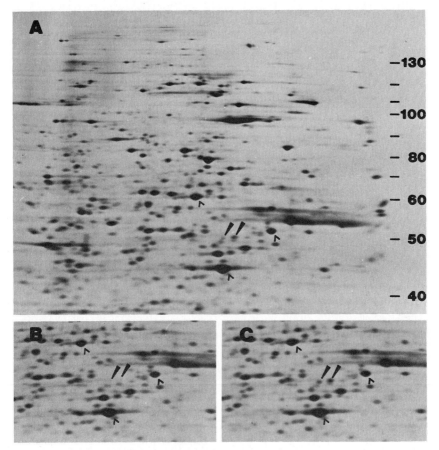

FIG. 2. 8-Bromo-cAMP-specific disappearance of R_I spots in cells chased for 8 hr after labeling with [^{35}S]methionine. S49 cells labeled for 1 hr with 100 μCi/ml [^{35}S]methionine were washed by centrifugation through a serum cushion then incubated for 8 hr in conditioned medium without drug (A), with 2 mM 8-bromo-cAMP (B), or with 1 mM dibutyryl-cAMP (C). Proteins (10^6 acid precipitable cmp) from whole cell extracts were subjected to two-dimensional gel electrophoresis, and autoradiograms were exposed for 6 days. Arrowheads indicate the two forms of R_I as in Fig. 1. The numbers to the right of A indicate approximate polypeptide molecular weights ($\times 10^3$).

chased for 8 hr in the absence (Fig. 2A) or presence of 8-bromo-cAMP (Fig. 2B) or dibutyryl-cAMP (Fig. 2C) to illustrate this approach. Label is easily detected in the positions of R_I in patterns from control and dibutyryl-cAMP-treated cells, but no label is detected in these positions from 8-bromo-cAMP-treated cells; the absence of labeled species in the positions of R_I in the patterns from 8-bromo-cAMP-treated cells argues that the radioactivity in these positions in gels from control and dibutyryl-cAMP-treated cells is entirely R_I. This procedure will not detect metabolically labile contaminants that might be a problem in gel patterns from pulse-labeled cells. A more general approach for detecting possible contaminants is to compare peptide maps of R_I species excised from gels of crude or affinity-purified material. Our experience suggests, however, that there is sufficient variability in the relative mobilities of R_I and potential contaminants that most contaminants will be discovered in the course of successive electrophoretic runs. Reproducible resolution of both forms of R_I from crude cell extracts requires great care in the preparation and running of gels plus a considerable measure of good luck.

Quantitation of Radioactivity in R_I

The gel approach described here is most useful for obtaining qualitative or semiquantitative data on R_I metabolism. For many experiments densitometry of gel autoradiograms is not necessary, as the human eye is very sensitive to differences in spot intensities. So long as gels are loaded with equal protein radioactivity, comparisons of R_I spots alone provide the relevant information on R_I metabolism. It is useful to make autoradiographic exposures of several lengths to aid in these comparisons; standardization wedges (above) provide a means of checking that spot intensities are linearly related to exposure and an aid in visual comparisons of spot intensities. It is more difficult to integrate the intensities of two or more spots by eye, so where comparisons involve gels where the distribution of label in the two forms of R_I is markedly different, densitometry might be necessary.

Two-dimensional gel autoradiograms are most conveniently scanned with a two-dimensional autodensitometer which successively measures the optical densities of contiguous small areas of film to generate a two-dimensional array of optical density values. Computer software has been described that can use such data to integrate radioactivity in individual protein species[10,11] and even to enhance resolution between partially over-

[10] J. I. Garrels, *J. Biol. Chem.* **254**, 7961 (1979).
[11] J. Bossinger, M. J. Miller, K.-P. Vo, E. P. Geiduschek, and N.-H. Xuong, *J. Biol. Chem.* **254**, 7986 (1979).

lapping species.[12] Spots can also be scanned with a one-dimensional microdensitometer by positioning the spot directly ahead of the light beam and scanning with a beam width just sufficient to include the entire spot in a single pass.[4] If different autoradiographic exposures must be used to obtain reliable optical density data for different points within an experiment, that data can be normalized by using standardization wedges or by scanning spots whose optical densities lie within the linear response range of the film in all exposures used.

Acknowledgments

The procedures described in this chapter were developed under Grants CA-14733 and AM-27916 from the National Institutes of Health.

[12] D. A. Agard, R. A. Steinberg, and R. M. Stroud, *Anal. Biochem.* **111**, 257 (1981).

[26] Phosphorylase Kinase from Rabbit Skeletal Muscle

By PHILIP COHEN

Glycogen phosphorylase b + ATP → Glycogen phosphorylase a + ADP
Glycogen synthase a + ATP → Glycogen synthase b + ADP

The phosphorylation of phosphorylase b takes place on a serine, 14 residues from the N-terminus of the polypeptide chain[1] and converts the enzyme from a form that requires AMP for activity to one that is almost fully active in the absence of this allosteric effector. The phosphorylation of glycogen synthase a occurs on a serine only seven residues from the N-terminus[2] and transforms the enzyme from a form that is almost fully active in the absence of glucose-6-phosphate to one that is more dependent on this activator.[2-6] The phosphorylation of glycogen phosphorylase and glycogen synthase occurs at comparable rates *in vitro*.

[1] K. Titani, A. Koide, J. Hermann, L. H. Ericsson, S. Kumor, R. D. Wade, K. A. Walsh, H. Neurath, and E. H. Fischer, *Proc. Natl. Acad. Sci. U.S.A.* **74**, 4762 (1977).
[2] N. Embi, D. B. Rylatt, and P. Cohen, *Eur. J. Biochem.* **100**, 339 (1979).
[3] P. J. Roach, A. A. DePaoli-Roach, and J. Larner, *J. Cyclic Nucleotide Res.* **4**, 245 (1978).
[4] A. A. DePaoli-Roach, P. J. Roach, and J. Larner, *J. Biol. Chem.* **254**, 4212 (1979).
[5] T. R. Soderling, A. K. Srivastava, M. A. Bass, and B. S. Khatra, *Proc. Natl. Acad. Sci. U.S.A.* **76**, 2536 (1979).
[6] K. Y. Walsh, D. B. Millikin, K. K. Schlender, and E. M. Reimann, *J. Biol. Chem.* **254**, 6611 (1979).

Assay Method

Principle

The conversion of phosphorylase b to a can be conveniently followed by assaying phosphorylase in the absence of AMP. The procedure is a modification of that described by Krebs.[7]

Reagents

Dilution Buffer. Phosphorylase kinase (EC 2.7.1.38) is diluted in 50 mM sodium glycerophosphate–0.1% (v/v) 2-mercaptoethanol pH 7.0 (buffer A) and this solution is stored on ice.

Phosphorylase b. This is prepared from rabbit skeletal muscle according to Fischer and Krebs[8] and recrystallized three times in the presence of AMP. The crystals are redissolved by warming at 30° in buffer A containing 2.0 mM EDTA, to give a final concentration of 30–50 mg/ml. The solution is treated several times with Dowex AGI-X4 resin to remove AMP,[9] and when the absorbance ratio has decreased to 0.53–0.54, the solution is dialyzed against buffer A containing 2.0 mM EDTA and 50% (v/v) glycerol pH 7.0. The AMP free phosphorylase b (~100 mg/ml) is stored at $-20°$, where it is stable for at least 2 years. Prior to use, aliquots are dialyzed overnight at 4° against buffer A to remove EDTA and glycerol, and diluted to 15 mg/ml in the same buffer, using an absorbance index, $A_{280\ nm}^{1\%}$, of 13.1 to measure the protein concentration.[10] The solution is stored on ice, and usually maintains its activity for several weeks. The specific activity of the phosphorylase b is 80 U/mg in the presence of 1.0 mM AMP using the assay described below, where 1.0 U of activity is that amount which releases 1.0 μmol of P_i per minute from glucose 1-phosphate.

Assay Buffers. Phosphorylase kinase is generally assayed at either pH 6.8 or 8.2. A solution containing 0.25 M Tris–0.25 M sodium glycerophosphate is adjusted to either pH 6.8 or 8.6 with 6 N HCl. The pH 6.8 buffer (5.0 ml) is mixed with 30 mM N'-(2-hydroxyethyl)ethylenediamine-N,N',N'-triacetic acid [HEDTA] (1.0 ml), 225 mM CaCl$_2$ (0.1 ml) and water (3.9 ml). The pH 8.6 buffer (5.0 ml) is mixed with 30 mM nitrilotriacetate [NTA] (1.0 ml), 225 mM CaCl$_2$ (0.1 ml), and water (3.9 ml). These buffers yield free Ca^{2+} concentrations in the final assay of 50 μM (6.8) and ~70 μM (8.2), which are saturating for the enzyme. The Ca-NTA buffer pH 8.6 slowly precipitates on storage and is therefore made

---

[7] E. G. Krebs, this series, Vol. 8, p. 543.
[8] E. H. Fischer and E. G. Krebs, *J. Biol. Chem.* **231**, 65 (1958).
[9] D. P. Gilboe, K. L. Larson, and F. Q. Nuttall, *Anal. Biochem.* **47**, 20 (1972).
[10] P. Cohen, T. Duewer, and E. H. Fischer, *Biochemistry* **10**, 2683 (1971).

up freshly each week. The assay buffers are stored frozen and thawed just before use.

Magnesium acetate (60 mM)–18 mM ATP pH 7.0. This is stored frozen, and thawed just before use.

Sodium maleate (0.1 M) pH 6.5 containing 1.0 mg bovine serum albumin/ml and 0.2% v/v 2-mercaptoethanol. The 2-mercaptoethanol is added just before use and the solution stored at 0°.

Glucose 1-phosphate (0.15 M)–rabbit liver glycogen (2% w/v)–sodium maleate pH 6.5 (0.1 M). The glycogen is treated with mixed bed resin before use.

Ammonium Molybdate Solution. This contains 14 ml of H_2SO_4 (36 N) 2.5 g of ammonium molybdate and 900 ml of water.

Aminonitrosulfonic Acid Reagent (ANSA). $Na_2S_2O_5$, (27.2 g), 1.0 g of anhydrous Na_2SO_3, and 0.5 g of ANSA are dissolved in 200 ml of water, by heating at 90°.

Procedure

The assay buffer, pH 6.8 or 8.6 (0.02 ml), is mixed with phosphorylase b (0.02 ml) and diluted phosphorylase kinase (0.01 ml) in a 10 × 1-cm glass test tube, and the mixture warmed at 30° for 3 min. The reaction is initiated with Mg-ATP (0.01 ml). The final pH of the assays are 6.8 or 8.2. After 5 min, 2.4 ml of ice cold sodium maleate pH 6.5 is added, which stops the reaction (solution A). The assays are performed in duplicate and control incubations are carried out in which phosphorylase kinase is replaced by dilution buffer. This corrects for traces of phosphorylase a activity in the phosphorylase b preparations, and for traces of phosphate present in any of the reagents.

Solution A (0.1 ml) is then added to 0.1 ml of the glucose-1-phosphate-glycogen solution that has been prewarmed for 3 min at 30° in a 12 × 1.2-cm test tube. After 5 min, the reaction is stopped by the addition of 4.55 ml of ammonium molybdate solution, followed by 0.25 ml of ANSA. After a further 5 min the absorbance at 620 nm is recorded and the phosphate released is calculated by reference to a standard curve.

The phosphorylase kinase is diluted so that no more than 20% of the phosphorylase b is converted to phosphorylase a. This corresponds to an absorbance increase of ~0.5 at 620 nm. The assay is very sensitive and even skeletal muscle extracts must be diluted 200-fold prior to assaying at pH 8.2.

Definition of Unit

One unit of phosphorylase kinase activity is that amount which catalyzes the conversion of 1.0 μmol of phosphorylase b to phosphorylase a

per minute in the standard assay at pH 8.2. In order to make this calculation the molecular weight of phosphorylase is taken as 97,400,[1] and the specific activity of phosphorylase a in the absence of AMP is taken as 70% of the specific activity of phosphorylase b in the presence of AMP.[10]

Alternative Assay

Phosphorylase kinase can also be assayed by measuring the incorporation of ^{32}P-radioactivity into phosphorylase b from [γ-^{32}P]ATP (5×10^7 cpm/μmol). The incubations are identical to those described above except that the ATP is reduced to 1.0 mM, and the reaction performed in a 1.5-ml plastic microcentrifuge tube. At the end of the assay the reaction is terminated by the addition of 1.0 ml of 5% w/v trichloroacetic acid, and processed as described elsewhere in this volume.[11] This assay is most convenient when glycogen synthase is used as the substrate. The final concentration of glycogen synthase in the assay is 0.6 mg/ml, and the procedure for purifying and storing this substrate is outlined in another chapter of this book.[11]

Purification Procedure

The preparation is modified from Brostrom *et al.*[12] All steps are performed at 0–4°. A lethal dose of sodium pentobarbitone (~5 ml at 100 mg/ml) is injected into the marginal ear vein of a large (4 kg) female New Zealand white rabbit, and the animal exsanguinated by cutting the jugular vein. The rabbit is skinned, and the muscle from the hind limbs and back rapidly removed and placed in ice. The muscle (about 800 g) is ground coarsely in an electric meat mincer, and homogenized in a Waring Blender at low speed for 30 sec with 2.5 vol of 4.0 mM EDTA–0.1% v/v 2-mercaptoethanol pH 7.0. The suspension is centrifuged at 6000 g for 45 min and the supernatant decanted through glass wool (step 1). The solution, which should be pH 6.7–6.8, is acidified to pH 6.1 by the addition of 1.0 N acetic acid and allowed to stand for 15 min. The turbid suspension is centrifuged at 6000 g for 45 min and the pH 6.1 supernatant used to prepare other enzymes, such as glycogen synthase kinase-3.[11] The precipitate is resuspended in 100 mM sodium glycerophosphate–4.0 mM EDTA–0.1% v/v 2-mercaptoethanol pH 8.2 (50 ml), using a hand homogenizer to disperse the pellet finely, and diluted to 140 ml with buffer A + 2.0 mM EDTA (step 2). The suspension is centrifuged at 78,000 g for 100 min and the supernatant carefully decanted through glass wool (step 3). One volume of cold neutral ammonium sulfate (475 g/ml) is added to 2 vol

[11] B. A. Hemmings and P. Cohen, (1982). this volume [36].
[12] C. O. Brostrom, F. L. Hunkeler, and E. G. Krebs, *J. Biol. Chem.* **246**, 1961 (1971).

TYPICAL PURIFICATION OF RABBIT MUSCLE PHOSPHORYLASE KINASE[a]

Step	Volume (ml)	Protein (mg)	Activity (units)	Specific activity (units/mg)	Purification (n-fold)	Yield (%)
1. Extract	10,200	204,000	12,240	0.06	1	100
2. pH 6.1 precipitation	800	19,200	11,120	0.58	9.7	91
3. 30,000 rpm supernatant	640	5,060	6,910	1.37	23	56
4. 30% $(NH_4)_2SO_4$ precipitate	22.4	1,434	4,230	2.95	49	35
5. Sepharose 4B eluate	225	210	2,150	10.2	170	18

[a] At steps 1–4 protein was measured by the procedure of Lowry et al. (J. Biol. Chem. **193**, 265, 1951). At step 5, the absorbance index, $A_{280\ nm}^{1\%} = 12.4$, found for purified phosphorylase kinase[13] was used; 4500 g muscle (six rabbits) was used in this preparation. The specific activity in muscle extracts ranges from 0.06 to 0.08 U/mg.

of supernatant, and after standing for 5 min, the suspension is centrifuged at 10,000 g for 10 min. The supernatant is discarded, and the precipitate is gently resuspended in a minimum volume of buffer A + 2.0 mM EDTA and dialyzed for 60 min against this buffer. The dialysate (2.5 ml) is clarified by centrifugation at 15,000 g for 2 min (step 4) and subjected to gel filtration on a column of Sepharose 4B (150 × 2 cm) equilibrated in buffer A + 2.0 mM EDTA. Three major peaks of 280 nm absorbance are eluted from the column.[13] The middle peak (phosphorylase kinase), which is eluted at a V_e/V_0 value of ~1.9, is collected (step 5), concentrated to ~4 ml by vacuum dialysis, further dialyzed against buffer A + 2 mM EDTA + 50% v/v glycerol, and stored at −20°. Under these conditions there is no loss of activity for at least a year. A summary of the purification is given in the table.

Comments on the Purification

The key step is the precipitation of phosphorylase kinase at pH 6.1 with the protein–glycogen complex, and it is therefore essential to minimize glycogenolysis during the excision of the muscle. This is achieved by cutting the hind limb muscles with sharp scissors along the bones, and by cutting the back muscle with a sharp scalpel along the vertebral column. Transverse cuts, which cause vigorous contraction of the muscle, are avoided until the last moment, and the muscle then plunged rapidly into ice, which stops further contraction. If the pH of the extract is below 6.5 excessive glycogenolysis has occurred, little phosphorylase kinase will precipitate at pH 6.1, and the extract should be discarded.

[13] P. Cohen, Eur. J. Biochem. **34**, 1 (1973).

Phosphorylase kinase has a strong tendency to aggregate and lose activity when kept in high concentrations of ammonium sulfate for prolonged periods, and step 4 should be carried out rapidly, as described under the purification procedure.

If homogeneous enzyme is required, only the leading half of the peak from Sepharose 4B should be pooled. The trailing half contains variable amounts of an impurity that has a subunit molecular weight of 100,000, slightly greater than that of phosphorylase.[13]

The purification procedure is highly reproducible and can be completed within 24 hr up to step 5. It can be scaled up sixfold without any increase in preparation time, simply by using a much larger Sepharose 4B column (150 × 5 cm). Enzyme (30–40 mg) is routinely isolated from a single rabbit (~800 g muscle) corresponding to an overall yield of 15–20%.

Properties

The leading half of the peak from Sepharose 4B is essentially homogeneous. Phosphorylase kinase is composed of four types of subunit α, β, γ, and δ with molecular weights of 145,000, 130,000, 45,000, and 17,000, respectively. The smallest active species has the structure $(\alpha\beta\gamma\delta)_4$.[14] The α- and β-subunits are components phosphorylated by cAMP-dependent protein kinase,[13] the γ-subunit appears to be the catalytic subunit,[15] while the δ-subunit is identical to the Ca^{2+}-binding protein calmodulin.[14,16] Phosphorylase kinase comprises 0.6–0.8% by weight of the soluble muscle protein (see the table), and its intracellular concentration is 2.5 μM.[17] The K_m for ATP is 0.4 mM.

Purified phosphorylase kinase is contaminated with traces of at least two other protein kinases, namely cAMP-dependent protein kinase and an enzyme that has been variously termed glycogen synthase kinase-5, casein kinase-2, casein kinase TS, casein kinase N-II, and casein kinase-G.[18] These two contaminants do not phosphorylate phosphorylase b, and therefore do not interfere with the routine assay. Both contaminating protein kinases phosphorylate glycogen synthase, however, and may affect studies of its phosphorylation when incubations are carried out for prolonged periods with high concentrations of phosphorylase kinase.[18]

---

[14] S. Shenolikar, P. T. W. Cohen, P. Cohen, A. C. Nairn, and S. V. Perry, *Eur. J. Biochem.* **100**, 329 (1979).

[15] J. R. Skuster, J. F. Jesse Chan, and D. J. Graves, *J. Biol. Chem.* **255**, 2203 (1980).

[16] R. J. Grand, S. Shenolikar, and P. Cohen, *Eur. J. Biochem.* **113**, 359 (1981).

[17] P. Cohen, *Curr. Top. Cell. Reg.* **14**, 117 (1978).

[18] P. Cohen, D. Yellowlees, A. Aitken, A. Donella-Deana, B. A. Hemmings, and P. J. Parker, *Eur. J. Biochem.* **124**, 21 (1982).

These problems can be prevented by adding the specific protein inhibitor of cAMP-dependent protein kinase to the assays, and by including heparin (2 μg/ml) which inhibits glycogen synthase kinase-5 specifically.[18]

The pH 6.8/8.2 activity ratio of the dephosphorylated form of phosphorylase kinase is 0.07 at 50–70 μM Ca^{2+}.[19] Lower activity ratios reported in the literature are due to the use of suboptimal concentrations of Ca^{2+} at pH 6.8. The addition of calmodulin (5 μg/ml) activates the enzyme fivefold at pH 6.8 and 1.5-fold at pH 8.2, and therefore increases the activity ratio to 0.23.[19] This is caused by the binding of a second molecule of calmodulin (termed the δ'-subunit) to phosphorylase kinase.[14,20] The δ'-subunit only binds to the enzyme in the presence of Ca^{2+} where it interacts with the α- and β-subunits.[14,20] This contrasts with the δ-subunit which remains tightly associated with the γ-subunit even in the absence of Ca^{2+}.[14,20]

Phosphorylation of the α- and β-subunits by cAMP-dependent protein kinase at 1.0 mM free Mg^{2+} activates the enzyme ~15-fold at pH 6.8 and ~1.5-fold at pH 8.2 and therefore increases the pH 6.8/8.2 activity ratio to ~0.7–0.8 at 50–70 μM Ca^{2+}. Phosphorylation also decreases the activation constant for Ca^{2+} from 10–20 μM to ~1 μM.[19]

Limited proteolysis by trypsin activates the enzyme ~20-fold at pH 6.8 and 1.3-fold at pH 8.2, and raises the pH 6.8/8.2 activity ratio to ~0.9 at 50–70 μM Ca^{2+}. Similar results are obtained with chymotrypsin. Activation by trypsin is accompanied by degradation of the α- followed by the β-subunit, while chymotrypsin degrades the α-subunit relatively selectively. The γ- and δ-subunits are unaffected by these proteinases at concentrations that produce maximal activation.[19] Neither the phosphorylated nor proteolytically modified forms of phosphorylase kinase are activated significantly by the δ'-subunit.[19]

Limited proteolysis by trypsin decreases the concentration of Ca^{2+} required for activation from 10–20 μM to <0.1 μM.[19] The dramatic effects of limited proteolysis on the kinetic parameters have important practical implications since even 2% proteolysis drastically affects plots of activity vs Ca^{2+} concentration, decreasing the apparent $A_{0.5}$ for Ca^{2+} by almost 10-fold.[19] Such minor proteolysis of the α- and β-subunits is not detectable by polyacrylamide gel electrophoresis or by measuring the pH 6.8/8.2 activity ratio at saturating Ca^{2+}. Trace proteolysis of phosphorylase kinase by endogenous proteinases is difficult, if not impossible, to avoid during the preparation. For this reason it is advisable to study the kinetic properties of the enzyme using freshly prepared material purified up to step 3. After the final two steps of purification the properties of the enzyme alter

---

[19] P. Cohen, *Eur. J. Biochem.* **111**, 563 (1980).
[20] C. Picton and P. Cohen, *Eur. J. Biochem.* **111**, 553 (1980).

slightly, in a way that suggests a 1–2% proteolysis of the α-subunit may have occurred.[19]

Other properties of phosphorylase kinase are reviewed in detail elsewhere.[18,21,22]

Acknowledgment

This work was supported by a Program Grant from the Medical Research Council, London, England.

[21] P. Cohen, C. B. Klee, C. Picton, and S. Shenolikar, *Ann. N.Y. Acad. Sci.* **356,** 151 (1980).
[22] C. Picton, C. B. Klee, and P. Cohen, *Cell Calcium* **2,** 281 (1981).

[27] Cardiac Phosphorylase Kinase: Preparation and Properties

By HEI SOOK SUL, BERNADETTE DIRDEN, KAREN L. ANGELOS, PATRICK HALLENBECK, and DONAL A. WALSH

Phosphorylase kinase (ATP: phosphorylase b phosphotransferase, EC 2.7.1.38) is a complex enzyme of molecular weight 1.3×10^6 and a subunit composition position of $\alpha_4 \cdot \beta_4 \cdot \gamma_4 \cdot \delta_4$.[1-5] In muscle, two major isozymes exist distingished only by the size of the α subunit. The molecular weights of the β-, γ-, and δ-subunits in all types of muscle so far examined are 125,000, 48,000, and 17,000, respectively. The α subunit in fast-twitch glycolytic muscle is distinct and slightly larger than that from oxidative red skeletal and cardiac muscle and this subunit in the latter two is designated α'. As is illustrated in Fig. 1, the α- and α'-subunits from rabbit skeletal muscle are each slightly larger than those from bovine skeletal muscle and when coelectrophoresed[6] all four can be separated (wells 4–6). For the same species, however, the α'-subunit of cardiac muscle is identical to the α'-subunit from skeletal muscle (wells 1–3). Bovine car-

[1] T. Hayakawa, J. P. Perkins, D. A. Walsh, and E. G. Krebs, *Biochemistry* **12,** 567 (1973).
[2] P. Cohen, *Eur. J. Biochem.* **34,** 1 (1973).
[3] P. Cohen, A. Burchell, J. G. Foukles, P. T. W. Cohen, T. C. Vanaman, and A. C. Nair, *FEBS Lett.* **92,** 287 (1978).
[4] A. Burchell, P. T. W. Cohen, and P. Cohen, *FEBS Lett.* **67,** 17 (1976).
[5] R. H. Cooper, H. S. Sul, T. F. McCullough, and D. A. Walsh, *J. Biol. Chem.* **255,** 11794 (1980).
[6] U. K. Laemmli, *Nature (London)* **227,** 680 (1970).

1. Bovine SK

2. Bovine SK plus Bovine Cardiac

3. Bovine Cardiac

4. Bovine SK

5. Rabbit SK

6. Bovine SK plus Rabbit SK

7. Bovine SK

8. Rabbit SK plus Bovine Cardiac

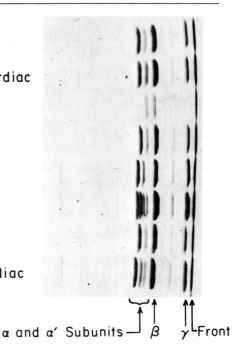

α and α' Subunits —┘ β γ—Front

FIG. 1. Gel electrophoresis of purified phosphorylase kinase from bovine heart, bovine skeletal muscle, and rabbit skeletal muscle. Bovine heart enzyme was purified as described in this text. Bovine and rabbit skeletal muscle enzymes were purified as described by Hayakawa et al.,[1] with the exception that the terminal chromatography on Sepharose 4B was replaced by zonal centrifugation using the conditions described in Step 6 for purification of the bovine heart enzyme. Electrophoresis was in accord with the method of Laemmli[6] using 5% gels.

diac α' subunit is also distinct from rabbit skeletal α' (well 8). The approximate molecular weights of these subunits are α rabbit skeletal, 145,000; α bovine skeletal, 142,000; α' rabbit skeletal, 139,000; α' bovine skeletal and heart, 135,000. These observed differences are not an artifact of preparation since identical results are obtained with enzyme isolated from initial cell extracts by immunoprecipitation. As is indicated in Fig. 1, the β-subunits in all types of bovine and rabbit muscle appear to be identical.

The γ-subunit of phosphorylase kinase contains the catalytic site[7] although there may be more than one type of catalytic site. The δ-subunit is identical to calmodulin and is the site of Ca^{2+}-dependent allosteric regulation, however, the fast-twitch skeletal isozyme (α), but not the α' isozyme, has a second site of Ca^{2+} sensitivity as a consequence of a readily

[7] K.-F. J. Chen and D. J. Graves, J. Biol. Chem. **257**, 5939, 5948, and 5956 (1982).

reversible interaction with either additional calmodulin or troponin C subunit (TNC); the latter two have also been referred to as being "phosphorylase kinase δ' subunit."[8] Both the α- (α') and β-subunits are phosphorylated, each by either the cAMP-dependent protein kinase or by autophosphorylation. Phosphorylation of either the α- (α') or β-subunits results in activation.[1,2,9,10]

In the preceding chapter in this volume[11] is a description of skeletal muscle phosphorylase kinase (α isozyme); this chapter reports upon the purification and properties of the enzyme isolated from bovine cardiac muscle and the procedure to measure the activation state of the cardiac enzyme in intact tissue.

Assay Method

Principle

Phosphorylase b kinase catalyzes the conversion of phosphorylase b to phosphorylase a by transferring the γ-phosphate of ATP to a single specific serine residue on each of the subunits. The initial assay,[12] based upon measurement of phosphorylase activity, has since been displaced by a determination of phosphorylase b phosphorylation; the latter is more sensitive, but investigators must be cautious to constantly monitor that the phosphorylase b used for this purpose has maintained its integrity. A partially denatured phosphorylase, as would be characterized by a decreased specific activity, may become a poorer substrate for phosphorylase kinase and/or a substrate for other protein kinases. The activity of the phosphorylase kinase is determined by measuring the amount of ^{32}P transferred from [γ-^{32}P]ATP to phosphorylase b. The protein is separated from the other components of the assay mixture by precipitation of it onto filter paper with trichloroacetic acid.

Reagents

Phosphorylase b. Skeletal muscle phosphorylase b (3× crystallized), prepared by the method of Fischer and Krebs,[13] may be stored conve-

[8] P. Cohen, *Eur. J. Biochem.* **111**, 562 (1980).
[9] H. S. Sul, R. H. Cooper, S. Whitehouse, and D. A. Walsh, *J. Biol. Chem.* **257**, 3484 (1982).
[10] H. S. Sul and D. A. Walsh, *J. Biol. Chem.* **257**, in press (1982).
[11] P. Cohen, this volume [26].
[12] E. G. Krebs, D. S. Love, G. E. Bratvold, K. A. Trayser, W. L. Meyer, and E. H. Fischer, *Biochemistry* **3**, 1023 (1964).
[13] E. H. Fischer and E. G. Krebs, *J. Biol. Chem.* **231**, 65 (1958).

niently as a powder obtained by direct lyophilization of the crystalline suspension within 3 days of crystal formation. This lyophilization requires particular care and, because of the high salt content of the crystalline suspension, must be done only from a very thin shell (e.g., 75 ml of crystalline suspension in a 2 liter flask). The lyophilized powder, stored at $-15°$, maintains full activity for at least 2 years.

To solubilize, 0.5 g of powder (i.e., phosphorylase plus salts) is dissolved by incubating at 30° for 1 min in 10 ml of 50 mM glycerol phosphate–45 mM 2-mercaptoethanol (pH unadjusted). Undissolved material is removed by centrifugation at 25° and then the solution is placed on ice for 3 hr. This induces crystallization of the phosphorylase due to the presence of 5'-AMP carried over from the initial crystallization. The crystals are collected by centrifugation, and phosphorylase is redissolved by suspending in 4 ml of 10 mM glycerol phosphate (pH 6.8)–45 mM 2-mercaptoethanol with 100 mg of acid-washed Norite and then incubating at 30° for 10 min. The Norite (plus some undissolved protein) is removed by centrifugation and the supernatant solution treated again for 3 min with 50 mg of Norite. The specific activity of phosphorylase obtained is in the range of 1350–1650 Cori[14] units/mg, its concentration, as prepared by this method, is between 60 and 70 mg/ml, it has an a/b ratio of 0.01, and an E_{280}/E_{260} absorbance ratio of between 1.7 and 1.8. This solution may be stored at 0° for a period of 2 to 4 weeks.

For phosphorylase kinase assay the stock solution is diluted in 10 mM glycerol phosphate–45 mM 2-mercaptoethanol (pH 6.8) to give a final assay concentration of 2.5 mg/ml. The optimum procedure is to incubate the stock solution at 15° for 20 hr and only add it to the diluent and then the preincubation mixture after both have reached a temperature of 20°. This protocol negates problems associated with cold-promoted changes in phosphorylase conformation.

Assay Buffer. 125 mM Tris–125 mM glycerol phosphate–50 μM CaCl$_2$, adjusted to either pH 8.6 or 6.8.

ATP-Mg^{2-}. 18 mM [γ-^{32}P]ATP (specific activity 10–50 cpm/pmol, prepared by the modification of the Glynn and Chappell procedure[15,16]) plus 60 mM magnesium acetate.

Preincubation Mixture. Two volumes of assay buffer, 2 volumes of phosphorylase b, and 1 volume of ATP-Mg^{2-} to give, in the final assay reaction mixture, concentrations of 41.67 mM Tris, 41.67 mM β-glycerol

---

[14] G. T. Cori, B. Illingworth, and P. J. Keller, this series, Vol. 1, p. 200.
[15] I. M. Glynn and J. B. Chappell, *Biochem. J.* **90**, 147 (1964).
[16] D. A. Walsh, J. P. Perkins, C. O. Brostrom, E. S. Hu, and E. G. Krebs, *J. Biol. Chem.* **246**, 1968 (1971).

phosphate, 16.7 μM CaCl$_2$, 3 mM ATP, 10 mM magnesium acetate, and 2.5 mg/ml phosphorylase b.

Tissue Extraction Buffer. 30 mM Tris (pH 7.5)–30 mM KCl–5 mM EDTA–100 mM NaF–1 mM phenylmethylsulfonyl fluoride.

Procedure

A 50-μl aliquot of preincubation mixture is preincubated at 30° for 2 min and the reaction then initiated by addition of 10 μl of diluted phosphorylase kinase. For determination of activity of pure protein the enzyme is diluted so that its concentration in the final assay reaction mixture is less than 2 μg/ml. The reaction is terminated after 10 min by transfer of a 50-μl aliquot onto squares (1 × 1 cm) of Whatman ET 31 filter paper and the paper immediately immersed in 10% (w/v) trichloroacetic acid. The papers are then washed for 30-min periods, once with 10% trichloroacetic acid at 4°, once with 5% trichloroacetic acid at 4°, twice with 5% trichloroacetic acid at 25°, and then once with 95% ethanol at 25° for 5 min; this method of washing is as described by Corbin and Reimann.[17] The papers are dried and then counted in a toluene-based scintillation fluor.

Units. One unit of enzyme activity is defined as that amount which catalyzes the incorporation of 1 μmol of ^{32}P from [γ-^{32}P]ATP into phosphorylase b per min. For enzyme specific activity, assays are at pH 8.2.[18] Activation of the enzyme can be expressed either as activity at pH 6.8 per mg of protein, or by the pH 6.8/pH 8.2 activity ratio.

Assay of Phosphorylase Kinase in Tissue Extracts and Determination of the Degree of Activation. To determine the activity of phosphorylase kinase in intact tissue, the heart must be rapidly frozen by the use of Wollenberger tongs and the tissue powdered at −70°. Between 150 and 200 mg of frozen powder at −70° and 0.75 ml of tissue extraction buffer are added in rapid succession to a Teflon-glass homogenizer (at 0°) and the homogenization continued for three or four strokes after the powder is thawed and dispersed. The extract is then centrifuged at 39,000 g for 15 min, the supernatant decanted through glass wool, and the filtrate diluted with an equal volume of 10 mM glycerol phosphate (pH 6.8)–5 mM EDTA–125 mM NaF–45 mM 2-mercaptoethanol. The assay of activity is identical to that described above except that the reaction mixture should contain 10 mg/ml glycogen, the phosphorylase b concentration should be increased to 5 mg/ml, and the reaction time should be for 30 min. The most reliable index of activation is the specific activity at pH 6.8; the pH 6.8/pH 8.2 activity ratio may also be used but this value is often subject to

---

[17] J. D. Corbin and E. M. Reimann, this series, Vol. 38, p. 287.
[18] Assay buffer at pH 8.6 results in a pH of 8.2 in the final reaction.

a wider dispersion of the data. Glycogen is used in this assay to increase the sensitivity. It stimulates the activity at pH 6.8 by 3- to 3.5-fold and at pH 8.2 by 1.6-fold.

Purification of Cardiac Phosphorylase Kinase

Reagents

Extraction Buffer. 30 mM Tris (pH 7.4)–50 mM NaF–30 mM NaCl–5 mM EDTA–0.1 mM phenylmethylsulfonyl fluoride–5 mM 2-mercaptoethanol.
Polyethylene Glycol. 50% (w/v) polyethylene glycol (6000 MW) dissolved in H_2O.
Buffer A. 30 mM Tris (pH 7.4)–50 mM NaF–30 mM NaCl–5 mM EDTA–10% sucrose (w/v) and 5 mM 2-mercaptoethanol.
Buffer B. 50 mM MES (pH 7.0)–50 mM NaCl–2 mM EDTA–5% (w/v) sucrose–5 mM 2-mercaptoethanol.
Buffer C. 50 mM glycerol phosphate (pH 6.8)–2 mM EDTA–5 mM 2-mercaptoethanol.

Procedure

Step 1, Tissue Extraction. Four beef hearts are obtained fresh from a local slaughterhouse and transported on ice to the laboratory. All subsequent steps are carried out at 4°. After removal of the pericardium and fat tissue, the heart muscle (4 kg wet weight) is cut into strips, passed through a meat grinder, and then homogenized in 2.5 volumes of extraction buffer in a Waring blender for 5 sec at low speed and 15 sec at high speed. After centrifugation at 7000 g for 15 min, the supernatant is decanted through glass wool.

Step 2, First Polyethylene Glycol Precipitation. The extract is brought to 3% (w/v, final concentration) polyethylene glycol by the addition of the 50% (w/v) stock solution. After stirring for 10 min, the precipitate is collected by centrifugation at 7000 g for 15 min and resuspended in 1 liter of buffer A using a Teflon-glass Potter-Elvejhem homogenizer. The above suspension (polyethylene glycol precipitate fraction) is immediately centrifuged for 90 min at 40,000 g and the supernatant solution then decanted through glass wool to remove coagulated lipid.

Step 3, Second Polyethylene Glycol Fractionation. To the supernatant obtained from the first polyethylene glycol precipitation is added polyethylene glycol to a concentration of 3% (w/v), the suspension is stirred for 5 min, and then centrifuged at 20,000 g for 30 min. The pellet is discarded and additional polyethylene glycol is added to the supernatant to a final concentration of 10% (w/v). The suspension is again stirred for 5 min and centrifuged at 20,000 g for 30 min. The white fluffy pellets are suspended in 100 ml of buffer B using a Teflon-glass homogenizer. This fraction is stored at 0° overnight. The next day, the solution is centrifuged again at 27,000 g for 30 min and the supernatant is used for further purification.

Step 4, First Zonal Sucrose Density Gradient Centrifugation. Sucrose density gradient centrifugation in a zonal rotor has proved to be an efficient and reproducible method for purification of the enzymes of glycogen metabolism. Further purification of the cardiac phosphorylase kinase obtained from polyethylene glycol fractionation (Step 3) is accomplished with a 7.5 to 35% sucrose gradient in the 1655-ml capacity Beckman Ti-15 zonal rotor. The gradient is produced with a two-chamber gradient maker with the output mixing chamber being closed and containing 800 ml of buffer C, and the open addition chamber containing 1100 ml of buffer C plus 40% (w/v) sucrose. The gradient is pumped into the outer rim of the zonal rotor, spinning at 3000 rpm, until the concentration in the core overflow line reaches 7.5% sucrose. The phosphorylase kinase solution obtained from the preceding Step 3 is then applied to the gradient via the core input and overlayed with 650 ml of buffer C. Finally, a 50-ml cushion of buffer C plus 40% sucrose is introduced via the rim inlet. Centrifugation is for 17 hr at 30,000 rpm. The rotor is unloaded via displacement with 40% sucrose. The profile of sedimentation is illustrated in Fig. 2; the

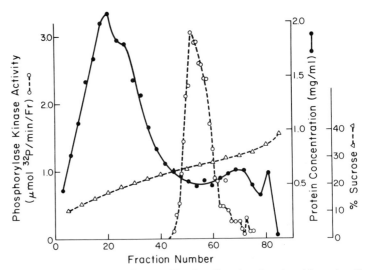

FIG. 2. First zonal sucrose gradient profile of cardiac phosphorylase kinase (purification, Step 4). Fraction volumes were 10 ml.

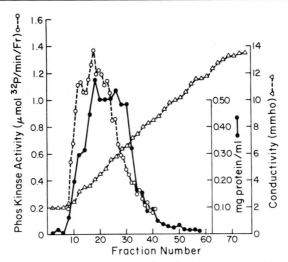

FIG. 3. DEAE-cellulose chromatography of cardiac phosphorylase kinase (purification, Step 5). Fraction volumes were 5 ml.

enzyme sediments as a fairly symmetrical peak, well separated from lower molecular weight contaminants.

Step 5, DEAE-Cellulose Chromatography. The fractions containing phosphorylase kinase (Fig. 2) are pooled (~100 ml) and applied to a column of DEAE-cellulose (1.5 × 10 cm) equilibrated with buffer C containing 10% sucrose. The column is washed with 100 ml of this buffer and then eluted with a 200 ml linear gradient of 0 to 0.5 M NaCl in the same buffer. The profile of elution is indicated in Fig. 3. Quite typically, phosphorylase kinase commences elution at 3 mmho, reaching a maximum peak at 4 mmho, but there is a distinctive trailing edge. Only the fractions making up the symmetrical main peak are pooled for subsequent purification.

Step 6, Second Zonal Sucrose Density Gradient Ultracentrifugation. The enzyme eluted from DEAE may be purified to near homogeneity by a second zonal centrifugation. The conditions employed are analogous to those of the first zonal centrifugation (Step 4) with the exception of the use of a Ti-14 rotor (650-ml capacity) and a gradient of 15 to 30% sucrose (w/v). No more than 50 ml of enzyme solution is applied to the gradient. Centrifugation is for 20 hr at 35,000 rpm. The profile of elution is indicated in Fig. 4. The enzyme in the pooled fractions may be concentrated 2- ~3-fold using an Amicon ultrafiltation cell with a XM-300 membrane and can then be stored at −70° for several months with little loss of activity.

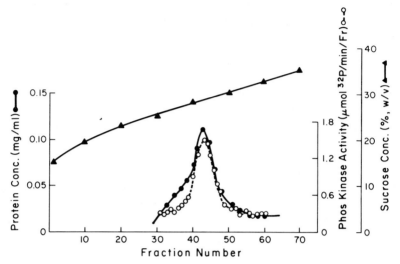

FIG. 4. Second zonal sucrose gradient profile of cardiac phosphorylase kinase (purification, Step 6). Fraction volumes were 5 ml. From Cooper et al.[5]

Summary of Purification

A summary of the purification is presented in the table. The enzyme obtained by this procedure is 90 ~ 95% pure with a specific activity in the range of 2 ~ 4 units per mg of protein and a pH 6.8/8.2 activity ratio of between 0.07 and 0.12.

PURIFICATION OF BOVINE CARDIAC PHOSPHORYLASE KINASE

Step	Specific activity (units/mg)	Purification (n-fold)	Yield (%)
1. Initial extract	0.00291	1	100
2. Precipitate of 3% polyethylene glycol	0.0353	12	63
3. Polyethylene glycol fractionation, 3–10%	0.0777	27	58
4. Zonal centrifugation (Ti-15)	0.953	328	10
5. DEAE-cellulose column	2.06	708	5
6. Zonal centrifugation (Ti-14)	3.68	1265	3

Properties of Cardiac Phosphorylase Kinase

The properties of the bovine cardiac enzyme have been presented.[5,9,10,19] With 80 μM phosphorylase b and 20 mM Mg^{2+} at pH 8.2 the K_m for ATP is 0.22 mM and with 3 mM ATP the K_m for phosphorylase b is 15 μM. Mn^{2+} can substitute for Mg^{2+} in the kinase reaction but is only 70% as effective. When the concentration of ATP exceeds that of divalent cations, kinase activity is totally inhibited. UTP and CTP do not serve as substrates and GTP and ITP show only marginal activity. The enzyme requires Ca^{2+} for activity with a K_a at pH 6.8 of 1.9 μM.

The enzyme can be activated two- to fourfold at pH 6.8 by cAMP-dependent protein kinase-catalyzed phosphorylation. As with the skeletal muscle enzyme, β-subunit phosphorylation precedes that of the α'-subunit but both β-subunit phosphorylation and α'-subunit phosphorylation cause enzyme activation. In contrast to skeletal muscle phosphorylase kinase (α isomer), the β-subunit can only be phosphorylated to a maximum level of 1 mol per four β-subunits, indicating strong site–site negative cooperativity. Mn^{2+} stimulates cAMP-independent phosphorylation of the α'-subunit, most likely due to autocatalysis, and this activates the enzyme three- to sixfold. The phosphorylation of both the α'- and β-subunits has been demonstrated in the intact perfused rat heart.[20]

[19] R. H. Cooper, H. S. Sul, and D. A. Walsh, *J. Biol. Chem.* **256**, 8030 (1981).
[20] T. E. McCullough and D. A. Walsh, *J. Biol. Chem.* **254**, 7336 (1974).

[28] Separation of the Subunits of Muscle Phosphorylase Kinase

By K.-F. JESSE CHAN and DONALD J. GRAVES

Phosphorylase kinase (ATP:phosphorylase b phosphotransferase, EC 2.7.1.38) is the key enzyme linking the hormonal and neural controls of glycogen metabolism.[1-3] The holoenzyme purified from rabbit skeletal muscle is a huge and complex macromolecule.[4,5] It has a molecular weight of ca. $1-3 \times 10^6$ and is composed of four different subunits with a stoichi-

[1] H. Nimmo and P. Cohen, *Adv. Cyclic Nucleotide Res.* **8**, 145 (1977).
[2] P. Cohen, *Curr. Top. Cell. Regul.* **14**, 117 (1978).
[3] G. M. Carlson, P. J. Bechtel, and D. J. Graves, *Adv. Enzymol.* **50**, 41 (1980).
[4] T. Hayakawa, J. P. Perkins, D. A. Walsh, and E. G. Krebs, *Biochemistry* **12**, 567 (1973).
[5] P. Cohen, *Eur. J. Biochem.* **34**, 1 (1973).

ometry of $(\alpha \beta \gamma \delta)_4$.[6–8] The γ-subunit has been identified as a catalytic subunit toward the conversion of phosphorylase b,[7–10] whereas the δ-subunit is identical to the ubiquitous Ca^{2+}-binding protein, calmodulin.[11] The α- and β-subunits can be phosphorylated and dephosphorylated and are important for regulation of catalytic activity.[2] Also, it has been suggested that the β-subunit contains a catalytic site for phosphorylase conversion.[12] It is now established that red soleus muscle and cardiac muscle, both functionally more oxidative than white skeletal muscle, contain the isozyme $(\alpha' \beta \gamma \delta)_4$.[13,14] The isozyme containing the α'-subunit is not stimulated by exogenous calmodulin and can be separated from the isozyme containing the α-subunit by affinity chromatography.[15] The molecular structure and topography of the holoenzyme, the modes of subunit-subunit interactions, and the means of biological information transfer among the different subunits are essentially unknown.

One approach to the study of the complex structure–function relationships of phosphorylase kinase is to isolate and characterize the individual subunits and their reconstituted entities. Through the dissociation of the holoenzyme with chaotropic salts, sodium dodecyl sulfate, and limited proteolysis, different subunit forms of the enzyme can be generated.

Methods to Monitor Dissociation

Enzyme Activity Assays

The standard assays consist of 0.25 M Tris/0.25 M sodium β-glycerophosphate/0.6 mM $CaCl_2$/60 mM magnesium acetate, adjusted to pH 6,8 or pH 8.6 with HCl (10 μl); phosphorylase b (6–30 mg/ml) in 40 mM sodium β-glycerophosphate/30 mM 2-mercaptoethanol, pH 6.8 (20 μl);

---

[6] S. Shenolikar, P. T. W. Cohen, P. Cohen, A. C. Nairn, and S. V. Perry, *Eur. J. Biochem.* **100**, 329 (1979).
[7] J. R. Skuster, K.-F. J. Chan, and D. J. Graves, *J. Biol. Chem.* **255**, 2203 (1980).
[8] K.-F. J. Chan and D. J. Graves, *J. Biol. Chem.* **257**, 5939 (1982).
[9] K.-F. J. Chan and D. J. Graves, *J. Biol. Chem.* **257**, 5948 (1982).
[10] K.-F. J. Chan and D. J. Graves, *J. Biol. Chem.* **257**, 5956 (1982).
[11] R. J. A. Grand, S. Shenolikar, and P. Cohen, *Eur. J. Biochem.* **113**, 359 (1981).
[12] E. H. Fischer, J. O. Alaba, D. L. Brautigan, W. G. L. Kerrick, D. A. Melencik, H. J. Moeschler, C. Picton, and S. Pocinwong, in "Versatility of Proteins" (C. H. Li, ed.), Proc. Symp. Proteins, Academic Press, New York, 1978.
[13] H. P. Jennissen and L. M. G. Heilmeyer, *FEBS Lett.* **42**, 77 (1974).
[14] A. Burchell, P. T. W. Cohen, and P. Cohen, *FEBS Lett.* **67**, 17 (1976).
[15] R. K. Sharma, S. W. Tam, D. M. Waisman, and J. H. Wang, *J. Biol. Chem.* **255**, 11102 (1980).

H_2O or effectors (10 μl); [γ-^{32}P]ATP (0.6–18 mM) (10 μl); and phosphorylase kinase (*vide infra*, usually between 0.6 and 12 μg/ml) (10 μl). The final pH of the reaction is 6.8 or 8.2. The reactions are carried out at 30° and are initiated with [γ-^{32}P]ATP after a 1-min preincubation. The enzyme activity is monitored by following ^{32}P incorporation into phosphorylase by using filter paper assays[16]; briefly, aliquots of the reaction mixture are pipetted onto Whatman No. 1 or phosphocellulose papers (2 × 2 cm) and are placed immediately into ice-cold 10% (v/v) trichloroacetic acid and 0.5% (w/v) sodium pyrophosphate (5 ml per paper). The papers are washed three more times with 5% (v/v) trichloroacetic acid and once in ethanol, and then dried; the radioactivity is counted by using liquid scintillation spectrometry. Alternatively, the *a* form of phosphorylase produced can be assayed in the direction of glycogen synthesis, essentially according to the method of Illingworth and Cori.[17] Under these conditions, the initial rate of reaction usually is linear for 5 min at pH 8.2, though a pronounced lag is observed at pH 6.8. The ratio of the activities at pH 6.8 to 8.2 usually is around 0.05. Upon dissociation of the holoenzyme, this ratio is increased to nearly unity, and the hysteresis at pH 6.8 is abolished. In addition, phosphorylase kinase loses its sensitivity to Ca^{2+} when dissociated, and up to 30% of the maximal activity at pH 8.2 can be obtained when assayed in the presence of EGTA.[5-8]

Physicochemical Characterization

Native phosphorylase kinase has a sedimentation coefficient of 23.6–26.1 S.[4,5] The appearance of lower molecular weight species can be detected by gel filtration or by density gradient ultracentrifugation. The gradients (5 ml) consisted of 5–20% sucrose, 1 mM dithioerythritol, 50 μM $CaCl_2$ in 50 mM 2-(*N*-morpholino)ethanesulfonic acid at pH 7.0. The samples (100 μl) were applied on top of the gradient, the tubes were placed into a SW 50.1 rotor and centrifuged at 4° in a Beckman LS-75 ultracentrifuge for 5 hr at 50,000 rpm, or 11 hr at 40,000 rpm. At the end of the run, 0.25-ml fractions were collected by using a Buchler densiflow apparatus and analyzed for kinase activity. A typical run is shown in Fig. 1. Aldolase, bovine serum albumin, and catalase were used as standards. The molecular composition of lower molecular weight species can then be analyzed by gel electrophoresis in the presence of SDS and by densitometry.

[16] E. M. Reimann, D. A. Walsh, and E. G. Krebs, *J. Biol. Chem.* **246**, 1986 (1971).
[17] B. Illingworth and G. T. Cori, *Biochem. Prep.* **3**, 1 (1953).

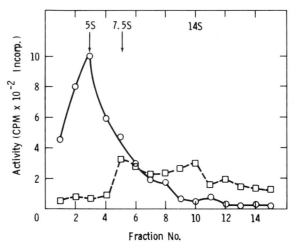

FIG. 1. Sucrose density gradient centrifugation of nonactivated phosphorylase kinase that had been incubated with either ATP or LiBr. Kinase (0.45 mg/ml) was incubated in a solution containing either 2.5% sucrose, 0.5 mM EDTA, 0.5 mM CaCl$_2$, 34 mM β-glycerophosphate, 32 mM Tris, 100 mM ATP, pH 8.6 at 0° for 33 hr or 2.5% sucrose, 2.5 mM β-glycerophosphate, 0.5 mM EDTA, 50 mM MES, 1 M LiBr, pH 7.0 at 0° for 9 hr. Kinase (0.009 mg) was applied to a 5-ml sucrose gradient (5 to 20%) containing 31 mM β-glycerophosphate, 31 mM Tris, 100 mM ATP, 10 μM CaCl$_2$, pH 8.6, and centrifuged at 4° for 11 hr at 40,000 rpm. The gradients were fractionated and assayed at pH 8.2. Phosphorylase kinase incubated with ATP (□); with LiBr (○). (From Skuster et al.[7])

Preparation of Subunit Forms

Treatment with Sodium Dodecyl Sulfate

Phosphorylase kinase is dissociated by using sodium dodecyl sulfate,[4] although no enzymic activity has been recovered following this treatment. Optimally, three different subunit forms—(α + α' + B), γ, and δ—can be isolated.

The holoenzyme (2–10 mg/ml) is first boiled in a water bath for 3–5 min, during which flocculation is observed. This mixture is chilled and centrifuged at 18,000 g for 10 min. The heat-stable δ-subunit is obtained in the supernatant with greater than 80% purity and can be further purified to apparent homogeneity by using ion-exchange chromatography (DEAE-cellulose or DEAE-Sephadex A-50) with a linear gradient of NaCl (0.1–0.7 M). The precipitate, which contains the other three subunits, is redissolved in 50 mM sodium β-glycerophosphate, pH 7.0, 1% sodium dodecyl sulfate, 1 mM dithioerythritol, and then subjected to gel filtration on a Sephadex G-200 (or Sephacryl S-300) column (1.3 × 105 cm) equilibrated with 50 mM sodium β-glycerophosphate, pH 7.0, 0.1% sodium dodecyl sulfate, and 0.5 mM dithioerythritol. Two protein peaks can be monitored at A_{280}. The first peak contains both the ($\alpha + \alpha'$)- and the β-subunits, whereas the second peak contains solely the γ-subunit. One advantage of this method is that large amount of materials can be purified, albeit no separation of the α- and the β-subunits is achieved. Further separation of these two subunits by other chromatographic techniques could be attempted. Alternatively, the α- and the β-subunits might be purified by using preparative sodium dodecyl sulfate gel electrophoresis.

Dissociation with LiBr

A large number of salts, NaCl, NaClO$_4$, KI, NH$_4$NO$_3$, NaNO$_3$, and RbBr, at concentrations in the molar range are unable to cause dissociation of nonactivated phosphorylase kinase.[18,19] At low temperature, low pH, and long incubation periods, ATP can partially dissociate both the trypsin-activated and the nonactivated forms of the enzyme.[7,18] However, LiBr (1.8 M) is the most effective agent in causing dissociation.[7–10] At least two monomeric and enzymatically active species, the $\alpha\gamma\delta$ and the $\gamma\delta$ complexes, could be isolated.

Preparation of the Active $\alpha\gamma\delta$ Complex

Nonactivated phosphorylase kinase is routinely purified through DE-52 column chromatography[5,8] and is stored in 50 mM sodium β-glycerophosphate, pH 7.0, 2 mM EDTA, 0.5 mM dithioerythritol, and 10% sucrose. Dissociation of the holoenzyme is initiated with dropwise additions of a 2.2-ml salt mixture (containing 1.455 ml of 10 M LiBr, 0.6 ml of 1.375 M MES, pH 7.0, 60 µl of 200 mM CaCl$_2$, and 85 µl of 1.0 M magnesium acetate) to a 6.0-ml enzyme solution (1.0–2.5 mg/ml) in a plastic vial, at 0°, and with constant stirring. The mixture is allowed to stand on ice for 6–7 hr. Turbidity usually is observed after 1–2 hr of incubation and is clarified by centrifugation at 18,000 g for 10 min. The precipitate, which is highly enriched with the β-subunit, can be solubilized in the presence of sodium dodecyl sulfate and used for the purification of this subunit by using a gel filtration technique.[8] The clear supernatant is applied to a Sephadex G-150 column (2.1 × 108 cm) equilibrated with 100 mM MES, pH 7.0, 1.0 mM dithioerythritol, and 0.1 mM CaCl$_2$ (buffer A). Typically, three activity peaks are eluted (Fig. 2). The most

---

[18] D. J. Graves, T. Hayakawa, R. A. Horvitz, E. Beckman, and E. G. Krebs, *Biochemistry* **12**, 580 (1973).
[19] G. M. Carlson and D. J. Graves, *Biochemistry* **15**, 4476 (1976).

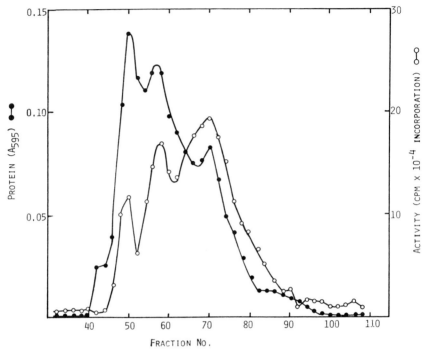

FIG. 2. Gel filtration of LiBr-treated phosphorylase kinase. Phosphorylase kinase (1.32 mg/ml) was incubated at 0° in a buffer containing 1.76 M LiBr, 10.4 mM Mg^{2+}, 100.8 mM MES, pH 7.0, 1.46 mM Ca^{2+}, 1.46 mM EDTA, 36.6 mM sodium β-glycerophosphate, 22 mM β-mercaptoethanol, and 7.3% sucrose for 6.5 hr. The incubation mixture was clarified by centrifugation at 18,000 g for 10 min and then was applied to a Sephadex G-150 column (2.1 × 108 cm) equilibrated with buffer A. Fractions of approximately 3 ml were collected and assayed for protein and kinase activity.

active fractions of the middle peak are pooled and immediately applied to a DE-52 column (0.7 × 7.4 cm) equilibrated with buffer A. The $\alpha\gamma\delta$ complex is eluted as a single peak with a 0–0.4 M NaCl gradient. After dialysis against 50% glycerol in buffer A, it is stored in aliquots at −20°. Figure 3 shows the gel electrophoresis patterns of the isolated $\alpha\gamma\delta$ complex and the holoenzyme of phosphorylase kinase. It seems essential that all procedures be carried out as fast as possible and that plastic tubes be used throughout. The $\alpha\gamma\delta$ complex prepared in this manner is stable for at least 2 months.

Preliminary studies on the dissociation of the $\alpha\gamma\delta$ complex by using EGTA and sucrose density-gradient ultracentrifugation were unsuccessful, though other means have not been attempted. The three individual subunits, however, can be isolated through treatment with SDS and subsequent gel filtration on a Sephacryl S-300 column.[8]

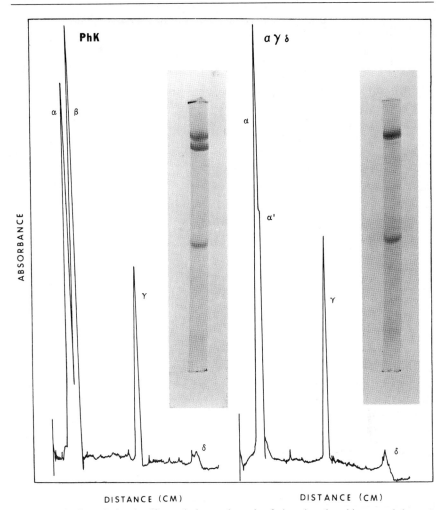

FIG. 3. Sodium dodecyl sulfate gel electrophoresis of phosphorylase kinase and the $\alpha\gamma\delta$ complex. Polyacrylamide gel electrophoresis of nonactivated phosphorylase kinase (PhK) and the $\alpha\gamma\delta$ complex in the presence of SDS were performed according to (35) with 10% gels. Densitometric tracings of these two gels are also shown (analysis of these results in the table).

Preparation of the Active $\gamma\delta$ Complex

Dissociation of phosphorylase kinase is carried out essentially as previously described, except that Mg^{2+} is omitted and the incubation period is extended to 18–24 hr.[7,8] It can also be prepared using peak 3 of Fig. 2 but lower yields are usually obtained. After clarification, the incubation mixture is applied to a Sephadex G-150 column equilibrated with buffer

MOLAR RATIOS OF THE SUBUNITS OF PHOSPHORYLASE
KINASE, $\alpha\gamma\delta$ AND $\gamma\delta$ COMPLEXES[a]

Enzyme forms	Molar ratios of subunits			
	$\alpha + \alpha'$	β	γ	δ
Phosphorylase kinase	1.00	1.07	1.08	0.88
$\alpha\gamma\delta$ complex	1.00	—[b]	1.16	1.19
$\gamma\delta$ complex	—[b]	—[b]	1.00	0.79

[a] The molar ratios were calculated from densitometric tracings of SDS–polyacrylamide gels of phosphorylase kinase, the $\alpha\gamma\delta$ complex, and the catalytically active $\gamma\delta$ species.[8]

[b] Molar ratios for these subunits were not calculated because only trace or negligible amounts were present.

A. Under these conditions, a major activity peak corresponding to a molecular weight of ca. 70,000 is observed. The peak fractions are pooled and immediately applied to a 3-ml Blue-Dextran Sepharose 4B (or Blue-Sepharose 4B, Pharmacia) column. Elution is carried out with a linear gradient (0–150 mM) of ATP in buffer A, pH 7.0, and the $\gamma\delta$ complex is obtained as a single peak between 30 and 60 mM ATP. Because of the presence of ATP, the kinase activity preferably should be monitored by assaying the amount of phosphorylase a generated. Otherwise, the ^{32}P incorporation into phosphorylase should be calculated according to the changes in specific radioactivity. The estimated molar ratios for the holoenzyme, the $\alpha\gamma\delta$ and $\gamma\delta$ complexes are summarized in the table.

Treatment of the $\gamma\delta$ complex with EGTA and subsequent sucrose density-gradient sedimentation indicated that an active γ-subunit may be generated.[10] Nevertheless, further investigation of this behavior is necessary.

Dissociation by Limited Proteolysis

Native phosphorylase kinase can be activated through limited proteolysis.[4,20] Nonactivated phosphorylase kinase (1–4 mg/ml) is incubated with trypsin in 50 mM glycerophosphate (pH 6.8) containing 2 mM EDTA and 10% sucrose.[18,20] With a weight ratio of kinase : trypsin of 1000 : 1, the enzyme is activated after 5 min of incubation. The reaction can be stopped by adding a sixfold excess of soybean trypsin inhibitor. With trypsin treatment, the activity ratio at pH 6.8/8.2 approaches unity[18] and the α- and β-subunits are partially degraded,[20] yet the sedimentation constant of trypsin-activated kinase (22.4 S) is close to that of the native

[20] T. Hayakawa, J. P. Perkins, and E. G. Krebs, *Biochemistry* **12**, 574 (1973).

enzyme.[18] However, in the presence of 20 mM ATP in the cold, or upon dilution lower molecular weight active species can be detected. A 6 S species is enriched in the γ-subunit. The experiments showed that the intact structure of phosphorylase kinase is not necessary for catalytic activity, but it was not possible to delineate which subunit was responsible for catalytic activity. By using limited chymotrypsin digestion and subsequent gel filtration of the partially degraded enzyme, a different conclusion was made, which suggested that the β-subunit may contain the catalytic site.[12]

There are a few drawbacks in these types of experiments. (1) It is difficult to reproduce the exact state of degradation every time. (2) Unless confirmed by sequence analysis or other means, the active species isolated may still represent a composite of large and small peptide fragments of unknown origin. Thus, it is difficult to draw unambiguous conclusions on the functional roles of the different subunits. (3) It also is possible that the degraded subunits have changed or lost some of their original properties. But limited proteolysis is useful in the study of phosphorylase kinase in conjunction with other techniques.

Concluding Remarks

Phosphorylase kinase is the first protein kinase discovered[21] which catalyzes covalent modification through protein phosphorylation, a process that is now recognized as a paramount control mechanism in biological systems. Although considerable literature exists on the studies of this enzyme,[2,22] many of its molecular details and its regulation by multisite phosphorylation and dephosphorylation, as exemplified through the interplay of Ca^{2+} and cyclic nucleotides, remain an intriguing challenge. The ability to isolate less complicated subunit forms of this complex macromolecule should pave the way for future investigation of these problems. Active complexes containing the subunits $\alpha\gamma\delta$ and $\gamma\delta$ have been derived from the holoenzyme. The individual subunits can be obtained after denaturation of the holoenzyme or the active complexes. Presently, it is not possible to obtain all of the individual subunits in the native state.

Acknowledgments

This is Paper No. J-10662 of the Iowa Agriculture and Home Economics Experiment Station, Ames, Iowa, Project No. 2120. The author acknowledges the support of the National Institutes of Health (Grant No. GM 09587) and the Experiment Station.

[21] E. G. Krebs, D. J. Graves, and E. H. Fischer, *J. Biol. Chem.* **234**, 2867 (1959).
[22] G. M. Carlson, P. J. Bechtel, and D. J. Graves, *Adv. Enzymol.* **50**, 41 (1979).

[29] Use of Peptide Substrates to Study the Specificity of Phosphorylase Kinase Phosphorylation

By DONALD J. GRAVES

Phosphorylase kinase catalyzes the phosphorylation of phosphorylase b, producing phosphorylase a, the activated form of glycogen phosphorylase.[1] The enzyme phosphorylates few other proteins; it has been reported, however, to phosphorylate itself,[2] glycogen synthase,[3-5] troponin I,[6] troponin T,[7] κ-casein,[8] histone Hl,[9] and protein from sarcoplasmic reticulum[10] and sarcolemma[11] and a brain protein involved in synaptic activation,[12] but phosphorylase seems to be the preferred substrate. The exact reasons for the variance in rates of protein phosphorylation is not known, but the differences must be related to the structure of the phosphorylatable region of the proteins. The phosphorylatable region can be defined as segments of the protein structure that interact with the phosphorylating enzyme. The region could be a simple linear segment of amino acids in the protein but also could contain other parts of the polypeptide chain or even a neighboring subunit in an oligomeric protein. The amino acid sequence around the phosphorylatable residue no doubt is important for phosphorylation, and one approach that can be used to learn about the specificity of the reaction is to use well-defined peptides containing the phosphorylatable residue as substrates. These substrates can be obtained by chemical or enzymatic fragmentation of the protein

[1] E. G. Krebs and E. H. Fischer, *Biochim. Biophys. Acta* **20**, 150 (1956).
[2] R. L. DeLange, R. B. Kemp, W. D. Riley, R. A. Cooper, and E. G. Krebs, *J. Biol. Chem.* **243**, 2200 (1968).
[3] P. J. Roach, A. A. DePaoli-Roach, and J. Larner, *J. Cyclic Nucleotide Res.* **4**, 245 (1978).
[4] T. R. Soderling, A. K. Srivastava, M. A. Bass, and B. S. Khatra, *Proc. Natl. Acad. Sci. U.S.A.* **76**, 2536 (1979).
[5] P. Cohen, *Curr. Top. Cell. Regul.* **14**, 117 (1978).
[6] J. T. Stull, C. O. Brostrom, and E. G. Krebs, *J. Biol. Chem.* **247**, 5272 (1972).
[7] S. V. Perry and H. A. Cole, *Biochem. J.* **141**, 733 (1974).
[8] A. A. Roach-DePaoli, E. W. Bingham, and P. J. Roach, *Arch. Biochem. Biophys.* **212**, 229 (1981).
[9] H. Tabuchi, E. Hashimoto, S. Nakomura, H. Yamamura, and Y. Nishizuka, *J. Biochem.* **89**, 1433 (1981).
[10] A. Schwartz, M. L. Entman, K. Komike, L. K. Lane, W. B. Van Wintle, and E. P. Bornet, *Biochim. Biophys. Acta* **426**, 57 (1976).
[11] P. J. St. Louis and P. V. Sulakhe, *Eur. J. Pharmacol.* **43**, 277 (1977).
[12] M. Browning, W. Bennett, and G. Lynch, *Nature (London)* **278**, 273 (1979).

substrates or by chemical synthesis. Studies are presented herein that show how experiments with peptide substrates can give information about the effect of size of the substrate, amino acid sequence, and organized structure on reactions catalyzed by phosphorylase kinase.

Preparation of Enzymes and Substrates

The source of the enzyme is rabbit skeletal muscle. Fresh tissue is used, and extraction and purification of nonactivated phosphorylase kinase are performed by following the procedure of Hayakawa et al.[13] The final step in the purification procedure utilizes DEAE-cellulose chromatography as described by Cohen.[14] From 2.5 kg of muscle, it is possible to obtain approximately 60 mg of enzyme. SDS–gel electrophoresis of this material shows that the enzyme contains equivalent amounts of α-, β-, γ-, and δ-subunits.[15] Activated phosphorylase kinase is prepared by incubating phosphorylase kinase with Mg ATP and the catalytic subunit of cAMP-dependent protein kinase. The activated kinase, when prepared according to a procedure described by Tessmer et al.,[16] contained 1.7–2.4 mol of phosphate/mol of α-subunit and 0.9–1.2 mol of phosphate/mol of β-subunit. The ratio of activities of the activated enzyme at pH 6.8 and 8.2 was between 0.3 and 0.4. The ratios of activities of the nonactivated enzyme at the same pH values was 0.05 or smaller.

A tetradecapeptide, Ser Asp Gln Glu Lys Arg Lys Gln Ile *Ser* Val Arg Gly Leu, containing the phosphorylatable serine of phosphorylase *b*, can be prepared from phosphorylase *a* by chymotryptic digestion.[17] The isolated phosphorylated peptide then is dephosphorylated by treatment with alkaline phosphatase. It was not possible to obtain the dephosphorylated peptide directly from phosphorylase *b* by chymotryptic digestion. Ion-exchange chromatography using Dowex-50 was used initially to purify the dephosphorylated peptide,[17] but poor yields were obtained with this ion exchanger. Better yields were obtained by using Bio-Rex 63, an intermediate cation-exchange resin containing a phosphoric acid as a functional group, in place of Dowex-50.[18]

The tetradecapeptide and other peptide substrates were also obtained by chemical synthesis by following the solid-phase methodology of Merri-

---

[13] T. Hayakawa, J. P. Perkins, D. A. Walsh, and E. G. Krebs, *Biochemistry* **12**, 574 (1973).
[14] P. Cohen, *Eur. J. Biochem.* **34**, 1 (1973).
[15] S. Shenolikar, P. T. W. Cohen, P. Cohen, A. C. Nairn, and S. V. Perry, *Eur. J. Biochem.* **100**, 329 (1979).
[16] G. W. Tessmer, J. R. Skuster, L. B. Tabatabai, and D. J. Graves, *J. Biol. Chem.* **252**, 5666 (1977).
[17] C. Nolan, W. B. Novoa, E. G. Krebs, and E. G. Fischer, *Biochemistry* **3**, 542 (1964).
[18] G. W. M. S. Tessmer, Thesis, Iowa State University.

field.[19] Approximately 300 mg of the crude synthetic tetradecapeptide was purified by ion-exchange chromatography on a (60 × 2.5 cm) column of Bio-Rex 63 utilizing a linear gradient between 0.2 M pyridine-acetate, pH 3.1, and 2.0 M pyridine-acetate, pH 5.0. The major peak was pooled, lyophilized, and dissolved in 1% acetic acid. Next, molecular sieve chromatography was done on a (90 × 2.5 cm) column of Sephadex G-15 in 1% acetic acid. The major peak again was pooled and lyophilized. The tetradecapeptide migrated as a single component on high-voltage electrophoresis and contained the exact proportions of amino acids found in the material obtained from phosphorylase by chymotryptic digestion. The overall yield of the pure tetradecapeptide was 36%.[16] Other peptides listed in Table I were purified by gel filtration utilizing Sephadex G-15 equilibrated with 1% acetic acid. Further purification when needed was accompanied by chromatography on Bio-Rex 63 followed by gel filtration on Sephadex G-15.

An amino-terminal segment containing the phosphorylatable site of phosphorylase b was obtained by CNBr fragmentation by a procedure adapted from Saari and Fischer.[20] In this case, CNBr fragmentation was done directly on 700 mg of unmodified phosphorylase b in 70% formic acid by utilizing a 100-fold excess of CNBr (1.7 g) over the methionine residues. After 24 hr at room temperature, the reaction was diluted 10-fold with H_2O and lyophilized. The residue was dissolved in 1 M formic acid and applied to a column (2.0 × 110 cm) of Sephadex G-50. Three major peaks were obtained, and the second was pooled, lyophilized, dissolved in 0.1 M formate, pH 2.9, containing 7 M urea and applied to a column (1.5 × 31 cm) of SP-Sephadex. Peptides were eluted with a gradient of 0.1–0.75 M sodium formate, pH 2.9, containing 7 M urea. The last peak contained the amino terminal segment, residues 1–99. After dialysis against 1 M formic acid, the peptide was lyophilized. The residue was then dissolved in 0.7 M formic acid and relyophilized. This procedure was repeated with 0.5 and 0.1 M formic acid and then with H_2O. The overall yield of the pure peptide was 45%.[21]

Assay for Phosphorylation

Reaction rates of phosphorylation were measured by following the incorporation of [^{32}P]phosphate into peptides by utilizing [γ-^{32}P]ATP. The reaction mixtures of 30° contained 30 μl of peptide, 10 μl of nonactivated or activated phosphorylase kinase (final concentration, 5–10 μg/ml), 10 μl

[19] R. B. Merrifield, *J. Am. Chem. Soc.* **86**, 304 (1964).
[20] J. C. Saari and E. H. Fischer, *Biochemistry* **12**, 5225 (1973).
[21] M. Hurst and D. J. Graves, in preparation.

TABLE I
K_m AND V_{max} VALUES FOR NONACTIVATED PHOSPHORYLASE KINASE WITH SYNTHETIC PEPTIDES[a]

Ser–Asp–Gln–Glu–Lys–Arg–Lys–Gln–Ile–Ser–Val–Arg–Gly–Leu[b]
5   6   7   8   9   10   11   12   13   14   15   16   17   18

Peptide number	Peptide length	Substitution	$K_m (\times 10^3\ M^{-1})$	V_{max} (μmol/min/mg)
1	5–18	None	1.2 ± 0.20	2.9 ± 0.27
2	9–18	None	0.9 ± 0.20	2.9 ± 0.60
3	10–18	None	0.9 ± 0.11	1.7 ± 0.13
4	11–18	None	1.5 ± 0.01	0.49 ± 0.004
5	5–16	None	0.2 ± 0.005	0.010 ± 0.0001
6	9–16	None	0.2 ± 0.03	0.038 ± 0.004
7	10–16	None	0.2 ± 0.003	0.029 ± 0.004
8	9–17	None	0.9 ± 0.07	0.88 ± 0.04
9	11–16	None		
10	9–18	10 Ala	0.9 ± 0.08	2.7 ± 0.19
11	10–16	10 Lys	0.8 ± 0.47	0.029 ± 0.013
12	9–18	11 Ala	0.9 ± 0.09	0.58 ± 0.05
13	9–18	10, 11 Ala	1.7 ± 0.45	0.55 ± 0.17
14	9–18	16 Ala	2.3 ± 0.04	0.18 ± 0.003
15	9–18	16 Gly	1.0 ± 0.12	0.12 ± 0.013
16	11–18	13, 15 Gly		
17	9–18	13 Gly	0.8 ± 0.04	0.52 ± 0.02
18	9–18	15 Gly	0.9 ± 0.03	0.25 ± 0.001
19	10–16	12 Glu	0.8 ± 0.12	0.019 ± 0.002
20	11–16	14 Thr		
21	10–16	14 Thr		
Phosphorylase b			0.27 ± 0.06	15.0 ± 0.82

[a] The specific activity measured at pH 8.2 with phosphorylase b as determined by the method of Brostrom *et al.*[27] is 85,000 units/mg. The kinetic constants were obtained by fitting the experimental data to a hyperbola function by the method of least squares; values are given as the mean ± standard deviation.

[b] The residues are numbered according to their actual sequence in phosphorylase b.[28]

of [γ-^{32}P]ATP-Mg^{2+}, 18–60 mM, pH 7.0, and 10 μl of 0.25 M glycerophosphate, 0.25 M Tris, 0.6 mM Ca^{2+} at pH 8.6. Analysis was accomplished by one of three procedures, depending on the size and charge of the peptide.[16] For most positively charged peptides shown in Table I, up to 50-μl aliquots of the reaction were spotted on 2.0 × 2.0-cm squares of Whatman cellulose phosphate paper P 81 (Whatman Inc., Clifton, NJ). The filter papers were dropped into a beaker containing 1 M acetic acid. For each paper, 10 ml of 1 M acetic acid was used. After 20 min, the acetic acid was decanted, and the process was repeated three more times. A final wash of 5 min in ethanol was used. The papers were air-dried and then placed in a toluene mixture for scintillation counting.[22] With this assay procedure, the progress curves were linear with respect to time and enzyme concentration. Aliquots from reaction mixtures could also be applied to a Pasteur pipet containing Dowex-50. ATP was removed by washing with pyridine acetate buffers.[16] The bound peptide on the resin was measured by scintillation counting. An assay developed by Kemp et al.[23] utilizing Dowex-1 resin to bind ATP also was used for the measurement of rates of phosphorylation.

Enzymatic Activity on Peptide Substrates of the Phosphorylase Sequence

In using peptides as substrates, it is important to establish that the specificity of protein phosphorylation is the same as it is in the native protein. Also, if the intent is to use peptide substrates as models of the natural reaction, a comparison of the enzyme properties on the different substrates is useful. The early experiments of Nolan et al.[17] suggested that the peptide, Ser Asp Gln Glu Lys Arg Lys Gln Ile Ser Val Arg Gly Leu, derived from phosphorylase by chymotryptic digestion, could be phosphorylated by phosphorylase kinase. Later, it was shown by Tessmer and Graves[24] that the seryl residue between Ile and Val, the seryl residue phosphorylated in phosphorylase b, is phosphorylated in the peptide with no phosphorylation occurring on the amino terminal serine. With nonactivated phosphorylase kinase and the peptide Ser Asp Gln Glu Lys Arg Lys Gln Ile Ser Val Arg Gly Leu, kinetic studies showed (1) a requirement for Ca^{2+}, (2) a lag in the catalytic reaction that could be overcome by preincubation with substrate, (3) low ratio in activities at pH 6.8/pH 8.2, (4) activation of the enzyme by trypsin or by phosphorylation with cAMP-dependent protein kinase.[25] All these features are characteristics of the reaction seen with phosphorylase b as substrate, suggesting that a peptide substrate can serve as a useful model for the natural substrate, phosphorylase b. Two features are different, however, about the reaction with the above peptide: (1) the K_m value is higher and the V_m is lower than with phosphorylase b; (2) phosphorylation of the seryl residue between Ile and

[22] G. A. Bray, *Anal. Biochem.* **7**, 279 (1960).
[23] B. E. Kemp, D. B. Bylund, T.-S. Huang, and E. G. Krebs, *Proc. Natl. Acad. Sci. U.S.A.* **72**, 3448 (1975).
[24] G. W. Tessmer and D. J. Graves, *Biochem. Biophys. Res. Commun.* **50**, 1 (1973).
[25] G. M. Carlson, L. B. Tabatabai, and D. J. Graves, in "Metabolic Interconversion of Enzymes, Fourth International Symposium, Arad, Israel, 1975" (S. Shaltiel, ed.), pp. 50–59. Springer-Verlag, Berlin and New York, 1976.

Val also is phosphorylated by cAMP-dependent protein kinase. Inasmuch as phosphorylase b cannot be phosphorylated by cAMP-dependent protein kinase,[26] these studies show that some aspects of the specificity of phosphorylation are lost with the peptide as substrate.

A series of peptides (~300 mmol of each) were chemically synthesized to evaluate what features in the primary structure were important for phosphorylation. The purified peptides were individually tested as substrates with a reaction mixture similar to that described in the section, Assay for Phosphorylation. Peptides were used at concentrations ranging from 1.5 to 3.0 mM. At different intervals, 10-μl aliquots were removed and spotted on Whatman 2MM paper for high-voltage electrophoresis. After electrophoresis, a autoradiogram was prepared to determine which peptides were phosphorylated. Peptides that showed significant phosphorylation were examined more closely in kinetic studies to evaluate K_m and V_m values.

Table I lists the results obtained with nonactivated phosphorylase kinase on various synthetic peptides. Little or no phosphorylation could be detected with peptides 9, 16, 20, and 21 in a high-voltage electrophoresis assay and therefore were not evaluated further. One of the major conclusions reached is that not all 14 amino acid residues of the reference peptide, peptide 1, are important for phosphorylase kinase recognition. The first five amino acid residues, starting from the amino terminus of the peptide, Ser, Asp, Gln, Glu, and Lys, have little influence on phosphorylation, and the sixth residue, Arg, has no large effect on the K_m or V_m value. On the other hand, the results obtained with peptide 5 (V_m value is reduced 300-fold) show that amino acids Leu and Gly on the carboxyl end are very important for optimal phosphorylation. A second conclusion is that substitution of residues in the sequence, Lys Gln Ile Ser Val Arg, has important effects on phosphorylase kinase. This is illustrated in Table I[27,28] and in a more complete study done later by Viriya and Graves[29] shown in Table II. Arginine is important, as illustrated by the effects of substitution by lysine and glutamic acid, peptide 10 and 11. Dramatic effects also were seen by substitution of serine by threonine, peptide 9, asparagine for glutamine, peptide 4 and glutamic acid for valine, peptide 8. The sequence, Lys Gln Ile Ser Val Arg, is found in human,[30] rat,[31]

[26] T. R. Soderling, J. P. Hickinbottom, E. M. Reimann, F. L. Hunkeler, D. A. Walsh, and E. G. Krebs, *J. Biol. Chem.* **245,** 6317 (1970).
[27] C. O. Brostrom, F. L. Hunkeler, and E. G. Krebs, *J. Biol. Chem.* **246,** 1961 (1971).
[28] K. Titani, A. Koide, J. Hermann, L. H. Ericsoon, S. Kumor, R. D. Wade, K. A. Walsh, H. Neurath, and E. H. Fischer, *Proc. Natl. Acad. Sci. U.S.A.* **74,** 4762 (1977).
[29] J. Viriya and D. J. Graves, *Biochem. Biophys. Res. Commun.* **87,** 17 (1979).
[30] R. C. Hughes, A. A. Yunes, E. G. Krebs, and E. H. Fischer, *J. Biol. Chem.* **237,** 40 (1962).
[31] C. L. Sevilla and E. H. Fischer, *Biochemistry* **8,** 2161 (1969).

TABLE II
K_m AND V_m VALUES FOR SYNTHETIC PEPTIDES

Ac–Lys–Gln–Ile–Ser–Val–Arg–Gly–Leu
 11   12   13   14   15   16   17   18[a]

Peptide number	Substitution	K_m (mM)	V_m (μmol/min/mg)
1	11 NH$_2$-Lys	1.48 ± 0.36	1.84 ± 0.25
2	None	1.56 ± 0.03	3.09 ± 0.11
3	11 Ac-Glu	3.86 ± 1.10	0.23 ± 0.06
4	12 Asn	0.21 ± 0.05	0.07 ± 0.01
5	13 Glu	2.18 ± 0.39	0.88 ± 0.42
6	13 Val	0.30 ± 0.06	0.41 ± 0.05
7	15 Ile	0.57 ± 0.12	2.33 ± 0.29
8	15 Glu	3.23 ± 0.65	0.06 ± 0.01
9	14 Thr	0.70 ± 0.30	0.04 ± 0.01
10	16 Lys	2.37 ± 0.44	0.14 ± 0.02
11	16 Glu	n.d.[b]	n.d.[b]

[a] The residues are numbered according to their actual sequence in phosphorylase b.
[b] Not determined because of the low rate of phosphorylation.

rabbit, and dogfish phosphorylase,[32] but not in yeast phosphorylase.[33] All these phosphorylases, except yeast, are phosphorylated by rabbit muscle phosphorylase kinase, further illustrating the significance of primary structure around the phosphorylatable serine for phosphorylation.

Studies of the specificity requirements and the kinetic parameters for reactions catalyzed by activated phosphorylase kinase (phosphorylated by cAMP-dependent protein kinase) also have been done.[16] The results are similar to those obtained with nonactivated kinase; e.g., the K_m value for the tetradecapeptide, Ser Asp Gln Gly Lys Arg Lys Gln Ile Ser Val Arg Gly Leu, and other peptide substrates are in the mM range. Substitutions of residues have similar effects, e.g., substitution of arginine on the carboxyl side of the phosphorylatable serine causes the K_m value to increase 2.6-fold and the V_m value to decrease about 8-fold as seen for nonactivated kinase (Table I). Studies on the effect of size show that a hexapeptide, Ile Ser Val Arg Gly Leu, can be phosphorylated but that the V_m value is only about 5% of the above tetradecapeptide.[34] Kinetic studies showed that V_m value for the reaction with the reference tetradecapeptide, peptide **1** of Table I, was half of that obtained with phosphorylase b.

[32] P. Cohen, J. C. Saari, and E. H. Fischer, *Biochemistry* **12**, 5233 (1973).
[33] K. Lerch and E. H. Fischer, *Biochemistry* **14**, 2009 (1975).
[34] L. B. Tabatabai, and D. J. Graves, *J. Biol. Chem.* **253**, 2196 (1978).

The K_m value, however, was more than 70-fold higher.[34] In this case, it seems that nonactivated and activated kinase differ. With nonactivated kinase, the K_m value for peptide substrates is not so different from that of phosphorylase b (Table I).

Several features that are important in a peptide substrate for assaying phosphorylase kinase activity are met by peptide 10, Lys Ala Lys Gln Ile Ser Val Arg Gly Leu (Table I). First, phosphorylation can be followed easily by incorporation of ^{32}P from [γ-^{32}P]ATP and by utilizing a filter paper assay with Whatman cellulose phosphate paper. Second, the substrate has a K_m and V_m value as good as the model tetradecapeptide (Table I). Third, the peptide is poorly phosphorylated by cAMP-dependent protein kinase in comparison with the peptide where Ala is replaced by Arg.[35] Thus, the peptide could be used in purified systems or in situations where some cAMP-dependent protein kinase is present along with phosphorylase kinase.

An alternative substrate that can be used for phosphorylase kinase studies is a CNBr fragment of the amino terminal end of phosphorylase b.[34] The fragment contains the first 99 residues of phosphorylase. The fragment is phosphorylated by phosphorylase kinase but cannot be phosphorylated by cAMP-dependent protein kinase.[21] Thus, the CNBr fragment, but not small peptides, has the same specificity of phosphorylation seen with native phosphorylase b. A K_m value intermediate between phosphorylase b and peptide 1 of Table I was obtained with activated phosphorylase kinase.[34] Because the fragment contains α-helical structure as measured by circular dichroism[35] and is consistent with structure established by X-ray crystallographic studies of phosphorylase b,[36] the studies suggest that structural features are important for the effectiveness of phosphorylation and the specificity of the process.

Enzymatic Activity on Peptide Substrates of the Glycogen Synthase Sequence

Several laboratories reported that glycogen synthase[37-40] can be phosphorylated by phosphorylase kinase, and kinetic studies with activated

---

[35] M. Hurst, J. K.-F. Chan, and D. J. Graves, unpublished results.
[36] L. N. Johnson, E. A. Stura, K. S. Wilson, M. S. P. Sansom, and I. T. Weber, *J. Mol. Biol.* **134**, 639 (1979).
[37] P. J. Roach, A. DePaoli-Roach, and J. Larner, *J. Cyclic Nucleotide Res.* **4**, 245 (1978).
[38] T. R. Soderling, A. K. Srivastava, M. A. Bass, and B. S. Khatra, *Proc. Natl. Acad. Sci. U.S.A.* **76**, 2536 (1979).
[39] A. A. DePaoli-Roach, P. J. Roach, and J. Larner, *J. Biol. Chem.* **254**, 4212 (1979).
[40] R. X. Walsh, D. M. Millikin, K. K. Schlender, and E. M. Reimann, *J. Biol. Chem.* **254**, 6611 (1979).

TABLE III
ALIGNMENT OF THE AMINO-TERMINAL SEQUENCES OF GLYCOGEN PHOSPHORYLASE AND GLYCOGEN SYNTHASE[a]

	1	2	3	4	5	6	7	8	9	10
Phosphorylase	Ac	Ser	Arg	Pro	Leu	Ser	Asp	Gln	Glu	Lys	Arg
Synthase	—	—	Pro	Leu	Ser	—	—	—	—	—

	11	12	13	14	15	16	17	18	19	20	21	22
Phosphorylase	Lys	Gln	Ile	Ser(P)	Val	Arg	Gly	Leu	Ala	Gly	Val	Glu
Synthase	Arg	Thr	Leu	Ser(P)	Val	Ser	Ser	Leu	Pro	Gly	Leu	Glu

[a] The residues are numbered according to their actual sequence in phosphorylase b.[28]

phosphorylase kinase showed that the K_m value was approximately the same as that determined with phosphorylase b but that the V_m value was about one-half.[38] Like phosphorylase, the site of phosphorylation of glycogen synthase is on the amino terminal region, and amino acid sequence analysis showed that Ser_7 in the protein is phosphorylated.[41,42] The sequence of the imino terminal peptide containing the phosphorylated serine is Pro Leu Ser Arg Thr Leu Ser(P) Val Ser Ser Leu Pro Gly Leu Glu, and as is shown in Table III, considerable homology exists with the phosphorylatable region of the phosphorylase sequence. Notable differences between the two structures are the gap of five residues, the substitution of serine for Arg_{16} and of lysine for Arg_{11}.

To evaluate differences in the reactions of phosphorylase kinase with glycogen synthase and phosphorylase, peptides of both sequences can be used as substrates. Table IV summarizes results of kinetic studies obtained with chemically synthesized peptides.[43] Peptide 1, representing the first 15 amino acid residues of glycogen synthase, is phosphorylated with a V_m value about threefold lower than that of a peptide of the phosphorylase sequence, peptide 2, but the K_m is somewhat lower. A shorter peptide, of the glycogen synthase sequence, peptide 3, which is missing Glu, Leu, Gly, and Pro, is not so good a substrate as peptide 1. The minimal length of peptide required for good phosphorylation has not been determined. The studies with peptides corroborate the kinetic results obtained with the natural protein substrates, i.e., glycogen synthase or peptides of its phosphorylatable region are reasonably good substrates for phosphorylase kinase even though an arginyl residue is not found on the carboxyl side of the phosphorylatable serine. Similar conclusions were

---

[41] T. R. Soderling, V. S. Sheorain, and L. H. Ericsson, *FEBS Lett.* **106**, 181 (1979).
[42] N. Embi, D. B. Rylatt, and P. Cohen, *Eur. J. Biochem.* **100**, 339 (1979).
[43] J. K.-F. Chan and D. J. Graves, *J. Biol. Chem.* **257**, 3655 (1982).

TABLE IV
KINETIC PARAMETERS FOR PHOSPHORYLASE KINASE WITH SYNTHETIC PEPTIDES OF THE GLYCOGEN SYNTHASE SEQUENCE[a]

	Phosphorylase kinase	
Peptide substrates[b]	K_m (mM)	V_m (μmol min^{-1} mg^{-1})
1 G. S. (1 → 15)	0.70	0.86
2 Phos. (5 → 18)	1.20	2.50
3 G. S. (1 → 11)	1.02	0.32
4 G. S. (1 → 11)Arg$_9$	0.81	2.25
5 G. S. (1 → 11)Lys$_4$	3.50	0.18

[a] Assays were carried out as described in the text.
[b] Peptide substrates:
  G. S. (1 → 15), Pro Leu Ser Arg Thr Leu Ser Val Ser Ser Leu Pro Gly Leu Glu
  Phos. (5 → 18), Ser Asp Gln Glu Lys Arg Lys Gln Ile Ser Val Arg Gly Leu
  G. S. (1 → 11), Pro Leu Ser Arg Thr Leu Ser Val Ser Ser Leu
  G. S. (1 → 11)Arg$_9$, Pro Leu Ser Arg Thr Leu Ser Val Arg Ser Leu
  G. S. (1 → 11)Lys$_4$, Pro Leu Ser Lys Thr Leu Ser Val Ser Ser Leu
  The numbers indicate the residue numbers found in the native proteins; Phos., phosphorylase; G. S., glycogen synthase.

drawn by Kemp and John.[44] It is clear, however, that by replacing serine by arginine on the carboxyl terminal side augments phosphorylation. The V_m value with peptide 4 is sevenfold greater than that of peptide 3. Substitution of lysine for arginine on the amino terminal side has the opposite effect. Here, the V_m with peptide 5 is lowered almost twofold, and the K_m is increased 3.5-fold from peptide 3.[43] Kemp and John[44] showed that substitution of arginine by leucine in a decapeptide of the glycogen synthase sequence resulted in a peptide that could not be phosphorylated. The sequence homology of the phosphorylatable site of glycogen synthase and phosphorylase no doubt is a major reason for the good rates. The gap of five residues in the synthase sequence from phosphorylase (Table III) seems to be of little consequence and is consistent with the results on phosphorylase types of peptides (Table I).

Peptide Inhibitors

Kinetic studies showed that the peptide, Arg Lys Gln Ile Thr Val Arg, is a competitive inhibitor with respect to a peptide substrate.[34] The nonlin-

[44] B. E. Kemp, and M. J. John, *Cold Spring Harbor Conf. Cell Proliferation* **8**, 331 (1981).

ear parabolic competitive kinetics obtained suggest that the inhibitor can add more than once to the enzyme. The peptide, Ile Ser Val Arg Gly Leu, is a substrate, but the peptide missing Leu is not.[45] It is, however, a competitive inhibitor with a K_i value about fivefold higher than the K_m value of peptide 1 of Table I. No inhibition could be obtained if the nitro blocking group on the arginyl residue was not removed.[45] The amino acid derivative, arginine methyl ester is a weak competitive inhibitor, and a K_i value of 33 mM was evaluated for inhibition of reactions with phosphorylase b.[24] Phosphorylase kinase can recognize peptides containing all D-amino acids. No phosphorylation occurs, but such a peptide containing D-amino acids, Leu Ser Tyr Arg Arg Tyr Ser Leu, was found to be a competitive inhibitor (K_i = 0.8 mM) with respect to peptide 1 of Table I.[43] Tetrapeptides containing D-amino acids, Arg Leu Ser Leu and Leu Ser Leu Arg are competitive inhibitors with K_i values of 1.6 and 3.3 mM, respectively.[46]

Concluding Remarks

Peptides containing the phosphorylatable sites of phosphorylase and glycogen synthase are phosphorylated at the same residue found in the natural protein substrates. They can provide a simple means for the assay of kinase activity. Although not stressed in this article, studies with peptide substrates along with protein substrates can be used to determine whether effectors influence the reaction by binding to the substrate, enzyme, or both. The use of different peptides provides useful information about the specificity of the enzymatic reaction and could lead to the design of specific inhibitors for various biological studies. The results suggest that phosphorylase kinase has multiple binding requirements because all residues in the immediate region of the phosphorylatable site are important for phosphorylation. It seems that higher orders of structure also are important for specificity and efficiency of phosphorylation. Peptide substrates can be effectively phosphorylated by phosphorylase kinase, but the K_m values are rather high in comparison with values obtained with the protein substrates.

Acknowledgments

This is Paper No. J-10610 of the Iowa Agriculture and Home Economics Experiment Station, Ames, Iowa, Project No. 2120. The author acknowledges the support of the National Institutes of Health (Grant No. GM 09587) and the Experiment Station.

[45] D. J. Graves, T. M. Martensen, J.-I. Tu, and G. W. Tessmer, *in* "Metabolic Interconversion of Enzymes, Third International Symposium, Seattle, 1973," pp. 53–61. Springer-Verlag, Berlin and New York, 1974.
[46] J. K.-F. Chan, Ph.D. dissertation, Iowa State University.

[30] Smooth Muscle Myosin Light Chain Kinase

By MICHAEL P. WALSH, SUSAN HINKINS, RENATA DABROWSKA, and DAVID J. HARTSHORNE

Assay

Principle

Myosin light chain kinase catalyzes the transfer of the γ-phosphate of ATP to a serine residue on a specific class of myosin light chains. In the case of smooth muscle myosin the two 20,000-dalton light chains are phosphorylated and the site of phosphorylation is serine-19.[1] The myosin light chain kinase from all muscle sources is absolutely dependent on Ca^{2+} and calmodulin (the ubiquitous Ca^{2+}-binding regulatory protein[2]), a feature that is particularly useful in assay procedures. The assay usually involves the quantitation of acid-stable phosphate incorporated into the myosin light chain using Mg^{2+}-[γ-^{32}P]ATP as the phosphate donor in the presence and absence of Ca^{2+} and calmodulin. Either the isolated 20,000-dalton light chain or the intact myosin molecule can be used as the substrate, i.e., phosphate acceptor. It should be noted, however, that the kinetics of phosphorylation of the two substrates are not identical.[3]

Reagents

[γ-^{32}P]ATP [the tetra(triethylammonium) salt] diluted with stock nonradioactive ATP, pH 7.5, to give an activity of 2000–5000 cpm/nmol ATP (using Čerenkov counting).
Tris–HCl, 100 mM, pH 7.5
$MgCl_2$, 16 mM
$CaCl_2$, 0.4 mM or EGTA, 4 mM, pH 7.5
  It is convenient to combine the above three solutions to give either a +Ca^{2+} or a −Ca^{2+} assay medium. Dilute times four in final assay mixture.
KCl, 3 M, used to adjust final KCl concentration in assay to approximately 60 mM.

---

[1] T. Maita, J.-I. Chen, and G. Matsuda, *Eur. J. Biochem.* **117**, 417 (1981).
[2] M. P. Walsh and D. J. Hartshorne, in "Biochemistry of Smooth Muscle" (N. L. Stephens, ed.). CRC Press, Cleveland, Ohio, in press.
[3] A. Persechini and D. J. Hartshorne, *Science* **213**, 1383 (1981).

Calmodulin, 1 mg/ml in distilled, deionized water
Myosin, ~10 mg/ml in 0.3 M KCl, 10 mM Tris–HCl, pH 7.5, 0.2 mM dithiothreitol.

Procedure

The reaction mixture (0.5 ml) contains 25 mM Tris–HCl, pH 7.5, 4 mM MgCl$_2$, 0.1 mM CaCl$_2$, 60 mM KCl, 15 μg/ml calmodulin, 0.5 mg/ml myosin, and myosin light chain kinase. Reaction at 25° is initiated by the addition of [γ-^{32}P]ATP to a final concentration of 0.75 to 1 mM. Reactions are quenched by addition of 0.45 ml of reaction mixture to 0.5 ml of 25% trichloroacetic acid (TCA), 2% sodium pyrophosphate in a plastic funnel (Isolab Quik-Sep column, code QS-P) fitted with a fiber glass disc. A cotton plug is also added to prevent blockage of the filter by precipitated protein. Excess [γ-^{32}P]ATP is washed out using a Millipore filtration unit with 5% TCA, 1% sodium pyrophosphate (four washes). The funnels are stoppered, 0.5 ml of 25% TCA, 2% sodium pyrophosphate added, and incubated in a water bath at 80° for 10 min (to remove acid-labile phosphate). The filters are again washed four times on the Millipore filtration apparatus with 5% TCA, 1% sodium pyrophosphate, and placed in plastic scintillation vials. Distilled, deionized water (15 ml) is added and protein-bound ^{32}P quantitated by Čerenkov counting.

Protein Purifications

Calmodulin

Calmodulin is purified from bull or ram testes by a modification of the procedure of Wallace *et al.*[4] Frozen testes (~250 g) are chopped and homogenized in a Waring blender at top speed for 2 min in 3 volumes of 25 mM Tris–HCl, pH 7.5, 1 mM EGTA, 0.02% NaN$_3$. The homogenate is centrifuged at 10,000 g for 60 min. The supernatant is filtered through glass wool and centrifuged at 150,000 g for 90 min. The supernatant is again filtered through glass wool and 1 M CaCl$_2$ added to a final concentration of 2 mM. The whole sample is applied (at a flow rate of ~20 ml/hr) to a column (1.5 × 30 cm) of 2-chloro-10-(3-aminopropyl)phenothiazine coupled to Sepharose 4B according to Jamieson and Vanaman[5] and previously equilibrated with 25 mM Tris–HCl, pH 7.5, 1 mM CaCl$_2$, 0.02%

---

[4] R. W. Wallace, E. A. Tallant, and W. Y. Cheung, *in* "Calcium and Cell Function, Volume 1, Calmodulin" (W. Y. Cheung, ed.), p. 13. Academic Press, New York, 1980.

[5] G. A. Jamieson, Jr. and T. C. Vanaman, *Biochem. Biophys. Res. Commun.* **90**, 1048 (1979).

NaN$_3$ (buffer A). The column is then washed with buffer A until A_{280} returns to baseline. Weakly bound proteins are eluted by washing the column for at least 12 hr with buffer A containing 0.5 M NaCl. Calmodulin is then eluted with 25 mM Tris–HCl, pH 7.5, 10 mM EGTA, 0.5 M NaCl, 0.02% NaN$_3$, dialyzed vs 10 mM NH$_4$HCO$_3$, pH 8.0 (2 × 10 liters), and lyophilized. The residue (pure calmodulin) is dissolved in distilled, deionized water at a concentration of ~15 mg/ml. It may be necessary to add one or two drops of NH$_3$ to dissolve the calmodulin. Yield ~80 mg/250 g testes. It is recommended that the affinity column be regenerated by washing with several column volumes of 6 M guanidine hydrochloride followed by buffer A after each use.

Smooth Muscle Myosin

Myosin is prepared from frozen turkey gizzards (250 g) by the method of Persechini and Hartshorne.[3] The absence of calmodulin and myosin light chain kinase is verified by appropriate myosin light chain kinase assays. Myosin prepared by this method is usually dephosphorylated.

Myosin Light Chain Kinase

Step 1. Homogenization. Turkey gizzards (fresh or stored frozen at −20°) are trimmed of fat, chopped, and minced to yield 250 g of mince. This is homogenized in a Waring blender at top speed for 3 × 5 sec in 4 volumes of 20 mM Tris–HCl, pH 7.5, 40 mM KCl, 1 mM MgCl$_2$, 1 mM EGTA, 1 mM dithiothreitol (DTT), 0.05% (v/v) Triton X-100 (buffer B) and centrifuged at 15,000 g for 15 min. The pellet is suspended and homogenized as before in 4 volumes of buffer B lacking Triton X-100 and centrifuged. The pellet is resuspended and homogenized as before in 4 volumes of buffer B lacking Triton X-100 and centrifuged.

Step 2. Kinase Extraction. The pellet from the previous step is suspended and homogenized in 4 volumes of 40 mM Tris–HCl, pH 7.5, 60 mM KCl, 25 mM MgCl$_2$, 1 mM EGTA, 1 mM DTT (Buffer C) and centrifuged at 15,000 g for 30 min. The supernatant (which contains the myosin light chain kinase) is filtered through glass wool.

Step 3. Reverse Ammonium Sulfate Fractionation. Solid ammonium sulfate is added slowly with stirring to the filtered supernatant to 60% saturation (363 g/liter). Buffer C is then added slowly to reduce the ammonium sulfate concentration to 40% saturation (226 g/liter). (The advantage of this reverse ammonium sulfate procedure[6] is that it results in more effective separation of protease activity. The proteases are largely re-

---

[6] The reverse ammonium sulfate procedure was suggested to us by Dr. S. Ebashi, University of Tokyo.

tained in the precipitate at 40% saturation whereas the myosin light chain kinase is soluble.) Centrifugation at 15,000 g for 30 min yields a supernatant containing the bulk of the kinase activity. In order to concentrate the kinase, ammonium sulfate is added to the supernatant, after filtering through glass wool, to 60% saturation (123 g/liter). The sample is centrifuged at 15,000 g for 30 min. The supernatant is discarded and the pellet is dissolved in ~50 ml of 20 mM K_2HPO_4, pH 8.0, 1 mM EGTA, 1 mM EDTA, 1 mM DTT, 0.02% NaN_3 (buffer D) with the aid of a hand-operated glass–glass homogenizer and dialyzed overnight vs buffer D (2 × 10 liters).

Step 4. Affi-Gel Blue Chromatography. The dialyzed sample is applied to a column of Affi-Gel Blue (Bio-Rad) previously equilibrated with buffer D. Excess protein is washed off the column with buffer D and bound proteins are then eluted with a linear NaCl gradient generated from

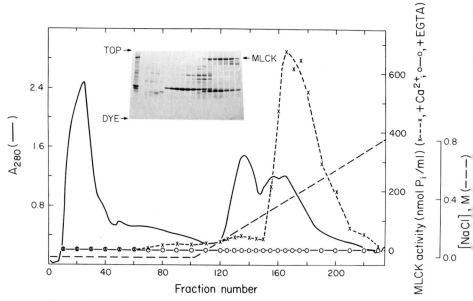

FIG. 1. Affi-Gel Blue chromatography of the dialyzed 40–60% ammonium sulfate fraction. The column (1.4 × 25 cm) was equilibrated with buffer D and eluted with a linear NaCl gradient (0–0.8 M) as described in the text at a flow rate of 14 ml/hr. Fractions of 3.3 ml were collected. Myosin light chain kinase activity was measured for 5 min. in the presence of 0.1 mM $CaCl_2$ (×---×) or 1 mM EGTA (○—○). Selected fractions were also subjected to SDS–PAGE (inset); fractions (from left to right): sample applied to column, 10, 20, 30, 40, 50, 120, 125, 130, 135, 140, 145, 150, 155, 160, 165, 170, 175, 180, 190. "TOP" indicates the top of the running gel (5% polyacrylamide stacking gel has been removed). "DYE" indicates the position of the tracking dye (bromphenol blue). Fractions 160–185 were pooled and dialyzed as described in the text.

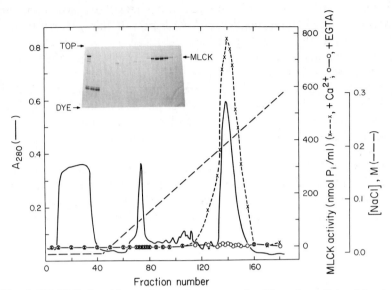

FIG. 2. DEAE-Sephacel ion-exchange chromatography. The column (1.4 × 25 cm) was equilibrated with buffer E and eluted with a linear salt gradient (0–0.3 M NaCl) as described in the text at a flow rate of 14 ml/hr. Fractions of 3.3 ml were collected and examined as described in the legend to Fig. 1. The inset shows a Coomassie blue-stained SDS–polyacrylamide gel of the following fractions (left to right): sample applied to column, 20, 30, 40, 60, 70, 75, 80, 90, 100, 110, 120, 125, 130, 135, 137, 140, 145, 150, 160. Fractions 132–146 were pooled and stored as described in the text.

200 ml each of buffer D and buffer D containing 0.8 M NaCl. Selected fractions are assayed for myosin light chain kinase activity in the presence and absence of Ca^{2+} and examined by 0.1% sodium dodecyl sulfate, 7.5–20% polyacrylamide gradient slab gel electrophoresis (SDS–PAGE) according to Laemmli.[7] A typical profile is shown in Fig. 1. The kinase elutes with a peak of activity at 0.4 M NaCl. Fractions are pooled on the basis of the activity measurements and the gel pattern, and dialyzed overnight vs 2 × 10 liters of 20 mM Tris–HCl, pH 7.5, 1 mM EDTA, 1 mM EGTA, 1 mM DTT (buffer E).

Step 5. DEAE-Sephacel Chromatography. The dialyzed sample is applied to a column of DEAE-Sephacel previously equilibrated with buffer E. Excess protein is washed off the column with buffer E and bound proteins are then eluted with a linear NaCl gradient generated from 250 ml each of buffer E and buffer E containing 0.3 M NaCl. Selected fractions are assayed for kinase activity in the presence and absence of Ca^{2+} and subjected to SDS–PAGE. Figure 2 shows a typical elution profile. Frac-

[7] U. K. Laemmli, *Nature (London)* **227**, 680 (1970).

TABLE I
PURIFICATION OF TURKEY GIZZARD MYOSIN LIGHT CHAIN KINASE[a]

Purification step	Total volume (ml)	Protein concentration[b] (mg/ml)	Total protein (mg)	Specific activity[c] (units/mg)	Total activity[c] (units)	Yield[d] (%)	Purification[d] (n-fold)
Mg^{2+} extract supernatant	1080	2.68	2894.4	0.061	176.6	100	—
40–60% ammonium sulfate	48	7.16	343.7	0.441	151.6	85.8	7.2
Dialyzed Affi-Gel Blue eluate	85.8	1.31	112.4	1.03	115.8	65.6	16.9
DEAE-Sephacel eluate	85.8	0.214	18.4	4.18	76.9	43.5	68.5

[a] Starting material: 250 g turkey gizzard.
[b] Determined according to R. F. Itzhaki and D. M. Gill, *Anal. Biochem.* **9**, 401 (1964), or T. Spector, *Anal. Biochem.* **86**, 142 (1978).
[c] One unit is defined as that amount of enzyme required to catalyze the incorporation of 1 μmol of P$_i$/min into the 20,000-dalton light chains of turkey gizzard myosin under conditions defined in the text.
[d] Based on the Mg^{2+} extract supernatant. Considerable purification of the kinase has already been achieved at this stage due to removal of essentially all the soluble proteins, but quantitation becomes feasible only at this step.

tions are pooled on the basis of the activity measurements and the gel pattern, made 5% (w/v) in sucrose and stored at −20 or −70° in plastic tubes in small aliquots (0.5–1.0 ml).

A summary of the purification procedure is shown in Table I. From these data, one can calculate a cellular content of myosin light chain kinase of 169 mg/kg or 211 mg/liter of intracellular water (assumed to be 80% of wet weight). This corresponds to a concentration of 1.6 μM assuming $M_r = 130,000$. This value is an underestimate since the kinase is not quantitatively extracted from washed myofibrils with a high Mg^{2+} concentration. The purified kinase is shown in Fig. 3. Densitometric scans of these and similar gels of other kinase preparations indicate >96% purity.

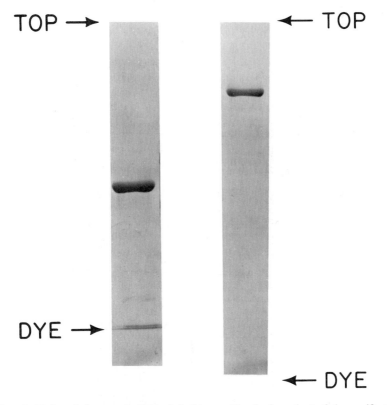

FIG. 3. Purity of the myosin light chain kinase. The final product of the purification procedure was electrophoresed in the presence of 0.1% SDS on 5–10% (left) or 7.5–20% (right) polyacrylamide gradient slab gels. Gels were stained with Coomassie brilliant blue R-250 and destained with 10% acetic acid.

General Comments and Properties

The advantages of this procedure over previously published methods of purification of myosin light chain kinases include (1) speed and ease; (2) elimination of the need for expensive mixtures of protease inhibitors, which appears to be due partly to the use of EGTA throughout (digestion of the kinase occurs in the presence of Ca^{2+}) and partly to the reverse ammonium sulfate step, and (3) elimination of the need for a calmodulin-Sepharose affinity column which requires the purification of large amounts of calmodulin.

The above procedure has been used to purify myosin light chain kinase from several smooth muscles, including chicken and turkey gizzard, bovine rumen, and bovine aorta. The preparation of stomach myosin light chain kinase includes the modification that a 0–1.5 M NaCl gradient is used with the Affi-Gel Blue chromatography since the kinase elutes at a higher ionic strength (0.7 M NaCl) than the gizzard kinase.[8] The difference in the positions of elution from Affi-Gel Blue could be due to the differences in molecular weight of the two enzymes: M_r = 155,000 for the stomach kinase, and M_r = 130,000 for the gizzard kinase. The myosin light chain kinase from bovine aorta was not homogeneous which probably reflects proteolytic degradation. The latter has been noted by others,[9] although DiSalvo et al.,[10] using a different procedure, reported the isolation of a single protein of M_r = 120,000.

Properties

Turkey gizzard myosin light chain kinase exists as a monomer of 130,000 daltons and is asymmetric (Stokes radius = 75 Å, sedimentation coefficient = 4.45 S, frictional coefficient = 1.85[11]). The amino acid compositions of various smooth muscle myosin light chain kinases are compared in Table II.

Substrate Specificity

Myosin light chain kinases appear to be absolutely specific for the phosphorylatable light chains of myosin. Myosin of the same tissue is the preferred substrate of a given kinase.

---

[8] M. P. Walsh, S. Hinkins, I. L. Flink, and D. J. Hartshorne, *Biochemistry* **21**, 6890 (1982).
[9] B. Vallet, A. Molla, and J. G. Demaille, *Biochim. Biophys. Acta* **674**, 256 (1981).
[10] J. DiSalvo, J. Miller, D. Blumenthal, and J. T. Stull, *Biophys. J.* **33**, 276a (1981).
[11] R. S. Adelstein and C. B. Klee, *J. Biol. Chem.* **256**, 7501 (1981).

TABLE II
AMINO ACID COMPOSITIONS OF SMOOTH MUSCLE MYOSIN LIGHT CHAIN KINASES[a]

Residue	Turkey gizzard No. 1	Turkey gizzard No. 2[b]	Chicken gizzard No. 1[c]	Chicken gizzard No. 2[d]	Bovine stomach[e]
Lysine	104.8 (80.6)	115.1 (92.1)	101.5 (96.7)	134.6 (103.5)	127.6 (82.3)
Histidine	16.9 (13.0)	13 (10.4)	15.9 (15.1)	14.7 (11.3)	18.8 (12.1)
Arginine	41.9 (32.2)	42.1 (33.7)	38.4 (36.6)	49.6 (38.2)	46.9 (30.3)
Aspartic acid	121.8 (93.7)	109.8 (87.8)	91.9 (87.5)	129.9 (99.9)	139.8 (90.2)
Threonine	67.2 (51.7)	68.5 (54.8)	61.0 (58.1)	80.1 (61.6)	98.2 (63.4)
Serine	89.7 (69.0)	86.4 (69.1)	86.4 (82.3)	108.6 (83.5)	102.8 (66.3)
Glutamic acid	155.0 (119.2)	155.2 (124.2)	139.1 (132.5)	190.3 (146.4)	169.6 (109.4)
Proline	62.1 (47.8)	58.0 (46.4)	47.2 (45.0)	78.3 (60.2)	108.6 (70.1)
Glycine	75.2 (57.8)	64.2 (51.4)	71.4 (68.0)	93.0 (71.5)	100.3 (64.7)
Alanine	81.5 (62.7)	85.1 (68.1)	73.2 (69.7)	91.5 (70.4)	98.0 (63.2)
Cysteine	22.0 (16.9)	27.9 (22.3)	16[f] (15.2)	ND	22.8 (14.7)
Valine	76.3 (58.7)	82.6 (66.1)	51.9 (49.4)	77.4 (59.5)	77.2 (49.8)
Methionine	21.6 (16.6)	21.7 (17.4)	19.7 (18.8)	33.0 (25.4)	17.1 (11.0)
Isoleucine	55.5 (42.7)	51.6 (41.3)	44.1 (42.0)	56.4 (43.4)	54.6 (35.2)
Leucine	71.4 (54.9)	66.9 (53.5)	59.2 (56.4)	80.6 (62.0)	99.0 (63.9)
Tyrosine	30.9 (23.8)	24.7 (19.8)	22.8 (21.7)	35.6 (27.4)	27.3 (17.6)
Phenylalanine	36.2 (27.8)	32.1 (25.7)	27.0 (25.7)	46.8 (36.0)	38.7 (25.0)
Tryptophan	ND	14.6	ND	ND	ND
Total residues	1130	1120	967	1300	1347

[a] Values are expressed as residues/mole assuming the following molecular weights: 130,000 (turkey gizzard No. 1), 125,000 (turkey gizzard No. 2), 105,000 (chicken gizzard No. 1), 130,000 (chicken gizzard No. 2), and 155,000 (bovine stomach). Values in parentheses are expressed as residues/100,000 daltons to facilitate comparisons.
[b] From R. S. Adelstein and C. B. Klee, *J. Biol. Chem.* **256,** 7501 (1981).
[c] From R. Dabrowska, D. Aromatorio, J. M. F. Sherry, and D. J. Hartshorne, *Biochem. Biophys. Res. Commun.* **78,** 1263 (1977).
[d] From H. Uchiwa, T. Kato, H. Onishi, T. Isobe, T. Okuyama, and S. Watanabe, *J. Biochem. (Tokyo)* **91,** 273 (1982).
[e] From Walsh et al.[8]
[f] From D. J. Hartshorne, R. F. Siemankowski, and M. O. Aksoy, in "Muscle Contraction. Its Regulatory Mechanisms" (S. Ebashi, K. Maruyama, and M. Endo, eds.), p. 287. Japan Scientific Society Press, Tokyo, and Springer-Verlag, Berlin and New York, 1980.

Regulation

Myosin light chain kinase activity is absolutely dependent on Ca^{2+} and calmodulin.[12] The active species contains kinase and calmodulin in 1 : 1

[12] R. Dabrowska, J. M. F. Sherry, D. K. Aromatorio, and D. J. Hartshorne, Biochemistry **17,** 253 (1978).

molar ratio in the presence of Ca^{2+}.[11] Kinase activity is also affected *in vitro* by phosphorylation catalyzed by cyclic AMP-dependent protein kinase.[13,14] In the absence of bound calmodulin, myosin light chain kinase of turkey gizzard is phosphorylated at two sites (A and B). In the presence of bound calmodulin, the kinase is phosphorylated only at site A. Site A phosphorylation has no observable effect on kinase activity, while site B phosphorylation results in a 5- to 20-fold decrease in the affinity of the kinase for calmodulin. Myosin light chain kinase activity *in vivo* may, therefore, be subject to concerted regulation by Ca^{2+} and cyclic AMP.

Various properties of myosin light chain kinases isolated from smooth and striated muscles and nonmuscle tissues have been compared.[15,16]

[13] R. S. Adelstein, M. A. Conti, D. R. Hathway, and C. B. Klee, *J. Biol. Chem.* **253**, 8347 (1978).
[14] M. A. Conti and R. S. Adelstein, *J. Biol. Chem.* **256**, 3178 (1981).
[15] M. P. Walsh, *Cell Calcium* **2**, 333 (1981).
[16] M. P. Walsh and D. J. Hartshorne, in "Calcium and Cell Function, Volume III" (W. Y. Cheung, ed.), p. 223. Academic Press, New York, 1982.

[31] Calcium-Activated, Phospholipid-Dependent Protein Kinase (Protein Kinase C) from Rat Brain

By USHIO KIKKAWA, RYOJI MINAKUCHI, YOSHIMI TAKAI, and YASUTOMI NISHIZUKA

Ca^{2+}-activated, phospholipid-dependent protein kinase (protein kinase C) is a new species of protein kinase that is activated by diacylglycerol in the presence of both membrane phospholipid and physiological concentrations of Ca^{2+}. Neither cyclic nucleotide nor calmodulin is involved in the enzyme activation. Protein kinase C was originally found as an inactive enzyme that can be activated by limited proteolysis with Ca^{2+}-dependent neutral thiol protease.[1,2] The enzyme was subsequently shown to be alternatively activated in a reversible manner by attachment to membrane phospholipid in the presence of Ca^{2+}.[3,4] Further analysis has

[1] Y. Takai, A. Kishimoto, M. Inoue, and Y. Nishizuka, *J. Biol. Chem.* **252**, 7603 (1977).
[2] M. Inoue, A. Kishimoto, Y. Takai, and Y. Nishizuka, *J. Biol. Chem.* **252**, 7610 (1977).
[3] Y. Takai, A. Kishimoto, Y. Iwasa, Y. Kawahara, T. Mori, and Y. Nishizuka, *J. Biol. Chem.* **254**, 3692 (1979).
[4] Y. Takai, A. Kishimoto, Y. Iwasa, Y. Kawahara, T. Mori, Y. Nishizuka, A. Tamura, and T. Fujii, *J. Biochem.* **86**, 575 (1979).

revealed that a small amount of diacylglycerol dramatically increases the affinity of the enzyme for Ca^{2+} as well as for phospholipid, and thereby causes activation of the enzyme at physiological concentrations of Ca^{2+}.[5,6] Diacylglycerol is not present in resting cell membranes in amounts sufficient to activate the enzyme. The neutral lipid seems to be provided by the breakdown of phosphatidylinositol by the action of phospholipase C when cells are stimulated by extracellular messengers. The diacylglycerol thus produced causes marked activation of protein kinase C and then disappears quickly for the resynthesis of phosphatidylinositol through phosphatidic acid and CDP-diacylglycerol.[7] This enclosed circle of phosphatidylinositol breakdown and resynthesis has been known as "phosphatidylinositol turnover" since its first discovery by Hokin and Hokin[8] in 1955. Diacylglycerol and/or phosphatidic acid may also supply arachidonic acid for prostaglandin synthesis.[9,10] The phospholipid breakdown is provoked by a wide variety of extracellular messengers, including epinephrine (α-effect), acetylcholine (muscarininc effect), histamine (H1 effect), several peptide hormones, growth factors, and many other biologically active substances[7]; these messengers usually increase the cytosolic Ca^{2+} concentration and do not increase cellular cyclic AMP levels. Protein kinase C may play some role in the control of cellular functions and proliferation by this group of extracellular messengers. In fact, the enzyme is widely distributed in many tissues and organs in mammals[11] as well as in other organisms,[12] and evidence strongly suggesting that this transmembrane signaling mechanism indeed operates in physiological processes has been obtained with thrombin-stimulated human platelets as a model system.[13,14] The present chapter describes the purification and properties of protein kinase C from a rat brain soluble fraction. The en-

[5] Y. Takai, A. Kishimoto, U. Kikkawa, T. Mori, and Y. Nishizuka, *Biochem. Biophys. Res. Commun.* **91**, 1218 (1979).
[6] A. Kishimoto, Y. Takai, T. Mori, U. Kikkawa, and Y. Nishizuka, *J. Biol. Chem.* **255**, 2273 (1980).
[7] R. H. Michell, *Biochim. Biophys. Acta* **415**, 81 (1975).
[8] L. E. Hokin and M. R. Hokin, *Biochim. Biophys. Acta* **18**, 102 (1955).
[9] R. L. Bell, D. A. Kennerly, N. Stanford, and P. W. Majerus, *Proc. Natl. Acad. Sci. U.S.A.* **76**, 3238 (1979).
[10] M. M. Billah, E. G. Lapetina, and P. Cuatrecasas, *J. Biol. Chem.* **256**, 5399 (1981).
[11] R. Minakuchi, Y. Takai, B. Yu, and Y. Nishizuka, *J. Biochem.* **89**, 1651 (1981).
[12] J. F. Kuo, R. G. G. Andersson, B. C. Wise, L. Mockerlava, I. Salomonsson, N. L. Brackett, N. Katoh, M. Shoji, and R. W. Wrenn, *Proc. Natl. Acad. Sci. U.S.A.* **77**, 7039 (1980).
[13] Y. Kawahara, Y. Takai, R. Minakuchi, K. Sano, and Y. Nishizuka, *Biochem. Biophys. Res. Commun.* **97**, 309 (1980).
[14] Y. Takai, K. Kaibuchi, T. Matsubara, and Y. Nishizuka, *Biochem. Biophys. Res. Commun.* **101**, 61 (1981).

zyme apparently lacks tissue and species specificities, at least in the physical and kinetic properties that are described in this chapter.

Assay Method

Principle. Protein kinase C catalyzes the transfer of the γ-phosphate of ATP to the seryl or threonyl residues of various protein substrates. The enzyme is routinely assayed by measuring the incorporation of ^{32}P from [γ-^{32}P]ATP into calf thymus histone H1 as a phosphate acceptor in the presence of Ca^{2+}, phospholipid, and unsaturated diacylglycerol. Histone H1 rather than whole histone mixture should be employed as it is the best model substrate so far tested. After incubation, the radioactive histone is separated from the reaction mixture by precipitation with 25% trichloroacetic acid followed by filtration on a nitrocellulose membrane. Basal activity is measured in the presence of ethylene glycol bis(β-aminoethyl ether)-N,N,N',N'-tetraacetic acid (EGTA) instead of Ca^{2+}, phospholipid, and diacylglycerol.

Reagents

Tris–HCl, 0.5 M, pH 7.5
Magnesium acetate, 25 mM
[γ-^{32}P]ATP, 0.25 mM, specific activity 5–15 × 10^4 cpm/nmol
Calf thymus histone H1, 5 mg/ml in H_2O. Calf thymus whole histone is prepared by the method of Johns.[15] Histone H1 is fractionated from the whole histone mixture by the method of Oliver *et al.*[16]
Calcium chloride, 2.5 mM
EGTA, 2.5 mM
Phospholipid, 10 mg/ml in chloroform. Total lipid is extracted from bovine brain or human erythrocyte ghosts by the method of Folch *et al.*[17] Phospholipid is fractionated from the total lipid by silicic acid column chromatography by the method of Rouser *et al.*[18]
Diolein (Nakarai Chemicals), 0.2 mg/ml in chloroform
Trichloroacetic acid, 25% (w/v)

Removal of Ca^{2+} from Reagents. For kinetic analysis it seems essential to remove as much Ca^{2+} as possible. Under the best conditions thus far tested the concentration of Ca^{2+} needed for half-maximal activity is in

---

[15] E. W. Johns, *Biochem. J.* **104**, 78 (1967).

[16] D. Oliver, K. R. Sommer, S. Panyim, S. Spiker, and R. Chalkley, *Biochem. J.* **129**, 349 (1972).

[17] J. Folch, M. Lees, and G. H. S. Stanely, *J. Biol. Chem.* **226**, 497 (1957).

[18] G. Rouser, G. Kritchevsky, and A. Yamamoto, *in* "Lipid Chromatographic Analysis" (G. V. Marinetti, ed.), Vol. 1, p. 99. Dekker, New York, 1967.

the 10^{-7} M range.[19] Ca^{2+}-free reagents are prepared by passing them through a column of Chelex 100, a resin specific for chelating divalent cations. Ca^{2+}-free water is prepared with a silica double distillation apparatus (Heraeus, Type Bi-18). Plastic columns, connections, and test tubes are used throughout these procedures. Spectrograde magnesium nitrate containing less than 100 ppm of Ca^{2+} (Merck, E. G.) may be suggested.

Procedure. The reaction mixture (0.25 ml) contains 5 μmol (10 μl) of Tris–HCl at pH 7.5, 1.25 μmol (50 μl) of magnesium acetate, 2.5 nmol (10 μl) of [γ-^{32}P]ATP, 50 μg (10 μl) of histone H1, 125 nmol (50 μl) of calcium chloride, 10 μg of phospholipid, 0.2 μg of diolein, and the enzyme fraction. Before being added, equal volumes of phospholipid and diolein in chloroform are mixed and the solvent is removed under nitrogen. The residue is then suspended in a small volume of 20 mM Tris–HCl at pH 7.5 by sonication with a Kontes sonifier K881440 for 5 min at 0°, and the resultant lipid micelles are added as specified above for assay. Basal activity is measured in the presence of 125 nmol (50 μl) of EGTA instead of calcium chloride, phospholipid, and diolein. The incubation is carried out for 3 min at 30°. When necessary Ca^{2+}-free reagents are employed and the reaction is performed in plastic tubes as described.[5,6] The reaction is stopped by the addition of 3 ml of 25% trichloroacetic acid. Acid-precipitable materials are collected on a nitrocellulose membrane filter (Toyo-Roshi membrane filter, comparable to Millipore filter; pore size, 0.45 μm) in a suction apparatus. The membrane filter is washed at least four times, each time with 3 ml of 25% trichloroacetic acid. The radioactivity of the acid-precipitable materials is determined using a Nuclear-Chicago Geiger Muller gas flow counter, Model 4338, or a Packard Tri-Carb liquid scintillation spectrometer, Model 3330.

Units. One unit of protein kinase C is defined as that amount of enzyme which incorporates 1 μmol of phosphate from ATP into histone H1 per minute at 30°. Specific activity is expressed as units per milligram of protein. Protein is determined by the method of Lowry *et al.*[20] with bovine serum albumin as a standard.

Application to Crude Extract. In general, with crude extracts it is rather difficult to demonstrate the absolute requirement of enzyme for Ca^{2+}, phospholipid, and diacylglycerol for several reasons: crude extracts normally contain several protein kinases, such as cyclic AMP-dependent protein kinases and Ca^{2+}- and calmodulin-dependent protein kinases, which may react with some undefined endogenous substrates. In addition, small quantities of Ca^{2+} and phospholipid are always present

---

[19] K. Kaibuchi, Y. Takai, and Y. Nishizuka, *J. Biol. Chem.* **256**, 7146 (1981).
[20] O. H. Lowry, N. J. Rosebrough, A. L. Farr, and R. J. Randall, *J. Biol. Chem.* **193**, 265 (1951).

in crude extracts. The stimulatory effect of diacylglycerol is observed only at physiological concentrations of Ca^{2+} below 10^{-5} M; kinetically, diacylglycerol increases the affinity of enzyme for Ca^{2+} as well as for phospholipid (see below). Thus, we suggest the use of DEAE-cellulose (DE-52) column (Step 2) to demonstrate or quantitate accurately the protein kinase C in crude extracts from most tissues.

Alternative Assays. The activity of protein kinase C can be measured by two alternative procedures.[1,2] First, because protein kinase C is able to phosphorylate protamine even in the absence of Ca^{2+}, phospholipid, and diacylglycerol due to a still unclarified reason, the activity is assayed with 100 μg of protamine as a substrate under the same conditions as described above except that none of the three activators is added and that the reaction is stopped by the addition of ice-cold 10% trichloroacetic acid. Second, protein kinase C is activated by limited proteolysis with Ca^{2+}-dependent neutral thiol protease, resulting in the formation of a catalytic fragment that is fully active in the absence of Ca^{2+}, phospholipid, and diacylglycerol. Thus, the enzyme may be quantitated by measuring the formation of this catalytic fragment.[2]

Purification Procedure

Protein kinase C is very susceptible to proteolysis, particularly by Ca^{2+}-dependent neutral thiol protease, to produce a catalytically active fragment that is totally independent of Ca^{2+}, phospholipid, and diacylglycerol. Thus, any tissue stored frozen may not be used as an enzyme source. In addition, relatively higher concentrations of EGTA and ethylenediaminetetraacetic acid (EDTA) are indispensable to keep the enzyme intact during the purification procedures. Rat brain soluble fraction is employed in the preparation of the enzyme simply because the enzymatic activity in this tissue is much higher (see below). However, the purification procedures up to Step 4 described below may also yield partially purified enzyme preparations from various mammalian tissues, including rat liver, rat intestinal smooth muscle, human platelets, and human peripheral lymphocytes.[11]

Step 1. Preparation of Crude Extract. Male Sprague–Dawley rats, weighing 150 to 200 g, are employed. Twenty-five rats are decapitated and the cerebra (approximately 25 g wet weight) are quickly removed and homogenized in a Potter-Elvehjem Teflon-glass homogenizer with 6 volumes of 20 mM Tris–HCl, pH 7.5, containing 0.25 M sucrose, 10 mM EGTA, and 2 mM EDTA. The homogenization and all subsequent operations are performed at 0–4°. The homogenate is centrifuged for 60 min at 100,000 g. The supernatant is employed as crude extract.

Step 2. DE-52 Chromatography. The crude extract (672 mg of protein) is applied to a DE-52 column (14 × 3 cm, Whatman) equilibrated with 20 mM Tris–HCl, pH 7.5, containing 50 mM 2-mercaptoethanol, 5 mM EGTA, and 2 mM EDTA. The column is washed with 300 ml of the same buffer solution, and then rewashed with 700 ml of 20 mM Tris–HCl, pH 7.5, containing 50 mM 2-mercaptoethanol, 1 mM EGTA, and 1 mM EDTA. The enzyme is eluted by application of a 1200-ml linear concentration gradient of NaCl (0 to 0.3 M) in the rewashing buffer at a flow rate of 75 ml/hr. Fractions of 18.6 ml each were collected. When protein kinase C is assayed, two peaks, major and minor, appear. The major peak is eluted between 40 and 100 mM NaCl (Fractions 11 through 22), and the minor peak between 140 and 200 mM NaCl (Fractions 30 through 38).[21] The major peak is collected and concentrated to 5 ml by an Amicon ultrafiltration cell equipped with a PM-10 filter membrane.

Step 3. Sephadex G-150 Chromatography. The enzyme solution (38.6 mg of protein) is subjected to a Sepadex G-150 column (90 × 2.6 cm, Pharmacia Fine Chemicals) equilibrated with 20 mM Tris–HCl, pH 7.5, containing 50 mM 2-mercaptoethanol, 0.5 mM EGTA, and 0.5 mM EDTA. Elution is carried out with the same buffer at a flow rate of 20 ml/hr. Fractions of 5.0 ml each are collected. When each fraction is assayed for protein kinase C, an apparently single peak appears in Fractions 46 through 54. These fractions are pooled and concentrated to 10 ml by ultrafiltration as described above.

Step 4. Isoelectrofocusing Electrophoresis. The enzyme solution of the previous step (5.36 mg of protein) is subjected to isoelectrofocusing electrophoresis using a 110-ml column (LKB instruments). The pH gradient (pH 5 to 7) is established during electrophoresis after the sequential addition of a carrier ampholyte solution (1.5%, w/v, LKB Ampholine pH 5–7) in a 0 to 50% (w/v) sucrose gradient[22] in the presence of 10 μM cyclic AMP[23] and 10 mM 2-mercaptoethanol. The enzyme solution is applied in the central ampholyte solution. Electrophoresis is continued at 600 V over a period of 42 hr for equilibration. After electrophoresis, 2.0-ml

---

[21] The minor peak is partially purified[2] and is shown to be indistinguishable from the major fraction in its physical, kinetic, and catalytic properties. The exact nature of this heterogeneity of protein kinase C is under investigation.

[22] The density gradient suitable for this isoelectrofocusing electrophoresis also may be prepared using glycerol [20 to 50% (w/v)] instead of sucrose. The enzyme at this step may be stored in 50% (w/v) glycerol at −20° after dialysis against the buffer solution containing 20% (w/v) glycerol followed by concentration by ultrafiltration.

[23] Cyclic AMP is effective to dissociate cyclic AMP-dependent protein kinase, which slightly contaminates the enzyme preparation, into its catalytic and regulatory subunits. The isoelectric points are 5.3 for the holoenzyme, 7.4 and 8.2 for the catalytic subunit, and 4.8 for the regulatory subunit of cyclic AMP-dependent protein kinase.

fractions are collected. When each fraction is assayed for protein kinase C, a major and two minor peaks appear in Fractions 14 through 20, Fractions 7 through 11, and Fractions 23 through 27, respectively. The major activity, which is focused at pH 5.6, is pooled and dialyzed overnight against 500 ml of 20 mM Tris–HCl, pH 7.5, containing 50 mM 2-mercaptoethanol, 0.5 mM EGTA, and 20% (w/v) sucrose.

Step 5. Blue-Sepharose Chromatography. The dialyzed enzyme solution (0.60 mg of protein) is mixed with 0.5 volume of 20 mM Tris–HCl, pH 7.5, containing 30 mM MgCl$_2$, 10 mM 2-mercaptoethanol, and 0.5 mM EGTA. The mixed solution is applied to a blue-Sepharose CL-6B column (3.5 × 1.2 cm, Pharmacia Fine Chemicals) equilibrated with 20 mM Tris–HCl, pH 7.5, containing 5 mM MgCl$_2$, 10 mM 2-mercaptoethanol, and 0.5 mM EGTA. After washing with 40 ml of the equilibrated buffer, the enzyme is eluted by application of a 40-ml linear concentration gradient of NaCl (0 to 2 M) in the same buffer at a flow rate of 30 ml/hr. Fractions of 0.8 ml each are collected. When protein kinase activity is measured, two peaks, major and minor, appear. The minor peak, which is eluted between 0.5 and 0.8 M NaCl (Fractions 12 through 19), is independent of Ca^{2+}, phospholipid, and diolein. The major peak, which is eluted between 0.9 and 1.5 M NaCl (Fractions 21 through 38), is dependent on these activators. The major peak is pooled.

Step 6. Phenyl-Sepharose Chromatography. The enzyme solution of the previous step (92 µg of protein) is applied to a phenyl-Sepharose CL-4B column (2.5 × 1.0 cm, Pharmacia Fine Chemicals) equilibrated with 20 mM Tris–HCl, pH 7.5, containing 10 mM 2-mercaptoethanol, 0.5 mM EGTA, 0.5 mM EDTA, and 1.5 M NaCl. The column is washed with 20 ml of the same buffer. The enzyme is eluted by decreasing the concentration of NaCl from 1.5 M linearly to 0 M in 20 ml of the buffer solution, followed by 15 ml of 20 mM Tris–HCl, pH 7.5, containing 10 mM 2-mercaptoethanol, 0.5 mM EGTA, and 0.5 mM EDTA. The flow rate is 30 ml/hr and 2-ml fractions are collected. A single sharp peak of protein kinase C is eluted between 0.4 and 0 M NaCl (Fractions 10 through 13).

A summary of the purification procedures is given in Table I. Protein kinase C is purified approximately 800-fold with an approximate yield of 5%. The purified protein kinase is apparently homogeneous as judged by sodium dodecyl sulfate (SDS)–polyacrylamide gel electrophoresis.

Properties

Stability. The homogeneous preparation of protein kinase C is unstable and may be stored at 0° for only a few days. The enzyme in the isoelectrofocusing electrophoresis step is relatively stable and may be

TABLE I
PURIFICATION OF PROTEIN KINASE C

Step	Protein (mg)	Activity (units × 10³)	Specific activity (units/mg × 10³)	Yield (%)	Purification (n-fold)
1. Crude extract	672	956	1.42	100	1
2. DE-52	38.6	604	15.6	63	11
3. Sephadex G-150	5.36	520	97.0	54	68
4. Isoelectrofocusing electrophoresis	0.60	307	512	32	361
5. Blue-Sepharose	0.092[a]	53.6	583	6	410
6. Phenyl-Sepharose	0.043[a]	47.3	1090	5	767

[a] Determined by gel scanning after polyacrylamide gel electrophoresis by the method of K. Weber, J. R. Pringle, and M. Osborn, this series, Vol. 26, p. 3.

stored at 0° for at least a few weeks in the presence of 0.5 mM EGTA. The enzyme is inactivated by freezing and thawing and by lyophilization.

Physicochemical Properties. The molecular weight of protein kinase C is estimated to be 82,000 by SDS-polyacrylamide gel electrophoresis. The sedimentation coefficient is estimated by sucrose density gradient analysis to be 5.1 S, which corresponds to a molecular weight of 77,000. The Stokes radius is estimated by a gel filtration analysis to be 42 Å, which corresponds to a molecular weight of 87,000. The enzyme is apparently composed of a single polypeptide chain with no subunit structure. The isoelectric point is approximately pH 5.6. A preliminary analysis indicates that protein kinase C has two functionally different, hydrophobic and hydrophilic domains. The hydrophobic domain seems to attach to membranes, and the hydrophilic domain carrying the catalytic site most likely faces the cytosolic side to exhibit enzymatic activity.

Mode of Activation. Protein kinase C is normally inactive, but is activated by diacylglycerol in the presence of Ca^{2+} and membrane phospholipid. The activation is very specific for Ca^{2+}, and other divalent cations are inactive. Among various phospholipids phosphatidylserine is indispensable. A small amount of diacylglycerol markedly increases the affinity of the enzyme for phosphatidylserine and sharply decreases the Ca^{2+} concentration to the 10^{-6} M range, which gives rise to the maximum enzyme activation. In addition, phosphatidylethanolamine, which is ineffective by itself, further increases the affinity of the enzyme for Ca^{2+} to the 10^{-7} M range when added together with phosphatidylserine and diacylglycerol.[19] Diacylglycerols that contain at least one unsaturated fatty acid at either position 1 or 2 are equally active, irrespective of the chain length of the other fatty acyl moiety, and those containing two saturated fatty acids are far less effective.[5,6,24] These phospholipids and diacylglycerol are most effective when first mixed in chloroform, evaporated to dryness under nitrogen, and then sonicated as described above before being added to the reaction mixture. Protein kinase C is free of calmodulin, and the enzymatic activity is not affected by exogenously added calmodulin nor by cyclic nucleotides.

Specificity. When tested *in vitro,* protein kinase C shows broad substrate specificity in a manner analogous to that of cyclic AMP-dependent protein kinase. The enzyme phosphorylates, to various extents, five species of calf thymus histones, protamine,[25] myelin basic proteins, microtubule-associated proteins, and many unidentified soluble and membrane-bound proteins.[26] Although a 40K protein in human platelets has been identified as a physiological substrate whose phosphorylation appears to be intimately related to serotonin release,[13,14] the precise intracellular targets for this enzyme remain to be clarified for most tissues. Evidence available at present indicates that the enzyme phosphorylates the seryl and threonyl residues but not the tyrosyl residue and that protein kinase C clearly differs from cyclic nucleotide-dependent protein kinases in its catalytic properties. For instance, with calf thymus histone H1 the enzyme phosphorylates the seryl and threonyl residues located in the carboxyl-terminal portion, whereas both cyclic AMP-dependent and cyclic GMP-dependent protein kinases phosphorylate a seryl residue (Ser-38) in the amino-terminal portion of this histone molecule.[27] Casein and phosvitin are practically inert as substrates for protein kinase C.

Kinetic Properties. The apparent K_m value for ATP and calf thymus histone H1 are 6 μM and 50 μg/ml, respectively. GTP does not serve as a phosphate donor. The optimum magnesium concentration is 5 to 10 mM. The optimum pH is 7.5 to 8.0 with 40 mM Tris–acetate as a test buffer.

[24] T. Mori, Y. Takai, B. Yu, J. Takahashi, Y. Nishizuka, and T. Fujikura, *J. Biochem.* **91**, 427 (1982).

[25] The dependence of the enzymatic reaction on Ca^{2+} and phospholipid differs slightly among histone species employed. With histone H1 the enzyme almost absolutely requires Ca^{2+} as well as phospholipid, whereas with histone H2B the reaction proceeds slowly in the absence of Ca^{2+} and phospholipid. Protamine appears to be exceptional, since it serves as a preferable substrate in the absence of Ca^{2+} and phospholipid even when a homogeneous enzyme preparation is employed. It is theoretically possible that the purified preparation is still a mixture of many enzymes, but plausible evidence strongly suggests that the enzyme is a single entity. Although the reason for this variation is not known, protamine may be used as a phosphate acceptor without the addition of Ca^{2+}, phospholipid, and diacylglycerol.

[26] Y. Nishizuka, *Mol. Biol. Biochem. Biophys.* **32**, 113 (1980).

[27] Y. Iwasa, Y. Takai, U. Kikkawa, and Y. Nishizuka, *Biochem. Biophys. Res. Commun.* **96**, 180 (1980).

Activator and Inhibitor. Tumor-promoting phorbol esters, such as 12-*O*-tetradecanoylphorbol 13-acetate, are able to substitute for diacylglycerol in the ng/ml range *in vitro,* and thereby directly activate protein kinase C in the presence of Ca^{2+} and phospholipid.[28] There is also evidence that this mechanism of activation may operate *in vivo.*[28] In contrast, various phospholipid-interacting drugs, such as chlorpromazine, dibucaine, imipramine, phentolamine, verapamil, and tetracaine, inhibit protein kinase C.[29] This effect is attributed to the inhibition of the activation process but not to interaction with the active site of the enzyme. This is supported by the fact that a catalytic fragment of protein kinase C, which is obtained by limited proteolysis with Ca^{2+}-dependent neutral thiol protease, is not susceptible to any of these drugs. Tumor-promoting phorbol esters do not affect this catalytic fragment. It is noteworthy that trifluoperazine, known as an inhibitor of calmodulin, is also a powerful inhibitor of protein kinase C.[12] Neither cyclic AMP-dependent nor cyclic GMP-dependent protein kinase is affected by the activators and inhibitors mentioned above.

Tissue Distribution. Protein kinase C appears to be widely distributed in most tissues and organs of mammals[11] as well as in lower organisms.[12] Table II shows the relative activities of protein kinase C and cyclic AMP-dependent protein kinase in some mammalian tissues. When assayed with calf thymus histone H1 as a model substrate, the activity of protein kinase C far exceeds that of the cyclic AMP-dependent enzyme in many tissues, e.g., platelets, brain, lymphocytes, and intestinal smooth muscle.

Subcellular Distribution. In most tissues, except brain, liver, and kidney, protein kinase C is almost totally recovered in the soluble supernatant fractions when tissues are homogenized in the presence of EGTA as described above. In brain, however, approximately one-third of the total activity is recovered in the soluble fraction, another one-third in the crude mitochondrial fraction, and the remainder in the microsomal and nuclear fractions as fractionated by a slightly modified method of Gray and Whittaker[30] in the presence of EGTA. Further fractionation analysis has revealed that most of the enzyme in the crude mitochondrial fraction is associated with synaptosomal membranes. This membrane-associated enzyme may be solubilized with 1% Triton X-100 (w/v) and is partially purified by DE-52 column chromatography and gel filtration on a Sephadex G-150 column under the conditions employed for the purification of

---

[28] M. Castagna, Y. Takai, K. Kaibuchi, K. Sano, U. Kikkawa, and Y. Nishizuka, *J. Biol. Chem.* **257,** 7847 (1982).

[29] T. Mori, Y. Takai, R. Minakuchi, B. Yu, and Y. Nishizuka, *J. Biol. Chem.* **255,** 8378 (1980).

[30] E. G. Gray and V. P. Whittaker, *J. Anat.* **96,** 79 (1962).

TABLE II
DISTRIBUTION OF PROTEIN KINASE C AND CYCLIC AMP-DEPENDENT PROTEIN KINASE

Tissue	Protein kinase C (units/mg × 10^3)	Cyclic AMP-dependent protein kinase[a] (units/mg × 10^3)
Platelets[b]	6.30	0.34
Brain	1.90	0.25
Lymphocytes[b]	1.06	0.32
Small intestinal smooth muscle	0.77	0.56
Granulocytes[b]	0.53	0.10
Lung	0.36	0.29
Kidney	0.28	0.15
Liver	0.18	0.13
Adipocyte	0.17	0.27
Heart	0.11	0.23
Skeletal muscle	0.08	0.11

[a] Cyclic AMP-dependent protein kinase is assayed under conditions similar to those for protein kinase C, except that 0.25 nmol of cyclic AMP is added instead of Ca^{2+}, phospholipid, and diolein.
[b] From human peripheral blood. Sprague–Dawley rats are employed for other tissues.

the soluble enzyme. The purified membrane-associated enzyme is indistinguishable from the soluble enzyme in its physical and kinetic properties.

Acknowledgments

This investigation was supported in part by research grants from the Scientific Research Fund of the Ministry of Education, Science and Culture, Japan (1979–1982), the Intractable Diseases Division, Public Health Bureau, the Ministry of Health and Welfare, Japan (1982), a grant-in-aid of New Drug Development from the Ministry of Health and Welfare, Japan (1979–1982), the Yamanouchi Foundation for Research on Metabolic Disorders (1977–1982), the Foundation for the Promotion of Research on Medicinal Resources, Japan (1977–1980), Takeda Pharmaceutical Science Foundation (1979–1981), and the Mitsuhisa Cancer Research Foundation (1981).

[32] Liver Calmodulin-Dependent Glycogen Synthase Kinase

By M. Elizabeth Payne and Thomas R. Soderling

Vasopressin, angiotensin II, and α-adrenergic agonists have been shown to promote the inactivation of glycogen synthase in rat liver,[1-5] presumably by a calcium-mediated process.[3,6] In addition, Garrison and co-workers[6] have reported that treatment of isolated hepatocytes with vasopressin or angiotensin II leads to a calcium-dependent increase in the phosphorylation of glycogen synthase. Calmodulin-dependent glycogen synthase kinase may mediate the hepatic effects on glycogen synthase of α-adrenergic agonists, vasopressin, and/or angiotensin II.

Calmodulin-dependent glycogen synthase kinase, which catalyzes the transfer of the γ-phosphate of ATP to serine seven residues from the amino terminus of glycogen synthase, has been isolated from liver.[7] This kinase is completely dependent on the presence of calmodulin (approx. $A_{0.5} = 0.1 \ \mu M$) and calcium for activity and completely inhibited by trifluoperazine ($I_{0.5} = 4 \ \mu M$) and EGTA.[7] The pH optimum is at 7.8 with maximal activity exhibited in HEPES buffer and at a magnesium ion concentration of 10 mM.[8]

From a Stokes radius of 69.9 Å, a sedimentation coefficient of 10.6 S, and an assumed partial specific volume of 0.725 cm^3/g, an approximate molecular weight of 305,000 g/mol can be calculated.[8] While an unequivocal subunit structure has not been defined, the major bands after polyacrylamide gel electrophoresis in the presence of sodium dodecyl sulfate correspond to a doublet at molecular weights of approximately 50,000 to 53,000 and minor bands at 63,000 and 67,000.[8]

Calmodulin-dependent glycogen synthase kinase is active toward both skeletal muscle glycogen synthase and liver glycogen synthase but does

---

[1] H. DeWulf and H. G. Hers, *Eur. J. Biochem.* **6**, 552 (1968).
[2] N. J. Hutson, F. T. Brumley, F. D. Assimacopoulos, S. C. Harper, and J. H. Exton, *J. Biol. Chem.* **251**, 5200 (1976).
[3] W. G. Strickland, P. F. Blackmore, and J. H. Exton, *Diabetes* **29**, 617 (1980).
[4] A. T. Hostmark, *Acta Physiol. Scand.* **88**, 248 (1973).
[5] D. A. Hems, L. M. Rodrigues, and P. D. Whitton, *Biochem. J.* **160**, 367 (1976).
[6] J. C. Garrison, M. K. Borland, V. A. Florio, and D. A. Twible, *J. Biol. Chem.* **254**, 7147 (1979).
[7] M. E. Payne and T. R. Soderling, *J. Biol. Chem.* **255**, 8054 (1980).
[8] M. E. Payne and T. R. Soderling, *J. Biol. Chem.* **258**, 2376 (1982).

not phosphorylate skeletal muscle phosphorylase b, liver phosphorylase b, skeletal muscle or cardiac myosin light chain, liver pyruvate kinase, casein, or histone IIA.[7–9] It catalyzes the rapid incorporation of up to 0.5 mol of ^{32}P/mol of synthase subunit resulting in partial inactivation of glycogen synthase.[7,8] Another 0.5 to 0.8 mol of ^{32}P/mol of synthase subunit can be gradually incorporated; however, little further inactivation occurs.

Materials and Methods

White, female, 3- to 5-kg New Zealand rabbits were used. [γ-^{32}P]ATP was prepared according to the method of Walseth and Johnson.[10] Calmodulin was purified from porcine brain as described by Watterson *et al.*[11] The calmodulin-Sepharose 4B was prepared basically using the method reported by March *et al.*[12] The heat-stable protein kinase inhibitor was purified from rabbit skeletal muscle according to the method of Walsh *et al.*[13]

Glycogen synthase *a* was purified from rabbit skeletal muscle as described previously.[14] The covalently bound phosphate has been estimated to be 0.1 to 0.5 mol P/mol synthase subunit.[15] In the absence of exogenous kinase, there was neither conversion from glycogen synthase *a* to *b* nor incorporation of ^{32}P into glycogen synthase—even when assayed in the presence of cAMP or calmodulin. When ATP was deleted from the reaction mixture, no change in the −G6P/+G6P activity ratio was observed.

Phosphorylation reaction mixtures of 50 μl contained 50 mM HEPES (pH 7.5 at 25°), 0.2 mM CaCl$_2$ (in excess over EDTA), 10 mM Mg(C$_2$H$_3$O$_2$)$_2$, 0.4 to 0.8 mg/ml of glycogen synthase *a* (or other protein substrate where indicated), kinase (10 μl), 0.5 mM [γ-^{32}P]ATP (400–850 cpm/pmol), and 10 μg/ml calmodulin (unless otherwise indicated). When necessary, the calmodulin-dependent glycogen synthase kinase was appropriately diluted just prior to the assay in 50 mM HEPES, 10% (v/v) ethylene glycol, 1 mg/ml crystalline bovine serum albumin, pH 7.5 at 5°.

[9] T. R. Soderling and M. E. Payne, *Cold Spring Harbor Conf. Cell Proliferation* **8**, 413 (1981).
[10] T. Walseth and R. Johnson, *Biochim. Biophys. Acta* **562**, 11 (1979).
[11] D. M. Watterson, W. G. Harrelson, Jr., P. M. Keller, F. Sharief, and T. C. Vanaman, *J. Biol. Chem.* **251**, 4501 (1976).
[12] S. C. March, I. Parikh, and P. Cuatrecasas, *Anal. Biochem.* **60**, 149 (1974).
[13] D. A. Walsh, C. D. Ashby, C. Gonzalez, D. Calkins, E. H. Fischer, and E. G. Krebs, *J. Biol. Chem.* **246**, 1977 (1971).
[14] T. R. Soderling, A. K. Srivastava, M. A. Bass, and B. S. Khatra, *Proc. Natl. Acad. Sci. U.S.A.* **76**, 2536 (1979).
[15] T. R. Soderling, J. P. Hickenbottom, E. M. Reimann, F. L. Hunkeler, D. A. Walsh, and E. G. Krebs, *J. Biol. Chem.* **245**, 6317 (1970).

Phosphorylation reactions were conducted at 30°. Reaction blanks were included in which glycogen synthase was omitted or in which the calmodulin-dependent glycogen synthase kinase was omitted. In both cases, the ^{32}P incorporation was negligible. The reactions were terminated in one of the following ways: (1) by spotting 10 µl on a 2-cm^2 Whatman 3 MM chromatography paper and immediately placing in cold 10% (w/v) trichloroacetic acid + 8% (w/v) sodium pyrophosphate; (2) by the addition of 20 µl of glacial acetic acid followed by spotting 40 µl on 2 × 2 cm^2 of Whatman 31 ET filter paper and placing in cold 10% (w/v) trichloroacetic acid + 8% (w/v) sodium pyrophosphate. After washing for 15 min, the filter papers were then washed, dried, and counted as described by Corbin and Reimann.[16]

Buffers Used in the Purification Procedure

Buffer A: 20 mM PIPES, 1 mM EGTA, 10% ethylene glycol, 1 mM DTT, 40 mM NaCl, 0.1 mM PMSF, 10 mg/liter TPCK, adjusted to pH 6.8 at 5°.

Buffer B: 50 mM HEPES, 1 mM EDTA, 10% ethylene glycol, 1 mM DTT, 0.1 mM PMSF, 10 mg/liter TPCK, adjusted to pH 7.5 at 5°.

Buffer C: 50 mM HEPES, 1 mM EDTA, 10% ethylene glycol, 1 mM DTT, 0.1 mM PMSF, 150 mM NaCl, pH 7.5, at 5°.

Buffer D: 50 mM HEPES, 0.2 mM CaCl$_2$, 1 mM DTT adjusted to pH 7.0 at 5°.

Purification of Liver Calmodulin-Dependent Glycogen Synthase Kinase

The major steps used in purifying liver calmodulin-dependent glycogen synthase kinase are outlined in Scheme 1. This procedure has been used to reproducibly purify the enzyme from rabbit liver, as detailed in this chapter, as well as from rat liver which contains much less of the calmodulin-dependent synthase kinase.[17]

After injection of a lethal dosage of Nembutal, the rabbits were bled from the jugular vein and their livers quickly excised. All subsequent steps were carried out at 4–5°. The livers from two rabbits were first ground coarsely in a meat grinder, and then homogenized (1 : 4, w/v) with a Teflon-glass homogenizer in 20 mM PIPES, 4 mM EDTA, 2 mM EGTA, 10% ethylene glycol, 1 mM dithiothreitol, 0.25 M sucrose, 0.2 mM PMSF, 1 mg/liter pepstatin A, 10 mg/liter TPCK, 1 mg/liter leupeptin adjusted to pH 6.8 at 5°. The homogenate was centrifuged at 10,000 g for 30 min. The

---

[16] J. D. Corbin and E. M. Reimann, this series, Vol. 38, p. 287.
[17] C. M. Schworer, M. E. Payne, A. T. Williams, and T. R. Soderling, *Arch. Biochem. Biophys.* **224**, in press (1983).

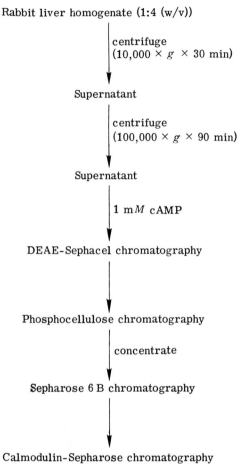

Scheme 1

pellet was discarded, and the supernatant was filtered through glass wool and then centrifuged at 100,000 g for 1.5 hr. After filtering the supernatant through glass wool, cAMP was added to a final concentration of 1 mM and incubated for 30 min at 4°. The cAMP was included to induce the dissociation of the regulatory and catalytic subunits of the cAMP-dependent protein kinase. The cAMP-treated supernatant was chromatogrammed on DEAE-Sephacel (Fig. 1). Using these conditions (see Fig. 1 legend), control experiments demonstrated that the catalytic subunit of the cAMP-dependent protein kinase did not bind to the DEAE-Sephacel. After assaying the fractions for glycogen synthase kinase activity in the absence and presence of calmodulin and in the absence and presence of

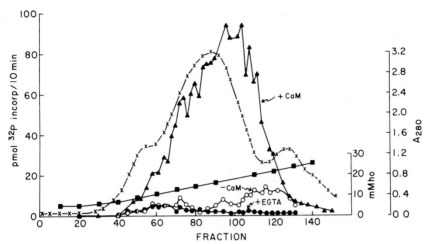

FIG. 1. DEAE-Sephacel chromatography of calmodulin-dependent glycogen synthase kinase. The cAMP-treated supernatant was mixed for 2 hr at 5° with 200 ml of DEAE-Sephacel previously equilibrated with buffer A without NaCl but containing 0.1 mM cAMP. The DEAE-Sephacel was washed extensively (20 bed volumes) with this buffer until the absorbance at 280 nm was below 0.05. Then, the resin was washed with buffer A until the absorbance at 259 nm was less than 0.05. After pouring the ion-exchange resin into a column, the column was developed beginning at fraction 1 with a 2 liter linear gradient of buffer A (40 mM NaCl) to buffer A made 0.4 M NaCl. The flow rate was 126 ml/hr and 11-ml fractions were collected. Fractions were assayed for glycogen synthase kinase activity in the presence of calmodulin (▲), in the absence of calmodulin (CaM) (○), or in the presence of EGTA (●). The absorbance (×) was monitored at 280 nm and the conductivity was determined. (■).

EGTA, the fractions containing the calmodulin-stimulated glycogen synthase kinase activity (from 11 to 22 mmho) were pooled and dialyzed overnight against 8 liters of buffer B. Using several different preparations of kinase, the profile observed was extremely reproducible. The assays for calmodulin-dependent glycogen synthase kinase were linear with time and with dilution.

Pepstatin A (1 mg/liter) and leupeptin (0.5 mg/liter) were added to the dialyzed, pooled fractions from the DEAE-Sephacel chromatography step, and this was applied to a 2.5 × 20-cm phosphocellulose column (Fig. 2). The peak of calmodulin-dependent glycogen synthase kinase activity eluted at a conductance of approximately 27 mmho. This activity was not affected by either 0.1 mM cAMP or the heat-stable inhibitor of the cAMP-dependent protein kinase. Approximately 60% of the protein applied to the phosphocellulose column did not bind. No kinase activity toward glycogen synthase was detected either in the absence or presence of

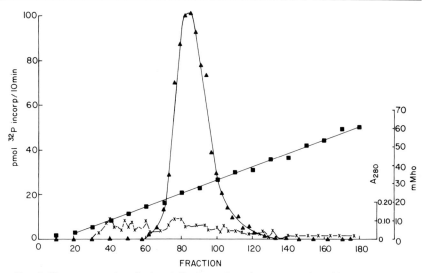

FIG. 2. Chromatography of calmodulin-dependent glycogen synthase kinase on phosphocellulose. The dialyzed sample obtained from the DEAE-Sephacel column step was applied to a column (2.5 × 20 cm) of phosphocellulose equilibrated with buffer B. After the addition of the enzyme, the column was washed with buffer B (about 250 ml) until the absorbance at 280 nm was less than 0.05. At fraction 1, the column was eluted with a 1.6-liter linear gradient from 0 to 0.8 M NaCl in buffer B. The flow-rate was 60 ml/hr. Glycogen synthase kinase activity (▲) was assayed in the presence of calmodulin. The absorbance (×) was monitored at 280 nm. The conductivity (■) was measured in every tenth fraction.

calmodulin in the flow-through fraction. Chromatography on phosphocellulose also separated the calmodulin-sensitive glycogen synthase kinase from the major peak of casein kinase activity which eluted at a conductance of about 60 mmho (not shown). The fractions containing the calmodulin-sensitive kinase activity (22–31 mmho) were pooled and concentrated with an Amicon ultrafiltration unit (PM-10 membrane) to about 5 ml.

The concentrated sample obtained after phosphocellulose chromatography was dialyzed overnight against buffer C and applied to a Sepharose 6B column. A single peak of calmodulin-dependent glycogen synthase kinase activity was observed at V/V_0 of 1.6 with a K_{av} = 0.424 (Fig. 3). No activity was detectable when the fractions were assayed either in the absence of added calmodulin or in the presence of EGTA. The fractions containing the calmodulin-dependent glycogen synthase kinase activity (fractions 82 to 102) were pooled.

Further evidence for the physical interaction of the calmodulin-dependent glycogen synthase kinase with calmodulin was obtained by forming a

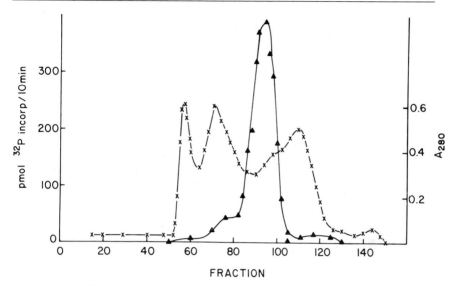

FIG. 3. Gel filtration chromatography of calmodulin-dependent glycogen synthase kinase. The dialyzed enzyme (5 ml) after phosphocellulose chromatography was applied to a column (1.5 × 87 cm) of Sepharose 6B equilibrated with buffer C. Fractions were monitored for absorbance at 280 nm (×) and assayed for glycogen synthase kinase activity in the presence of calmodulin (▲). The flow rate was 5 ml/hr, and 1-ml fractions were collected. The void volume of the column was at fraction 56.

reversible, calcium-dependent complex with calmodulin coupled to Sepharose 4B. Calmodulin-dependent glycogen synthase kinase was eluted from the column using buffer containing EGTA (Fig. 4). The fractions were collected in polypropylene tubes to which ethylene glycol and calcium had been added to final concentrations of 10% (v/v) and 1.4 mM, respectively. The fractions eluted with EGTA plus NaCl were immediately pooled and dialyzed against several changes of 50 mM HEPES, 1 mM EDTA, 10% ethylene glycol (v/v), 50% glycerol (v/v), 1 mM DTT adjusted to pH 7.5 at 5°. This sample was then further concentrated with an Amicon ultrafiltration unit (PM-10 membrane) and stored at −20°. Although in Fig. 4 it appears that about half or more of the calmodulin-dependent kinase was not retained by the affinity column, this is misleading since the kinase assays were not linear with respect to time. The total calmodulin-dependent kinase activity toward glycogen synthase in the flow-through fractions was actually about 20% of that eluted with EGTA plus NaCl. No additional calmodulin-stimulated glycogen synthase kinase activity adsorbed when the flow-through fractions were reapplied to a different calmodulin-Sepharose column using the same conditions—thus

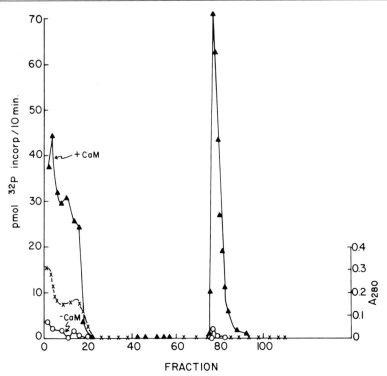

FIG. 4. Calmodulin affinity column chromatography of calmodulin-dependent glycogen synthase kinase. A calmodulin-Sepharose column (0.7 × 23.5 cm) was equilibrated with buffer D. The dialyzed enzyme obtained from the Sepharose 6B column was slowly applied to the column, and the flow-through fractions were immediately reapplied to the column. The column was washed with buffer D, and at fraction 40 the equilibration buffer was supplemented with 0.2 M NaCl. Beginning with fraction 70, buffer E (containing 1.2 mM EGTA) was applied to the column. Aliquots of the fractions were assayed for glycogen synthase kinase activity in the presence (▲) and absence of (○) of calmodulin. The absorbance (×) was monitored at 280 nm. The assays for kinase activity were not linear with respect to ^{32}P incorporation into glycogen synthase.

indicating it was not due to an overloading of the column. Furthermore, no additional calmodulin-stimulated glycogen synthase activity adsorbed when the flow-through fractions were applied to calmodulin-Sepharose columns varying the concentration of calcium from 0.1 to 1.0 mM or varying the pH from 6.8 to 8.5.

Because numerous protein kinases which are able to phosphorylate glycogen synthase are present in the early stages of purification, reliable quantitations of recoveries or fold purifications are difficult to obtain. In

PURIFICATION OF CALMODULIN-DEPENDENT GLYCOGEN SYNTHASE KINASE

Fraction	Total protein (mg)	Total volume (ml)	Total activity (μmol/min)	Specific activity (μmol/min/mg)	Yield (%)	Purification (n-fold)
Homogenate	15,046	898	7.48	4.9×10^{-4}	100	1.0
10,000 g supernatant	7,607	737	2.77	3.6×10^{-4}	37	0.7
100,000 g supernatant	6,108	667	2.77	4.5×10^{-4}	37	0.9
After DEAE-Sephacel	524	672	2.44	4.7×10^{-3}	33	10
After phosphocellulose	18	4.4	0.85	4.8×10^{-2}	11	96
After Sepharose 6B	3	3.4	0.28	9.5×10^{-2}	4	200
After CaM-Sepharose	0.02	0.4	0.04	2.2	0.6	4500

order to estimate the recoveries and fold purifications, phosphorylation of glycogen synthase was assayed in the presence of the heat-stable protein inhibitor of cAMP-dependent protein kinase, calcium, and either trifluoperazine or calmodulin. The calmodulin-dependent glycogen synthase kinase activity was calculated as the difference between the activities with calmodulin and trifluoperazine. In order to minimize the effect of possible endogenous inhibitors, the crude extracts were diluted 20- to 30-fold. The reactions were linear with time and dilution. The table summarizes the recovery, purification, and activity of the enzyme at the various stages of purification. The calmodulin-dependent glycogen synthase kinase was purified about 4500-fold from rabbit liver with a 0.6% recovery. The data presented in steps 1–3 represent the combined and averaged values from five different preparations, while steps 4 and 5 are from three different preparations. Subsequent purification was carried out with two preparations which were pooled.

[33] Casein Kinase I

By GARY M. HATHAWAY, POLYGENA T. TUAZON, and
JOLINDA A. TRAUGH

Casein kinase I is a multipotential, cyclic nucleotide-independent, Ca^{2+}-independent protein kinase which phosphorylates acidic substrates including casein. The type I casein kinases have been identified and described in a variety of eukaryotes including yeast, plants, fowl, and mammals.[1] The enzyme is present in nuclei and cytoplasm and found associated with membranes, ribosomes, and mitochondria.[1a] The type I casein kinases have been termed multipotential enzymes since they phosphorylate a number of different substrates.[1] The name, casein kinase, is derived from the substrate used for purification and characterization of the enzyme. Casein is the substrate of choice due to its ready availability. The type I casein kinase is distinct from both the type II casein kinases,[2] which are also present in many eukaryotic tissues, and the mammary gland enzymes, which function in the physiological phosphorylation of casein.[3]

Casein kinase I from rabbit reticulocytes has a sedimentation coefficient of 3.2 S and a molecular weight of around 37,000 as determined by polyacrylamide gel electrophoresis.[1,4] Thus, the native enzyme appears to consist of a single subunit. In the phosphorylation reaction, casein kinase I utilizes only ATP as the phosphoryl donor and phosphorylates seryl residues. The recognition sequence in casein has been determined to be Glu–X–Ser.[1,5] Substrates which have been identified for the enzyme include translational initiation factors, messenger ribonucleoprotein particles, nuclear acidic proteins including RNA polymerase, red cell membrane proteins such as spectrin and band 3, and glycogen synthase (for review see Ref. 1). In addition, the enzyme is autophosphorylated.[4] Although casein kinase I has been shown to modify a number of different substrates, the functional role for the enzyme in metabolism remains to be

[1] G. M. Hathaway and J. A. Traugh, *Curr. Top. Cell. Regul.* **21**, 101 (1982).
[1a] The type I casein kinase has also been identified as nuclear kinase NI, casein kinase S, and casein kinase A.
[2] G. M. Hathaway and J. A. Traugh, this volume [34].
[3] R. M. Whitney, J. R. Brunner, K. E. Ebner, H. M. Farrell, Jr., R. V. Josephson, C. V. Morr, and H. E. Swaisgood, *J. Dairy Sci.* **59**, 795 (1976).
[4] G. M. Hathaway and J. A. Traugh, *J. Biol. Chem.* **254**, 762 (1979).
[5] P. T. Tuazon, E. W. Bingham, and J. A. Traugh, *Eur. J. Biochem.* **94**, 497 (1979).

elucidated. It has been postulated that the casein kinases function to integrate metabolism due to the large number and variety of substrates identified for the enzymes.[1]

Assay of Casein Kinase I

The standard assay for casein kinase I is identical to that described for casein kinase II[2]; 50 mM Tris–HCl, pH 7.2, 10 mM MgCl$_2$, 150 mM KCl, 0.10 mM ATP, and 5 mg/ml mixed dephosphocasein (see below). However, optimal activity is not observed at the same KCl concentrations which give optimal activity with casein kinase II. Like casein kinase II, the concentration of KCl required for maximum activity is dependent on the concentration of casein as shown in Fig. 1A. When mixed casein is used and increased to 5 mg/ml, the salt optimum increases to about 300 mM. At lower concentrations of casein (0.15–1.0 mg/ml), the KCl optimum is about 150 mM KCl. When β-casein A is substituted for mixed casein at 0.5 mg/ml, the KCl optimum is 50 mM (Fig. 1B). The observation that the salt optimum is both a function of casein concentration as well as homogeneity of the casein substrate suggests that these salt effects may reflect changes in the state of the substrate in solution rather than a direct effect of monovalent cations on the enzyme itself. MgCl$_2$ is saturating at 5.0 mM at all KCl concentrations and addition of higher concentrations (up to 20 mM) has little effect on activity (Fig. 2). Incorporation of

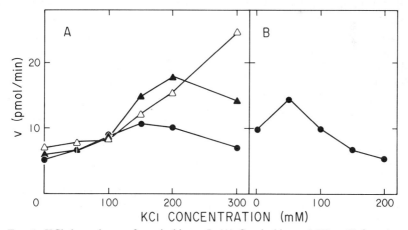

FIG. 1. KCl dependence of casein kinase I. (A) Casein kinase I (12 × 10^{-6} e.u.) was assayed in a 0.1-ml reaction volume containing 50 mM Tris–HCl, pH 7.2, 0.10 mM ATP (180 cpm/pmol), 10 mM MgCl$_2$, KCl as indicated, and 0.50 (●), 1.0 (▲), and 5.0 mg/ml (△) mixed casein. (B) Casein kinase I (12 × 10^{-6} e.u.) was assayed under the same conditions with 0.5 mg/ml β-casein A.

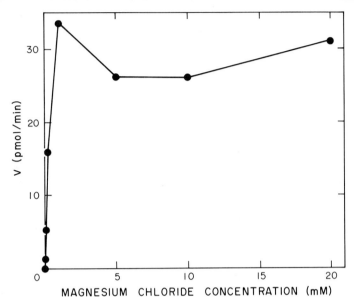

FIG. 2. Magnesium optimum for casein kinase I. Casein kinase I was assayed using standard conditions at various concentrations of $MgCl_2$.

phosphate is linear with the addition of up to 120 μunits/assay. One unit of enzyme activity equals 1 μmol ^{32}P incorporated into casein per minute.

Purification of Casein Kinase I

DEAE and Phosphocellulose Chromatography

The initial steps in the purification of casein kinase I are identical to those described for casein kinase II. The two enzymes are separated on DEAE-cellulose (see Ref. 2). Casein kinase I elutes prior to casein kinase II and at a KCl concentration between 0.05 and 0.10 M. After assaying, peak fractions are pooled and the sample is adjusted to the column buffer and applied directly to a 75 ml column of phosphocellulose. The phosphocellulose column (3.7 × 7.0 cm), equilibrated with 25 mM potassium phosphate, pH 6.8, 1.0 mM EDTA, 10 mM 2-mercaptoethanol, and 0.02% NaN_3, is developed with a 500 ml, 0 to 1.25 M NaCl gradient. Aliquots of column fractions are diluted 20-fold with cold, double distilled water so that a maximum of 10–20 enzyme μunits are added into the assay. The enzyme elutes from phosphocellulose at approximately 0.5 M NaCl. At this stage, casein kinase I is not pure, but is essentially free of casein kinase II, the hemin controlled repressor, protease activated kinase I and II, and the type I and II cAMP-dependent protein kinase. Casein kinase I is stable for at least a year at 4° when sodium azide (0.02%) is present, but labile when subjected to freezing. Casein kinase I may be purified further by chromatography on sulfopropyl-Sephadex (Pharmacia) and hydroxylapatite (BioGel HT, Bio-Rad), DNA cellulose, Affi-Gel blue Sepharose, or Sephadex G-150.

Chromatography on Sulfopropyl-Sephadex and Hydroxylapatite

Casein kinase I can be purified to apparent homogeneity (as determined by gel electrophoresis in sodium dodecyl sulfate) by chromatography on sulfopropyl-Sephadex followed by chromatography on hydroxylapatite.[4,6] Since we have experienced considerable losses of the enzyme by dialyzing, pooled fractions are routinely diluted to a salt concentration about 0.1 M to achieve conditions compatible with binding to these resins. Ice cold double distilled water or dilute (5 mM potassium phosphate, pH 6.8) buffer is used as diluent and the salt concentration is monitored by conductance at 0°. The diluted enzyme is pumped onto the column (2 × 2.6 cm for sulfopropyl-Sephadex and 1.6 × 5 cm for hydroxylapatite). A 400-ml, linear gradient of 0 to 1.25 M NaCl in 25 mM potassium phosphate, pH 6.8, is used to develop the sulfopropyl-Sephadex column; the enzyme elutes at 0.3 to 0.5 M. A 300 ml linear gradient of 0.025 to 0.5 M potassium phosphate, pH 6.8 containing 1 mM EDTA, 10 mM 2-mercaptoethanol, and 0.02% sodium azide is applied to the hydroxylapatite column. The enzyme elutes at about 250 mM phosphate[4,6] and has a specific activity of 0.5 units/mg.

Gel Filtration

Following phosphocellulose chromatography, casein kinase I was purified to apparent homogeneity in one additional step, gel filtration on Sephadex G-150. Casein kinase I was concentrated by precipitation in 70% $(NH_4)_2SO_4$ and resuspended in column buffer to a final volume of 9 ml. The sample was applied to a 170 ml Sephadex G-150 column equilibrated in 20 mM Tris, pH 7.4, 100 mM KCl, 1 mM EGTA, 10 mM 2-mercaptoethanol, and 0.02% sodium azide and fractions of 3 ml were collected at a flow rate of 27 ml/hr. The majority of the protein eluted in the void volume while casein kinase I was greatly retarded due to the small size of the protein. The highly purified enzyme was concentrated immediately to retain activity (see Enzyme Concentration below).

---

[6] G. M. Hathaway, T. S. Lundak, S. M. Tahara, and J. A. Traugh, this series, Vol. 60, p. 495.

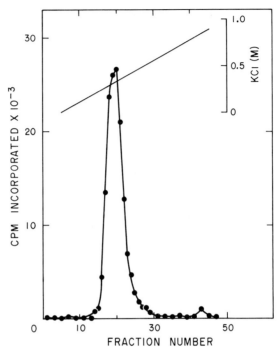

FIG. 3. DNA-cellulose chromatography of casein kinase I. Casein kinase I (0.15 e.u.), purified by chromatography on phosphocellulose, was applied to a 5 ml column of DNA-cellulose and assayed with casein as substrate. The salt gradient was determined by conductivity.

Chromatography on DNA Cellulose

Casein kinase I can be purified fivefold by chromatography on DNA cellulose. DNA cellulose is prepared from phenol-extracted, calf thymus DNA according to the procedure of Potuzak and Dean.[7] After resuspension in 50 mM Tris–HCl, pH 7.1 containing 1 mM EDTA, 10 mM 2-mercaptoethanol, and 0.02% sodium azide, the resin is defined once. Casein kinase I, purified through the phosphocellulose chromatography step, is diluted fourfold to a NaCl concentration of 0.15 M and applied to a 1.6 × 2.5 cm column of DNA cellulose. After washing with three column volumes (15 ml) of starting buffer, the column is developed with an 80 ml gradient of 0 to 1.0 M KCl in the same buffer; 2-ml fractions are collected. Casein kinase I activity elutes at about 0.3 M KCl (Fig. 3).

[7] H. Potuzak and P. D. G. Dean, *FEBS Lett.* **88,** 161 (1978).

Chromatography on Affi-Gel Blue Sepharose

Casein kinase I binds tightly to Affi-Gel blue Sepharose and is purified approximately sixfold on this resin. In the absence of ATP, 2.0 M potassium phosphate fails to elute the enzyme activity (Hathaway and Traugh, unpublished results). However, like the enzyme from calf thymus,[8] casein kinase I is eluted by a gradient of NH_4Cl if 10 mM ATP is present. Casein kinase I purified through the phosphocellulose step is diluted to achieve an ion strength of 0.3 M and applied directly to a 3.0 ml column of Affi-Gel blue Sepharose (Bio-Rad Laboratories). After washing with 3 column volumes of buffer (50 mM Tris–HCl, pH 7.9, 0.1 mM EDTA, 0.2 mM DTT), an 80 ml gradient of 0 to 1 M NH_4Cl, containing all the components of the buffer, plus 10 mM ATP, is applied. Enzyme activity elutes at 0.4 M NH_4Cl. $MgCl_2$ is not included in the gradient as the enzyme adheres more tightly to the resin and a lower recovery is observed when Mg^{2+} is present. Nucleotide can be removed by diluting the column fractions and concentrating the enzyme with a small hydroxylapatite column as outlined below.

Enzyme Concentration

It is important to note that following phosphocellulose chromatography, further purification results in a preparation which loses activity rapidly unless it is quickly concentrated. This is readily accomplished with greater than 95% recovery using hydroxylapatite. Small columns (<1 ml volume) of hydroxylapatite are made from a stock solution containing 2 g cellulose powder (Sigma-cell 50) in 25 ml of hydroxylapatite (Bio-Rad HT), 5 mM potassium phosphate, pH 6.8. Fractions containing casein kinase from the previous purification step are pooled and diluted 5- to 10-fold with double distilled water to reduce the salt concentration. The enzyme is pumped through hydroxylapatite at a flow rate of 100–150 ml/hr and the bound enzyme is eluted with 0.5 M potassium phosphate, pH 6.8, containing 1 mM EDTA and 0.01% sodium azide. Small fractions (0.3–0.5 ml) are collected and assayed for activity and protein concentration using the procedures of Lowry et al.[9] or Bradford.[10] Active fractions are pooled to attain a final protein concentration of at least 1.0 mg/ml.

[8] M. E. Dahmus, *J. Biol. Chem.* **256**, 3319 (1981).
[9] O. H. Lowry, N. J. Rosebrough, A. L. Farr, and R. J. Randall, *J. Biol. Chem.* **193**, 265 (1951).
[10] M. Bradford, *Anal. Biochem.* **72**, 248 (1976).

Inhibitors of Casein Kinase I

A number of inhibitors have been described for casein kinase II including heparin, 2,3-bisphosphoglycerate, inositol hexasulfate, and pyridoxal 5'-phosphate[1,2]; however, these compounds either have no effect on casein kinase I or are considerably less inhibitory than observed with casein kinase II. Casein kinase I is not inhibited by concentrations of heparin up to 7 μg/ml.[11] Inositol hexasulfate, pyridoxal 5'-phosphate, and 2,3-bisphosphoglycerate are somewhat inhibitory with 50% inhibition observed at concentrations of 0.1, 1.5, and 15 mM, respectively under standard assay conditions. With 2,3-bisphosphoglycerate, a sevenfold higher concentration is required to inhibit casein kinase I relative to casein kinase II. The K_i for the inhibitor with casein kinase I is 3.0 mM. This suggests that 2,3-bis-phosphoglycerate is of less importance as a regulatory compound for casein kinase I than for casein kinase II.

Analysis of the Casein Kinases with Mixed Casein and Purified Casein Variants

Casein consists of three major phosphoproteins, α_{S1}-, β-, and k-caseins. These caseins contain phosphoryl groups added endogenously by the mammary gland casein kinase. Several variants of each phosphoprotein have been identified and the amino acid sequences have been determined (see Ref. 3 for review). Mixed casein, readily available from a number of commercial sources, is ideal for routine assays during purification of the casein kinases. It is dephosphorylated prior to use in order to obtain a more homogeneous substrate and for solubilization of the substrate. For more sophisticated kinetic analyses and examination of site specificity by phosphopeptide mapping, homogeneous substrates are required and purified variants of α_{S1}- and β-caseins are utilized. For our studies, these variants have been generously supplied by Dr. Elizabeth Bingham, USDA, Philadelphia, PA. These variants, when fully phosphorylated by the mammary gland enzyme, or completely dephosphorylated, contain a single phosphate-acceptor site for casein kinase I and a different site for casein kinase II. The presence or absence of endogenous phosphate has no effect on the specific sites phosphorylated by casein kinase I[5] whereas partially dephosphorylated casein contains additional sites for casein kinase II.[1] Thus it is important to utilize fully phosphorylated or fully dephosphorylated casein for studies with casein kinase II. Either the α_{S1}- or β-casein variants can be used as defined substrates for kinetic analyses or examination of site specificity. The latter is an impor-

[11] G. M. Hathaway, T. H. Lubben, and J. A. Traugh, *J. Biol. Chem.* **255**, 8038 (1980).

tant tool for the complete identification of casein kinases isolated from sources other than reticulocytes.

Preparation of Dephosphorylated Casein

Mixed casein (Matheson, Coleman and Bell) is dephosphorylated using a modification of the procedure of Reimann et al.[12] Dephosphorylation is carried out in our laboratory by suspending 10 g of casein in 100 ml of 50 mM Tris, pH 9.5; the solution is brought to 100° and kept at that temperature for 10 min. Sodium hydroxide (0.1 N) is added as necessary to maintain the pH at 9.5. The casein is extensively dialyzed against 40 volumes of 50 mM Tris–HCl, pH 7.5, 1 mM EGTA, and 5 mM EDTA, then against water, changed three times daily for 48 hr. The solution may be stored frozen, or lyophilized and stored as the solid over a suitable desiccant at $-20°$.

Determination of Phosphoserine and Phosphothreonine

The phosphoacceptor amino acid in casein is determined by partial acid hydrolysis of the $^{32}P_i$-labeled substrate. Radiolabeled casein is precipitated by addition of 100% cold trichloroacetic acid to the standard reaction mixture to a final concentration of 25%. The precipitate is collected by centrifugation at 10,000 g for 10 min, washed once with 5% trichloroacetic acid and once with ethanol/ether (2/1), and dissolved in a sufficient amount of 0.1 M NH_4HCO_3, pH 8.6, to bring the concentration of casein to 1–2 mg/ml. Following dialysis against several changes of water at 4°, the sample is dissolved in 4 M HCl and partial acid hydrolysis is carried out in a sealed evacuated tube at 110° for 4 hr. The hydrolysate is lyophilized and the residue resuspended in 0.02 ml of water. Phosphorylated residues are analyzed by thin-layer electrophoresis; the entire sample is spotted, together with phosphoserine and phosphothreonine as internal standards, on a single lane. Standards of phosphoserine, phosphothreonine, and $^{32}P_i$ are spotted in separate lanes. Electrophoresis is conducted at pH 1.9 in 1.8% formic acid : 7.3% acetic acid at 500 V for 2 hr at 4°. Radioactivity is visualized by autoradiography and the standards are identified by staining with ninhydrin. Radioactivity is quantitated by scraping the indicated spots from the plate and counting them in toluene scintillant.

Determination of Proteolytic Phosphopeptides

Casein kinase I modifies a single residue in the α_{S1}-casein variants, serine-41, and a single residue in the β-casein variants, serine-22.[5] These

[12] E. M. Reimann, D. A. Walsh, and E. G. Krebs, J. Biol. Chem. **246**, 1986 (1971).

sites are not modified endogenously nor are they modified by casein kinase II.[13] Casein kinase II phosphorylates threonine-49 in α_{S1}-casein and threonine-41 in β-casein. The phosphopeptides resulting from chymotryptic digestion of β-casein are unique for each casein kinase and can be identified by two-dimensional fingerprinting using thin-layer electrophoresis in the first dimension and ascending chromatography in the second dimension followed by autoradiography. Casein is incubated with sufficient enzyme to maximally phosphorylate the substrate (10 μe.u. per 0.4 mg casein for 8–10 hr). Acid precipitation, resuspension, and dialysis of the casein are carried out as described above and the casein is incubated with chymotrypsin (ratio of 25:1) at 37° for 12–16 hr. The digests are lyophilized and the residues dissolved in a small volume of water. Approximately 180 μg of casein hydrolysate is applied to a thin-layer cellulose plate (Eastman, 20 × 20 cm) and electrophoresis is carried out for 2 hr at 400 V in pyridine:acetic acid:H_2O (10:0.4:90) at pH 6.4 in a cooled apparatus. The dried plates are developed with 1-butanol:acetic acid:H_2O (3:1:1) by ascending chromatography. The peptides are visualized with ninhydrin spray and the phosphopeptides by autoradiography on X-ray film. The chymotryptic peptides obtained following phosphorylation of casein kinase I and casein kinase II are shown in Fig. 4. The resulting phosphopeptides are distinct and different for each enzyme.

Identification of Casein Kinase I in Cell Lysates and Homogenates

The two major casein kinase activities, type I and II, are observed in extracts of almost all eukaryotic cell types when casein is used as substrate. However, the majority of the casein kinase activity in cell extracts may be inhibited. For instance, in reticulocyte lysates, total casein kinase activity is greater than 80% inhibited.[2] This inhibition can be released by dilution or by gel filtration and the total casein kinase activity quantified. Thus, in some cases, it may be essential to remove inhibitors from cell extracts to obtain an accurate indication of total activity.

The amounts of the individual casein kinase activities can be determined in diluted or gel filtered preparations of cell extract or lysate with heparin. Heparin has been shown to completely inhibit casein kinase II but has no effect on casein kinase I.[11] In reticulocyte lysates, addition of 0.7 μM heparin is sufficient to eliminate all casein kinase activity due to the type II enzyme.[2] Casein kinase I activity is obtained by difference and represents 20–25% of the total activity. This is corroborated by chromatography of the postribosomal supernate on DEAE-cellulose.[4] Since 1.0 ml of packed reticulocytes contains about 0.040 enzyme units of total

[13] P. T. Tuazon and J. A. Traugh, *J. Biol. Chem.* **253**, 1746 (1978).

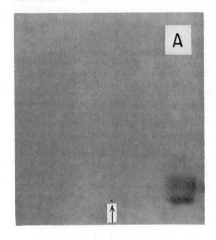 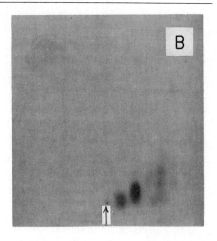

CASEIN KINASE I CASEIN KINASE II

FIG. 4. Phosphopeptide maps of chymotryptic digests of β-casein A^2 phosphorylated by casein kinase I and II. Casein β-A^2 was maximally phosphorylated with casein kinase I (A) or II (B) and digested with chymotrypsin. The products were analyzed by two-dimensional fingerprinting and autoradiography. The origin is indicated with an arrow on the autoradiograms. Taken from Tuazon et al.[5]

casein kinase activity, about 0.008–0.010 enzyme units/ml of activity in packed cells is casein kinase I.

Acknowledgments

We are indebted to Dr. Elizabeth W. Bingham for the generous supply of casein variants and Linda S. Collins for excellent technical assistance. The research described in this article was supported by Public Health Service Grants GM21424 and GM26738.

[34] Casein Kinase II

By GARY M. HATHAWAY and JOLINDA A. TRAUGH

A second, multipotential casein kinase, casein kinase II, has been identified in both the nucleus and cytoplasm of a wide variety of eukaryotes.[1,1a] Casein kinase II is structurally and chromatographically dis-

[1] G. M. Hathaway and J. A. Traugh, *Curr. Top. Cell. Regul.* **21,** 101 (1981).
[1a] The enzyme has been isolated from numerous tissues and has been described variously as nuclear kinase NII, glycogen synthase kinase 5, PC 0.7, casein kinase TS, casein kinase G, and eIF-2β kinase.

tinct from casein kinase I, but like the latter enzyme, phosphorylates acidic substrates. Although many of the proteins phosphorylated by the two enzymes are the same, the sites modified by the two enzymes differ. Casein kinase II phosphorylates both seryl and threonyl residues; the recognition sequence for the enzyme requires two acidic residues following the phosphorylatable residue and has been identified as Ser(Thr)–Glu(Ser-P)–Asp(Glu).[1,2] Casein kinase II utilizes both ATP and GTP as phosphoryl donors in the phosphotransferase reaction. The apparent K_m for ATP is 10 μM, 40 μM for GTP.[3] The enzyme has an $\alpha_2\beta_2$ structure with subunit molecular weights of approximately 42,000 and 24,000, respectively.[1] The α-subunit has been shown to contain the ATP binding site and is presumably the catalytic subunit.[4] The β-subunit is the acceptor site for autophosphorylation.[1,3]

Regulation of the type II casein kinase appears to be quite complex. Monovalent cations stimulate casein kinase activity with most if not all substrates[1] and polyamines have been shown to increase activity.[5,6] A number of inhibitory compounds for the enzyme have been identified including heparin,[7] inositol hexasulfate,[8] pyridoxal 5'-phosphate,[9] and 2,3-bisphosphoglycerate.[1,10,11] The amount of inhibitor required for 50% inhibition of the casein kinase ranges from 9.6 nM for heparin to 1.0 mM for 2,3-bisphosphoglycerate. Both heparin and 2,3-bisphosphoglycerate could have physiological significance in different cell types. All of the inhibitors are competitive with respect to the substrate, casein, and not with ATP.[1,7–10] 2,3-Bisphosphoglycerate has also been shown to act competitively with respect to the endogenous substrate, initiation factor 2.[12] Since the inhibitor compounds are structurally disimilar from the protein substrates, the data suggest an anionic binding site, which recognizes phosphate and/or sulfate groups in a closely associated configuration.[9] In recent studies with 2,3-bisphosphoglycerate, we have shown that the enzyme is allosterically regulated and this regulation is affected by the Mg^{2+} concentration.[10]

[2] P. T. Tuazon, E. W. Bingham, and J. A. Traugh, *Eur. J. Biochem.* **94**, 497 (1979).
[3] G. M. Hathaway and J. A. Traugh, *J. Biol. Chem.* **254**, 762 (1979).
[4] G. M. Hathaway, M. J. Zoller, and J. A. Traugh, *J. Biol. Chem.* **256**, 11442 (1981).
[5] C. Cochet, D. Job, F. Pirollet, and E. M. Chambaz, *Endocrinology* **106**, 750 (1980).
[6] P. H. Mäenpää, *Biochim. Biophys. Acta* **498**, 294 (1977).
[7] G. M. Hathaway, T. H. Lubben, and J. A. Traugh, *J. Biol. Chem.* **255**, 8038 (1980).
[8] G. M. Hathaway and J. A. Traugh, *Fed. Proc. Fed. Am. Soc. Exp. Biol.* **40**, 391 (1981) (Abstr.).
[9] G. M. Hathaway and J. A. Traugh, *Fed. Proc. Fed. Am. Soc. Exp. Biol.* **41**, 3591 (1982) (Abstr.).
[10] G. M. Hathaway and J. A. Traugh, *Fed. Proc. Fed. Am. Soc. Exp. Biol.* **42**, 2857 (1983) (Abstr.).
[11] R. Kumar and M. Tao, *Biochim. Biophys. Acta* **410**, 87 (1975).
[12] M. I. Gonzatti-Haces and J. A. Traugh, *J. Biol. Chem.* **257**, 6642 (1982).

Casein kinase II has been shown to phosphorylate translational initiation factors, mRNP particles, spectrin, glycophorin, RNA polymerase, high-mobility-group protein 17, glycogen synthase, troponin T (for review see Ref. 1) and the regulatory subunit of the type II, cAMP-dependent protein kinase.[13] Although casein kinase II has been shown to modify a number of different endogenous substrates, no function for any of these phosphorylation events has been identified at this time. It has been postulated that casein kinase II may function to integrate total cell metabolism by regulating phosphorylation of a number of proteins in different metabolic pathways.[1]

Assay of Casein Kinase II

Enzyme activity is measured by monitoring the incorporation of radioactive phosphate from [γ-^{32}P]ATP into trichloroacetic acid-precipitable substrate. A convenient substrate for purposes of enzyme purification and routine assay is the commercially prepared whole casein which has been dephosphorylated according to a modification[14] of the procedure of Reimann et al.[15] However, for critical kinetic work, either a homogeneous variant of the α_{S1}- or β-caseins or purified endogenous substrate is mandatory as discussed previously.[14] Standard assays are set up in 1.0 ml polystyrene tubes[15a] (4.2 × 0.85 cm) and contain 50 mM Tris–HCl buffer, pH 7.1, 150 mM KCl, 10 mM MgCl$_2$, 0.100 mM [γ-^{32}P]ATP, casein kinase II (10–20 enzyme μunits), and 5 mg/ml mixed casein in a final volume of 0.1 ml. It is convenient to prepare a stock solution of buffer, KCl, MgCl$_2$, and casein at a 2.5-fold higher concentration so that these components can be added in a single 0.040-ml aliquot. This "pre-mix" can be frozen indefinitely. ATP, enzyme, and inhibitor (or water), added in 0.020-ml aliquots, complete the 0.1 ml reaction and either ATP or enzyme may be added last to initiate the reaction. Alternatively, ATP and "pre-mix" can be combined separately and added as an 0.06 or 0.08 ml "hot mix." This "hot-mix" is stable at −15° for at least a week. Operationally, tubes containing all components except the enzyme are allowed to equilibrate in a 30° waterbath for 3 min. The concentrated enzyme is diluted in ice-cold, 5 mM Tris–HCl, pH 7.2, and used immediately, as the enzyme is labile upon dilution. To monitor background levels, control samples contain an equal volume of buffer in place of enzyme. Each reaction is initiated at 15-sec

---

[13] R. L. Geahlen, D. F. Carmichael, and E. G. Krebs, *Fed. Proc. Fed. Am. Soc. Exp. Biol.* **41,** 661 (1982). (Abstr.)

[14] G. M. Hathaway, P. T. Tuazon, and J. A. Traugh, this volume [33].

[15] E. M. Reimann, D. A. Walsh, and E. G. Krebs, *J. Biol. Chem.* **246,** 1986 (1971).

[15a] Obtained from Milian Instruments, 26 Bd. Helvetique, 1207 Geneve, Switzerland, Catalog No. TPS-10.

intervals by addition of the enzyme. After 15 min at 30°, 0.08 ml of sample is removed from each assay at 15-sec intervals, spotted on a 1.6 × 2.6-cm rectangle of Whatman 31 ET filter paper, and dropped into 10% trichloroacetic acid. Papers are washed with slow stirring at room temperature for 10 min in a volume of trichloroacetic acid equal to 10 ml per sample. The solution is decanted to radioactive waste and replaced with an equal volume of 5% trichloroacetic acid. This wash is repeated twice; the trichloroacetic acid is extracted for 10 min in a volume of 95% ethanol equal to 5 ml per sample. Papers are dried under a heat lamp, sealed into 2 × 4-cm polyester bags[15b] with 0.2 ml of scintillation fluid, and counted. Control counts are subtracted from experimentals. The resultant cpm are corrected by a factor 100/80 and divided by the specific activity of the [γ-^{32}P]ATP and the total time of reaction for expression as picomoles incorporated per minute at 30°. One enzyme unit is defined as the amount of enzyme which catalyzes the incorporation of 1 μmol of ^{32}P$_i$ per minute under these standard conditions.

Figure 1 shows a time course obtained with the substrate, β-casein B. Casein was reduced to low levels (60–750 μg/ml) and the time of assay extended to 20 min to show the linearity of the reaction. Linearity was maintained for 20 min even at the lowest casein concentration. Figure 2 shows the results obtained when increasing amounts of casein kinase II are added to the standard assay. Product formation is linear with the addition of up to 20 enzyme μunits, but becomes increasingly nonlinear at higher concentrations of enzyme. Calculations reveal that the source of nonlinearity is not due to depletion of ATP or casein, but probably results from product inhibition. ADP has been found to inhibit the forward reaction with a K_i of 0.018 mM (Hathaway and Traugh, unpublished results).

Assay Parameters

Casein kinase II is active over a broad pH range (Fig. 3). Optimal activity is observed between pH 8.0 and 9.0, a 1.5-fold increase in velocity over that observed at 7.2. However, the assay is standardized at pH 7.2, which is closer to the intracellular pH.

The concentration of KCl required to give optimal activity is not constant, but varies with the concentration of the protein substrate.[10] Therefore, account must be taken of this effect when the casein concentration is varied in the reaction. When the KCl optimum is examined over a 10-fold range of casein concentrations, it is observed to vary between 100 to 200

[15b] Prepared from 3-ml Filmware tubes, Catalog No. 500-0030, Sybron/Nalge, Rochester, NY 14602.

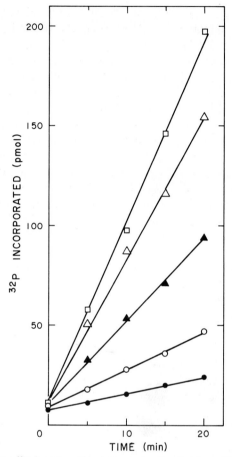

FIG. 1. Time course $^{32}P_i$ incorporation into casein. Casein kinase II (11×10^{-6} e.u.) was added to the 0.1 ml reaction containing 50 mM Tris–HCl, 0.10 mM ATP, 10 mM MgCl$_2$, and 120 mM KCl. The substrate, β-casein B, was 0.060 (●), 0.13 (○), 0.26 (▲), 0.50 (△), and 0.75 mg/ml (□). Assay conditions differed slightly from standard conditions but were optimized for this substrate.

mM KCl depending on the amount of casein (Fig. 4). Under standard conditions, the enzyme is assayed with saturating substrates at 150 mM KCl. Under these conditions, the enzyme displays a Mg^{2+} optimum between 5 and 15 mM (Fig. 5). While the enzyme is inactive in the absence of Mg^{2+}, concentrations higher than 15 mM are inhibitory. The concentrations of KCl, MgCl$_2$, and casein quoted here should be taken as a starting point for the assay of the casein kinase. Optimal concentrations may vary depending upon the source and purity of the protein substrate.

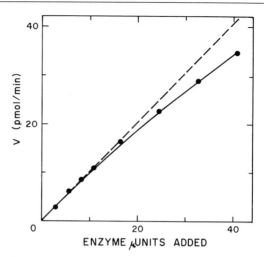

Fig. 2. Range of linear response of the standard assay with increasing concentrations of casein kinase II. Increasing amounts of casein kinase II were added to the standard assay. The deviation from linearity at 20×10^{-6} e.u. was 7.5%.

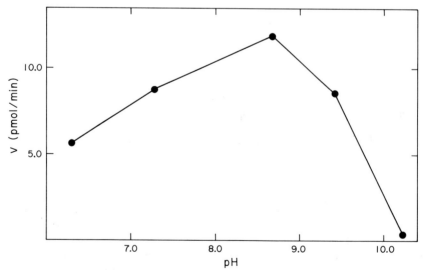

Fig. 3. Casein kinase assay, pH dependence. Standard assay conditions were used except that the following buffers were employed: 12.5 mM potassium phoshate, pH 6.3 or 25 mM Tris–HCl, pH 6.3; 12.5 mM potassium phosphate, 12.5 mM Tris–HCl, pH 7.3; 25 mM Tris–HCl, pH 8.7; 25 mM Tris–HCl, pH 9.4; 25 mM Tris–HCl, pH 10.25.

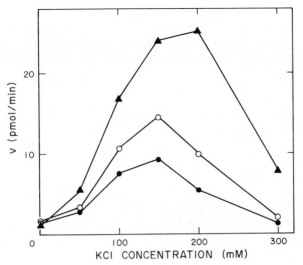

FIG. 4. Dependence of casein kinase II activity on KCl and casein concentrations. Casein kinase II (25×10^{-6} e.u.) was assayed under standard conditions at 0.5 mg/ml (●), 1.0 mg/ml (○), 5.0 mg/ml (▲) mixed casein.

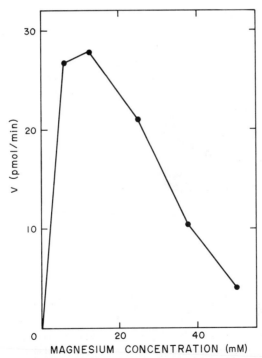

FIG. 5. Magnesium chloride optimum for casein kinase II. The concentration of Mg^{2+} was varied under standard assay conditions with 28×10^{-6} e.u. of casein kinase II.

Therefore, conditions should be optimized with respect to these salts for each new substrate or batch of dephosphorylated whole casein.

Purification of Casein Kinase II

Casein kinase II is routinely purified from 2 liters of postribosomal supernate from rabbit reticulocytes. For the preparation of postribosomal supernate, washed reticulocytes[16] are lysed in 2 mM MgCl$_2$ at two times the packed cell volume for 4 min. Cell lysis is terminated by addition of one packed cell volume of 0.15 M KCl. All procedures are carried out at 0–4°. The lysate is centrifuged at 15,000 g for 30 min. The pellet is discarded and the supernate is centrifuged 3 hr at 360,000 g. The supernate is decanted and brought to 20 mM Tris–HCl, pH 7.4, 2 mM EGTA, 5 mM EDTA, 10 mM 2-mercaptoethanol, and 0.02% NaN$_3$ (buffer A) using a 10-fold concentrated solution. The ribosomal pellet may be saved and utilized for other purposes. Phenylmethylsulfonyl fluoride is added to a concentration of 0.5 mM by the dropwise addition of a 0.1 M solution in isobutanol. The supernate is stirred at 4° and the protease inhibitor is added over a period of 10 min. DE-53 or DE-52 (800 ml settled volume), which has been preequilibrated with buffer A and degassed to remove CO$_2$, is added to the supernate and the mixture is stirred periodically with a glass rod. After 15 min the slurry is transferred to a large Buchner funnel and filtered by suction through two thicknesses of Whatman No. 1 filter paper. The resin is not allowed to dry, but is washed continuously with starting buffer until essentially free of hemoglobin. The protein-bound resin is transferred to a 4-liter beaker; buffer A is added to increase the total volume by about 40% and the slurry is poured into a 1-liter column (5.0 × 60 cm). After washing with 600 ml of buffer A, a 2.5-liter linear gradient ranging from 0 to 0.5 M KCl in buffer A is started and 20-ml fractions are collected. Casein kinase II elutes between 0.175 and 0.22 mM KCl. Aliquots of every other fraction are diluted 1/20 and assayed as described previously. Fractions containing the activity are pooled. An aliquot of the pooled fraction is taken at each step of the purification procedure and analyzed for activity. A second aliquot is stored at 4° until the end of the purification, whereupon all samples are analyzed for protein content by the procedures of either Lowry *et al.*[17] or Bradford.[18]

The pool from the DEAE-cellulose column is adjusted to 25 mM potassium phosphate, pH 6.8, 20 mM 2-mercaptoethanol, 10 mM EDTA, and 5 mM EGTA. Solid ammonium sulfate is added slowly to 80% saturation at 0° along with dilute NH$_4$OH, as needed, to maintain the pH at 7.0.

[16] J. A. Traugh and S. B. Sharp, this series, Vol. 60, p. 534.
[17] O. H. Lowry, N. J. Rosebrough, A. L. Farr, and R. J. Randall, *J. Biol. Chem.* **193**, 265 (1951).
[18] M. Bradford, *Anal. Biochem.* **72**, 248 (1976).

The precipitate is collected by centrifugation and dissolved in a minimal volume of 25 mM phosphate buffer, pH 6.8, containing 10 mM 2-mercaptoethanol, 1 mM EDTA, 2 mM EGTA, and 0.02% NaN$_3$ (buffer B) and dialyzed exhaustively against buffer B. The sample is pumped onto a 50-ml column (2.5 × 10 cm) of phosphocellulose equilibrated with buffer B. Under these conditions, greater than 80% of the casein kinase II passes through the column while casein kinase I binds. The flow-through containing casein kinase II is adjusted to 0.25 M NaCl with solid NaCl and applied to a 100 ml phosphocellulose column (2.5 × 20 cm) equilibrated with buffer B containing 0.25 M NaCl. Under these conditions, casein kinase II adheres to the phosphocellulose. A linear, 400 ml gradient ranging from 0.25 to 1.25 M NaCl in buffer B is applied. Casein kinase II elutes at about 0.7 M NaCl. While the enzyme is not pure at this stage, the preparation is free of other protein kinases and may be stored up to a year at 4° without loss of activity if sodium azide (0.02%) or another suitable fungicide or bacteriostatic agent is added.

Further purification of the enzyme is achieved by chromatography on hydroxylapatite as described previously,[3,19] or by chromatography on a third phosphocellulose column under the same conditions as the second column. The resulting enzyme has high specific activity and the preparation is greater than 90% pure as shown by polyacrylamide gel electrophoresis in sodium dodecyl sulfate. The enzyme from the final phosphocellulose column is concentrated on hydroxlyapatite. The pooled activity is diluted with double distilled water to a conductivity equivalent to 0.1 M potassium phosphate and applied to a 0.7-ml column of hydroxylapatite equilibrated at pH 6.8 with 25 mM potassium phosphate, 10 mM 2-mercaptoethanol, and 0.02% sodium azide. Casein kinase II is eluted in a single step when potassium phosphate is increased to 400 mM. Fractions of 0.2 to 0.4 ml are collected, assayed, and combined so that a final activity of at least 2 units per ml (about 1.0 mg/ml protein) is obtained. In the best preparations, recovery is 30–40% of that from the DEAE-cellulose step. The purified enzyme is stable at 4° for up to a year under these conditions if azide (0.02%) is present. Lower protein concentrations lead to a rapid loss of enzyme activity; freezing inactivates the enzyme.

Autophosphorylation of Casein Kinase II

When casein kinase II is incubated at 30° in the presence of [γ-^{32}P]ATP, radioactivity is incorporated into the enzyme. Analysis of the product by polyacrylamide gel electrophoresis in sodium dodecyl sulfate

[19] G. M. Hathaway, T. S. Lundak, S. M. Tahara, and J. A. Traugh, this series, Vol. 60, p. 495.

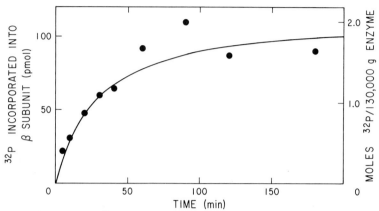

FIG. 6. Time course of $^{32}P_i$ incorporation into casein kinase II. Casein kinase II (14.5 μg) was incubated at 30° for the indicated times and analyzed by gel electrophoresis.

reveals that virtually all of the $^{32}P_i$ is incorporated into the 24,000 dalton β-subunit.[3] Casein kinase II (14 μg) is autophosphorylated in 0.30 ml in 50 mM Tris-HCl, 50 mM potassium phosphate, pH 6.9, 10 mM MgCl$_2$, and 0.20 mM ATP. The components are preincubated for 5 min at 30° prior to initiation of the reaction by the addition of 0.025 ml of enzyme. Incubation is at 30°. Aliquots of 0.030 ml are removed at timed intervals and sample buffer containing sodium dodecyl sulfate (0.02 ml) is added to stop the reaction.[19] After heating for 15 min at 65° the samples are loaded onto 10% polyacrylamide gels.[19,20] Following electrophoresis, casein kinase II is located by staining with Coomassie blue and the radioactivity is identified by autoradiography. Radioactive gel bands are excised along with controls (areas of the gel containing no protein) and incubated overnight at room temperature in counting vials containing 0.5 ml of 1% SDS. Following addition of 7.5 ml of Triton-toluene (1 : 2) scintillation mixture or 4.0 ml of Scint-A (Packard Industries), the radioactivity is monitored in a scintillation counter. The enzyme incorporates up to 2 mol of phosphate per 130,000 g of protein or 1 mol per mol of β-subunit (Fig. 6). When the pH dependency of the autophosphorylation reaction is examined, a pH optimum around 7.5 is observed (data not shown).

Inhibition of Casein Kinase II

A number of inhibitors of casein kinase II have been identified including heparin, 2,3-bisphosphoglycerate, inositol hexasulfate, and pyridoxal

[20] U. K. Laemmli, *Nature (London)* **227**, 680 (1970).

5'-phosphate.[4,7–11] All of these inhibitors are competitive with respect to the protein substrate.[7–10]

Heparin

Of the several inhibitors studied, heparin has proved to be both the most potent and the most specific for the type II casein kinase. Inhibition of casein kinase II is measured under standard conditions. Porcine heparin (Sigma) is made up gravimetrically in double distilled water. Owing to the competitive nature of the inhibition, 0.15 μg/ml of heparin (9.6 nM) is necessary to yield 50% inhibition at saturating casein concentrations as shown in Table I. An apparent K_i of 1.4 nM has been obtained with heparin using mixed casein as substrate.[7] It should be noted that while heparin is a potent inhibitor of casein kinase II, it interacts with a number of endogenous proteins including translational initiation factors, ribosomes, RNA polymerase, etc. In the cases that have been examined, higher levels of heparin are required to inhibit the enzyme when these proteins are used as substrates.[6] This is probably due to competition for heparin between the enzyme and the substrate.

2,3-Bisphosphoglycerate

Solid 2,3-bisphosphoglycerate (Sigma), either as the pentacyclohexylammonium or the Tris salt, is dissolved in water and the pH is brought to 7.0 with 0.1 M HCl or 0.1 M NaOH. The volume is adjusted to give a final concentration of 25 mM and diluted to give working solutions of 2.5, 5.0, 10.0, 15.0, and 20 mM. Additions of 0.020 ml of these solutions are made to the assay. With 2,3-bisphosphoglycerate, the inhibition has been shown

TABLE I
INHIBITORS OF CASEIN KINASES

Inhibitor[a]	Casein kinase I $I_{50}{}^a$ (mM)	Casein kinase II I_{50} (mM)
Heparin	—[b]	9.3×10^{-6}
Inositol hexasulfate	0.1	0.04
Pyridoxal 5'-phosphate	1.5	0.4
2,3-Bisphosphoglycerate	15	2.1

[a] I_{50} is defined as the concentration of inhibitor required to give 50% inhibition under standard assay conditions.
[b] No inhibition was observed at 50 times the amount which gave 50% inhibition with the type II enzyme.

to be cooperative and the degree of inhibition is dependent upon the Mg^{2+} concentration.[9,10] Because of the cooperative nature of the inhibition, it has not been possible to report a K_i for 2,3-bisphosphoglycerate. However, at 0.5 mg/ml β-casein B and 10 mM $MgCl_2$, 1 mM 2,3-bisphosphoglycerate gives about 50% inhibition. Similar results are found when initiation factor 2 is the substrate.[12] At saturating levels of whole casein (5.0 mg/ml) the I_{50} is 2.1 mM. This compares with values of 0.01 μM for heparin, 40 μM for inositol hexasulfate, and 400 μM for pyridoxal 5'-phosphate (Table I).

Inositol Hexasulfate and Pyridoxal 5'-Phosphate

In a similar manner to that described for 2,3-bisphosphoglycerate, inositol hexasulfate and pyridoxal 5'-phosphate are made up at pH 7.0 at concentrations of 10 mM. Solutions of pyridoxal 5'-phosphate are made fresh daily and kept in the dark. They are quantitated in 50 mM potassium phosphate at pH 7.0 by absorbance at 388 nm (ε = 4900 M^{-1} cm^{-1}).[21]

Activation of Casein Kinase II by Polyamines

The polyamine, spermine, has been shown to stimulate casein kinase II.[5,6] The degree of stimulation is dependent on several parameters including KCl and Mg^{2+} concentrations (Hathaway and Traugh, unpublished results). At limiting substrate (0.5 mg/ml β-casein A) and conditions for optimal enzymatic activity (100 mM KCl, 10 mM $MgCl_2$), a 1.5-fold stimulation of activity is observed at 2 mM spermine. However, when Mg^{2+} or KCl are varied in the reaction, the response of the enzyme toward spermine is markedly variable. At low concentrations of Mg^{2+} (2.0 mM) the enzyme is stimulated threefold, while at high concentrations of Mg^{2+} (>20 mM) or KCl (>150 mM) inhibition by spermine is observed. The amount of spermine required to give 50% of the stimulatory response decreases as Mg^{2+} decreases. At 10 mM Mg^{2+}, the concentration of spermine which gives 50% maximal stimulation is 0.63 mM; when Mg^{2+} is reduced to 2 mM, 0.24 mM spermine is required.

Identification of Casein Kinase II in Cell Lysates or Extracts

The strong inhibition obtained with heparin, together with the fact that the inhibition is specific for the type II casein kinase, allows measurement of this enzyme in mixtures of protein kinases such as cell lysates and

[21] C. A. Storvick, E. M. Benson, M. A. Edwards, and M. Woodring, *Biochem. Biophys. Res. Commun.* **106**, 131 (1982).

TABLE II
EFFECT OF DILUTION ON QUANTITATION OF CASEIN KINASE II
IN LYSATE

Tissue	Sample[a]	Dilution	Enzyme activity (pmol/min)
Reticulocyte	Lysate	0	22.3
		1:2	23.8
		1:10	12.3
		1:20	6.3
Reticulocyte	Postribosomal supernate	0	21.3
		1:2	23.4
		1:10	10.2
		1:20	5.6
Erythrocyte	Lysate	0	9.8
		1:2	8.9
		1:10	2.2
		1:20	1.2

[a] Enzyme was added as lysate or postribosomal supernate (0.014 ml/0.100 ml reaction) and assayed using standard conditions.

extracts. It is important to note that casein kinase II is largely inhibited in reticulocyte lysate, as shown by a lack of correlation between enzyme activity and dilution (Table II). It should be emphasized that the nonlinearity is not due to substrate depletion nor product formed during the assay since the activity is well within the limits imposed by the assay. This nonlinearity can be eliminated by dilution, dialysis, or gel filtration on Sephadex G-25 (Hathaway and Traugh, unpublished results), presumably by removing a low-molecular-weight inhibitor which may vary from cell type to cell type.

When increasing concentrations of heparin are added to an assay mixture containing either reticulocyte lysate or postribosomal supernate, total casein kinase activity is inhibited. Figure 7 shows the titration of casein kinase activity in reticulocyte lysate by porcine heparin. Maximal inhibition is observed at 0.7 μg/ml heparin. The percentage of the total activity inhibited by heparin is observed to be 80 ± 3%. This value is independent of whether lysate or postribosomal supernate is chosen as the enzyme source as long as dilution or gel filtration preceded the measurement. Since heparin is highly specific for casein kinase II, it follows that this enzyme must account for about 80% of the casein kinase activity in lysates. This is in accord with the distribution of casein kinase activity observed following DEAE-cellulose chromatography.[3] Since total casein

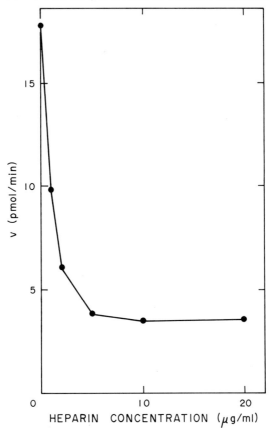

FIG. 7. Inhibition of casein kinase activity by heparin in a reticulocyte lysate. Rabbit reticulocyte lysate was desalted by gel filtration on Sephadex G-25. The lysate was then further diluted 1 : 3 to yield a total dilution of 1 : 6; 0.014 ml was used as the enzyme source in the standard assay. Heparin was added at concentrations up to 20 μg/ml.

kinase activity is calculated to be about 0.040 e.u. per ml of packed cells, this suggests about 0.032 e.u. are casein kinase II.

Identification of Casein Kinase II

Casein kinase II can be distinguished from casein kinase I by numerous criteria.[1] The enzymes are chromatographically distinct on DEAE-cellulose and on phosphocellulose. Casein kinase I elutes at 0.05 M KCl from DEAE-cellulose and 0.5 M NaCl from phosphocellulose whereas casein kinase II elutes at 0.20 and 0.7 M salt from these columns, respectively. Casein kinase I utilizes only ATP as the phosphoryl donor while casein kinase II effectively utilizes both ATP and GTP. The identity of the casein kinases can be confirmed by examining the sites of phosphorylation using casein variants. Casein kinase I phosphorylates only seryl residues while casein kinase II modifies threonyl residues in fully phosphorylated and fully dephosphorylated casein and seryl residues in partially dephosphorylated casein.[1,2] Two-dimensional phosphopeptide maps obtained following incubation of casein kinase II with native or dephosphorylated casein variants give identical results. These phosphopeptide patterns are distinct from those observed with casein kinase I (for methodology and results, see Ref. 14). Finally, heparin can be used to distinguish between the two enzymes. Heparin specifically inhibits casein kinase II whereas no effect on casein kinase I is observed with a 50-fold excess of the inhibitor (7 μg/ml).

Acknowledgments

We wish to thank Dr. Elizabeth Bingham for generously supplying the casein variants used in these studies. The research described here was supported by Grant GM 26738 from the U.S. Public Health Service.

[35] Pyruvate Dehydrogenase Kinase from Bovine Kidney

By FLORA H. PETTIT, STEPHEN J. YEAMAN, and LESTER J. REED

$$\text{Dephosphoenzyme } (E_1) + \text{ATP} \xrightarrow{\text{kinase}} \text{phosphoenzyme} + \text{ADP}$$

In eukaryotic cells the pyruvate dehydrogenase complex is located in mitochondria, within the inner membrane-matrix compartment. The complex is composed of multiple copies of three major components: pyruvate dehydrogenase (E_1), dihydrolipoamide acetyltransferase (E_2), and dihydrolipoamide reductase (E_3). The E_2 component forms a structural core, composed of 60 subunits arranged in a pentagonal dodecahedron-like particle, to which E_1 and E_3 are bound by noncovalent bonds. The complex also contains small amounts of two regulatory enzymes, a kinase and a phosphatase, that modulate the activity of E_1 by phosphorylation (inactivation) and dephosphorylation (activation), respectively.[1] The kinase (EC 2.7.1.99) is tightly bound to E_2 and copurifies with the complex, whereas the phosphatase is loosely associated with the complex.[2]

[1] T. C. Linn, F. H. Pettit, and L. J. Reed, *Proc. Natl. Acad. Sci. U.S.A.* **62**, 234 (1969).
[2] T. C. Linn, J. W. Pelley, F. H. Pettit, F. Hucho, D. D. Randall, and L. J. Reed, *Arch. Biochem. Biophys.* **148**, 327 (1972).

Assay Method

Principle. Kinase activity in preparations of the pyruvate dehydrogenase complex can be detected by monitoring inactivation of the complex in the presence of ATP and Mg^{2+} or incorporation of ^{32}P-labeled phosphoryl groups from [γ-^{32}P]ATP into the complex, i.e., into its E_1 component. Quantitative assay of kinase activity is based on measurement of the initial rate of incorporation of ^{32}P-labeled phosphoryl groups into crystalline E_1 in the presence of E_2. E_2 stimulates the rate of phosphorylation of E_1 by the kinase three- to fivefold. The method is modified from that of Linn *et al.*[2]

Reagents

Assay buffer: 0.02 M potassium phosphate, pH 7.0, 1 mM $MgCl_2$, 0.1 mM EDTA, 2 mM dithiothreitol

Dihydrolipoamide acetyltransferase, prepared from pyruvate dehydrogenase complex as described below

Pyruvate dehydrogenase, crystalline, prepared from pyruvate dehydrogenase complex as described below

[γ-^{32}P]ATP (Amersham, Arlington Heights, Illinois), 5 mM, in assay buffer, with a specific radioactivity of about 10,000 cpm/nmol

Trichloroacetic acid, 10%

Aqueous counting scintillant (Amersham)

Procedure. The reaction mixture contains 0.3 mg of crystalline pyruvate dehydrogenase and 0.2–2 units of kinase in 0.18 ml of the assay buffer.[3] After equilibration of the solution at 30° for 30 sec, 0.02 ml of the radioactive ATP solution is added to start the reaction. Three 0.05-ml aliquots are withdrawn at 20-sec intervals and applied to 2.2 cm (diameter) disks of Whatman 3MM chromatography paper. The paper disks are placed immediately in cold 10% trichloroacetic acid. The disks are washed four times (10 min each) with cold 10% trichloroacetic acid, twice (5 min each) with 95% ethanol, and once (5 min) with diethyl ether. The papers are air-dried and placed in vials containing 5 ml of scintillant; radioactivity is determined in a scintillation counter.

Units. Units are expressed as micromoles of ^{32}P incorporated per minute at 30°, and specific activities as units per milligram of protein. Protein is determined by the biuret method[4] with crystalline bovine serum albumin as standard.

---

[3] When free kinase, i.e., kinase not complexed with E_2, is assayed, 50 μg of the latter enzyme is included in the assay mixture. When possible, kinase is added to the assay mixture from concentrated solution to minimize dilution of the enzyme. If dilution is necessary, assay buffer containing 0.3% bovine serum albumin is used.

[4] A. G. Gornall, C. J. Bardawill, and M. M. David, *J. Biol. Chem.* **177,** 751 (1959).

PURIFICATION OF PYRUVATE DEHYDROGENASE KINASE FROM BOVINE KIDNEY

Fraction	Volume (ml)	Protein (mg)	Specific activity[a]	Recovery (%)
Pyruvate dehydrogenase complex	10	347	0.012	100
E_2-kinase subcomplex	4.3	78	0.028	52
Kinase	1.0	4	0.33[b]	32

[a] Micromoles of ^{32}P incorporated into pyruvate dehydrogenase per minute per mg protein.
[b] Acetyltransferase (E_2) (50 µg) was added to the assay mixture.

Purification Procedure

Pyruvate dehydrogenase kinase is tightly bound to the acetyltransferase (E_2) component of the pyruvate dehydrogenase complex. The kidney complex contains 3–5 times as much kinase activity as the heart complex. Therefore, the highly purified pyruvate dehydrogenase complex from bovine kidney is the preferred source for isolation of the kinase. The complex is separated into E_1, E_3, and an E_2-kinase subcomplex. The subcomplex is then resolved into E_2 and the kinase. The purification procedure is modified from that of Linn et al.[2] and is summarized in the table.[5] Unless specified otherwise, all operations are performed at 4°.

Step 1. Resolution of Pyruvate Dehydrogenase Complex at pH 9.0. To a concentrated solution of the highly purified pyruvate dehydrogenase complex (about 300 mg in <10 ml) from bovine kidney[6] is added solid NaCl and 1 M glycine (adjusted to pH 9.5 with NaOH) to make the final concentrations 1 and 0.1 M, respectively. The volume is adjusted to 10 ml; the pH of the solution is 9.0. The solution is kept in an ice bath for at least 30 min and then applied to a column (2.5 × 50 cm) of Sepharose 6B that has been equilibrated with buffer A (0.1 M glycine, pH 9.0; 1 M NaCl; 1 mM $MgCl_2$; 0.1 mM EDTA). The column is developed with this buffer. The flow rate is about 8 ml/hr, and fractions of about 2.6 ml are collected. Two major protein peaks are eluted, as determined by monitoring A_{280} of the fractions. The first peak to emerge from the column contains the E_2-kinase subcomplex. The second peak contains E_1 and E_3.

Step 2. Ammonium Sulfate Precipitation. Fractions containing the E_2-kinase subcomplex are combined, and solid ammonium sulfate (0.12 g/ml) is added with stirring. The solution is stirred for 15 min, then the precipitate is collected by centrifugation at 30,000 g for 15 min. The

---

[5] The procedure is also applicable to resolution of the pyruvate dehydrogenase complex from bovine heart.
[6] F. H. Pettit and L. J. Reed, this series, Vol. 89, p. 376.

precipitate is dissolved in about 3 ml of 0.05 M potassium phosphate, pH 7.5, containing 1 mM MgCl$_2$ and 0.1 mM EDTA, and the solution is dialyzed for about 30 hr with stirring against two 500-ml portions of this same buffer.

Step 3. Treatment with p-Hydroxymercuriphenyl Sulfonate. The dialyzed protein from step 2 is diluted to 5 mg/ml with buffer A containing 2 mM dithiothreitol and dialyzed against this same buffer for 6 hr.[7] The dialyzed solution is placed in a centrifuge tube, and 0.01 volume of 0.25 M monosodium p-hydroxymercuriphenyl sulfonate (Sigma Chemical Co., St. Louis, Missouri) is added. After mixing, the preparation is kept on ice for 30 min. If no precipitation occurs, another 0.03 ml of the mercurial is added. After 30 min, 0.01-ml aliquots of the mercurial are added at 10-min intervals until no further local precipitation occurs. The mixture is kept overnight on ice, and the precipitate is collected by centrifugation at 30,000 g for 20 min. The precipitate contains E_2, and the supernatant fluid contains the kinase. The precipitate is dissolved overnight in a minimal amount of 0.05 M potassium phosphate, pH 7.5, containing 30 mM dithiothreitol, 0.1 mM MgCl$_2$, 0.01 mM EDTA; the solution is clarified by centrifugation at 30,000 g for 10 min. The yield of acetyltransferase is about 50 mg.[8] The kinase solution is dialyzed for 16 hr, with stirring, against two 1-liter portions of buffer B (0.01 M imidazole, adjusted to pH 7.3 with solid DL-asparagine, 0.1 mM MgCl$_2$, 0.01 mM EDTA). The solution is centrifuged twice at 144,000 g for 2.5-hr periods to remove trace amounts of E_2. The supernatant fluid is concentrated by vacuum dialysis against buffer B in a Bio-Molecular Dynamics concentrator (Bio-Molecular Dynamics, Beaverton, Oregon).

Step 4. DEAE-Cellulose Chromatography. Kinase preparations from step 3 are usually at least 90% pure as judged by sodium dodecyl sulfate (SDS)–polyacrylamide gel electrophoresis. Preparations of lesser purity are purified further by chromatography on Whatman DE-52. The enzyme is applied to a column (0.9 × 1.5 cm) of DE-52 that has been equilibrated with buffer B. The column is washed with 3 ml of this buffer, and then developed with 3-ml portions of this buffer containing 0.05, 0.1, and 0.2 M NaCl; 0.5-ml fractions are collected. The kinase elutes at 0.2 M NaCl. The active fractions are combined and dialyzed against buffer B. The recovery of kinase activity in this step is about 85%.

Purification and Crystallization of Pyruvate Dehydrogenase. The E_1-containing fractions from step 1 are made 10 mM with respect to dithiothreitol and kept on ice for 30 min. Solid ammonium sulfate (0.142 g/

---

[7] Treatment of the E_2-kinase subcomplex with dithiothreitol at alkaline pH is essential for subsequent resolution of the subcomplex in the presence of mercurial.

[8] Protein is determined after precipitation with trichloroacetic acid to remove dithiothreitol.

ml) is added with stirring. The suspension is stirred for 5 min, then the precipitate is removed by centrifugation at 30,000 g for 15 min; the pellet is discarded. A second portion of solid ammonium sulfate (0.107 g/ml) is added with stirring. The precipitate is collected by centrifugation. It contains E_1; the supernatant fluid contains E_3.[9] The precipitate is dissolved in about 2.2 ml of buffer C (0.02 M potassium phosphate, pH 7.0, 0.1 mM $MgCl_2$, 0.01 mM EDTA, 2 mM dithiothreitol) and dialyzed overnight against the same buffer. The solution is dialyzed over a 3-day period against buffer C that has been adjusted consecutively to pH 6.8, 6.7, and 6.6. The dialyzed preparation is warmed to 30° for 10 min and then clarified by centrifugation for 15 min at 20°. The solution contains about 100 mg of protein. It is kept on ice for about 2 weeks. The needle-like crystals of E_1 are collected by centrifugation and redissolved in 1.2 ml of buffer C by warming at 30° for 15 min. Any insoluble material is removed by centrifugation at 20°. The yield of crystalline pyruvate dehydrogenase is 20–66%. The solution is divided into small portions and stored at −50°.

Properties

Pyruvate dehydrogenase kinase obtained from step 3 or step 4 is at least 90% pure as judged by analytical ultracentrifugation. The sedimentation velocity pattern shows a major peak with a sedimentation coefficient (s_{20}) of about 5.5 S in 0.01 M imidazole buffer, pH 7.5.[2] The kinase consists of two apparently different subunits with molecular weights of about 48,000 and 45,000 as estimated by SDS–polyacrylamide gel electrophoresis.[10] The kinase doublet is difficult to detect on SDS–polyacrylamide gels of the kidney pyruvate dehydrogenase complex because of the small amount of kinase present in the complex (about 5% by weight). The kinase exhibits a pronounced tendency to aggregate in buffers other than buffer B. The kinase is labile to freezing and thawing. It loses about 50% of its activity during storage for 1 month at 4°. Sodium azide (0.02%) is usually included in the storage buffer.

Catalytic Properties.[11] Pyruvate dehydrogenase kinase requires a divalent cation (Mg^{2+} or Mn^{2+}). Its activity is not affected by cyclic AMP or cyclic GMP. This kinase catalyzes transfer of the γ-phosphoryl group of ATP to three serine residues in the α subunit (M_r 41,000) of pyruvate

---

[9] Highly purified dihydrolipoamide reductase (5–8 mg) is collected between 0.45 and 0.65 ammonium sulfate saturation.

[10] F. H. Pettit, L. R. Stepp, S. J. Yeaman, and L. J. Reed, unpublished results.

[11] F. Hucho, D. D. Randall, T. E. Roche, M. W. Burgett, J. W. Pelley, and L. J. Reed, *Arch. Biochem. Biophys.* **151**, 328 (1972).

dehydrogenase (E_1).[12,13] Phosphorylation proceeds markedly faster at site 1 than at sites 2 and 3, and phosphorylation at site 1 correlates closely with inactivation of E_1. The apparent K_m for $MgATP^{2-}$ is about 20 μM. The apparent K_m for E_1 is about 20 μM. In the presence of the acetyltransferase (E_2) this value is decreased to about 0.6 μM, and the rate of phosphorylation of E_1 is increased three- to fivefold. This acceleration in rate could be due to favorable topographical positioning of E_1 with respect to its kinase or possibly to a change in conformation of either E_1, the kinase, or both enzymes, induced by binding to E_2.

Kinase activity is stimulated by acetyl-CoA and by NADH if K^+ or NH_4^+ ions are present,[14] and kinase activity is inhibited by ADP, pyruvate, and dichloroacetate.[15] ADP is competitive with respect to ATP, and this inhibition apparently requires the presence of a monovalent cation.[16] These effectors apparently act directly on the kinase.[17] On the other hand, the coenzyme thiamin pyrophosphate inhibits kinase activity by acting on the protein substrate (E_1) rather than on the kinase.[18]

Kinase activity is also inhibited by N-ethylmaleimide and by certain disulfides.[19] Inhibition by disulfides is highly specific and is reversed by thiols. 5,5'-Dithiobis(2-nitrobenzoic acid) is the most potent inhibitor, showing significant inhibition at a concentration as low as 1 μM. The kinase apparently contains a thiol group (or groups) that is involved in maintaining a conformation of the enzyme that facilitates phosphorylation and inactivation of E_1.

Pyruvate dehydrogenase kinase appears to be highly specific for pyruvate dehydrogenase. It exhibits little activity, if any, toward rabbit skeletal muscle phosphorylase b or glycogen synthase a, various histones, or casein.

---

[12] C. R. Barrera, G. Namihira, L. Hamilton, P. Munk, M. H. Eley, T. C. Linn, and L. J. Reed, *Arch. Biochem. Biophys.* **148**, 343 (1972).

[13] S. J. Yeaman, E. T. Hutcheson, T. E. Roche, F. H. Pettit, J. R. Brown, D. C. Watson, and G. H. Dixon, *Biochemistry* **17**, 2364 (1978).

[14] F. H. Pettit, J. W. Pelley, and L. J. Reed, *Biochem. Biophys. Res. Commun.* **65**, 575 (1975).

[15] S. Whitehouse, R. H. Cooper, and P. J. Randle, *Biochem. J.* **141**, 761 (1974).

[16] T. E. Roche and L. J. Reed, *Biochem. Biophys. Res. Commun.* **59**, 1341 (1974).

[17] L. J. Reed, *Curr. Top. Cell. Regul.* **18**, 95 (1981).

[18] J. R. Butler, F. H. Pettit, P. F. Davis, and L. J. Reed, *Biochem. Biophys. Res. Commun.* **74**, 1667 (1977).

[19] F. H. Pettit, J. Humphreys, and L. J. Reed, *Proc. Natl. Acad. Sci. U.S.A.* **79**, 3945 (1982).

[36] Glycogen Synthase Kinase-3 from Rabbit Skeletal Muscle

By BRIAN A. HEMMINGS and PHILIP COHEN

Glycogen synthase a + 3ATP → Glycogen synthase b + 3ADP

MgATP-dependent protein → MgATP-dependent protein
phosphatase (inactive)   phosphatase (active)

Glycogen synthase kinase-3 is one of the five glycogen synthase kinases that have been identified in skeletal muscle,[1] and is of major importance in determining the kinetic properties of glycogen synthase *in vivo*.[2,3] It catalyzes the phosphorylation of three serine residues on glycogen synthase,[4-6] converting the enzyme from a form that is almost fully active in the absence of glucose-6P, to one that is largely dependent on this allosteric activator.[7] Glycogen synthase kinase-3 also has a second activity that is not shared by any other protein kinase, namely, the ability to activate an enzyme termed the MgATP-dependent protein phosphatase.[6] The importance of this second activity *in vivo* is not yet understood.

Assay Methods

Phosphorylation of Glycogen Synthase

This assay measures the incorporation of ^{32}P radioactivity from [γ-^{32}P]ATP into glycogen synthase.

Reagents

All reagents are stored on ice.

---

[1] P. Cohen, D. Yellowlees, A. Aitken, A. Donella-Deana, B. A. Hemmings, and P. J. Parker, *Eur. J. Biochem.* **124,** 21 (1982).
[2] P. J. Parker, N. Embi, P. Cohen, and F. B. Caudwell, *Eur. J. Biochem.* **124,** 47 (1982).
[3] P. J. Parker, F. B. Caudwell, and P. Cohen, *Eur. J. Biochem.* **130,** 227 (1983).
[4] D. B. Rylatt, A. Aitken, T. Bilham, G. D. Condon, N. Embi, and P. Cohen, *Eur. J. Biochem.* **107,** 529 (1980).
[5] C. Picton, A. Aitken, T. Bilham, and P. Cohen, *Eur. J. Biochem.* **124,** 37 (1982).
[6] B. A. Hemmings, D. Yellowlees, J. C. Kernohan, and P. Cohen, *Eur. J. Biochem.* **119,** 443 (1982).
[7] N. Embi, D. B. Rylatt, and P. Cohen, *Eur. J. Biochem.* **107,** 519 (1980).

Glycogen synthase a is isolated from rabbit skeletal muscle[8,9] and passed through phosphocellulose equilibrated in 25 mM sodium glycerophosphate–1.0 mM EDTA–0.1% (v/v) 2-mercaptoethanol–10% (v/v) glycerol pH 7.0 (solution A) to remove traces of contaminating glycogen synthase kinase activities.[10] It is stored at $-20°$ at a concentration of ~6 mg/ml and aliquots are diluted to 1.5 mg/ml with solution A just before use. The concentration of glycogen synthase is determined using an absorbance index, $A_{280\ nm}^{1\%}$ of 13.4.[8]

The specific protein inhibitor of cyclic AMP-dependent protein kinase (termed "the inhibitor protein") is partially purified[11] and stored at $-20°$ as a concentrated solution in 50 mM Tris–HCl, pH 7.0–0.1% (v/v) 2-mercaptoethanol (solution B) containing 1.0 mM EDTA. Aliquots are diluted 100-fold in 0.01 mg/ml heparin just before use.

Dilution of glycogen synthase kinase-3. The enzyme is stored at $-20°$ in solution B containing 1.0 mM EDTA and 50% glycerol. Aliquots are diluted in solution B containing 1.0 mM EDTA and 1.0 mg bovine serum albumin/ml just before use.

15 mM magnesium acetate–1.0 mM [γ-^{32}P]ATP (5×10^7 cpm/μmol), pH 7.0.

Procedure. Glycogen synthase a (0.04 ml) is incubated with 2.0 mM EGTA (0.01 ml), inhibitor protein containing heparin (0.02 ml) and glycogen synthase kinase-3 (0.01 ml) at 30° for 3 min in a 1.5-ml plastic microcentrifuge tube. The reaction is initiated with MgATP (0.02 ml) and stopped after 2–10 min by the addition of 1.0 ml of 5% (v/v) trichloroacetic acid. After standing for 10 min, the suspension is centrifuged at 15,000 g for 2 min, the supernatant is discarded, and the precipitate is washed three times with 1.0-ml portions of 25% (w/v) trichloroacetic acid. Finally, the centrifuge tube is inserted into a scintillation vial, and the ^{32}P radioactivity is analyzed by Cerenkov counting. The assays are carried out in duplicate. Control incubations are included, in which glycogen synthase kinase-3 is omitted, to correct for any trace glycogen synthase kinase activity present in purified glycogen synthase. At early stages of the purification further control incubations are essential in which glycogen synthase is omitted, to correct for phosphorylation of proteins present in partially purified glycogen synthase kinase-3.

EGTA, inhibitor protein and heparin are included in the incubations to inhibit any phosphorylase kinase, cyclic AMP-dependent protein kinase,

---

[8] H. G. Nimmo, C. G. Proud, and P. Cohen, *Eur. J. Biochem.* **68**, 21 (1976).
[9] F. B. Caudwell, J. F. Antoniw, and P. Cohen, *Eur. J. Biochem.* **86**, 511 (1978).
[10] T. R. Soderling, A. K. Srivastava, M. A. Bass, and B. S. Khatra, *Proc. Natl. Acad. Sci. U.S.A.* **76**, 2536 (1979).
[11] G. A. Nimmo, and P. Cohen, *Eur. J. Biochem.* **87**, 341 (1978).

and glycogen synthase kinase-5,[1] respectively, that may be trace contaminants in glycogen synthase or glycogen synthase kinase-3. Glycogen synthase may also contain traces of a modified form of phosphorylase kinase that has lost its sensitivity to regulation by calcium ions, and is therefore no longer inhibited by EGTA.[12] This is largely removed by passing glycogen synthase through phosphocellulose.[10]

Definition of Unit. One unit of glycogen synthase kinase-3 is that amount which catalyzes the incorporation of 1.0 µmol of phosphate into glycogen synthase per minute from [γ-^{32}P]ATP. In making this calculation the subunit molecular weight of glycogen synthase is taken as 86,000.[5]

Activation of the MgATP-Dependent Protein Phosphatase

The MgATP-dependent protein phosphatase is completely inactive until it is incubated with glycogen synthase kinase-3 and MgATP.[6] Glycogen synthase kinase-3 can therefore be assayed by quantitating its ability to activate the MgATP-dependent protein phosphatase under conditions where the rate of increase in protein phosphatase activity is directly proportional to the amount of glycogen synthase kinase-3 added.[6] The MgATP-dependent protein phosphatase has a broad substrate specificity and is conveniently assayed by the release of ^{32}P-inorganic phosphate from ^{32}P-labeled phosphorylase *a*.[13]

Reagents

All reagents are stored on ice.

The MgATP-dependent protein phosphatase is partially purified from rabbit skeletal muscle[13] and stored at −20° in solution B containing 1.0 mM EDTA and 50% (v/v) glycerol at a concentration of 200 units/ml. Aliquots are diluted 100-fold in solution B containing 1.0 mg bovine serum albumin/ml just prior to assay.

Dilution of glycogen synthase kinase-3. This is diluted into solution B containing 1.0 mg bovine serum albumin/ml just prior to assay.

5.0 mM MgCl$_2$–0.5 mM ATP, pH 7.0.

^{32}P-labeled phosphorylase *a* (10^8 cpm/µmol). The phosphorylase *a* is prepared from rabbit skeletal muscle phosphorylase *b* using phosphorylase kinase and [γ-^{32}P]ATP,[14] and stored as a crystalline suspension at 30 mg/ml in solution B containing 1.0 mM EDTA and 250 mM NaCl. Immediately prior to use, the crystals are dissolved by warming at 30° for a few minutes, and diluted to 3.0 mg/ml with solution B.

[12] N. Embi, D. B. Rylatt, and P. Cohen, *Eur. J. Biochem.* **100**, 339 (1979).
[13] A. A. Stewart, B. A. Hemmings, P. Cohen, J. Goris, and W. Merlevede, *Eur. J. Biochem.* **115**, 197 (1981).
[14] J. F. Antoniw, H. G. Nimmo, S. J. Yeaman, and P. Cohen, *Biochem. J.* **162**, 423 (1977).

Procedure. The reaction is carried out in a 1.5-ml plastic microcentrifuge tube. MgATP-dependent protein phosphatase (0.02 ml; 0.04 units) is incubated at 30° for 5 min with 0.01 ml of diluted glycogen synthase kinase-3, and the activation reaction is initiated with 0.01 ml of MgATP. After 5 min, 0.02 ml of phosphorylase *a* is added to initiate the protein phosphatase reaction. After a further 5 min, the reactions are terminated by the addition of 0.1 ml of 20% (w/v) trichloroacetic acid. The suspension is allowed to stand for 10 min and centrifuged at 15,000 *g* for 2 min. The supernatant (0.1 ml) is transferred to another microcentrifuge tube containing 1.0 ml of dioxane-based scintillant, which is inserted into a scintillation vial and counted. Control incubations are included in which MgATP is omitted.

Definition of Unit. One unit of MgATP-dependent protein phosphatase is that amount which catalyzes the release of 1.0 μmol of phosphate/min from phosphorylase *a*. One unit of glycogen synthase kinase-3 is that amount which increases the phosphorylase phosphatase activity of the MgATP-dependent protein phosphatase by 1.0 U/min. The activation of the MgATP-dependent protein phosphatase is linear up to 0.001 mU of glycogen synthase kinase-3 added to the assay, corresponding to a 20% conversion of the protein phosphatase to its activated form during the incubation.[6]

Purification of Glycogen Synthase Kinase-3

Preparation of Glycogen Synthase Agarose

Affi-Gel 15 (12.5 ml; Bio-Rad) is washed with water and resuspended in 5 ml of 100 mM sodium bicarbonate, pH 8.0, containing 15 mg of glycogen synthase *a*. After mixing for 16 hr at 4°, 10.0 ml of 1.0 M glycine, pH 8.0, is added. After a further 60 min at 4°, the gel is washed with 40 ml of 25 mM Tris–HCl, pH 7.0–1.0 mM EDTA–0.1% (v/v) 2-mercaptoethanol–5% (v/v) glycerol–0.5 M NaCl and then equilibrated in the same buffer without the NaCl. Of the glycogen synthase 95% is covalently attached to Agarose by this procedure, and can be used repeatedly without any loss of binding capacity.

Procedure. All operations are carried out at 4°. The starting point for the preparation is the pH 6.1 supernatant obtained from a phosphorylase kinase preparation, which is described elsewhere in this book.[15] The pH 6.1 supernatant (~10,000 ml) obtained from 5000 g muscle (six rabbits) is adjusted to pH 7.0 with 10 M ammonium hydroxide, and taken to 33%

---

[15] P. Cohen, this volume [26].

saturation by the addition of solid ammonium sulfate (192 g/liter). After standing for 60 min, the suspension is centrifuged at 6000 g for 45 min, and the supernatant decanted and taken to 55% saturation by the addition of more ammonium sulfate (172 g/liter). After standing for a further 60 min, the suspension is again centrifuged at 6,000 g for 60 min, and the supernatant discarded. The precipitate is resuspended in 10 mM Tris–HCl, pH 7.0 (30°) containing 1.0 mM EDTA and 0.1% (v/v) 2-mercaptoethanol, and dialyzed for 15 hr against four changes of the same buffer. Denatured protein is removed by centrifugation at 20,000 g for 30 min and the supernatant (step 2) applied to a 15 × 9 cm column of DEAE—cellulose (Whatman DE-52) equilibrated in the dialysis buffer. Glycogen synthase kinase-3 is not retained by the column under these conditions, but it is necessary to wash the column with three column volumes of equilibration buffer to elute all of the enzyme. The active fractions (step 3) are adjusted to 40 mM Tris–HCl, pH 7.0–1.0 mM EDTA–0.1% (v/v) 2-mercaptoethanol–5% (v/v) glycerol, and all solutions after this step contain these three stabilizing agents. The enzyme is applied to a 14 × 4-cm column of phosphocellulose (Whatman P-11) equilibrated with 50 mM Tris–HCl pH 7.0, and washed extensively with the same buffer containing 0.1 M NaCl. Glycogen synthase kinase-3 is eluted with equilibration buffer containing 0.2 M NaCl (step 4). It should be noted that the bulk of the glycogen synthase kinase-3 activity is eluted after the major protein peak, resulting in a purification of 30- to 50-fold at this step.[6]

The eluate from phosphocellulose is applied directly to a 15 × 3.5-cm column of Affi-Gel blue (Bio-Rad) equilibrated in 50 mM Tris–HCl, pH 7.0–0.2 M NaCl. The column is washed with equilibration buffer until the absorbance of the effluent at 280 nm is zero and then developed with a 1000-ml linear salt gradient from 0.2 to 1.0 M NaCl. Glycogen synthase kinase-3, which is eluted at 0.4 M NaCl, is pooled (step 5), concentrated to ~10 ml by vacuum dialysis, dialyzed against 25 mM Tris–HCl, pH 7.0, and chromatographed on a glycogen synthase Agarose column (9.0 × 1.5 cm). The column is developed with a 200-ml linear salt gradient from 0 to 0.3 M NaCl in the equilibration buffer. Glycogen synthase kinase-3 which is eluted at 0.2 M NaCl is pooled (step 6), concentrated to ~5 ml by vacuum dialysis, dialyzed against 25 mM Tris–HCl, pH 7.0, and rechromatographed on glycogen synthase Agarose as described above (step 7). The active fractions are concentrated, dialyzed against 50 mM Tris–HCl, pH 7.0–1.0 mM EDTA–50% (v/v) glycerol and stored at −20°. Under these conditions there is little or no loss of activity for one year. A summary of the purification is given in the table.

[16] M. M. Bradford, *Anal. Biochem.* **72,** 248 (1976).

PURIFICATION OF GLYCOGEN SYNTHASE KINASE-3 FROM RABBIT SKELETAL MUSCLE[a]

Step	Protein (mg)	Activity (mU) Method A	Activity (mU) Method B	Specific activity (mU/mg) Method A	Specific activity (mU/mg) Method B	Activity ratios, Methods B/A	Purification (n-fold)	Yield (%)
1. Extract	170,000							
2. Ammonium sulfate (33–55%)	14,000							
3. DEAE-cellulose	2240	1870	33,000	0.83	14.7	18	1	100
4. Phosphocellulose	63	1805	27,000	28.6	430	15	34	96
5. Affi-Gel blue	12	1240	20,000	103	1700	16.5	128	66
6. First glycogen synthase agarose	0.4	436	6,500	1090	17,000	15.6	1310	23
7. Second glycogen synthase agarose	0.05	107	1,350	2060	26,000	12.6	2500	6

[a] Muscle (5000 g) (six rabbits) was used in this preparation. Protein was measured by the method of Bradford.[16] Since the activity cannot be measured accurately at the first two steps (see the text), the purification and yield were calculated relative to the values at step 3. The results are taken from Hemming et al.[6] Activity measurements were carried out by measuring the phosphorylation of glycogen synthase (Method A) or activation of the Mg ATP-dependent protein phosphatase (Method B).

Comments on the Purification

When muscle extracts are adjusted to pH 6.1 and centrifuged to sediment the protein–glycogen complex,[15] virtually all the glycogen synthase kinase-3 remains in the pH 6.1 supernatant,[6] and this fraction is therefore a convenient starting point for the preparation.

Glycogen synthase kinase-3 cannot be measured with any certainty at the first two stages of purification. For example, the presence of glycogen synthase kinase-4, which is unaffected by EGTA, heparin, or the inhibitor protein, interferes with estimation of the activity by Method A. The purification and yield are therefore given relative to step 3. However, based on the protein that is eliminated at these first three stages (98–99%, see the table), the overall purification should be at least 50,000-fold, even if the yield through step 3 was as low as 25%. Enzyme (~50–100 μg) is isolated from 5000 g muscle (see the table) corresponding to an overall yield of 6–12% from step 3. The preparation is highly reproducible and can be completed in about 7 days.

Virtually all the phosphorylase kinase activity and 90–95% of the glycogen synthase kinase-4 activity is removed by step 2. Remaining traces disappear at steps 3 and 4, which also eliminate glycogen synthase kinase-5. The catalytic subunit of cyclic AMP-dependent protein kinase follows glycogen synthase kinase-3 through step 4 and is only partially removed at step 5. This activity is usually eliminated by step 7, but some preparations are still contaminated with traces of this activity. This is easily inactivated by the inhibitor protein and hence does not interfere with studies of the phosphorylation of glycogen synthase by glycogen synthase kinase-3.

The activity ratio Method B/Method A remains constant throughout the purification (see the table) indicating that the phosphorylation of glycogen synthase and activation of MgATP-dependent protein phosphatase are catalyzed by the same enzyme. It should be noted, however, that at least 100-fold higher dilution of the enzyme is necessary when assaying by Method B.

Properties

Despite the extensive (~80,000-fold) purification, the preparations of glycogen synthase kinase-3 are not completely homogeneous. The most active fractions show one major protein staining band M_r approx. = 51,000 and several minor components.[6] If the major band is glycogen synthase kinase-3, then the preparations are 50–60% pure. Incubation with [γ-^{32}P]ATP-Mg leads to the phosphorylation of the M_r = 51,000 component and up to 4 mol of phosphate can be incorporated per mole.[6] This suggests that glycogen synthase kinase-3 undergoes an autophosphorylation reaction, like many protein kinases. Gel filtration on Sephadex G-100 equilibrated in solution B containing 10% (v/v) ethanediol and 0.01% Brij 35 gives an M_r approx. of 57,000 for enzyme purified through step 4 indicating that the purified glycogen synthase kinase-3 is a monomer.[6] However, if step 3 is omitted, and gel filtration performed in solution B containing 1.0 mM EDTA and 0.5 M NaCl, the M_r approx. is 72,000 with a shoulder at $M_r = 57,000$, while omission of NaCl causes aggregation to a form that is eluted in the void volume on Sephadex G-150.[1] It is not yet known whether glycogen synthase kinase-3 interacts with another protein in these fractions, particularly at low ionic strength, or whether the M_r approx. = 57,000 species is an active fragment generated from a larger form by limited proteolysis.

The K_m for glycogen synthase in Assay Method A is 0.3 mg/ml at 1.0 mM ATP and 3.0 mM Mg^{2+}, and the K_m values for ATP and GTP at 0.6 mg/ml glycogen synthase are 0.025 and 0.4 mM, respectively. The V_{max} at saturating concentrations of GTP is 65% of that at saturating ATP. The K_m values for ATP and GTP are similar in Assay Methods A and B. The activity is unaffected by cyclic AMP, cyclic GMP, Ca^{2+}, or calmodulin in either assay.[6,7]

Glycogen synthase kinase-3 catalyses the phosphorylation of the R$_{II}$-subunit of cyclic AMP-dependent protein kinase (6 μM) at 10–15% of the rate at which glycogen synthase (6 μM) is phosphorylated.[17] Partially dephosphorylated and hydrolyzed casein (100 μM) is phosphorylated at 3% of the rate of glycogen synthase (6 μM).[6] Glycogen synthase kinase-3 also phosphorylates a minor component present in variable amounts in commercial preparations of phosvitin. This minor component (M_r approx. = 35,000) has been purified, and its N-terminal sequence (nine residues) is identical to that of the major component (M_r approx. = 40,000) (A. Chan and B. A. Hemmings, unpublished experiments). Preliminary evidence suggests that the $M_r = 35,000$ protein may be a less highly phosphorylated form of the $M_r = 40,000$ species.[1] Glycogen synthase kinase-3 does not phosphorylate a large number of proteins that are phosphorylated by other protein kinases, such as glycogen phosphorylase, phosphorylase kinase, protein phosphatase inhibitor-1, L-type pyruvate kinase, acetyl-CoA carboxylase, ATP-citrate lyase, and histones H1 and H2B.[1,6]

The three serines in glycogen synthase phosphorylated by glycogen synthase kinase-3 are all within nine residues in the polypeptide chain.[4,5]

---

[17] B. A. Hemmings, A. Aitken, P. Cohen, M. Rymond, and F. Hofmann, *Eur. J. Biochem.* **127**, 473 (1982).

The amino acid sequence in this region is

Arg–Tyr–Pro–Arg–Pro–Ala–Ser(P)–Val–
                    Pro–Pro–Ser(P)–Pro–Ser–Leu–Ser(P)–Arg

The order of phosphorylation of these sites is unknown.

Glycogen synthase kinase-3 phosphorylates two serines in the R_{II}-subunit located 44 and 47 residues from the N-terminus.[17] The sequence is

$\quad\quad\quad\quad\quad\quad\quad$ 44 $\quad\quad\quad\quad\quad\quad$ 47
Leu–Arg–Glu–Ala–Arg–Ser(P)–Arg–Ala–Ser(P)–Thr–Pro–Pro–Ala–

The sequences Ser(P)-X-Pro-Pro and Ser(P)-Arg are therefore common to both substrates and may be important in determining the substrate specificity of the enzyme.

Although activation of the MgATP-dependent protein phosphatase requires the presence of glycogen synthase kinase-3 and MgATP, there is currently no evidence that this is a phosphorylation reaction. It has been suggested that glycogen synthase kinase-3 catalyzes the insertion of Mg^{2+} or MgATP into the protein phosphatase.[18,19]

Acknowledgments

This work was supported by a Program Grant from the Medical Research Council, London, England, and by the British Diabetic Association.

Note Added in Proof

Recent work indicates that activation of the Mg-ATP-dependent protein phosphatase does involve a phosphorylation reaction [R. A. Hemmings, T. R. Resink, and P. Cohen, *FEBS Lett.* **150,** 319 (1982)].

[18] S. D. Yang, J. R. Vandenheede, J. Goris, and W. Merlevede, *J. Biol. Chem.* **255,** 11759 (1980).
[19] S. D. Yang, J. R. Vandenheede, and W. Merlevede, *FEBS Lett.* **126,** 57 (1981).

[37] Double-Stranded RNA-Dependent eIF-2α Protein Kinase

By RAY PETRYSHYN, DANIEL H. LEVIN, and IRVING M. LONDON

The inhibition of protein synthesis initiation in reticulocyte lysates by low levels of double-stranded RNA (dsRNA) (1–20 ng/ml)[1–3] is largely due to the activation of a cAMP-independent protein kinase (dsI) that phosphorylates the α-subunit (38K) of the initiation factor eIF-2 (eIF-2α).[4,5] Similar inhibition of protein synthesis and phosphorylation of eIF-2α occurs in heme-deficient reticulocyte lysates[6–8] and intact reticulocytes.[9,10] The inhibition of protein synthesis by dsRNA or by heme deficiency is characterized by (1) biphasic kinetics of inhibition, (2) polysome disaggregation, (3) depletion of the [40 S · Met-tRNA$_f^{Met}$]–ribosomal complex, (4) potentiation of inhibition by ATP, and (5) reversal and prevention of inhibition by the addition of exogenous eIF-2 or high levels of cAMP or GTP (for reviews, see Refs. 11, 12). In addition, both modes of inhibition result in the phosphorylation of proximal or identical serine site(s) of eIF-2α which are the same *in vitro* and in lysates.[13] However, several criteria indicate that the two eIF-2α kinases are distinct molecular entities. These include differences in (1) intracellular localization, (2) modes of activation,[4,5] (3) immunogenic properties,[14] (4) phosphoprotein

[1] E. Ehrenfeld and T. Hunt, *Proc. Natl. Acad. Sci. U.S.A.* **68,** 1075 (1971).
[2] T. Hunt and E. Ehrenfeld, *Nature (London)* **230,** 91 (1971).
[3] T. Hunter, T. Hunt, R. J. Jackson, and H. D. Robertson, *J. Biol. Chem.* **250,** 409 (1975).
[4] P. J. Farrell, K. Balkow, T. Hunt, R. J. Jackson, and H. Trachsel, *Cell* **11,** 187 (1977).
[5] D. H. Levin and I. M. London, *Proc. Natl. Acad. Sci. U.S.A.* **75,** 1121 (1978).
[6] D. H. Levin, R. S. Ranu, V. Ernst, and I. M. London, *Proc. Natl. Acad. Sci. U.S.A.* **73,** 3112 (1976).
[7] G. Kramer, G. M. Cimadevilla, and B. Hardesty, *Proc. Natl. Acad. Sci. U.S.A.* **73,** 3078 (1976).
[8] V. Ernst, D. H. Levin, and I. M. London, *Proc. Natl. Acad. Sci. U.S.A.* **76,** 2118 (1979).
[9] P. J. Farrell, T. Hunt, and R. J. Jackson, *Eur. J. Biochem.* **89,** 517 (1978).
[10] A. Leroux and I. M. London, *Proc. Natl. Acad. Sci. U.S.A.* **79,** 2147 (1982).
[11] *Cold Spring Harbor Conf. Cell Proliferation* **8,** 931 (1981).
[12] D. H. Levin, V. Ernst, A. Leroux, R. Petryshyn, R. Fagard, and I. M. London, *in* "Protein Phosphorylation and Bio-regulation" (G. Thomas, E. J. Podesta, and J. Gordon, eds.), p. 130. Karger, Basel, 1980.
[13] V. Ernst, D. H. Levin, A. Leroux, and I. M. London, *Proc. Natl. Acad. Sci. U.S.A.* **77,** 1286 (1980).
[14] R. Petryshyn, H. Trachsel, and I. M. London *Proc. Natl. Acad. Sci. U.S.A.* **76,** 1575 (1979).

profiles,[8] (5) molecular weights, and (6) chromatographic properties during purification of the eIF-2α kinases.[15-17] The mechanism by which the phosphorylation of eIF-2α results in the inhibition of protein synthesis is not fully understood. Previous studies have shown that the phosphorylation of eIF-2α diminishes its interaction with other initiation components[18-22] and results in an apparent impairment in the ability of eIF-2 to recycle. The inhibition of protein synthesis and the phosphoprotein profiles produced by dsRNA are not unique to reticulocytes, but are also observed in dsRNA-treated extracts from interferon-sensitized nonerythroid cells.[23-26]

We and others have described the purification of an inactive (latent) form of dsI and an activated form of dsI from reticulocyte lysates.[16,17,27-30] This chapter describes the purification and characterization of the latent form of the reticulocyte double-stranded RNA activated eIF-2α kinase (dsI). At all stages of purification of latent dsI, activation was dependent on dsRNA and ATP and was accompanied by a dsRNA-dependent phosphorylation of a polypeptide doublet (67K/68.5K) of dsI. Increase in the phosphorylation of the 68.5K component of dsI is concomitant with increased dsRNA-dependent eIF-2α kinase activity (dsI). In the purified state the enzyme is activated by an autokinase mechanism. It is not known whether this autokinase mechanism operates under physiologic conditions.

[15] H. Trachsel, R. S. Ranu, and I. M. London, *Proc. Natl. Acad. Sci. U.S.A.* **75,** 3654 (1978).
[16] D. H. Levin, R. Petryshyn, and I. M. London, *Proc. Natl. Acad. Sci. U.S.A.* **77,** 832 (1980).
[17] R. Petryshyn, D. H. Levin, and I. M. London, *Biochem. Biophys. Res. Commun.* **94,** 1190 (1980).
[18] C. de Haro, A. Datta, and S. Ochoa, *Proc. Natl. Acad. Sci. U.S.A.* **75,** 1148 (1978).
[19] R. S. Ranu and I. M. London, *Proc. Natl. Acad. Sci. U.S.A.* **76,** 1079 (1979).
[20] A. Das, O. Ralston, M. Grace, R. Roy, P. Ghosh-Dastidar, H. K. Das, B. Yaghmai, S. Palmieri, and N. K. Gupta, *Proc. Natl. Acad. Sci. U.S.A.* **76,** 5076 (1979).
[21] J. Siekierka, K. I. Mitsui, and S. Ochoa, *Proc. Natl. Acad. Sci. U.S.A.* **78,** 220 (1981).
[22] R. S. Ranu, I. M. London, A. Das, A. Majumdar, R. Ralston, R. Roy, and N. K. Gupta, *Proc. Natl. Acad. Sci. U.S.A.* **75,** 745 (1978).
[23] A. Zilberstein, P. Federman, L. Schulman, and M. Revel, *FEBS Lett.* **68,** 119 (1976).
[24] P. J. Farrell, C. G. Sen, M. F. Dubois, L. Ratner, E. Slattery, and P. Lengyel, *Proc. Natl. Acad. Sci. U.S.A.* **75,** 5893 (1978).
[25] J. R. Lenz and C. Baglioni, *J. Mol. Biol.* **253,** 4219 (1978).
[26] C. Samuel, *Proc. Natl. Acad. Sci. U.S.A.* **76,** 600 (1979).
[27] D. H. Levin, R. Petryshyn, and I. M. London, *J. Biol. Chem.* **256,** 7638 (1981).
[28] H. Grosfeld and S. Ochoa, *Proc. Natl. Acad. Sci. U.S.A.* **77,** 6526 (1980).
[29] R. S. Ranu, *Biochem. Biophys. Res. Commun.* **97,** 2529 (1980).
[30] H. K. Das, A. Das, P. Ghosh-Dastidar, R. O. Ralston, B. Yaghmai, R. Roy, and N. K. Gupta, *J. Biol. Chem.* **256,** 6491 (1981).

There are several advantages in purifying and studying the latent form of dsI rather than the activated form. (1) Latent dsI is significantly more stable and can be readily stored for prolonged periods. (2) Examination of the molecular mechanism of activation and the function of dsRNA in this process is facilitated. (3) The use of latent dsI also facilitates studies on the regulation of dsI activation and activity by other cellular components, particularly protein phosphatases.

Reagents

[γ-^{32}P]ATP (20–40 Ci/mmol), Bolton-Hunter reagent (^{125}I, 2200 Ci/mmol), and [^{14}C]leucine (350 mCi/mmol) (New England Nuclear).
Sepharose 6B, Sephadex G-200, and Sephacryl S-200 (Pharmacia Fine Chemicals).
DEAE-cellulose (DE-52) and Phosphocellulose P11 (Whatman).
Agarose-poly(I) · poly(C) (P-L Biochemicals).
Poly(I) · poly(C), dithiothreitol (DTT), Tris (Trizma Base), and HEPES (Sigma Chemical Co.).

Buffers and Stock Solutions

KCl (2 M), EDTA (0.2 M), Mg(OAc)$_2$ (1 M), (NH$_4$)$_2$SO$_4$, saturated and neutralized to pH 7.0 with concentrated NH$_4$OH, Tris–HCl, pH 7.5–7.8 (1 M).

Buffer A. 50 mM Tris–HCl, pH 7.5, 80 mM KCl, 2 mM Mg(OAc)$_2$, 0.1 mM EDTA, 1 mM DTT, 10% (v/v) glycerol.

Buffer B. 25 mM Tris–HCl, pH 7.7, 50 mM KCl, 0.1 mM EDTA, 1 mM DTT, 10% (v/v) glycerol.

Buffer C. 10 mM Tris–HCl, pH 7.7, 50 mM KCl, 0.1 mM EDTA, 2 mM DTT, 5% (v/v) glycerol.

Biological Materials

Rabbit reticulocyte lysate is prepared as described in detail elsewhere.[31,32] In addition, large batches were purchased from two sources: (1) Grand Island Biological Inc. and (2) Pel-Freeze Inc. *P. chrysogenum* mycophage dsRNA was a gift of Dr. Hugh Robertson (Rockefeller University). Highly purified eIF-2 was kindly provided by Dr. W. C. Merrick (Case Western Reserve University) and Dr. H. Trachsel (M.I.T.). Reovirus dsRNA and eIF-2 were also prepared by one of us (D.H.L.).

[31] T. Hunt, G. Vanderhoff, and I. M. London, *J. Mol. Biol.* **66**, 471 (1972).
[32] V. Ernst, D. H. Levin, and I. M. London, *J. Biol. Chem.* **253**, 7163 (1978).

Column Chromatography

DEAE-cellulose, phosphocellulose, agarose-poly(I) · poly(C), Sepharose 6B, Sephadex G-200, and Sephacryl S-200 columns were prepared in the cold as suggested by the suppliers, and equilibrated with starting buffer prior to use. All dextran gel columns including Sepharose 6B, Sephadex G-200, and Sephacryl S-200 were primed prior to use by filtration of small amounts of lysate followed by equilibration.

Assay Procedures

Two assays are utilized to determine the activation and activity of dsI obtained at each step of purification. They are (1) inhibition of protein synthesis in reticulocyte lysates, and (2) the phosphorylation of eIF-2α *in vitro* (protein kinase assay).

Inhibition of Protein Synthesis Assay

Aliquots of dsI activation assays are added directly to protein synthesizing reticulocyte lysates to measure the amounts of inhibitor activated. It is important to note that high poly(I) · poly(C) (20–25 μg/ml) which does not activate dsI or significantly inhibit protein synthesis during a 30-min incubation is added to all protein synthesis assayed to prevent further dsI activation by carryover dsRNA.[3,5] Protein synthesis assays (25 μl) containing 12.5 μl of reticulocyte lysates and optimal levels of hemin (20 μM) are carried out as described in detail elsewhere.[31,32] Unless otherwise stated, all incubations are at 30°; aliquots (3–5 μl) are removed at the indicated intervals and spotted on filter paper discs (grade 596, Schleicher and Scheull).[32] The extent of protein synthesis is determined by processing these discs as previously described.[32] A unit of inhibitory activity is defined in the table.

Protein Kinase Assay

Activation and activity of dsI are often assayed in one step and are based on the incorporation of ^{32}P into both dsI and eIF-2α. Protein kinase assays (20 μl) containing 25 mM Tris–HCl, pH 7.5 (or 10 mM HEPES, pH 7.2), 2 mM Mg(OAc)$_2$, 20–60 mM KCl, 20–50 μM [γ-^{32}P]ATP, dsRNA (20 ng/ml), eIF-2 (2–10 pmol), and appropriate levels of latent dsI are performed at 30° as previously described.[5,6] Other additions or changes are indicated in the figure legends. After incubation the reactions are terminated by the addition of SDS-denaturing buffer[33] and heated at

---

[33] W. F. Studier, *J. Mol. Biol.* **79**, 237 (1973).

PURIFICATION OF LATENT dsI[a]

Step	Protein (mg)	Total activity (units)	Specific activity (units/mg)
1. Lysates	25,600	—	—
2. Sepharose 6B	550	—	—
3. Ammonium sulfate	54.01	33,700	624
4. DEAE-cellulose	24.05	29,800	1,239
5. Phosphocellulose	2.15	10,000	4,651
6. Glycerol gradient	0.29	2,900	10,000

[a] One unit of inhibitory (IU) activity is defined as the amount of dsI required to reduce [^{14}C]leucine incorporation by 3 pmol in 30 min. A reduction of 1000 cpm (~3 pmol [^{14}C]leucine or 0.6 pmol globin) equals ~5% inhibition of protein synthesis in an average lysate. The reason for this measure is that there is a linear proportionality between the amount of dsI and the percentage inhibition which persists to about 50% inhibition. Further inhibition requires increased nonproportional levels of dsI. A reduction in incorporation of [^{14}C]leucine by 30 pmol (~50% inhibition) therefore represents 10 IU. All activity values have been corrected for any dsRNA-independent inhibition. Protein concentration was determined by the method of Lowry et al. (*J. Biol. Chem.* **193**, 265, 1951).

95° for 2 min (or 85° for 5 min). The proteins are separated by one-dimensional SDS–polyacrylamide gel electrophoresis (SDS–PAGE, 0.1% SDS–10% acrylamide–0.26% BIS) as described by Laemmli.[34] Gels are stained and dried and subjected to autoradiography to obtain phosphoprotein profiles.[32]

Activation of Latent dsI

To follow the purification of latent dsI at any given stage, column (or gradient) fractions are assayed for the activation of dsI in 25 μl incubations containing 25 mM Tris–HCl, pH 7.7, 2 mM Mg(OAc)$_2$, 1 mM DTT, 80 mM KCl, 5 mM creatine phosphate, 2.5 μg creatine phosphokinase, 1 mM ATP, and latent dsI aliquots. dsRNA (20 ng/ml) is added and the incubations are carried out at 30° for 30 min. When more pure fractions of dsI are used, creatine phosphate and creatine phosphokinase are omitted. Aliquots from the activation incubations are added to protein synthesis assays to determine dsI activity. Omission of dsRNA from the activation

[34] U. K. Laemmli, *Nature (London)* **227**, 680 (1970).

mixtures served to determine any dsRNA-independent inhibition of protein synthesis.

Purification of Latent dsI

Unless otherwise stated, all purification procedures are performed in the cold.

Sepharose 6B Chromatography (Steps 1 and 2 in the Table)

An 800 ml batch of rabbit reticulocyte lysate (stored at $-76°$) containing 10–15 A_{260} units of ribosomes per ml is rapidly thawed, supplemented with hemin (2 μM), and clarified by centrifugation for 10 min at 10,000 g at 4° to remove debris if present. Lysate volumes of 80–100 ml are separately filtered through a Sepharose 6B column (6 × 40 cm) preequilibrated with buffer A. The ribosomal fractions ($A_{260}/A_{280} = 1.7-1.8$) which are eluted in the void volume are collected and combined. This ribosomal fraction contains all of the latent dsI of the lysate and is used for further purification. A double-stranded RNA-dependent inhibitor (dsI) is associated exclusively with the ribosomes and can be clearly resolved from the heme-regulated inhibitor (HRI) which is present in the soluble protein fraction or high-speed supernatant (S100) (Fig. 1). Sepharose 6B fractions are eluted after the ribosomes are collected and stored at $-76°$; this material is useful for the purification of HRI. An alternative procedure for preparing reticulocyte ribosomes for dsI purification employs overnight sedimentation of lysates at 120,000 g in a Spinco 42.1 rotor at 4° through a 5 ml cushion of 50% glycerol in buffer A. The ribosome pellet is resuspended in buffer A at a concentration of 60–80 A_{260} units/ml and is used for further dsI purification. This procedure for ribosome preparation is more rigorous than Sepharose 6B chromatography due to the stresses of prolonged centrifugation and Dounce homogenization of the ribosomal pellet. However, it provides a more purified ribosome preparation which retains all of the latent dsI in the lysate. This preparation of ribosomes is also completely free of the heme-regulated eIF-2α kinase and other soluble proteins.

Preparation of Ribosomal Salt-Wash

The ribosome fraction (8000 A_{260} units in 800 ml) is brought to 0.49 M KCl by the slow addition of 4 M KCl. After stirring for 20 min on ice, the ribosomes are removed by sedimentation for 4 hr at 39,000 g at 4° in a Spinco 42.1 rotor. Latent dsI is quantitatively extracted by this procedure.

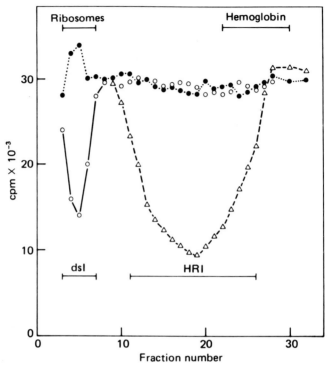

FIG. 1. Distribution of dsI and HRI in lysates fractionated on Sepharose 6B. Fresh lysate was fractionated in Sepharose 6B and each fraction was monitored for its capacity to form dsI (○) or HRI (△) as described in the text. Portions (5 μl) of each fraction in the activation profiles were added to separate protein synthesis assays (25 μl) containing 20 μg/ml of poly(I) · poly(C); untreated aliquots of the Sepharose fractions (●) were similarly assayed. After 30 min at 30°, 5-μl aliquots were assayed for protein synthesis.

Fractionation with Ammonium Sulfate (Step 3 in the Table)

The salt-wash is fractionated by ammonium sulfate precipitation at 0–39% (Fraction I) and 39–55% (Fraction II) saturation. After each fractionation step the salt-wash is stirred on ice for 30 min and the precipitate is collected by centrifugation at 10,000 g for 30 min at 4°. Each precipitate is suspended in 21.5 ml of buffer B and is dialyzed extensively against the same buffer. Fraction I (54.1 mg protein) contains the highest specific activity for dsRNA-dependent eIF-2α kinase activity (dsI) and inhibition of protein synthesis. Fraction II (30 mg protein) contains significant dsI activity as well as the initiation factor eIF-2. To achieve the highest yield it is recommended that a 0–55% ammonium sulfate fraction be used.

Chromatography on DEAE-cellulose (Step 4 in the Table)

Fraction I (54.1 mg protein) is applied to a DEAE-cellulose column (1.5 × 12 cm) preequilibrated with buffer B. The column is eluted with buffer B until the A_{280} in the effluent is less than 0.01. The remaining proteins are eluted by stepwise addition of buffer B containing increasing KCl concentrations (0.2, 0.3, and 0.5 M). Latent dsI does not bind to DEAE-cellulose under these conditions, and all the activity elutes in the 0.05 M KCl wash [dsI(DC)]. The protein fraction containing latent dsI is concentrated by ultrafiltration with a UM-10 membrane (Amicon) and nitrogen pressure at 4°. dsI(DC) displays on SDS–PAGE a marked dsRNA-dependent phosphorylation of eIF-2α and a broad phosphoprotein of approximately 67,000 daltons (dsI) (Fig. 2). Concomitant with

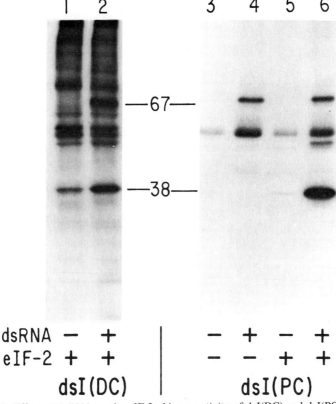

FIG. 2. Effect of dsRNA on the eIF-2α kinase activity of dsI(DC) and dsI(PC). Protein kinase assays (20 μl) contained 3 μg of latent dsI(DC) (lanes 1, 2), or 1 μg of latent dsI(PC) (lanes 3–6), dsRNA (20 ng/ml), and 0.5 μg of eIF-2 where indicated. Protein kinase assays and SDS–PAGE were as described in the text. The figure is an autoradiogram.

these dsRNA-dependent phosphorylations is a capacity to inhibit protein synthesis (Fig. 3).

Chromatography on Phosphocellulose (Step 5 in the Table)

The dsI(DC) (24.1 mg protein) is applied to a phosphocellulose column (1.5 × 6 cm) that is preequilibrated with buffer B. Proteins are eluted with buffer B until the A_{280} is less than 0.01; further elution is by stepwise addition of buffer B containing 0.2, 0.3 and 0.5 M KCl. The major portion of the latent dsI (70–75%) is eluted with 0.3 M KCl; however, some activity (25–30%) is observed in the 0.2 M KCl eluate. The 0.3 M KCl [dsI(PC)] fraction is concentrated to 1 ml (2.15 mg protein) by ultrafiltration. The phosphoprotein profile of dsI(PC) shows a marked dsRNA-dependent phosphorylation of eIF-2α as well as phosphorylation of a protein band of approximately 67,000 daltons (Fig. 2). Furthermore, the same dsI preparation demonstrates significant dsRNA-dependent inhibition of protein synthesis in lysates when it is activated (Fig. 3).

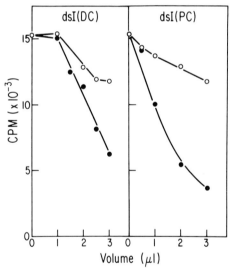

FIG. 3. Effect of dsI(DC) and dsI(PC) on protein synthesis. Latent dsI(DC) (20 μg) and latent dsI(PC) (21.5 μg) were incubated in 25 μl for 30 min at 30° in the absence (○—○) and presence (●—●) of dsRNA (20 ng/ml) as described in Assay Procedures. The reactions were chilled in ice, and aliquots were removed as indicated and assayed for their effect on protein synthesis. All protein synthesis assays were carried out as described in Assay Procedures. Each point represents a 5-μl aliquot at 30 min.

Glycerol Gradient Centrifugation (*Step 6 in the Table*)

Latent dsI(PC) (1 mg protein) is layered over an 11.5-ml 10–40% glycerol gradient in buffer B (containing 0.1 M KCl) and centrifuged for 40 hr at 35,000 rpm at 2° in a Spinco SW 41 rotor. Fractions of 0.5 ml are collected and assayed directly for dsRNA-dependent eIF-2α activity (Fig. 4) and inhibition of protein synthesis (Fig. 5). The autoradiogram (Fig. 4) indicates that dsI resolves into two phosphoproteins of 67,000 and 68,500 daltons during purification. There is increased activity with increased phosphorylation of the upper band. Active fractions are combined and

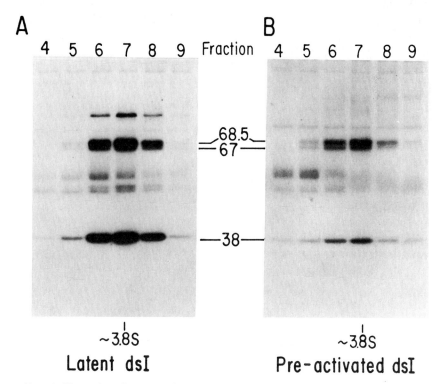

FIG. 4. Glycerol gradient centrifugation of latent and activated dsI(PC). For latent dsI (A), protein kinase assays containing 10 μl from each glycerol gradient fraction were incubated with eIF-2 (1 μg) and dsRNA (20 ng/ml) and analyzed by SDS–PAGE. Only assays from fractions 4–9 are shown. For activated dsI (B), latent dsI(PC) (215 μg) was activated and labeled in a 200 μl reaction volume containing 25 mM Tris–HCl, pH 7.7, 100 mM KCl, 2 mM Mg(OAc)$_2$, 20 ng/ml dsRNA, and 200 μM [γ-^{32}P]ATP (1.8 Ci/mmol). After 30 min at 30°, the reaction mixture was chilled and layered on a 10–40% glycerol gradient and centrifuged as described. Protein kinase assays on the preactivated dsI contained eIF-2 (1 μg) and gel electrophoresis were carried out as described in A except that dsRNA was omitted. The figure is an autoradiogram.

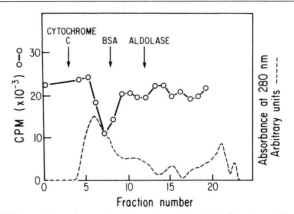

FIG. 5. Inhibition of protein synthesis by glycerol gradient fractions of latent dsI. Fractions from the glycerol gradient of latent dsI (Fig. 4) were assayed for inhibition of protein synthesis after activation of dsI. Aliquots (10 µl) from each fraction were incubated with dsRNA and ATP as described and 4.5-µl aliquots were assayed for their effect on protein synthesis. Each point represents a 5-µl aliquot assayed at 30 min for ^{14}C incorporation.

concentrated by vacuum dialysis and stored in liquid nitrogen at a concentration of 0.28 mg/ml protein. Storage of latent dsI in this manner results in no significant loss of activity; however, some loss of activity occurs with repeated freezing and thawing.

A stable latent dsI with similar purity is obtained by filtration on Sephacryl S-200 instead of the glycerol gradient step. Latent dsI(PC) (0.5 ml) is applied to a Sephacryl S-200 column (1.7 × 50 cm) preequilibrated with buffer C. Fractions (0.5 ml) are collected and assayed for dsRNA-dependent eIF-2α kinase activity and inhibition of protein synthesis. Filtration on Sephacryl S-200 yields good recovery, a five- to eight-fold increase in purification, and is more rapid.

Chromatography on Agarose-Poly(I) · Poly(C)

Latent dsI obtained from Step 6 is further purified extensively by affinity chromatography on agarose-poly(I) · poly(C). Protein (0.11 mg) is applied to a 100 µl bed volume of agarose-poly(I) · poly(C) in a 1 ml plastic pipette. Protein is eluted by stepwise fractionation in buffer B with increasing concentrations of KCl. Three consecutive 100-µl fractions, collected at 0.05 M (wash), 0.2 M, and 1.0 M KCl, remove all detectable proteins except dsI. dsI is eluted in one fraction (100 µl) of 2 M KCl plus 100 µM ATP (Fig. 6) and is concentrated by vacuum dialysis at 0° against buffer B containing 80 mM KCl and 100 µM ATP. The tight binding of dsI to this matrix results in a dsI preparation of near homogeneity (>95% as

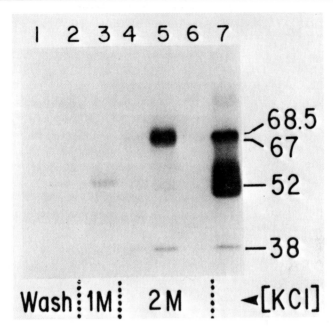

FIG. 6. Chromatography of dsI on agarose-poly(I) · poly(C). Aliquots (3 μl) were assayed for eIF-2α kinase activity at 37° for 10 min in the presence of 0.33 μg of eIF-2 (2 pmol). Assays were analyzed by SDS–PAGE. The figure is an autoradiogram of dsI assays of the wash (track 1), and consecutive fractions of the 1 M KCl (tracks 2 and 3) and 2 M KCl plus 100 μM ATP (tracks 4–6) eluates. Track 7 is an assay of the starting latent dsI.

judged by protein labeling with ^{125}I)[35] (Fig. 7). However, protein recovery is too low to determine specific activity or to detect a Coomassie blue-stained band. Approximately 50% of the eIF-2α kinase activity is recovered in this step and there is some loss of activity during storage at −76°. One consequence of this purification step is that low concentrations of poly(I) · poly(C) are eluted in all effluent fractions, and thus provide exogenous dsRNA for activation.

Properties of dsI

It is essential to note that in the purification described here, the dsRNA-dependent eIF-2α kinase remains in a latent state. At all stages in the purification, activation requires dsRNA and ATP. The dsRNA-dependent phosphorylation of the α-subunit (38K) of added eIF-2 and of a phosphoprotein doublet (67K/68.5K) is observed throughout purification.

[35] A. E. Bolton and W. M. Hunter, *Biochem. J.* **133**, 529 (1973).

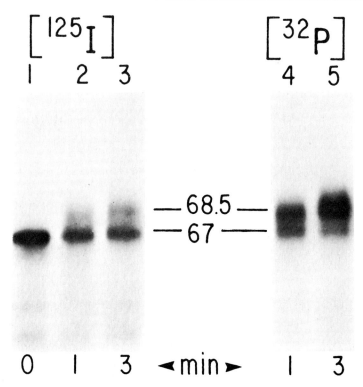

FIG. 7. ^{125}I- and ^{32}P-labeling of highly purified dsI. Protein kinase assays (20 μl), containing 8 μl of dsI purified by affinity chromatography (see Fig. 6), 10 mM HEPES, pH 7.2, 2 mM Mg(OAc)$_2$, 60 mM KCl, 20 ng/ml of dsRNA, and 50 μM unlabeled ATP (tracks 1–3) or 20 μM [γ-^{32}P]ATP (25 Ci/mmol) (tracks 4 and 5), were incubated at 30° for the indicated intervals. Assays 1–3 were stopped with 8 mM EDTA and then reacted for 30 min at 0° with benzene-free Bolton–Hunter reagent (~100 μCi of ^{125}I, 2200 Ci/mmol). All assays were separated by SDS–PAGE. The figure is an autoradiogram.

In particular, fractionation of latent dsI by glycerol gradient centrifugation (Figs. 4 and 5) and by Sephadex G-200 chromatography (Fig. 8) demonstrates a dsRNA-dependent phosphorylation of eIF-2α that is coincident with phosphorylation of the 67–68.5K doublet (Figs. 4 and 8) and with protein synthesis inhibitory activity. The table summarizes the purification procedure of latent dsI to the glycerol gradient step. In using commercially prepared lysates as starting material, it was not possible to determine accurately dsI activity in crude fractions (Steps 1 and 2) due to the presence of other inhibitors. Lysates prepared in our laboratory, however, consistently show a potent and selective inhibitory response to added dsRNA (Fig. 1). Losses of activity are observed in Steps 5 and 6;

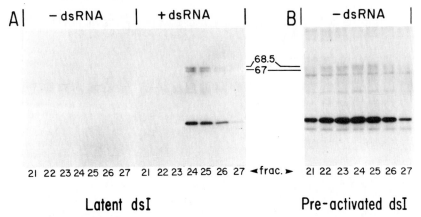

FIG. 8. Sephadex G-200 filtration of latent and activated dsI. Latent dsI from a gradient preparation (75 μg in 150 μl) was filtered through a Sephadex G-200 column (A). Protein kinase assays containing 0.5 μg of eIF-2 and 10 μl of column fractions 21–27 were carried out in the absence and presence of dsRNA as indicated and were analyzed by SDS–PAGE (A). For activation of dsI(PC) (B), latent dsI (51 μg) was first activated in a 300 μl volume as described in Fig. 4B, except that the [γ-^{32}P]ATP was 105 μM (1 Ci/mmol). Protein kinase assays of this profile (B) were analyzed by SDS–PAGE as in A. The figure is an autoradiogram.

however, the minimum purification from crude lysate to Step 6 is estimated at about 1000 to 2000-fold. Glycerol gradient centrifugation of dsI(PC) that is activated with dsRNA and [γ-^{32}P]ATP prior to centrifugation exhibited a sedimentation coefficient (3.7–3.8 S) similar to that of latent dsI, and again the prelabeled 67K/68.5K doublet was coincident with eIF-2α kinase activity (Fig. 4). These results suggest that activation of dsI does not affect its sedimentation properties or result in extensive molecular changes, a suggestion further supported by sizing comparisons on Sephadex G-200 (Fig. 8). In previous preliminary experiments, we reported a molecular weight for dsI of approximately 120,000 by analysis on Sephadex G-200; however, recent results with Sephacryl S-200 indicate that dsI has a molecular weight of approximately 68,000 (data not shown) in agreement with that reported by others.[29,30]

Nature of the dsI Phosphoprotein Doublet

The activation of latent dsI by an autophosphorylation mechanism occurs at a low temperature (10°) but at a reduced rate. This property facilitates a more detailed analysis of the kinetics of phosphorylation of the 67K/68.5K doublet. We have observed an asymmetric ^{32}P-labeling pattern in which the 67K component is initially phosphorylated followed by appearance of increasing label in the 68.5K component (Fig. 9, tracks

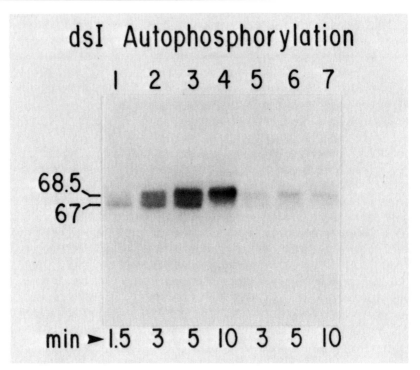

FIG. 9. Kinetics of the dsRNA-dependent phosphorylation of the 67K/68.5K polypeptide doublet. Protein kinase assays (20 µl) contained 10 mM HEPES, pH 7.2, 2 mM Mg(OAc)$_2$, 60 mM KCl, 20 ng/ml of dsRNA, 20 µM [γ-^{32}P]ATP (25 Ci/mmol); and 0.25 µg of glycerol gradient latent dsI (step 6). Incubation was at 10° for the indicated intervals. For three assays (tracks 5–7), 1 mM unlabeled ATP was added at 1.5 min, and incubation was continued for the indicated intervals. All assays were analyzed by SDS–PAGE as described. The figure is an autoradiogram.

1–4). However, if the ^{32}P-label in the 67K is chased at an early time (1.5 min, track 1) by the addition of unlabeled ATP, the radiolabel in the ^{32}P-67K component shifts into the 68.5K region (Fig. 9, tracks 5–7). These data suggest that the differences in the migrations are probably due to differences in the state of phosphorylation of dsI. This view is further supported by studies with highly purified dsI that has been labeled with ^{125}I, which exhibits a migration in SDS–PAGE at 67K (Fig. 7, track 1). If the dsI preparation is allowed to autophosphorylate briefly with unlabeled ATP prior to labeling with ^{125}I, a lightly labeled second component appears at 68.5K (Fig. 7, tracks 2 and 3). The two ^{125}I-labeled (and phosphorylated) bands are coincident with the ^{32}P-labeled 67K/68.5K doublet (Fig. 7, tracks 4 and 5). The shift in label to the 68.5K component is associated with increased eIF-2α kinase activity (Fig. 10), an indication that dsI is more active in a multiply phosphorylated state.

Concluding Remarks

The purification procedure described above provides a useful method for obtaining latent dsI. The dsRNA-dependent inhibition of protein synthesis is concomitant with the phosphorylation of eIF-2α and of the 67K/68.5K doublet of dsI. The dsRNA-dependent phosphorylation of dsI occurs by an autokinase mechanism *in vitro*. Activated (and

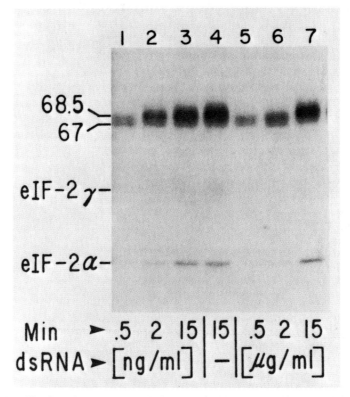

FIG. 10. Kinetics of autophosphorylation and eIF-2α kinase activity of highly purified dsI. dsI assays (20 μl) contained 3 μl of purified dsI (see Fig. 7), 10 mM HEPES, pH 7.2, 3 mM Mg(OAc)$_2$, 0.27 μg of eIF-2 (1.9 pmol), 74 μM [γ-^{32}P]ATP (6.7 Ci/mmol), and dsRNA (20 ng/ml) (tracks 1–3) or poly(I) · poly(C) (20 μg/ml) (tracks 5–7). A consequence of purification of dsI by agarose-poly(I) · poly(C) is that low levels of poly(I) · poly(C) elute in all effluent fractions, which eliminates the need for exogenous dsRNA for activation. This is demonstrated in track 4, which contains no added dsRNA. Incubation was at 37° for the indicated times. Assays were analyzed by SDS–PAGE. The figure is an autoradiogram.

phosphorylated) dsI displays chromatographic properties on DEAE-cellulose, phosphocellulose, and agarose-poly(I) · poly(C)[16,27] that are different from those of latent dsI although the molecular weight of dsI is unchanged by activation. The purification of a latent form of dsI facilitates further studies on the regulation of dsI activation and activity.

Acknowledgments

This work was supported in part by USPHS Grants AM-16272 and GM24825.

[38] Rhodopsin Kinase

By HITOSHI SHICHI, ROBERT L. SOMERS, and KATSUHIKO YAMAMOTO

The light-dependent phosphorylation of the visual pigment rhodopsin was discovered independently by three laboratories.[1-3] Rhodopsin kinase, the enzyme involved in the reaction, catalyzes the transfer of the terminal (γ) phosphate group of ATP to the opsin protein.

$$\text{Rhodopsin} + \text{ATP} \xrightarrow{\text{light}} \text{opsin-P} + \text{ADP}$$

Assay Method[4]

Principle. Bovine rod outer segments are treated with 5 M urea to prepare the substrate. This treatment denatures membrane-associated rhodopsin kinase but has little effect on rhodopsin.[4] The substrate is then incubated in the light with rhodopsin kinase, [γ-^{32}P]ATP, and [*adenine*-^{3}H]ATP. The inclusion of [^{3}H]ATP is to estimate the extent of binding of ATP itself. The kinase activity is determined by subtracting the amount of [^{3}H]ATP bound from the total ^{32}P incorporated in the opsin protein. At ATP concentrations 20 to 30 times the concentration of rhodopsin, the initial reaction rate follows the pseudo-first-order kinetics ($k \simeq 0.06$ min^{-1}) for at least 10 min.[4] In an alternative method, rod membranes are incubated with [γ-^{32}P]ATP either in the dark or in the light. The kinase activity is determined by subtracting the ^{32}P incorporated in the dark from

[1] H. Kühn and W. J. Dreyer, *FEBS Lett.* **20**, 1 (1972).
[2] D. Bownds, J. Dawes, J. Miller, and M. Stahlman, *Nature (London) New Biol.* **236**, 35 (1972).
[3] R. N. Frank, H. D. Cavanagh, and K. R. Kenyon, *J. Biol. Chem.* **248**, 596 (1973).
[4] H. Shichi and R. L. Somers, *J. Biol. Chem.* **253**, 7040 (1978).

the ^{32}P incorporated in the light. The phosphorylated protein is identified as rhodopsin after separation of the labeled protein by chromatographic[4] and electrophoretic methods.[1]

Reagents

0.30 M Tris–HCl, pH 7.4

1.0 mM ATP Na$_2$ containing [γ-^{32}P]ATP (200 to 600 dpm/pmol) and [*adenine*-2,8-^{3}H]ATP (400 to 600 dpm/pmol)

40 mM MgCl$_2$

Urea-treated rod membranes containing rhodopsin (see below for preparation)

10% trichloroacetic acid containing 5 mM H$_3$PO$_4$

Flood lamp (100–150 W)

Wratten gelatin filter, No. 21

Gelman HA filters or Millipore HA filters: pore size, 0.45 μm; diameter, 25 mm

Procedure. The reaction mixture (0.25 ml) is prepared by mixing 0.05 ml of Tris–HCl (pH 7.4, 12.5 μmol), 0.025 ml of ATP Na$_2$ (0.025 μmol) containing [γ-^{32}P]ATP and [*adenine*-2,8-^{3}H]ATP, 0.025 ml of MgCl$_2$ (1 μmol), 0.05 ml of enzyme (2 to 25 μg protein), and 0.10 ml of urea-treated membranes (1.5 to 2.0 nmol of rhodopsin). The reaction mixture is placed in a shaker-water bath adjusted at 37°. The water bath is covered with a lid which has a large light-intake area covered with a Wratten filter. The reaction is initiated by irradiation of sample with a flood lamp. The intensity of light reaching the sample is about 100 footcandles. The reaction is stopped by the addition of 1 ml of 10% prechilled trichloroacetic acid containing 5 mM H$_3$PO$_4$. Samples are then filtered on membrane filters under mild suction, washed five times by filtration with 2 ml of 10% trichloroacetic acid–5 mM H$_3$PO$_4$ for each wash. The filter is dried, dissolved in 1.0 ml of NCS tissue solubilizer, neutralized with 50 μl of glacial acetic acid, and mixed with 10 ml of toluene-based mixture. The ^{3}H and ^{32}P radioactivities of samples are counted in a scintillation counter. Rhodopsin kinase activity is determined by subtracting the amount of ATP bound (from ^{3}H radioactivity) from the total amount of phosphate incorporated (from ^{32}P radioactivity). One unit of activity is the amount (milligrams) of enzyme protein that transfers 1 μmol of phosphate per minute. Specific activity is defined as activity in units per milligram of enzyme protein. To prepare urea-treated membranes (substrate for rhodopsin kinase), rod outer segments (from 100 bovine retinas) are suspended in 100 ml of 50 mM Tris–HCl (pH 8.0) containing 5 mM EDTA and 5.0 M urea, sonicated for 3 min at 20 KC at 0°, diluted with 2 volumes of 50 mM Tris–HCl (pH 7.4), and centrifuged for 45 min at 100,000 g. The pellet is washed four times with 100 ml of Tris buffer for each wash by suspending and centrifuging, lyophilized, and stored at $-20°$ in the dark. The membranes are suspended in Tris–HCl (pH 7.4) and sonicated before use.

Purification Procedure[4]

Fresh bovine eyes are purchased from a local slaughterhouse and brought to the laboratory on ice in a light-tight container.

Step 1. Extraction. All operations are carried out in the dark or under dim red light. Two hundred retinas are suspended in 300 ml of 20% sucrose–0.1 M sodium phosphate (pH 6.5) containing 2 mM $MgCl_2$ and 0.1 mM EDTA (i.e., eight retinas in 12 ml of buffer per centrifuge tube) and shaken vigorously with a Vortex for 30 sec and centrifuged at 300 g for 5 min. The supernatant is collected and centrifuged at 20,000 g for 10 min to sediment rod outer segments. The enzyme is extracted from the outer segments either with 1.0 M NH_4Cl or with EDTA buffer; that is, the enzyme is extracted from rod membranes by homogenization with 1.0 M NH_4Cl in 50 mM Tris–HCl, pH 7.4 (1 ml/retina). The extract is fractionated with a saturated solution of ammonium sulfate (pH 7.4) at 3° at pH 7.4 to collect the enzyme precipitated at 20 to 45% saturation. The precipitate is dissolved in 50 mM Tris–HCl buffer, pH 7.4, containing 0.1 M KCl and 10^{-4} M dithiothreitol and dialyzed overnight at 3° against the buffer. Alternatively, the outer segments are suspended in 20 ml of 50 mM Tris–HCl buffer (pH 7.4) containing 1 mM EDTA and 0.1 mM dithiothreitol and dialyzed overnight against 1 liter of the same buffer at 3° in the dark. The suspension is then centrifuged at 105,000 g for 20 min to collect the extracted enzyme in the supernatant.

Step 2. Gel Filtration on Sephacryl S-200. The crude extract is loaded on a Sephacryl S-200 column (2.5 × 88 cm) which has been preequilibrated with 50 mM Tris–HCl (pH 7.4) containing 0.1 M KCl and 10^{-4} M dithiothreitol (elution buffer A). Four-milliliter fractions are collected at a flow rate of 96 ml/hr.

Step 3. Affinity Chromatography on Blue Sepharose CL6B. Active fractions from the Sephacryl column are pooled and placed on a Blue Sepharose CL6B column (1.6 × 9.5 cm) which has been preequilibrated with elution buffer A. Two-milliliter fractions are collected at a flow rate of 24 ml/hr. The column is eluted first with 80 ml of elution buffer A and then with a linear gradient of 0.1 to 1.0 M KCl in elution buffer A (120 ml each). Fractions containing kinase activity (eluted at about 0.2 M KCl) are pooled and dialyzed against elution buffer B (elution buffer A containing 15% glycerol). The Blue Sepharose column is preequilibrated with elution buffer B and the dialyzed sample is loaded on the column and eluted

Purification of Rhodopsin Kinase

Step	Total protein (mg)	Total units	Units/mg of protein	Yield (%)
1.0 M NH$_4$Cl extraction	437	0.131	0.00030	100
1 mM EDTA extraction	481	0.154	0.00032	118
Sephacryl S-200	9.5	0.082	0.00860	63
First Blue Sepharose	1.0	0.033	0.0330	25
Second Blue Sepharose	0.15	0.0042	0.0280	3.2

successively with 48 ml of elution buffer B, and 100 ml of elution buffer B containing 0.2 M KCl. Four-milliliter fractions are collected. The yield of purified enzyme is 1–10 μg from 100 retinas. The purification procedure is summarized in the table.

Properties

Nature of Product. Phosphorylated membranes are incubated with 9-*cis*-retinal (or 11-*cis*-retinal) in the dark and the regenerated pigment isorhodopsin (or rhodopsin) is extracted, purified, and identified.[4]

Purity of Enzyme. Purified enzyme is homogeneous as judged by electrophoresis in sodium dodecyl sulfate-containing polyacrylamide gel. The enzyme has a molecular weight of 52,000[4]–67,000[5] depending on the method of purification.

Specificity. The enzyme is highly specific for rhodopsin; phosvitin, casein, histone, and protamine are not phosphorylated.[4] In a strict sense, the substrate is not rhodopsin but a bleaching intermediate of rhodopsin formed following light absorption. Light has no effect on the enzyme itself. The enzyme does not catalyze the dephosphorylation of phosphorylated rhodopsin. ATP (K_m = 8 μM, V_{max} = 40 nmol/mg/min) is the preferred phosphate donor to GTP (K_m = 400 μM, V_{max} = 2 nmol/mg/min).

Inhibitors. The activity (with ATP) is inhibited 90% by 1 mM Zn^{2+}, 50% by 1 mM AMP, and 50% by 1 mM adenosine but not by cyclic AMP or cyclic GMP. Na^+ (100 mM) inhibits 90% of activity, while K^+ (100 mM) is without effect. Digitonin and Emulphogene BC720 at 0.1% inhibits about 60% of activity. Spermidine (1 mM) has no effect.

Phosphorylation Sites. Serine (and threonine) residues are phosphorylated.[1] Under optimal conditions, five to seven phosphate groups are incorporated per mol of rhodopsin.[4,6] The phosphorylation sites are pri-

---

[5] H. Kühn, *Biochemistry* **17**, 4389 (1978).
[6] U. Wilden and H. Kühn, *Biochemistry* **21**, 3014 (1982).

marily localized in the carboxyl terminal region (residues 1–20) of opsin peptide.[7]

pH Optima. Two pH optima are determined at about 6.0 and 7.5

Enzyme Stability. The enzyme stored in 0.1 M KCl at 3° loses 50% of its activity within a week. Monovalent cations such as K^+ and NH_4^+ and 15% glycerol stabilize the enzyme to some extent.

Distribution. Rhodopsin kinase is located on the external surface of the disk membrane.[8] The affinity of membrane for the enzyme increases upon the photobleaching of rhodopsin.[5]

Phosphorylation of Enzyme. There is evidence that rhodopsin kinase itself is phosphorylated by ATP and the reaction is neither stimulated by cyclic nucleotides nor affected by light or dark conditions.[9]

[7] P. A. Hargrave, S.-L. Fong, J. H. McDowell, M. T. Mas, D. R. Curtis, J. K. Wang, E. Juszczak, and D. P. Smith, *Neurochemistry* **1**, 231 (1980).
[8] H. Shichi and R. L. Somers, *Photochem. Photobiol.* **32**, 491 (1980).
[9] R. H. Lee, B. M. Brown, and R. N. Lolley, *Biochemistry* **21**, 3303 (1982).

[39] Polyamine-Dependent Protein Kinase and Phosphorylation of Ornithine Decarboxylase in *Physarum polycephalum*

By VALERIE J. ATMAR and GLENN D. KUEHN

$$ATP^{4-} + \text{ornithine decarboxylase} \rightarrow ADP^{3-} + H^+ + \text{phospho-ornithine decarboxylase}^{2-}$$
ATP : protein phosphotransferase (EC 2.7.1.37)

Major metabolic control of the biosynthesis of the polyamines, putrescine, spermidine, and spermine, has long been recognized to center at the reaction catalyzed by ornithine decarboxylase (OrnDCase) (L-ornithine carboxylyase, EC 4.1.1.17).[1] The reaction yields the diamine, putrescine, through pyridoxal phosphate-dependent decarboxylation of L-ornithine. A rapid increase in OrnDCase activity is an invariant element in the response of all eukaryotic cells stimulated to proliferate.[2] Intensive investigation of this property has identified two properties that make OrnDCase a unique catalyst among eukaryotic enzymes. First, many different agents have the capacity to produce marked enhancement of the enzyme

[1] D. R. Morris and R. H. Fillingame, *Annu. Rev. Biochem.* **43**, 303 (1974).
[2] D. B. Maudsley, *Biochem. Pharmacol.* **28**, 153 (1979).

activity.[1,3] These reagents include tumor-promoting agents, viral transformations, virtually all hormone administrations, and mechanical growth stimulation. Other reagents, such as the polyamines and interferon, rapidly inactivate OrnDCase in a variety of cell cultures. Second, OrnDCase has been observed to exhibit the shortest apparent half-life of any known eukaryotic enzyme, i.e., ~10 min.[4,5] In most cases, conclusions regarding the inducibility and rapid turnover (synthesis and degradation) rate of OrnDCase were derived from experiments on the rates of increase or decay of enzyme activity following treatment with pharmacological reagents. The precise loci of action for most of these reagents are unknown, thus leaving the interpretation of such an approach ambiguous.

Many mechanisms have been proposed to explain these rapid variations in OrnDCase activity.[6] A conservative appraisal of the available evidence for these proposals supports a stronger case for some type of posttranscriptional control for OrnDCase in contrast to a transcriptional control. Indeed, the original observation that OrnDCase exhibited the extraordinarily short half-life of ~10 min hinted of an enzymatically catalyzed process involving posttranslational modification. Moreover, there is increasing evidence that the polyamines, spermidine and spermine, exert a key regulatory role in some type of negative feedback process which is antagonized by putrescine.[7,8] Below, we describe the preparation of a protein kinase from nuclei or nucleoli of the slime mold, *Physarum polycephalum* which phosphorylates OrnDCase.[9,10] The phosphorylation reaction is absolutely dependent on spermidine and spermine. Putrescine antagonizes activation of the protein kinase by spermidine and spermine. Phosphorylation of OrnDCase inhibits its capacity to catalyze decarboxylation of L-ornithine.

In general, the regulation of several enzymes, which clearly function in a biosynthetic capacity, through reversible phosphorylation by protein kinases and phosphoprotein phosphatases, has been shown to be a negative feedback control mechanism.[11] However, the current example of

[3] A. Raina and J. Jänne, *Med. Biol.* **53**, 121 (1975).
[4] D. H. Russell and S. H. Snyder, *Mol. Pharmacol.* **5**, 253 (1969).
[5] C. A. Manen and D. H. Russell, *Life Sci.* **17**, 1769 (1975).
[6] P. P. McCann, *in* "Polyamines in Biomedical Research" (J. M. Gaugas, ed.), p. 109. Wiley, New York, 1980.
[7] T. J. Paulus and R. H. Davis, *J. Bacteriol.* **145**, 14 (1981).
[8] P. S. Mamont, A. M. Joder-Ohlenbusch, M. Nussli, and J. Grove, *Biochem. J.* **196**, 411 (1981).
[9] G. R. Daniels, V. J. Atmar, and G. D. Kuehn, *Biochemistry* **20**, 2525 (1981).
[10] V. J. Atmar and G. D. Kuehn, *Proc. Natl. Acad. Sci. U.S.A.* **78**, 5518 (1981).
[11] P. Cohen, "Recently Discovered Systems of Enzyme Regulation by Reversible Phosphorylation. Molecular Aspects of Cellular Regulation," Vol. 1. Elsevier, Amsterdam, 1981.

OrnDCase is unique. The end products of the biosynthetic pathway, spermidine and spermine, control the capacity of the protein kinase to phosphorylatively inactivate the first biosynthetic enzyme of the pathway, OrnDCase. This is a newly recognized role for protein kinases. Other examples of this phenomenon should be anticipated in general metabolism.[12]

Assay Method

Principle. The assay for polyamine-dependent protein kinase requires the demonstration that transfer of [^{32}P]HOPO$_3^{2-}$ from [γ-^{32}P]ATP to OrnDCase is spermidine and spermine dependent. The procedure involves measurement of acid-stable [^{32}P]HOPO$_3^{2-}$ incorporated into OrnDCase following termination of the protein kinase reaction by a cold solution of EDTA containing bovine serum albumin.

Reagents for Assay

Buffer reagent containing 0.3 M β-glycerol phosphate disodium salt, 30 mM Mg(CH$_3$CO$_2$)$_2$ · 4H$_2$O, 3 mM ethylene glycol-bis(β-aminoethyl ether)-N,N'-tetraacetic acid (EGTA). The final pH is adjusted to 6.0 with HCl at 25°.

[γ-^{32}P]ATP · Tris reagent, 100 mM, with specific radioactivity of approximately 5 × 10^9 cpm/μmol or 2250 mCi/mmol, prepared by a published procedure.[13]

Termination reagent containing 100 mM EDTA, 50 mg/ml bovine serum albumin.

Spermidine/spermine solution containing 3 mM spermidine (free base) and 3 mM spermine (free base).

HClO$_4$, solutions of 5 and 10% are needed.

Scintillation cocktail containing 1 liter of toluene, 1 liter of 2-ethoxyethanol, and 8 g of Omnifluor.

Procedure. The procedure described below has successfully measured the activity of polyamine-dependent protein kinase in four different eukaryotic cells including *P. polycephalum*, bovine spermatozoa, rat liver nuclei, and Ehrlich ascites tumor cells. Incubation mixtures at pH 6.8 contain 50 μl of buffer reagent, 10 μl of [γ-^{32}P]ATP · Tris reagent, 10 μl of spermidine/spermine solution, 10 μl of 0.7 M NaCl containing 10–100 μg of OrnDCase, and 40 μl of enzyme preparation in appropriate buffer (see Purification section). The phosphate-acceptor protein, OrnDCase, is pre-

---

[12] B. Lent and K.-H. Kim, *J. Biol. Chem.* **257**, 1897 (1982).
[13] I. M. Glynn and J. B. Chappell, *Biochem. J.* **90**, 147 (1964).

pared as previously described.[14] All components are preincubated at 30° for 5 min. [γ-^{32}P]ATP · Tris reagent is added to initiate the reaction at zero time. After 20 min, the reaction is stopped by addition of 15 μl of termination reagent. The reaction tube is placed on ice. Zero-time control reactions are treated similarly except that [γ-^{32}P]ATP is added to the reaction mixture after addition of the termination reagent. From the reaction tube, a sample of 0.1 ml is immediately spotted uniformly over the area of a disc, 2.5 cm diameter, of Whatman 3 MM paper. Discs are prerinsed in distilled water before use to remove loose fibers and are dried at room temperature. A maximum of 30 spotted discs are collectively washed successively in a beaker for 30 min in 10% $HClO_4$ at 5°, again for 30 min in 10% $HClO_4$ at 5°, and lastly in 5% $HClO_4$ for 30 min at room temperature. One liter of each $HClO_4$ solution is used. When these washings are complete, the disc is blotted on adsorbent paper to remove excess $HClO_4$ and is dried at room temperature or under a heat gun. Each dried disc is suspended in 5 ml of nonaqueous scintillation cocktail and is counted for [^{32}P]phosphate content. The specific radioactivity of the stock [γ-^{32}P]ATP solution is determined by counting a sample of the stock concurrently with the dried disc.

The incorporation of 1 μmol of [γ-^{32}P]$HOPO_3^{2-}$ into OrnDCase per minute at 30° under the conditions specified, defines one enzyme unit. Units per milligram of protein is the specific activity.

Purification from Physarum polycephalum

Organism and Isolation of Nuclei or Nucleoli. P. polycephalum used in this isolation is strain M_3cV originally isolated at the University of Wisconsin. The culture has been maintained since 1970 at New Mexico State University on a glucose–yeast extract–tryptone medium[15] with subculture every three days.

Nucleoli are isolated in bulk quantities from microplasmidia grown in 400-ml shake-cultures using Percoll gradient density centrifugation to remove troublesome mucopolysaccharides.[16] Nuclei are isolated from microplasmodia of *P. polycephalum* by the method of Mohberg and Rusch.[17]

Precolumn Preparation. All procedures are conducted near 4°. Freshly isolated nucleoli or nuclei are suspended in an equal volume of 50 m*M* Tris–HCl (pH 7.5) containing 0.3 *M* NaCl. The suspension is passed

---

[14] G. D. Kuehn, H. U. Affolter, V. J. Atmar, T. Seebeck, U. Gubler, and R. Braun, *Proc. Natl. Acad. Sci. U.S.A.* **76**, 2541 (1979).
[15] B. Chin and I. A. Bernstein, *J. Bacteriol.* **96**, 330 (1968).
[16] G. D. Kuehn, V. J. Atmar, and G. R. Daniels, this series, Vol. 94, p. 147.
[17] J. Mohberg and H. P. Rusch, *Exp. Cell Res.* **66**, 305 (1971).

twice through a cold French pressure cell at 1390 kg/cm². The extruded suspension is centrifuged for 2 hr at 200,000 g. To the supernatant fraction is added 20 mg of Bio-Rex 70 (Na⁺ form, 400 mesh, Bio-Rad Laboratories, Richmond, CA) per milligram of total protein. The slurry is stirred slowly for 30 min and is then centrifuged at 6,000 g for 15 min. The Bio-Rex resin is discarded. The supernatant fraction is treated twice more with Bio-Rex 70 by this same procedure. After the third treatment, the final protein solution, containing an enriched fraction of acidic nonhistone proteins, is concentrated to about 5 ml in an Amicon concentrator cell model 200 equipped with a PM10 membrane, under N_2 gas pressure. The concentrate is dialyzed to equivalent conductivity against 50 mM Tris–HCl (pH 7.5) containing 0.3 M NaCl.

A 1-cm i.d. × 15 cm column of phosphocellulose (Sigma, fine mesh, product no. C2258) is prepared and equilibrated in 50 mM Tris–HCl (pH 7.5) containing 0.3 M NaCl. Detailed procedures for the preparation and storage of the phosphocellulose were recently published in this series of methods.[16]

Isolation of Polyamine-Dependent Protein Kinase. The concentrate of dialyzed, nonhistone nucleolar or nuclear proteins derived from Bio-Rex 70 treatment is applied to the column. The total protein in the concentrate should not exceed 120 mg. Polyamine-dependent protein kinase binds strongly to phosphocellulose. Thus, the column is thoroughly eluted with 10–20 column volumes of original equilibration buffer at a flow rate of about 10–15 ml/hr. At the end of this procedure, the ultraviolet absorption at 280 nm of the eluant is essentially zero. Next, the elution buffer is changed to a 300-ml, linear gradient from 0.3–1.0 M NaCl in 50 mM Tris–HCl (pH 7.5). Fractions of 3 ml are collected. Polyamine-dependent protein kinase consistently elutes from the column when the NaCl concentration is 0.7–0.8 M. A representative elution profile is shown in Fig. 1. Tubes which contain the polyamine-dependent protein kinase are conveniently located by measuring the conductivity of the column fractions and assaying those which have a NaCl concentration in the region of 0.7 M NaCl.

Detection of the protein kinase exploits its novel property of autophosphorylation in the absence of OrnDCase.[9,10] Column fractions suspected of containing the enzyme are assayed for activity in the presence and the absence of spermidine/spermine. No OrnDCase substrate protein is added to these assays. Phosphorylation of endogenous substrate protein will be observed in an early and a late fraction (Fig. 1) from the column.

The early fraction (peak a, Fig. 1) contains the polyamine-dependent protein kinase, free of its natural phosphate-acceptor substrate, OrnD-

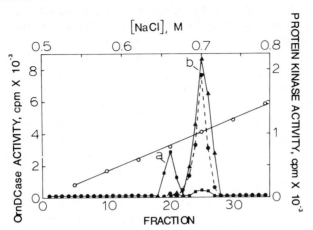

FIG. 1. Cofractionation of polyamine-dependent protein kinase (●,■) and OrnDCase (▲) enzymatic activities from nuclear proteins by phosphocellulose chromatography. Partially purified nonhistone proteins, 120 mg, are applied to a phosphocellulose column. After exhaustively washing the column to remove unbound proteins, a gradient of NaCl in 50 mM Tris–HCl (pH 7.5) is applied. Fractions of 3 ml are collected. Each fraction is tested for protein kinase activity capable of phosphorylating an endogenous protein substrate concurrently in the fraction in the presence (●) or absence (■) of a mixture of spermidine and spermine, 0.5 mM each. Each fraction was also tested for OrnDCase activity. Reprinted from V. J. Atmar and G. D. Kuehn, *Proc. Natl. Acad. Sci. U.S.A.* **78,** 5518 (1981).

Case. In this state, the protein kinase autophosphorylates and the phosphorylation is inhibited by spermidine/spermine. Autophosphorylation yields a phosphopeptide of M_r 26,000 on polyacrylamide gels in sodium dodecyl sulfate.[9] If OrnDCase is provided as substrate and the protein kinase reaction is conducted in >150 mM NaCl (see below), the protein kinase phosphorylates the 70,000 M_r subunit of OrnDCase in a reaction that is spermidine/spermine dependent.

The late fraction (peak b, Fig. 1) contains polyamine-dependent protein kinase copurified and complexed with its phosphate-acceptor substrate protein, OrnDCase. Sodium chloride at >150 mM concentration is required to dissociate the catalytically inactive complex to yield enzymatically active spermidine/spermine-dependent protein kinase or active OrnDCase.[10] In the presence of OrnDCase, the protein kinase phosphorylates the M_r 70,000 subunit of OrnDCase in a reaction that is absolutely dependent on spermidine/spermine (Fig. 1).[10]

The total amount of the protein kinase isolated by this method is very small. An upper limit of 100 μg of protein can be purified from 40 ml of wet-packed isolated nuclei. Electrophoretograms of purified preparations of the polyamine-dependent protein kinase show a single polypeptide of

M_r 26,000 in both 8 and 12.5% sodium dodecyl sulfate polyacrylamide gels.[9] The specific activity of the purified protein kinase ranges in the vicinity of 10–15 mU/mg of protein when OrnDCase is the substrate. Polyamine-dependent protein kinase activity cannot be demonstrated in crude extracts prepared from nuclei or nucleoli. It is only after partial fractionation through Bio-Rex 70 treatment that the activity can be detected. High phosphoprotein phosphatase activity, which is inhibitable by Na_2MoO_4, appears to be responsible for this effect. Phosphocellulose chromatography separates the phosphatase from the protein kinase. For these reasons, a purification summary table would not be meaningful.

General Properties of the Protein Kinase. Preparations of the protein kinase are stable for weeks when stored at 4° in buffered 0.7 M NaCl solution. The enzyme that copurifies with OrnDCase begins to lose its capacity to be activated by spermidine/spermine after about 3 weeks.

Putrescine reverses the capacity of spermidine/spermine to activate the protein kinase reaction. The mechanism for how this is accomplished is unknown.

OrnDCase is the only endogenous substrate yet identified for the polyamine-dependent protein kinase.[9] With OrnDCase as substrate, the K_m for ATP and Mg^{2+} are 10 and 53 μM, respectively. The kinase will phosphorylate casein at approximately 0.1 the rate observed with OrnDCase. However, the casein kinase activity is not affected by spermidine/spermine. No modifiers, other than the polyamines, have been found to affect the catalytic capacity of this enzyme.[9] Furthermore, preliminary interpretation of early experiments hints that the polyamines affect the protein substrate as opposed to the kinase itself.[9]

Purification from Other Sources

The procedures described have also successfully purified polyamine-dependent protein kinase from bovine spermatozoa,[18] nuclei from cultures of Ehrlich ascites tumor cells as well as the cytoplasmic compartment of this tumor cell line,[19] and from nuclei of rat liver (unpublished work). In the case of the ascites cells, it has recently been discovered that mouse fibroblast interferon induces polyamine-dependent protein kinase activity over 120-fold in the nucleus.[19] This activity was inseparable from previously described, double-stranded RNA-dependent protein kinase on phosphocellulose or on poly(I) · poly(C)-agarose affinity columns.

---

[18] V. J. Atmar, G. D. Kuehn, and E. R. Casillas, *J. Biol. Chem.* **256**, 8275 (1981).
[19] V. Sekar, V. J. Atmar, M. Krim, and G. D. Kuehn, *Biochem. Biophys. Res. Commun.* **106**, 305 (1982).

[40] Characterization of the Abelson Murine Leukemia Virus-Encoded Tyrosine-Specific Protein Kinase

By JEAN YIN JEN WANG and DAVID BALTIMORE

Abelson murine leukemia virus (A-MuLV) was isolated from a tumor induced by Moloney-MuLV (M-MuLV) in a glucocorticoid-treated mouse. A-MuLV causes a pre-B cell lymphoma which is different from the thymus-dependent disease induced by M-MuLV. The genome of A-MuLV is a derivative of M-MuLV with the nucleic acid sequence from the c-*abl* region of normal mouse genome replacing most of the M-MuLV genome. As a result A-MuLV is a replication-defective virus in which the v-*abl* sequence confers on the virus the ability to transform pre-B cells and fibroblasts *in vitro*. There are several different strains of A-MuLV producing proteins of different molecular weights: the smallest has a molecular weight of about 90,000 [A-MuLV (P90)] and the largest has a molecular weight of 160,000 [A-MuLV (P160)]. All of the A-MuLV proteins contain 30,000 daltons of the M-MuLV *gag* protein sequence, with the remaining 60,000 to 130,000 daltons of polypeptide chain being encoded by the v-*abl* sequence. The *gag* portion of the A-MuLV protein is not necessary for the viral transforming activity.[1]

Evidence that the A-MuLV Protein Is a Tyrosine-Specific Protein Kinase

The A-MuLV protein can be recognized by antibodies against the M-MuLV *gag* protein determinants. When the A-MuLV protein is immunoprecipitated by anti-*gag* sera and the pellet resuspended in buffer containing [γ-^{32}P]ATP, incorporation of ^{32}P into the A-MuLV protein can be observed. Phosphoamino acid analysis of the labeled A-MuLV protein shows the exclusive presence of phosphotyrosine.[2] Control experiments demonstrate that this tyrosine-phosphorylation reaction is specific to cell lysates containing the A-MuLV proteins and the correct antisera.[2] This result implies that the A-MuLV protein is capable of autophosphorylation on tyrosine residues, although the presence of contaminating protein kinase in the immunoprecipitates cannot be ruled out.

Direct evidence for the tyrosine-specific protein kinase activity comes

---

[1] R. Prywes and D. Baltimore, unpublished results (1982).
[2] O. N. Witte, A. Dasgupta, and D. Baltimore, *Nature (London)* **283**, 826 (1980).

from the expression of A-MuLV coding sequence in *E. coli* by recombinant DNA techniques.[3] The v-*abl* sequence from A-MuLV (P90) is placed downstream of the P_R promoter of λ phage and this stretch of coding information gives rise to a 62,000-dalton protein, as expected. Phosphoproteins containing phosphotyrosine can be found in *E. coli* cells expressing the 62,000-dalton protein, although *E. coli* ordinarily contains no phosphotyrosine.[3] The 62,000-dalton polypeptide contains phosphotyrosine itself, indicating that the A-MuLV protein can indeed carry out autophosphorylation reactions. At least seven other bacterial proteins are also phosphorylated on tyrosine residues in *E. coli* cells containing the 62,000-dalton A-MuLV protein.

Assay of the A-MuLV Protein Kinase Activity

Autophosphorylation Assay in an Immunoprecipitate

Materials

Lysis buffer: 10 mM Na phosphate, pH 7.5, 100 mM NaCl, 1.0% Triton X-100, 0.5% Na deoxycholate, 0.1% SDS, 1 mM EDTA, 5 mM phenylmethylsulfonyl fluoride (all detergent concentrations are percentages w/v).

Phosphorylation buffer: 20 mM Tris–HCl, pH 8.0, 10 mM MnCl$_2$, 1 μCi [γ-^{32}P]ATP at 2000–3000 Ci/mmol.

SDS buffer: 200 mM Tris–HCl, pH 6.8, 30 mM EDTA, 6% SDS, 1% (v/v) 2-mercaptoethanol, 30% glycerol, 0.05% bromphenol blue.

Goat anti-Moloney-MuLV sera.

Formaldehyde-fixed *Staphylococcus aureus*.

Procedure: Immunoprecipitation. A-MuLV-transformed cells (2 × 10^6) are washed free of medium with phosphate-buffered saline and lysed in 1 ml of lysis buffer. A 25-μl aliquot of a 50% suspension of *S. aureus* previously washed with the Lysis buffer containing 0.5% (w/v) bovine serum albumin is added to the cell lysate; after 1 hr of incubation on ice, the *S. aureus* is removed by centrifugation at 13,000 *g* for 15 min. This centrifugation step is usually carried out in an Eppendorf Microfuge at 4°. Five hundred microliters of the clarified lysate is then mixed with 5 μl of antiserum and incubated on ice for 6 to 18 hr. The antigen–antibody complex is collected onto *S. aureus*: usually 50 μl of a washed, 50% suspension is used for 5 μl of serum. After a 1-hr incubation on ice, the immunoprecipitate on *S. aureus* is pelleted by centrifugation at 13,000 *g* for 30 sec.

Phosphorylation Reaction. The immunoprecipitate is washed twice with cold 20 mM Tris–HCl (pH 8) by repeated resuspension and centrifu-

[3] J. Y. J. Wang, C. Queen, and D. Baltimore, *J. Biol. Chem.* **257**, 13181 (1982).

gation. Then 50 µl of the phosphorylation buffer is added to resuspend the pellet and the mixture is incubated at 30° for 10 min. The phosphorylation reaction is terminated by centrifuging the mixture, removing the supernatant, adding 50 µl of SDS buffer, and boiling for 10 min. This treatment solubilizes the immunocomplex and the supernatant of this SDS buffer extraction can be loaded onto a SDS–polyacrylamide gel for electrophoresis. The ^{32}P-labeled protein bands are visualized by autoradiography.

Comment. The autophosphorylation of the A-MuLV protein in an immunoprecipitate is not a regular measurement of the enzyme activity. The phosphorylation occurs at nanomolar concentrations of ATP, the amount of label incorporated reaches a plateau in 1 min at 30 or 0°, and only a small percentage of the immunoprecipitated A-MuLV protein is phosphorylated. However, this is a useful method for a qualitative determination of the presence or absence of protein kinase activity of the A-MuLV proteins. The phosphorylation on immunoprecipitates occurs intermolecularly but only when both molecules are coprecipitated. For example, one can mix a heat-denatured cell lysate containing the A-MuLV P120 protein with a nondenatured P160-lysate, immunoprecipitate both proteins, incubate the pellet with [γ-^{32}P]ATP, and observe the phosphorylation of both P120 and P160 although the heat-denatured P120-immunoprecipitate alone does not retain any kinase activity. This result shows that intermolecular phosphorylation does occur in this assay. However, if the heat-denatured P120-lysate is added to an immunoprecipitate of nondenatured P160 during the phosphorylation reaction, only the P160 protein can be labeled. Thus, the coprecipitation of both proteins is crucial to the intermolecular phosphorylation reaction. We interpret these results as an indication that only some preformed complexes of the A-MuLV proteins can incorporate phosphate in this assay. Thus, the immunoprecipitated form of the A-MuLV protein is not the equivalent of an immobilized enzyme. In fact, the immunoprecipitated A-MuLV protein does not phosphorylate a 36,000-dalton mouse cellular protein whereas it can do so in solution (see below).

A General Assay for the Phosphorylation of Tyrosine Residues in Proteins

Principle. The phosphoryl linkage of phosphotyrosine is much more resistant to alkaline conditions than that of phosphoserine or phosphothreonine, the two most commonly found phosphoamino acids. A quantitative assay for the specific phosphorylation of tyrosine residues takes advantage of this stability at high pH.

Procedures. The protein samples to be assayed for phosphotyrosine content are precipitated by the addition of 9 volumes of 10% trichloroacetic acid (TCA), 0.1 M Na-pyrophosphate. The pellet is collected by centrifugation, dissolved in the original sample volume of 1 N NaOH, and incubated at 37° for 1 hr. The NaOH is neutralized by adding an equal volume of 1 N HCl and an amount of bovine serum albumin is added to give a 2 mg/ml final concentration, then the proteins are precipitated again by adding a volume of 50% TCA to give a final TCA concentration of 10%. The protein pellet is washed three times with 10% TCA, 0.1 M Na-pyrophosphate and air-dried, and the radioactivity is measured by scintillation counting.

The standard error of this assay is within 2%. The alkaline treatment does not eliminate all of the phosphothreonine counts. However, a significant reduction of the nonphosphotyrosine counts can be achieved. An example of this assay result is shown in the table. After pulse-labeling with $^{32}P_i$, E. coli cells with or without the abl-encoded 62,000-dalton protein were collected, washed, and sonicated, and the cell extracts were subjected to the TCA precipitation and NaOH treatment described above. The difference between the two samples in total TCA-precipitable counts was twofold, whereas after NaOH treatment the difference becomes fivefold. This assay will be very useful for measuring tyrosine-specific protein kinase activity when a substrate of the A-MuLV protein becomes available. We are in the process of searching for such a substrate protein.

Partial Purification of the 120,000 Dalton A-MuLV Protein

Two batchwise fractionation procedures have been used to enrich for the P120 A-MuLV protein.[2] The behavior of the P120 protein during fractionation was followed by immunoprecipitation with anti-Moloney-MuLV sera. Because of the lack of a quantitative assay, the purification of the recovery of P120 through these steps could not be accurately determined. However, the autophosphorylating activity of P120 is retained

QUANTITATION OF NaOH-RESISTANT, TCA-PRECIPITABLE ^{32}P IN PULSE-LABELED E. coli CELLS[a]

Cell type	Total TCA-precipitable counts (cpm/mg protein)	NaOH-resistant counts (cpm/mg protein)
E. coli without A-MuLV coding sequence	4300	517
E. coli with A-MuLV coding sequence	9567	2410

[a] E. coli cells were pulse-labeled in minimal salt buffer with carrier-free $^{32}P_i$ as described by Wang et al.[3]

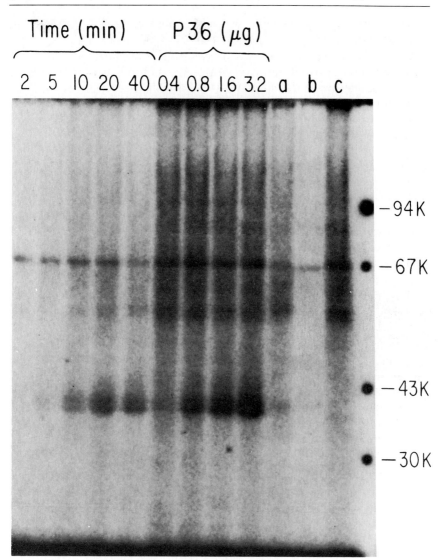

FIG. 1. Tyrosine phosphorylation of a 36,000-dalton mouse cellular protein by partially purified A-MuLV P120 protein. The phosphorylation reaction was carried out at 30° in 20 mM Tris–HCl, pH 8, 10 mM $MnCl_2$, 10 mM NaF, 5 mM 2-mercaptoethanol, and 0.5 mM [γ-^{32}P]ATP (500 cpm/pmol). The time of incubation or the amount of P36 present is indicated in the figure. (a) Reaction in the presence of 0.1% SDS; (b) reaction at 0°; (c) the A-MuLV P120-containing fraction alone without the purified P36 at 30° for 40 min.

through the fractionations and this activity can be stable for at least 6 weeks at $-20°$ in the partially purified form.[2]

Briefly, A-MuLV (P120) strain transformed lymphoid cells are extracted with the lysis buffer (see Autophosphorylation Assay) and mixed with DEAE-cellulose equilibrated with the same buffer. The unbound material and the washings of the resin are combined and added to phosphocellulose also equilibrated with the lysis buffer. The phosphocellulose resin is washed and the adsorbed material is eluted with lysis buffer plus 0.5 M NaCl. The batch-eluted material is concentrated by adsorption onto a small phosphocellulose column and reeluted with 0.5 M NaCl-lysis buffer without the 0.1% SDS but containing 25% glycerol.

The phosphocellulose-bound fraction can phosphorylate a 36,000-dalton protein (P36) purified from mouse Erhlich ascites cells. The purification of P36 is essentially by the method of Erikson et al. who have purified a homologous P34 from chicken cells.[4] Using antisera raised against the purified P36, it is shown that this protein is present in normal fibroblasts in an unphosphorylated form and it becomes phosphorylated in A-MuLV-transformed fibroblasts.[5] Purified P36 does not incorporate phosphate by itself, but it can be phosphorylated on tyrosine residues when the partially purified A-MuLV P120 (the phosphocellulose fraction) is present. As shown in Fig. 1, the phosphorylation of P36 increases with time and it is also a function of the P36 concentration. A doublet of P36 is observed, the lower band corresponding to the Coomassie blue-stained band. However, only 0.1% of the input P36 is phosphorylated and the rate of phosphorylation is not proportional to the amount of enzyme added. Therefore, this preparation of P36 is not an ideal substrate for the assay of A-MuLV protein kinase. The phosphorylation of P36 is sensitive to 0.1% SDS, it does not proceed at 0°, and it does not happen when the immunoprecipitates of the A-MuLV proteins are used as a source of enzyme.

Summary

There is sufficient evidence that the A-MuLV protein is a tyrosine-specific protein kinase. There are methods for detecting this kinase activity and this kinase has been expressed in *E. coli*. Because the information coding for the tyrosine-specific protein kinase is present in normal mouse cells, such an enzyme must have a normal physiologic function. The elucidation of this physiologic function and the understanding of the role of this enzyme activity in neoplastic transformation is the challenge of the future.

[4] E. Erikson and R. L. Erikson, *Cell* **21,** 829 (1980).
[5] J. Y. J. Wang and D. Baltimore, unpublished results (1981).

[41] Purification of the Receptor for Epidermal Growth Factor from A-431 Cells: Its Function as a Tyrosyl Kinase

By STANLEY COHEN

Murine epidermal growth factor (EGF) is a single polypeptide chain of 53 amino acid residues (MW 6045) containing three disulfide bonds, but devoid of alanine, phenylalanine, and lysine; its sequence is known.[1] EGF binds to specific plasma membrane receptors in target cells and initiates and maintains a complex program of biochemical and morphological events leading to cell growth and multiplication. Specific saturable receptors for EGF are demonstrable using ^{125}I-labeled EGF and a wide variety of cells and tissues. References may be found in a number of reviews.[1-3]

The human epidermoid carcinoma cell line A-431 contains an extraordinarily high concentration of membrane receptors for EGF, approximately 2 to 3 × 10^6 receptors per cell. The binding of EGF to membranes prepared from these cells results in an enhancement of the ability of the preparation to phosphorylate specific endogenous membrane proteins at tyrosine residues[4-6] in the presence of [γ-^{32}P]ATP. The major endogenous membrane protein whose phosphorylation is enhanced by the presence of EGF appears to be the EGF receptor itself.[7,8] Evidence has recently been obtained, using 5'-p-fluorosulfonylbenzoyladenosine to affinity label the EGF-stimulable kinase, which suggests that the EGF binding site, the EGF-stimulable protein kinase, and the site of phosphorylation all reside in the same polypeptide chain.[9] We have taken advantage of this observation by employing EGF covalently linked to Affi-Gel for the purification of the EGF-related tyrosine-specific protein kinase from detergent-solubilized material.

One comment is necessary concerning the isolation procedures. The assay for protein kinase activity relies on the ability of the crude membrane preparations and purified kinase to "autophosphorylate." The

---

[1] G. Carpenter and S. Cohen, *Annu. Rev. Biochem.* **48**, 193 (1979).
[2] M. D. Hollenberg, *Vitam. Horm.* **37**, 69 (1979).
[3] G. Carpenter and S. Cohen, *Recept. Recognition Ser.* B **13**, 43 (1981).
[4] G. Carpenter, L. King, and S. Cohen, *J. Biol. Chem.* **254**, 4884 (1979).
[5] L. King, G. Carpenter, and S. Cohen, *Biochemistry* **19**, 1524 (1980).
[6] H. Ushiro and S. Cohen, *J. Biol. Chem.* **255**, 8363 (1980).
[7] S. Cohen, G. Carpenter, and L. King, *J. Biol. Chem.* **255**, 4834 (1980).
[8] S. Cohen, H. Ushiro, C. Stoscheck, and M. Chinkers, *J. Biol. Chem.* **257**, 1523 (1982).
[9] S. A. Buhrow, S. Cohen, and J. V. Staros, *J. Biol. Chem.* **257**, 4019 (1982).

usual kinetic parameters of enzyme activity are not applicable and precise calculations of yield and specific activity are not yet possible.

Preparative Procedures

Preparation of EGF-Affi-Gel

Murine EGF was isolated by published procedures.[10] EGF was covalently linked to Affi-Gel 10 (an *N*-hydroxysuccinimide ester of a succinylated aminoalkyl BioGel A support) obtained from Bio-Rad. The gel was washed with 3 bed volumes of cold distilled water. EGF (20 mg) was dissolved in 3 ml of 0.1 M NaHCO$_3$ buffer, pH 8.1, and added to 6 g of wet gel cake. The reaction was allowed to proceed for 6 hr at room temperature with gentle agitation. Unreacted sites were blocked by the addition of 1 M ethanolamine adjusted to pH 8 with HCl (0.1 ml of the ethanolamine solution per milliliter of suspension). The reaction mixture was stirred at room temperature for 1 hr and the gel was then washed extensively using, successively, 4 liters of water, 2 liters each of 1 M NaCl, 0.1 M acetic acid, and 1 M urea, 4 liters of water, and finally, 2 liters of 20 mM HEPES buffer, pH 7.4. In control experiments, the extent of coupling under these conditions was estimated by including ^{125}I-labeled EGF in the reaction mixture and it was found that approximately 30% of the EGF was incorporated into the gel.[7] The above washing procedure also was employed to regenerate the gel after use.

Standard Phosphorylation Assays

For intact membranes the reaction mixtures contained A-431 membrane (20 to 80 μg of protein); HEPES buffer (20 mM, pH 7.4); MnCl$_2$ (1 mM); [γ-^{32}P]ATP (15 μM, 6 to 12 × 10^5 cpm); and EGF (100 ng) in a final volume of 60 μl. The reaction tubes were placed on ice and preincubated for 10 min in the presence or absence of EGF. The reaction was initiated by the addition of labeled ATP and incubation at 0° was continued for the desired times (usually 5–10 min). The reaction was terminated by pipetting 50-μl aliquots onto squares (2 cm) of Whatman No. 3MM filter paper which were dropped immediately into a beaker of cold 10% trichloroacetic acid containing 0.01 M sodium pyrophosphate. The filter papers were washed extensively with the trichloroacetic acid/pyrophosphate solution at room temperature, extracted with alcohol and ether, and dried, and the radioactivity was measured in a Nuclear Chicago gas flow counter.

[10] C. R. Savage and S. Cohen, *J. Biol. Chem.* **247**, 7609 (1972).

For Triton X-100-solubilized membranes, the procedure was essentially identical, except the reaction mixtures were preincubated in the presence or absence of EGF for 10 min at room temperature and then chilled on ice for 10 min prior to initiating the reaction by the addition of the labeled ATP. The reaction mixtures contained Triton X-100 at a final concentration of 0.2%.

When the phosphorylation reaction was carried out for short periods of time (under 3 min), the reaction was stopped by the addition of 5 μl of 5 N HCl prior to the removal of the 50-μl aliquot for measurement of radioactivity.

Gel Electrophoresis and Autoradiography

The A-431 membrane components were separated by SDS–gel electrophoresis (7.5% acrylamide) by the method of Laemmli.[11] Commercial molecular weight standards were used for calibration of gels. The slab gels were fixed, stained with Coomassie blue, dried under vacuum, and autoradiography (1 to 3 days) was performed using Kodak RP Royal X-Omat film.[5]

Hydrolysis and Identification of [^{32}P]Phosphotyrosine

The ^{32}P-labeled membrane material was hydrolyzed in 1 ml of 6 N HCl at 100° for 1.5 to 4 hr in tubes sealed under vacuum. The hydrolysates were dried in a rotary evaporator and dissolved in 25 μl of water, and 5-μl aliquots (to which the standard phosphorylated amino acids were added) were subjected to electrophoresis on Whatman No. 3MM paper at 20° for 5 to 8 hr at 800 V. Two buffer systems were employed: pyridine–acetic acid, pH 3.5 (20 ml of glacial acetic acid and 2.3 ml of pyridine in a final volume of 3 liters) and 6.9% formic acid. The standards (phosphotyrosine, phosphoserine, and phosphothreonine) were located by ninhydrin; the radioactive material was located by autoradiography using X-ray film. Phosphotyrosine may also be identified by chromatographic procedures[6] and high performance liquid chromatography.[12]

Growth of A-431 Cells and Preparation of Membrane and Vesicle Fractions

The A-431 cells were grown in 100-mm Falcon dishes or in roller bottles containing Dulbecco's modified Eagle's medium (GIBCO) supplemented with 5 or 10% calf serum (Flow Laboratories) and gentamycin (Microbiological Associates).

[11] U. K. Laemmli, *Nature (London)* **227,** 680 (1970).
[12] G. Swarup, S. Cohen, and D. L. Garbers, *J. Biol. Chem.* **256,** 8197 (1981).

Two procedures were used to obtain fractions enriched with plasma membranes. A membrane fraction was prepared[4] by the procedure of Thom et al.[13] In brief, the cells were scraped from dishes, concentrated into a small volume by centrifugation, and lysed by dilution into a hypotonic borate/EDTA buffer, pH 10.2. The lysed cells were filtered through a nylon screen and a crude membrane fraction was obtained by centrifugation at 25,000 g for 30 min. The pelleted material was resuspended, layered over 35% (w/w) sucrose, and centrifuged (40,000 g for 45 min) in a swinging bucket rotor. The material at the buffer–sucrose interface was collected with a Pasteur pipet, suspended in 10 mM HEPES, pH 7.4, and recentrifuged (75,000 g for 30 min). The final membrane preparation was resuspended in the HEPES buffer (approximately 10 mg of protein/ml), divided into small aliquots, frozen on dry ice, and stored at $-70°$. The EGF receptor/protein kinase isolated from this membrane preparation (see later section) appeared as a M_r 150,000 protein on SDS–polyacrylamide gels.

An alternate method[8] for the preparation of shed membrane vesicles was employed to obtain the "native" EGF receptor/protein kinase. Roller bottles containing A-431 cells were washed three times with 50 ml of Dulbecco's phosphate-buffered saline (PBS) at room temperature and once with 50 ml of Ca^{2+}, Mg^{2+}-free hypotonic PBS (1 volume of Ca^{2+}, Mg^{2+}-free PBS plus 19 volumes of water). Hypotonic PBS (150 ml) was then added and the bottles were rotated (3 rpm) for 15 min at room temperature. During this period, the cells swelled, but did not rupture. The hypotonic PBS was discarded, and the cells were washed one time with 50 ml of vesiculation buffer (100 mM NaCl, 50 mM Na_2HPO_4, 5 mM KCl, and 0.5 mM $MgSO_4$, pH 8.5). Then 50 ml of vesiculation buffer was readded and the bottles were rolled for 20 min at room temperature and for 1 hr at 37°. During this period, plasma membrane vesicles are shed into the vesiculation buffer and are easily seen under the microscope. The vesiculation buffer was then decanted through nylon screen (500 μm) into a flask in ice. The vesicle suspension was centrifuged at 150 g for 5 min at 4° and the pellet (debris) was discarded. The supernatant fluid was centrifuged at 20,000 g for 30 min at 4°. The vesicle pellets were resuspended in 40 ml of 10 mM HEPES buffer, pH 7.4, and recentrifuged for 1 hr at 80,000 g. The final pellets were gently resuspended in 10 mM HEPES, pH 7.4, using a glass homogenizer and stored in 125-μl aliquots at $-70°$ at a protein concentration of approximately 20 mg/ml. The final vesicle yield from one roller bottle (approximately 2×10^8 cells) was 5 mg of protein. The protein content of these preparations was quantitated by the proce-

---

[13] D. Thom, A. J. Powell, C. W. Lloyd, and D. A. Rees, *Biochem. J.* **168**, 187 (1977).

dure of Bradford[14] using γ-globulin as a standard. The receptor/protein kinase isolated from this vesicle preparation (see later section) appeared mainly as a M_r 170,000 protein on SDS–polyacrylamide gels.

Net phosphorylation using vesicles was approximately twice as great as compared to the membrane preparation, both in the presence and absence of EGF. Since phosphatases are present in the samples, no precise interpretation of these data can be drawn other than that both preparations possess an EGF-stimulated kinase system (and both are capable of binding EGF[8]). In both systems the major phosphorylated amino acid residue was identified as phosphotyrosine by the procedures outlined above.

Major differences in the EGF–receptor–kinase complexes present in the vesicle and membrane preparations are apparent when the phosphorylated preparations are examined by SDS–acrylamide electrophoresis and autoradiography. The major phosphorylated component in the vesicle preparations had a molecular weight of 170,000 with a minor component of 150,000; the major phosphorylated component in the membrane preparation had a molecular weight of 150,000 with a minor component in the 170,000 region. In both instances, phosphorylation was enhanced in the presence of EGF. The lower M_r protein appears to be a proteolytic degradation product of the "native" receptor/kinase.[8]

Solubilization and Purification Procedures

Standard Method for Membrane Solubilization[8]

Aliquots of the A-431 membrane or vesicle preparations were suspended in 20 mM HEPES buffer, pH 7.4, containing 5% Triton X-100 and 10% glycerol. The final membrane concentration was 6 mg of protein/ml. The suspension was allowed to stand at room temperature for 20 min and the mixture was centrifuged at 100,000 g for 60 min at 4°. The solubilized preparation could be stored for several weeks at −70° with little loss in receptor or kinase activity.

A comparison of the effects of EGF on phosphorylation using intact and solubilized membranes indicated that much (over 60%) of the basal activity and almost all (over 90%) of the EGF responsive activity are recoverable after solubilization.

The solubilization may be carried out at 0° or room temperature; slightly higher yields of phosphorylating activity (about 10% higher) are observed using room temperature solubilization. A variety of nonionic

[14] M. M. Bradford, *Anal. Biochem.* **72**, 248 (1976).

detergents were effective in solubilizing the kinase activity[7]; extracts prepared with either sodium cholate or sodium deoxycholate were inactive in the phosphorylation assay.

EGF-Affi-Gel Purification of EGF Receptor/Kinase

Aliquots of the solubilized membranes or vesicles (800 μl, 3.6 mg of protein) were added to 200 μl of packed EGF-Affi-Gel beads in a 1-ml Eppendorf pipet tip sealed with glass wool. The mixture was stirred at room temperature for 30 min. The tips were then centrifuged (1 min at 1000 g) to remove nonadsorbed material and the gel was washed three times with 1 ml of cold buffer (10% glycerol and 0.2% Triton X-100 adjusted to pH 7.2 with NaOH). To the final gel cake were added 300 μl of elution buffer (10% glycerol, 0.2% Triton X-100, and 5 mM ethanolamine, final pH 9.7). The mixture was stirred for 30 min at 0° and centrifuged to obtain the eluted receptor material, which was immediately neutralized with acetic acid. The addition of traces of phenol red as an internal indicator facilitated the neutralization. The protein content of these Triton-containing preparations was determined by the method of Peterson[15] using bovine albumin as a standard. The eluted material (300 μl) contained approximately 20–30 μg of protein using either the membrane or vesicle preparation. This preparative procedure may be easily scaled up fivefold by substituting a Bio-Rad disposable chromatographic column (9 ml capacity) for the 1-ml Eppendorf pipette tip, and increasing the volume of all reagents fivefold; other conditions were unchanged.

The affinity purified membrane and vesicle preparations both have the capacity to bind ^{125}I-labeled EGF and both contain EGF-stimulable protein kinase activity.[8] However, on SDS gels, the single major protein isolated from the membrane preparation had an apparent molecular weight of 150,000 whereas the vesicle preparation yielded a major 170,000 M_r protein. Both proteins are phosphorylated in the presence of ATP and the phosphorylation of both is enhanced in the presence of EGF (Fig. 1).

The specific activity as measured by the intrinsic kinase reaction was approximately 5- to 10-fold greater for the M_r 170,000 preparation as compared with the M_r 150,000 preparation (see the table). This 5- to 10-fold difference in specific activities of the two preparations was also observed when the concentration of ATP in the reaction mixture was increased from 15 μM (as in the table) to 150 μM. In view of the observation that when challenged by exogenous substrates (tubulin or certain antisera) the M_r 170,000 preparation contained less kinase activity than the M_r 150,000 preparation, it was suggested that the M_r 170,000 protein is in

[15] G. L. Peterson, *Anal. Biochem.* **83**, 346 (1977).

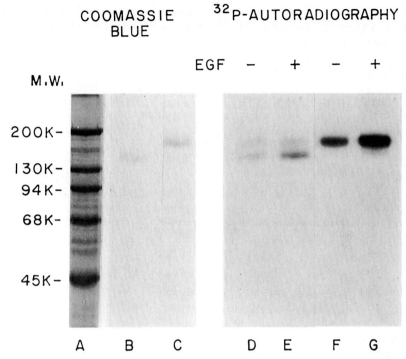

FIG. 1. Electrophoresis and autoradiography of solubilized A-431 membranes and vesicles purified by affinity chromatography following incubation with [γ-^{32}P]ATP. A-431 membranes or vesicles were solubilized in Triton and purified by affinity chromatography as described. Aliquots (1.2 μg of protein) of the purified material were subjected to the standard phosphorylation procedure. The reactions were stopped and analyzed by SDS–gel electrophoresis, Coomassie blue staining, and autoradiography. A to C, Coomassie blue stain. A, Molecular weight standards; B, purified membrane preparation; C, purified vesicle preparation. D to G, ^{32}P autoradiographs; D and E, purified membrane preparation phosphorylated in the absence (D) or presence (E) of EGF; F and G, purified vesicle preparation phosphorylated in the absence (F) or presence (G) of EGF. (From Cohen et al.[8])

a more favorable configuration for autophosphorylation than is the M_r 150,000 protein.

Figure 2 shows the time course of the incorporation of [^{32}P]ATP into the affinity purified M_r 170,000 preparation at 0° in the presence and absence of EGF. A stimulation of the endogenous phosphorylation by EGF is noted at all times examined (10 sec to 16 min). A comparison of these results with those obtained using intact membranes[4] or Triton-solubilized membranes[7] indicates that the effects of EGF were similar in all of the preparations. We have calculated from the data in Fig. 2 that after 16

COMPARISON OF BASAL AND EGF-ACTIVATED PHOSPHORYLATION
USING AFFINITY-PURIFIED RECEPTOR PREPARATIONS FROM
VESICLES AND MEMBRANES[a]

		^{32}P incorporated (cpm/µg protein)	
Preparation	Experiment	−EGF	+EGF
Membrane purified	1	436	1,212
	2	320	1,414
Vesicle purified	1	4,260	7,740
	2	5,840	10,320

[a] Affinity-purified receptor preparations (0.5 µg of protein) were assayed for phosphorylating activity in the presence and absence of EGF per 10 min at 0°.

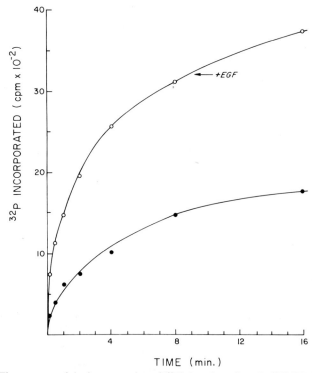

FIG. 2. Time course of the incorporation of [^{32}P]phosphate from [γ-^{32}P]ATP into purified vesicle preparations in the presence and absence of EGF. The standard phosphorylation assay for solubilized receptor was performed as described. ●---●, Minus EGF; ○---○, plus EGF. (From Cohen et al.[8])

min of incubation at 0° in the presence of EGF, approximately 0.3 to 0.4 μmol of ^{32}P was incorporated/μmol of receptor, assuming a molecular weight of 170,000. In other experiments, under the same conditions, but with different receptor preparations, we have observed up to 0.8 to 1.0 μmol of ^{32}P incorporated/μmol of receptor. However, tryptic digestions indicate that more than one site is phosphorylated (unpublished observation).

The presence of phosphotyrosyl-protein phosphatase activity in the starting membrane preparations and the absence of efficient exogenous substrates for the EGF receptor/kinase have retarded progress in understanding the kinetic properties of this enzyme. Although tubulin[8] and pp60^{src} antisera[16,17] may be used as substrates, they are not very effective. Recently, however, a series of pp60^{src}-related peptides was synthesized and the peptides were found to be substrates for the EGF receptor/kinase.[18] In addition, it has been observed[19] that vanadate inhibits membrane phosphotyrosyl-protein phosphatase activity without inhibiting the activity of the EGF receptor/kinase. The exploitation of these findings should help our understanding of this hormone-dependent enzyme system.

[16] M. Chinkers and S. Cohen, *Nature (London)* **290**, 516 (1981).
[17] J. E. Kudlow, J. E. Buss, and G. N. Gill, *Nature (London)* **290**, 519 (1981).
[18] L. J. Pike, B. Gallis, J. E. Casnellie, P. Bornstein, and E. G. Krebs, *Proc. Natl. Acad. Sci. U.S.A.* **79**, 1443 (1982).
[19] G. Swarup, S. Cohen, and D. L. Garbers, unpublished observations.

[42] Detection and Quantification of Phosphotyrosine in Proteins

By JONATHAN A. COOPER, BARTHOLOMEW M. SEFTON, and TONY HUNTER

Protein kinases can be differentiated according to their amino acid specificity. Casein kinases of type II and the cAMP-dependent kinases phosphorylate serine, while casein kinases of type I phosphorylate threonine and, to a lesser extent, serine. Recently, another group of kinases with strict specificity for tyrosine has been described.[1]

[1] T. Hunter and B. M. Sefton, *in* "The Molecular Actions of Toxins and Viruses" (P. Cohen and S. Van Heyningen, eds.), p. 333. Elsevier, Amsterdam, 1982.

There are thus three phosphorylhydroxyamino acids in cell proteins: phosphoserine, phosphothreonine, and phosphotyrosine.[2] Other phosphorylated residues identified in proteins include thioesters of cysteine, phosphoramidates of lysine and histidine, and acid anhydrides of glutamic and aspartic acid.[3] Some of these are intermediates in enzymatic reaction mechanisms. Others appear to be stable modifications, although the details of their synthesis are poorly understood. Of all the potential phosphate esters of the natural amino acids, only the three phosphorylhydroxyamino acids have reasonable chemical stability to extremes of pH.[4,5] For this reason they, and the enzymes which form them, are amenable to study.

Most tyrosine-specific protein kinase activities detected to date are implicated in cell growth control. They include the kinases associated with the cell surface receptors for two polypeptide growth factors, epidermal growth factor (EGF) and platelet-derived growth factor (PDGF), and with the transforming proteins of at least five genetically distinct groups of retroviruses.[6-12] Identification of proteins containing phosphotyrosine is thus of interest both from an enzymological standpoint and for an understanding of mechanisms which may be involved in malignant transformation. The procedures described here are also useful for identifying phosphodiester linkages through tyrosine, such as are found in some nucleotidyl-proteins[13] and some proteins bound covalently to nucleic acid.[14]

The activity of a tyrosine protein kinase in the cell is often apparent from an increase in the gross level of phosphotyrosine in cell proteins.[15] This is a consequence of the normal scarcity of phosphotyrosine. The induction of a cellular tyrosine protein kinase by growth factors or the

---

[2] Free phosphotyrosine has also been found in *Drosophila*. H. K. Mitchell and K. D. Lunan, *Arch. Biochem. Biophys.* **106**, 219 (1964).
[3] G. Taborsky, *Adv. Prot. Chem.* **28**, 1 (1974).
[4] R. H. A. Plimmer, *Biochem. J.* **35**, 461 (1941).
[5] Phosphohydroxyproline has not yet been found in proteins.
[6] W. Eckhart, M. A. Hutchinson, and T. Hunter, *Cell* **18**, 925 (1979).
[7] T. Hunter and B. M. Sefton, *Proc. Natl. Acad. Sci. U.S.A.* **77**, 1311 (1980).
[8] S. Cohen, G. Carpenter, and L. King, Jr., *J. Biol. Chem.* **255**, 4834 (1980).
[9] M. S. Collett, A. F. Purchio, and R. L. Erikson, *Nature (London)* **285**, 167 (1980).
[10] A. D. Levinson, H. Oppermann, H. E. Varmus, and J.M. Bishop, *J. Biol. Chem.* **255**, 11973 (1980).
[11] O. N. Witte, N. E. Rosenberg, and D. Baltimore, *Nature (London)* **281**, 396 (1979).
[12] B. Ek, B. Westermark, A. Wasteson, and C.-H. Heldin, *Nature (London)* **295**, 419 (1982).
[13] B. M. Shapiro and E. R. Stadtman, *J. Biol. Chem.* **243**, 3769 (1968).
[14] P. G. Rothberg, T. J. R. Harris, A. Nomoto, and E. Wimmer, *Proc. Natl. Acad. Sci. U.S.A.* **75**, 4868 (1978).
[15] B. M. Sefton, T. Hunter, K. Beemon, and W. Eckhart, *Cell* **20**, 807 (1980).

introduction of a viral tyrosine protein kinase by infection can raise the proportion of acid-stable protein bound phosphate present as phosphotyrosine 5–10 times. In normal cells this proportion ranges from 0.02 to 0.06%, and in transformed and growth factor-treated cells it ranges from 0.10 to 0.50%. Specific phosphoproteins of potential biological relevance can be examined for phosphate esterified to tyrosine by the experimental steps of protein purification, partial acid hydrolysis, and identification of the phosphoamino acids.

It is also possible to screen cellular phosphoproteins for candidates likely to contain phosphotyrosine, and to perform phosphoamino acid analysis on these. Candidates include proteins whose phosphate content increases under conditions where total cellular phosphotyrosine levels are elevated, detected by comparing one- or two-dimensional gel electrophoretograms of ^{32}P-labeled cell proteins. Such comparisons can be facilitated by incubation of the gel in alkali (Table I).[16,17] Since the phosphate ester linkages to tyrosine and threonine are stable to high pH, the pattern of labeled products becomes enriched for phosphates linked to tyrosine and to threonine. Unfortunately, not all phosphotyrosyl residues are resistant, presumably because of their local environment. Proteins identified in this way can be purified directly from the alkali-treated gels or from parallel untreated gels. Additional candidate proteins for analysis can be recognized by immunological procedures. Antisera have been raised against phosphotyrosine and its analogs, and used for protein purification by immunoprecipitation and immunoaffinity chromatography.[18] Proteins containing phosphotyrosine can be identified by "Western" blotting with such antisera and then purified by conventional chromatography procedures, or by one- or two-dimensional gel electrophoresis.

Whatever the source of the proteins for analysis, consideration must be given to the possibilities of (1) artifactual losses of phosphotyrosine and (2) artifactual transfer of phosphate to tyrosine. Both problems can be minimized by working rapidly in the cold and denaturing the proteins as soon as possible. Unscheduled tyrosine protein kinase activity during sample preparation can be minimized by chelating divalent cations essential for kinase activity. In the case of radiolabeled proteins, this problem can be circumvented by reducing the specific activity of precursor ATP with unlabeled ATP. Phosphatases specific for tyrosine are as yet poorly characterized, but may be inhibited by traces of Zn^{2+} (although this is incompatible with the use of chelating agents and buffers containing de-

---

[16] J. A. Cooper and T. Hunter, *Mol. Cell Biol.* **1**, 165 (1981).
[17] Y.-S. E. Cheng and L. B. Chen, *Proc. Natl. Acad. Sci. U.S.A.* **78**, 2388 (1981).
[18] A. H. Ross, D. Baltimore, and H. Eisen, *Nature (London)* **294**, 654 (1981).

TABLE I
Alkali Digestion of Radioactive Proteins Separated in Polyacrylamide Gels[a]

Step 1	After electrophoresis, the gel is stained if necessary and destained in 10% acetic acid, 10% isopropanol. Unstained gels should be fixed in this same solution to precipitate the separated proteins. The stacking gel is removed, if used
Step 2	The gel is rinsed in water for 10–15 min and dried under heat and vacuum onto a piece of Whatman 3MM paper or similar. This removes the acidic solution and allows rapid equilibration with the alkali in the next step. At this stage the positions of radioactive proteins can be detected by autoradiography
Step 3	The dried gel is immersed in a freshly made solution of 66 g 85% KOH (Analytical Grade) per liter of H_2O (i.e., 1 M), allowing about 300 ml of solution per 30 ml gel. The gel shrinks and peels away from the paper on which it was dried. The paper is removed. The container is covered and placed in an oven at 55° for 2 hr with occasional gentle agitation
Step 4	The alkali is then decanted from the gel. Care should be taken with the hot alkali and the ammonia vapor released during hydrolysis of the polyacrylamide. Hydrolysis causes the gels to swell and become fragile, and they must be handled with great care. The alkali is replaced with 10% acetic acid/10% isopropanol solution and the container gently swirled at room temperature for 2 hr. The acidic solution should be changed three or four times
Step 5	The gel should have shrunk to approximately its original size. After 10 min in H_2O the gel is dried as before

[a] This procedure has been applied to proteins contained in many different types of gel matrix, including 10, 15, and 20% polyacrylamide gels cross-linked with 0.13, 0.09, and 0.065% bisacrylamide, respectively, and 10% polyacrylamide gels cross-linked with 0.33% diallyltartardiamide. Polypeptides remain entrapped in the gel but many phosphate esters are hydrolyzed and eluted. Residual phosphate is contained in phosphoserine, phosphothreonine, and phosphotyrosine in the relative ratios 78, 21, and 1%, compared with 94, 6, and 0.2% before treatment [J. A. Cooper and T. Hunter, *Mol. Cell. Biol.* **1**, 165 (1981)].

oxycholate), or by adding free phosphotyrosine or analogs.[19,20] Probably the simplest strategy is, where possible, to lyse cells directly into a denaturing solution, or to stop an *in vitro* reaction with denaturing agents.

Physical and Chemical Properties of Phosphotyrosine

Synthesis of Phosphotyrosine. O^4-Phosphotyrosine is readily synthesized according to the protocol of Rothberg *et al.*[14] which involves heating L-tyrosine with P_2O_5 in H_3PO_4 to 100° for 72 hr. Phosphotyrosine is puri-

[19] J. G. Foulkes, R. F. Howard, and A. Ziemiecki, *FEBS Lett.* **130**, 197 (1981).
[20] D. L. Brautigan, P. Bornstein, and B. Gallis, *J. Biol. Chem.* **256**, 6519 (1981).

fied by ion exchange chromatography. Yields of the free acid are between 60 and 70%. Phosphotyrosine is commercially available from Sigma Chemical Company and Calbiochem-Behring Corporation.

Physical Properties. The molecular weight of the free acid is 261. The ultraviolet absorption spectrum of the free acid in H_2O (pH about 3) has a peak at 265 nm and a distinct shoulder at 270 nm with a molar extinction coefficient of 500. The absorption peak is not significantly shifted at high pH (λ_{max} 268, and 274 nm at pH 12). The spectrum of tyrosine (λ_{max} in H_2O 275 nm, with a shoulder at 281 nm, ε_m 1350) is distinct from that of phosphotyrosine and is significantly shifted in alkali. Phosphotyrosine has a lower fluorescence yield than tyrosine. When excited at 265 nm a solution of phosphotyrosine in H_2O emits maximally at 293 nm with a molar yield of 13,000 (cf. tyrosine in H_2O, excitation 274 nm, λ_{max} 302 nm, ε_m 26,000).

The ionization constants are very similar to those of phosphoserine and phosphothreonine: $pK_1 < 2$ (phosphate), pK_2 2.4 (COOH), pK_3 5.8 (phosphate), pK_4 9.4 (NH_2).

Chemical Properties. Phosphotyrosine is soluble in H_2O at pH 7 at concentrations up to at least 1 M at 25°. Tyrosine is much less soluble (concentration of a saturated solution at 25° is about 2 mM). Phosphotyrosine is reasonably stable as the free acid in aqueous solution at 4°, slowly hydrolyzing to phosphate and tyrosine. It is less stable in strong acid (50% hydrolyzed in 1 M HCl, 5 hr at 100°) than in strong alkali (1% hydrolyzed in 1 M NaOH, 5 hr at 100°).[4] The stabilities in weak acid may be effected by the presence of particular divalent cations.[21,22]

Detection of Phosphotyrosine

Labeled or Unlabeled Protein? Potentially, phosphotyrosine may be assayed in nonradioactive samples through separation by HPLC followed by reaction with fluorescamine or *o*-phthalaldehyde,[23,24] or, in principle, by radioimmunoassay with antiphosphotyrosine sera. Currently, the most sensitive methods involve radioactive labeling although there are situations where this is not practical.

Labeling the Tyrosine Moiety. Appreciable amounts of [^{3}H]phosphotyrosine accumulate in the proteins of cells labeled with [5,6-^{3}H]tyrosine, but the acid-catalyzed proton exchange during hydrolysis means that

[21] E. Bamann, H. Trapmann, and A. Schnegraf, *Chem. Ber.* **88**, 1726 (1955).
[22] T. A. Sundararajan, K. S. V. Sampath Kumar, and P. S. Sarma, *Biochim. Biophys. Acta* **28**, 148 (1958).
[23] J. C. Yang, J. M. Fujitaki, and R. A. Smith, *Anal. Biochem.* **122**, 360 (1982).
[24] G. Swarup, S. Cohen, and D. L. Garbers, *J. Biol. Chem.* **256**, 8197 (1981).

much of the label is lost prior to amino acid analysis. The specific activity of [^{14}C]tyrosine is too low to be a practical alternative. Phosphorylated tyrosines are not likely to be good substrates for chemical or enzymatic radioiodination, due to the electron-withdrawing effect of the phosphate group on the benzene ring.

Labeling the Phosphate Moiety. Proteins can be labeled with ^{32}P or ^{33}P *in vivo* or *in vitro*. The latter isotope is little used. While ^{33}P is more expensive and has somewhat lower specific activity, its lower energy emissions make for higher resolution and lower radiation exposure for the investigator. ^{33}P may be detected by autoradiography, or with increased efficiency by fluorography. For the latter procedure, gels can be impregnated with PPO or thin-layer plates with 2-methylnaphthalene.[25,26] ^{32}P emits energetic β-particles which can be detected efficiently with a fluorescent intensifying screen.[27] For any fluorographic procedure, film should be preexposed to yellow light and exposed at $-70°$ to achieve a linear increase in film absorbance with radioactivity.[25]

Phosphate Labeling in Vivo. Labeling intact cells with radioactive orthophosphate can lead to distorted estimates of the abundance of phosphotyrosine unless care is taken to equilibrate fully the intracellular pools of exchangeable phosphate in ATP, metabolic intermediates, and macromolecules. In general, it appears that tyrosine phosphates in protein turn over more rapidly than serine or threonine phosphates, since short labeling times lead to relatively more incorporation of ^{32}P into total cell phosphotyrosine than into phosphoserine or phosphothreonine, when compared with labeling to "steady state."[28] The apparent abundance of phosphotyrosine, as a percentage of total acid-stable phosphoamino acids, was 1.1% after 2 hr of labeling and 0.4% after 18 hr of labeling of Rous sarcoma virus-transformed cells.[28] [These percentages are not absolute (see below).] Equilibration of the pool is particularly important in experiments where the turnover of the pool of ATP is likely to change during the labeling period, e.g., when treating cells with EGF or serum. When ^{32}P$_i$ is added together with serum, the specific activity of ATP can change 2.4-fold within 1 hr, and the apparent phosphate content of many phosphoproteins changes similarly.[29,30]

For routine analysis we generally label growing cells in 35 mm dishes for 16–18 hr with ^{32}P$_i$ at a concentration of 1 mCi/ml of Dulbecco's modi-

---

[25] W. M. Bonner and R. A. Laskey, *Eur. J. Biochem.* **46**, 83 (1974).
[26] W. M. Bonner and J. D. Stedman, *Anal. Biochem.* **89**, 247 (1978).
[27] R. A. Laskey and A. D. Mills, *FEBS Lett.* **82**, 314 (1977).
[28] T. Hunter, B. M. Sefton, and K. Beemon, *in* "Animal Virus Genetics" (B. Fields, R. Jaenisch, and C. F. Fox, eds.), p. 499. Academic Press, New York, 1980.
[29] M. Nilsen-Hamilton, W. R. Allen, and R. T. Hamilton, *Anal. Biochem.* **115**, 438 (1981).
[30] D. D. Cunningham and A. B. Pardee, *Proc. Natl. Acad. Sci. U.S.A.* **64**, 1049 (1969).

fied Eagle's medium lacking phosphate but supplemented with 4% calf serum (not dialyzed). Serum has a measured [P_i] of 1.5 mM, so our labeling medium has about 6% of the [P_i] in complete Eagle's medium (1 mM P_i). The medium specific activity is thus about 17 Ci/mmol, but this probably changes during equilibration. For preparing individual proteins for analysis, the amount of added $^{32}P_i$ can be raised to 3 mCi/ml but the problem of radiolytic toxicity becomes more serious under these conditions. For a qualitative result with particularly scarce proteins it may be advantageous to label for 2–4 hr in serum-free medium since this reduces the proportion of label which is incorporated into nucleic acid. Disproportional labeling of the phosphoamino acids occurs under these conditions and precise quantification is not possible.

Phosphate Labeling in Vitro. [γ-^{32}P]ATP, [^{35}S]γ-thioATP, or other nucleoside triphosphates can be used as donors to label substrates in vitro. Generally the requirements are Mg^{2+} or Mn^{2+}, neutral pH buffer, and ATP. Monovalent cations appear to be unnecessary. Most tyrosine protein kinases can function at 4° and can use micromolar concentrations of ATP, so the radioactive ATP need not be diluted with cold ATP. A buffer of 20 mM 1,4-piperazinediethanesulfonic acid, pH 7.0, and 10 mM $MnCl_2$ has proven suitable for all applications we have tested including immunoprecipitates of viral transforming proteins and membranes containing EGF and PDGF receptors. Reactions are stopped by chelating agents and immediate preparation for gel electrophoresis.

Sample Preparation: Total Cellular Proteins. In order to estimate the proportion of total cellular protein phosphate esterified to tyrosine, no purification of protein is necessary, since it is possible to separate phosphotyrosine from all other phosphorylated compounds generated by hydrolysis (see below). However, to reduce the risk of overloading the separation system it is advisable to remove the bulk of the nucleic acids, phospholipids, and low-molecular-weight compounds prior to hydrolysis.

For a 35-mm dish of ^{32}P-labeled cells:

1. Wash twice with cold buffered saline and drain the dish thoroughly.
2. Add 0.3 ml RIPA [a solution of 1% Nonidet P-40, 1% sodium deoxycholate, 0.1% sodium dodecyl sulfate, 0.15 M NaCl, 0.01 M sodium phosphate pH 7.0, 1% Trasylol (Mobay Chemical Corporation)] supplemented with 2 mM EDTA.
3. Scrape with a rubber policeman, leave dish at 4° for 10 min to solubilize adherent structures completely, scrape again, and transfer to a 1.5-ml snap-cap tube.
4. Centrifuge at 14,000 rpm in the SS24 rotor (Sorvall) or JA21 rotor (Beckman) (20,000 g) at 2° for 20 min.
5. Transfer supernatant to a 1.5-ml snap-cap tube containing 0.4 ml

of NTE (a solution of 0.1 M NaCl, 0.01 M Tris–HCl, pH 7.5, 0.001 M EDTA) and 0.4 ml buffer-saturated phenol (redistilled), at room temperature.
6. Vortex (full speed) for 30 sec. Centrifuge for 1 min (microcentrifuge) at room temperature.
7. Discard aqueous layer. Reextract phenol layer once with 0.8 ml NTE. Discard aqueous layer, keeping the interface.
8. Transfer phenol layer to a 30-ml glass tube (Corex). Add 13 ml water and 2 ml 100% TCA. Mix well and stand at 0° for 1 hr. Carrier protein is unnecessary since the nonionic detergent from the RIPA buffer acts as a carrier for TCA precipitation.
9. Centrifuge at 10,000 rpm in the HB4 rotor (Sorvall) or JS13 rotor (Beckman), or 8,000 rpm in the SS34 angle rotor (Sorvall), for 10 min at 2°. Decant supernatant from the pellet of protein and detergent.
10. Extract pellet with 5 ml $CHCl_3$/MeOH (2:1) (room temperature) with gentle shaking. Collect protein precipitate with a 10 min centrifugation, as before but at room temperature. Decant supernatant from the translucent protein pellet.
11. Air dry the pellet. Radiation shielding is advisable until this stage. Dissolve the dry pellets in 200 μl of 5.7 M HCl (see below) at 100° for 2 min and transfer to 13 × 100-mm screw-cap tubes (Pyrex) with 2, 100 μl washes of 5.7 M HCl. This is now ready for acid hydrolysis. One microliter can be sampled for counting; 3 × 10^6 dpm releases about 500 dpm of phosphotyrosine (from normal cells) after partial acid hydrolysis (see below for maximum permissible protein loading for two-dimensional separation on thin layer plates).

Purification of Individual Proteins. Individual proteins can be purified from SDS gels of, e.g., immunoprecipitates or *in vitro* kinase reactions, or from two-dimensional gels, by standard procedures of homogenization and TCA precipitation.[31] As few as 100 cpm (Cerenkov) of [^{32}P]phosphoprotein can be analyzed successfully. We normally use 10–20 μg of γ-globulin as carrier for the precipitation. The precipitated protein is washed with cold ethanol or a freshly made 1:1 cold mixture of ethanol and ether, and dried, then dissolved in 5.7 M HCl. Small volumes can be transferred to Reactivials (Pierce), or can be hydrolyzed in 1.5-ml snap-cap tubes, providing the caps are clamped closed. At least 50 μl of HCl should be used.

^{32}P-labeled proteins may also be extracted from gels which have been

---

[31] K. Beemon and T. Hunter, *J. Virol.* **28**, 551 (1978).

incubated in alkali to reveal proteins containing phosphotyrosine and phosphothreonine according to the procedure described in Table I. This treatment causes cleavage of some peptide bonds and the polypeptide is not TCA-precipitable, so isolation of the ^{32}P-labeled peptides from other material extracted from the gel matrix is less straightforward. The following protocol works satisfactorily. It involves the hydrolysis of unpurified ^{32}P-labeled peptides with acid in the gel piece followed by the separation of the released phosphoamino acids from contaminants.[32]

1. Cut a piece from alkali-treated gel. Incubate for 1 hr (see below) in 1 ml 5.7 M HCl at 110° in a 13 × 100-mm screw-cap glass tube.
2. Add 4 ml H$_2$O, centrifuge at 2000 rpm in bench-top centrifuge for 10 min, transfer supernatant, freeze, and lyophilize over NaOH.
3. Dissolve yellow residue in 10 ml H$_2$O. Add 0.4 ml of 50% (v/v) slurry of Dowex AG1-X8 (Cl$^-$ form). Adjust pH of mixture to 7.5–8.5 with 10 mM NH$_4$OH.
4. Add 1 μl of a solution containing 1 mg/ml of each phosphoserine, phosphothreonine, and phosphotyrosine (unlabeled). These will allow estimation of the recoveries of phosphoserine, phosphothreonine, and phosphotyrosine during the subsequent steps, and act as internal markers for the phosphoamino acid separation.
5. Shake at room temperature for at least 4 hr.
6. Allow resin to settle. Wash the resin into a small "column" made from a blue pipette tip fitted with a 6-mm porous plastic sinter (Omnifit, Atlantic Beach, New York). Wash twice with 0.5 ml H$_2$O. Count the column plus contents by Cerenkov radiation in a scintillation counter to check that most of the ^{32}P has bound to the resin.
7. Elute the phosphoamino acids with two applications of 0.5 ml of 0.1 M HCl. The "column" may be inserted in a 1.5-ml snap-cap tube and centrifuged briefly to remove all liquid.
8. Clarify the eluate if necessary, count, and lyophilize.

Hydrolysis. Phosphoamino acids may be released by partial acid, enzymatic, or alkaline hydrolysis. We have little experience with enzymatic hydrolysis but suspect that it rarely goes to completion, possibly because the peptide bonds adjacent to phosphoamino acids are poor substrates. We routinely use acid hydrolysis, but proteolytic digestion followed by acid hydrolysis, or alkaline hydrolysis alone, may give superior yields of phosphotyrosine.[33] Complete hydrolysis is not satisfactory since the phosphate esters are labile to both acid and base. Recovery varies with the nature of individual residues. Our experience suggests that artifactual

---

[32] D. Templeton, personal communication.
[33] T. M. Martensen, and R. L. Levine, this volume [43].

TABLE II
RECOVERY OF [^{32}P]PHOSPHOAMINO ACIDS BY ACID HYDROLYSIS[a]

Hydrolysis time (hr)	Phosphotyrosine	Phosphothreonine	Phosphoserine
1	0.24	5.67	94.13
2	0.15	8.25	91.70
4	0.08	13.89	86.03

[a] Rous sarcoma virus-transformed chicken embryo cells were labeled for 18 hr with ^{32}P$_i$ and their phosphoproteins were hydrolyzed for 1, 2, or 4 hr in 5.7 M HCl at 110°. The recovery of each phosphoamino acid is expressed as a percentage of the total radioactivity recovered in all three phosphoamino acids, which was 21, 24, and 18%, respectively, of the total radioactivity in the sample [T. Hunter and B. M. Sefton, in "Protein Phosphorylation and Bio-Regulation" (G. Thomas, E. J. Podesta, and J. Gordon, eds.), p. 193. Karger, Basel, 1980)].

transfer of phosphate to or from phosphotyrosine does not occur during acid hydrolysis.

Hydrolysis in 5.7 M HCl at 110° for 2 hr releases an estimated 25% of total protein phosphotyrosyl residues as phosphotyrosine. The remainder is found as free P$_i$ (50%) and phosphotyrosyl peptides (25%). Shorter times of hydrolysis (1 hr) favor the recovery of phosphotyrosine, so we commonly use 1 hr hydrolysis when working with minute amounts of labeled protein. Longer times of hydrolysis (4 hr) increase the relative yield of phosphothreonine (Table II),[34] and decrease the yield of phosphotyrosine and an unidentified compound running close to phosphotyrosine on thin-layer electrophoresis (see below).

Acid Hydrolysis Procedure

1. To the dry protein, contained in a 13 × 100-mm glass tube (Pyrex) or Reactivial (Pierce) or 1.5-ml snap-cap tube, add a total of 50–400 μl of constant-boiling HCl (redistilled, or by dilution of analytical grade concentrated HCl to 5.7 M).
2. Dissolve if necessary by 1–2 min of incubation at 100° and vortexing. Spin down all liquid to bottom of tube.
3. It is not necessary to hydrolyze under an inert atmosphere. Cap the tubes tightly (clamp snap-cap tube caps closed) and place in 110°C oven for desired time.

[34] T. Hunter and B. M. Sefton, in "Protein Phosphorylation and Bio-Regulation" (G. Thomas, E. J. Podesta, and J. Gordon, eds.), p. 193. Karger, Basel, 1980.

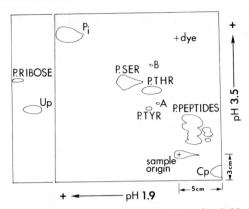

FIG. 1. Separation of phosphoamino acids by two-dimensional thin-layer electrophoresis. Diagram of a thin-layer plate on which a partial acid hydrolyzate of partially-purified ^{32}P-labeled cell proteins (see text) has been analyzed as described in Table III. At left are shown phosphate-containing compounds present in the sample but which migrate off the plate during the pH 1.9 electrophoresis. A and B are radiolabeled compounds of unknown structure. After elution from the cellulose and further hydrolysis they generate phosphoserine.

4. After hydrolysis, centrifuge liquid to bottom of tube. Evaporate to dryness under vacuum over NaOH pellets. It is not possible to freeze constant-boiling HCl, but it will evaporate without "bumping" if not cooled. A Speedivac (Savant) may be used.

Hydrolysis with alkali (5 M KOH at 155° for 1 hr) gives almost quantitative release of phosphotyrosine[33] but completely destroys phosphoserine and phosphothreonine due to base-catalyzed β-elimination. The hydrolyzates are suitable for HPLC analysis or a regular amino acid analyzer, but for thin-layer electrophoresis the KOH needs to be removed by established procedures.[35]

Separation of Phosphoamino Acids. Phosphoamino acids may be separated by HPLC or by thin-layer or paper electrophoresis. We will only concern ourselves with the separation of the three phosphorylhydroxyamino acids. Since hydrolyzates of proteins from ^{32}P-labeled cells frequently contain degradation products of nucleic acids and phospholipids, consideration must be given to the behavior of these compounds in the separation system. The chief contaminants appear to be 3'-UMP, 3'-CMP, ribose and deoxyribose phosphates, as well as phosphopeptides and free P_i. Lesser quantities of two unidentified compounds are also detected (see Fig. 1). Both give rise to [^{32}P]phosphoserine upon subse-

[35] H. Gould and H. R. Matthews, *in* "Laboratory Techniques in Biochemistry and Molecular Biology" (T. S. Work and E. Work, eds.), p 226. Elsevier, Amsterdam, 1976.

quent analysis. One of these compounds has similar electrophoretic mobilities to phosphotyrosine, so a highly resolving separation system is necessary.

In all separations of radioactive phosphoamino acids, unlabeled phosphoamino acid markers should be included as internal markers. Lyophilized hydrolyzates are generally dissolved in pH 1.9 buffer (Table III) containing 0.3–0.5 µl of a solution of 1 mg/ml each phosphoserine, phosphothreonine, and phosphotyrosine.

HPLC Systems. Phosphotyrosine is resolved from phosphoserine and phosphothreonine, but runs with acidic amino acids, on the cation-exchange column of the amino acid analyzer (isocratic elution 0.1 M citrate pH 2.5).[36] All three phosphorylhydroxyamino acids can be resolved on anion-exchange columns.[23,24]

Thin-Layer Systems. Phosphotyrosine differs significantly from phosphoserine and phosphothreonine in its hydrophobic benzene ring, and its greater mass. These enable separation on cellulose thin layer plates (100 µm, E. M. Laboratories, or similar) by chromatography and by electrophoresis at pH 3.5, where all three phosphorylhydroxy amino acids have a similar charge. Phosphotyrosine also presumably differs from phosphoserine and phosphothreonine in its first pK_a since they can be separated at pH 1.7,[37] but we have been unable to measure this ionization accurately. Phosphotyrosine cannot be separated from phosphothreonine at pH 1.9, the pH used traditionally for separation of phosphoserine and phosphothreonine. A summary of buffer compositions, running conditions, and mobilities is given in Table III.

When phosphate, phosphopeptides, and phosphoamino acids are the only labeled compounds present, e.g., in hydrolyzates of proteins labeled *in vitro* kinase reactions, one-dimensional separation by electrophoresis at pH 3.5 is adequate.

When proteins labeled *in vivo* are analyzed, separation by pH 3.5 electrophoresis alone is inadequate, since 3′-UMP comigrates with phosphotyrosine.[28,38] Two-dimensional separation by electrophoresis at pH 1.9 followed by electrophoresis at pH 3.5 is preferable. (In principle, it would be possible to perform the electrophoresis dimensions in the other order, but it is better to do the pH 1.9 dimension first because this buffer has a greater capacity.) A schematic representation of the mobilities of the phosphoamino acids and nucleotides is shown (Fig. 1). Whole cell protein samples should be analyzed on separate plates, but hydrolyzates of four

---

[36] P. Böhlen, personal communication.
[37] H. Ushiro and S. Cohen, *J. Biol. Chem.* **255**, 8363 (1980).
[38] M. Manai and A. J. Cozzone, *Anal. Biochem.* **124**, 12 (1982).

TABLE III
THIN-LAYER CELLULOSE SEPARATIONS OF PHOSPHOAMINO ACIDS[a]

Buffer	Voltage (kV)	Time (min)	Current (mA)	Mobilities (cm)		
				Phospho-tyrosine	Phospho-threonine	Phospho-serine
1.7[b]	1.5	30	50	1.7	1.3	3.2
1.9[c,d]	2.5	25	50	4.0	4.0	6.5
3.5[d,e]	2.0	20	35	6.0	8.0	9.0
1.9[c,f]	1.5	20	30	2.0	2.0	3.0
3.5[e,f]	1.3	16	25	3.0	3.8	4.0
Chromatography[g]	—	—	—	5.7	5.1	4.5

[a] Information given is for 20 × 20 cm 100 μm cellulose thin-layer plates, either plastic or glass backed. The electrophoresis apparatus used was a custom built cold-plate apparatus, fitted with an air bag device to apply pressure to the top surface of the thin-layer plate to prevent wicking of the buffer and consequent overheating. The voltages may need to be decreased and the times of electrophoresis increased for other apparatus designs. On our apparatus each end of the glass plate is linked to an electrode chamber by about 10 cm of Whatman 3MM paper (double thickness) wetted with buffer, and this should be taken into account in choosing suitable voltages.
[b] pH 1.7 buffer is 7% formic acid.
[c] pH 1.9 buffer is 50 : 156 : 1794 88% formic acid : glacial acetic acid : H_2O.
[d] Conditions for electrophoresis of one sample per plate (Fig. 1).
[e] pH 3.5 buffer is 10 : 100 : 1890 pyridine : glacial acetic acid : H_2O.
[f] Conditions for electrophoresis of four samples per plate (Fig. 2).
[g] Chromatography buffer is 5 : 3 isobutyric acid : 0.5 M ammonium hydroxide.

purified individual proteins can be analyzed on one plate if the origins are in the configuration shown in Fig. 2.

General Procedures for Thin Layer Electrophoresis

1. Sample application. Samples are dissolved in a minimum volume of pH 1.9 buffer (Table III) containing 0.3–0.5 μg of each unlabeled phosphorylhydroxyamino acid if the entire sample is to be loaded, or proportionately more if only a fraction is to be loaded. Samples are spotted onto the origin in 0.3- to 0.5-μl aliquots from a microliter syringe or micropipetter (Oxford). Each application is dried with a stream of cold air.
2. A dye marker (0.5 μl) containing 1 mg xylene cyanol FF and 5 mg of εDNP-lysine per ml can be applied to an origin distant from the sample (Fig. 1).
3. Wetting the plate. For one-dimensional electrophoresis and the second dimension of a two-dimensional separation, where the ori-

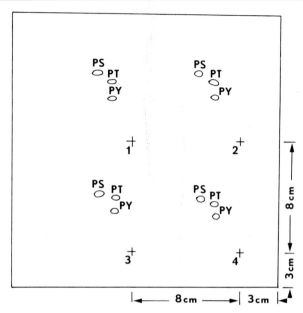

FIG. 2. Analysis of four samples by two-dimensional electrophoresis on a single thin-layer plate. The positions of nonradioactive phosphoamino acid standards, located with ninhydrin, are shown (PS, phosphoserine; PT, phosphothreonine; PY, phosphotyrosine).

gins are on a line, the plate is wetted on both sides of the origin line with two pieces of Whatman 3MM filter paper. The pieces are wetted in electrophoresis buffer, excess buffer removed, laid on both sides of the origin about 5–10 mm distant, and the buffer is allowed to creep up to the origin line. This concentrates the samples. For two-dimensional electrophoresis, where the origin(s) is one or several spots, a "blotter" is constructed from Whatman 3MM filter paper.[39] Two sheets 25 × 25 cm are sewn together, around the perimeter, and holes are made with a 0.5-in. cork borer at positions corresponding to the sample(s) and dye marker. The blotter is wetted in pH 1.9 buffer, blotted to remove excess buffer, and applied to the plate. The buffer should converge as an even circle on to the origin.

4. The wetted plate should appear dark but without puddles of liquid. It is promptly assembled into the electrophoresis apparatus and electrical contact made with the buffer tanks.

[39] Whatman 3MM or similar filter paper is preferable to high-rag blotting paper, since the latter appears to contain compounds which affect the phosphoamino acid separation.

5. After electrophoresis, the plate is dried quickly at room temperature if further separation is required, or in a 65° oven. Unlabeled markers are located on the dry plate by spraying with 0.2% ninhydrin in acetone and returning to the oven to develop the color (10–15 min).

Paper Electrophoresis Systems. Paper electrophoresis has certain advantages. It allows higher protein loads than thin-layer electrophoresis. It also has the advantage that multiple two-dimensional separations can be run in parallel after electrophoresis at the first pH. The phosphoamino acid regions from several samples can be cut out and sewn on to a fresh paper sheet for the second dimensions.[40] The same buffers are used as with thin-layer electrophoresis (Table III).

Quantitation of Phosphoamino Acids Separated on Cellulose Thin-Layer Plates. [^{32}P]Phosphoamino acids can be located on thin-layer plates by autoradiography. Overloading is manifested by streaking in the first dimension. Not more than 50 μg of protein hydrolysate should be loaded on a single origin. Care must be taken to distinguish the phosphotyrosine spot from the nearby spot A (Fig. 1). With low levels of radioactivity, quantitative information can be derived by densitometry of the film. For highly radioactive samples, the areas corresponding to the individual phosphoamino acids can be marked with pencil, after aligning the plate and film using radioactive ink marks. The cellulose is then simultaneously scraped and collected with a blue pipet tip fitted with a 6-mm plastic sinter (Omnifit). The tip is connected to a vacuum line and used like a vacuum cleaner. The tip is then stood (narrow end up) in a 1.5-ml snap-cap tube and 100 μl pH 1.9 buffer is added. The liquid is centrifuged through the cellulose into the tube, and the elution repeated with two further applications of buffer. The contents of the tube are then transferred to a scintillation vial and 0.2 ml of H_2O and 10 ml of aqueous scintillant are added.

Expression of Quantitative Data. Quantitation of the abundance of phosphoamino acids in a total cell protein or individual protein sample is not absolute because of differential stabilities and recoveries of the phosphoamino acids. We express the radioactivity in each phosphoamino acid as a percentage of the total radioactive phosphoamino acids recovered. In this way one normalizes for losses. Duplicate samples, particularly in the case of whole cell protein analysis, yield very similar results.

Sequencing Phosphorylated Peptides. Peptides containing [^{32}P]phosphotyrosine can be subjected to automated spinning cup sequence analysis (Beckman Sequenator, 0.3 M Quadrol program).[41] The cyclization and

[40] R. Martinez, K. D. Nakamura, and M. Weber, *Mol. Cell. Biol.* **2**, 653 (1982).
[41] T. Patschinsky, T. Hunter, F. Esch, J. A. Cooper, and B. M. Sefton, *Proc. Natl. Acad. Sci. U.S.A.* **79**, 973 (1982).

cleavage of the phosphotyrosine residue to form a ATZ-phosphotyrosine derivative occurs normally. (This is in contrast to what occurs with phosphoserine and phosphothreonine where cyclization results in β-elimination of the phosphate.) The ATZ-phosphotyrosine, however, is inefficiently extracted by the butyl chloride, since the phosphate moiety is ionized under the acid conditions of extraction. Usually only 2-5% of the ATZ-phosphotyrosine is extracted at the appropriate cycle. A similar fraction is extracted at each subsequent cycle. Thus the radioactive sequence profile for a peptide containing a single phosphotyrosine shows a peak with a long trail, which could mask a second phosphotyrosine residue. The ATZ-phosphotyrosine derivative can be converted to the PTH-phosphotyrosine derivative by heating at 80° in 0.1 N HCl for 10 min. The nature of the extracted radioactivity at each cycle can thus be confirmed, following conversion, by electrophoresis at pH 3.5 on a thin-layer cellulose plate. A PTH-phosphotyrosine marker can be synthesized according to conventional procedures.[42]

[42] J. Sjöquist, *Biochim. Biophys. Acta* **41**, 20 (1960).

[43] Base Hydrolysis and Amino Acid Analysis for Phosphotyrosine in Proteins

By TODD M. MARTENSEN and RODNEY L. LEVINE

The phosphorylation of certain protein tyrosyl residues by special kinases is closely linked to retrovirus-induced cell transformation[1,2] as well as the response of cells to growth factors[3-5] from several sources. The tyrosine phosphate (Tyr-P) content of normal cellular protein is less than 0.1% of the sum of serine phosphate and threonine phosphate; special techniques are required to identify and quantify this rare phos-

[1] J. E. Smart, H. Opperman, A. P. Czernilofsky, A. F. Purchio, R. L. Erikson, and J. M. Bishop, *Proc. Natl. Acad. Sci. U.S.A.* **78**, 6013 (1981).
[2] T. Patschinsky, T. Hunter, F. S. Esch, J. A. Cooper, and B. M. Sefton, *Proc. Natl. Acad. Sci. U.S.A.* **79**, 973 (1982).
[3] H. Ushiro and S. Cohen, *J. Biol. Chem.* **255**, 8363 (1980).
[4] F. H. Reynolds, Jr., G. J. Todaro, C. Fryling, and J. R. Stephenson, *Nature (London)* **292**, 259 (1981).
[5] B. Ek, B. Westermark, A. Wasteson, and C. H. Heldin, *Nature (London)* **295**, 419 (1982).

phoamino acid.[6] Taking advantage of the fact that Tyr-P is relatively stable to base,[7] a method employing rapid digestion of protein,[8] by which 80% of the Tyr-P in proteins can be recovered and quantified by means of an amino acid analyzer,[9] has been developed.

Methodology

Protein Hydrolysis and Amino Acid Analysis. Protein hydrolyses are carried out at 1–10 mg/ml (bovine serum albumin is added as carrier, if necessary) by dilution of protein solutions in water with an equal volume of 10.0 N KOH or by the addition of 5.0 N KOH to lyophilized samples. The KOH solution is stored in a tightly sealed plastic bottle. Hydrolyses with volumes greater than 1 ml are carried out in Wheaton 1-dram borosilicate vials (No. 224882). Hydrolyses with smaller volumes are carried out in 0.3 or 1.0-ml Reacti-Vials (Pierce). Sample volume should be at least 20% of the total volume of the vial. Cover the vials with unlined caps (PGC Scientific, Rockville, Md.) containing a Teflon/silicone cap liner (Pierce No. 12712). If the usual rubber-lined caps are used, remove the liner. (A narrow metal spatula blade permits easy removal of the rubber liner. The caps should then be boiled in water to remove loosely adhering particles.) The cap should be seated firmly, without overtightening. Place the capped vials in a heating block (Pierce Reacti-Block C for the 1-dram vial, and Reacti-Block A for the 1-ml and 0.3-ml vials) seated in a standard small bench top heater set at 155°. After incubation at 155° for 35 min in the *covered* heating block, remove the vials with forceps and place them in an unheated block to cool. Transfer the contents of each sample to conical test tubes (1–3 ml; Bellco Glass), and rinse each vial with distilled water equal to the sample volume. Place the tube containing the combined hydrolysate and wash on ice, and add 5.0 N HClO$_4$ equivalent to the volume of 5.0 N KOH, mix; then leave the tube on ice for a few minutes. Check the pH of the sample by spotting ~1 μl on pH indicator paper. If the pH is not 1.5–8, adjust it with a few microliters of 5.0 N HClO$_4$ or 5.0 N KOH. Centrifuge the sample at ~1000 g for 5 min at 0–4° to remove insoluble perchlorate and silicate. Remove an aliquot of the supernatant and mix it with one-tenth its volume of 2.0 N sodium formate, pH 2.0. Samples are normally stored at 4° prior to amino acid analysis. Samples put on ice or frozen will throw out more precipitate which will settle out by tapping the sample tube but may necessitate recentrifugation. Amino

---

[6] J. A. Cooper, B. M. Sefton, and T. Hunter, this volume [42].
[7] R. H. A. Plimmer, *Biochem. J.* **35,** 461 (1941).
[8] R. L. Levine, *J. Chromatogr.* **236,** 499 (1982).
[9] T. M. Martensen, *J. Biol. Chem.* **257,** 9648 (1982).

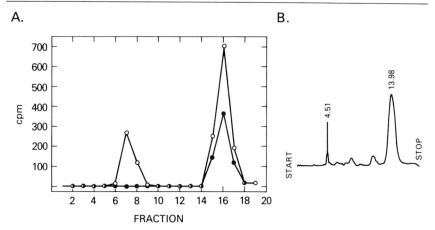

FIG. 1. Elution profile of a hydrolysate of phosphotyrosylglutamine synthetase applied to an amino acid analyzer. A sample of [^{32}P]phosphotyrosylglutamine synthetase (0.35 mg) was digested in 0.15 ml of 5.0 N KOH containing 0.12 μM [^{14}C]Tyr-P for 35 min at 155°, neutralized with HClO$_4$, and applied to the amino acid analyzer. Upon injection, a fraction collector was simultaneously activated to collect the eluate from the analyzer. (A) ^{32}P (○) and ^{14}C (●) radioactivity found in eluate from analyzer (~0.45 ml) collected at 1.0-min intervals. (B) Recorder tracing from analyzer. From T. M. Martensen, *J. Biol. Chem.* **257**, 9648 (1982).

acid analyses are done with an amino acid/peptide analyzer (Dionex) equipped with a stainless-steel microbore column (0.4 × 15 cm) packed with DC-5A resin (Dionex). The column is eluted with 67 mM sodium citrate buffer, pH 2.0, or preferably with 0.2 M sodium formate, pH 2.0, at a flow rate of 18 ml/hr. The eluate is mixed with *o*-phthalaldehyde in borate buffer[10] which enters the reaction coil at 9 ml/hr. The fluorescent amino acid derivatives are detected in a fluorometer (Gilson) connected to a recording integrator (Shimadzu). Tyr-P is measured after injection of the neutralized hydrolysate onto the column preequilibrated for 15 min with buffer. After elution for 15 min the column is washed with 0.1 N NaOH, 0.1 N NaCl, 3.5 mM EDTA for 10 min prior to reequilibration. The detection system can reliably measure 50–100 pmol.

Discussion

The half-life for phosphoryl bond hydrolysis of Tyr-P in 3 N HCl and 5 N KOH at 155° is 7.0 and 116 min, respectively.[9] Complete hydrolysis of proteins under these conditions requires minimally 20 min;[10] the recovery of Tyr-P after 30–35 min of base digestion is ~80%. Constant recoveries

[10] P. E. Hare, this series, Vol. 47, p. 3.

are obtained only when the sample volume is ≥20% of the volume of the hydrolysis vial. Protein concentration of hydrolysis samples is kept >1 mg/ml to minimize the risk of destruction of Tyr-P during the hydrolysis and workup. Also, since dilution of the sample from hydrolysis to analysis is 1:3.1, a lyophilization step may be avoided by using small volumes with high protein concentrations. Hydrolysates must be neutralized to pH 7–9 before lyophilization to prevent destruction of Tyr-P by $HClO_4$.

Figure 1 illustrates the utility of using an amino acid analyzer for the detection and quantification of Tyr-P in proteins. The amount of Tyr-P in [^{32}P]phosphotyrosylglutamine synthetase was quantified from the area of the peak which emerged at 13.98 min. All of the normal amino acids are retained on the analyzer resin at pH 2.0. Phosphoserine, phosphothreonine, cysteic acid and UMP would emerge at approximately the position of $^{32}P_i$ shown in Fig. 1 eluting well before Tyr-P. The major ^{32}P-labeled product of the base digest was shown to coelute with authentic [^{14}C]Tyr-P. Inclusion of trace amounts of [^{14}C]Tyr-P as an internal standard makes quantification easier, and supports the identification of ^{32}P-labeled material emerging from the analyzer as Tyr-P. The detection of [^{32}P]Tyr-P in base hydrolysates of proteins eluted from sodium dodecyl sulfate gels was possible by this technique.[9] The detection of Tyr-P in trichloroacetic acid precipitates of crude cell extracts incubated with [^{32}P]ATP by this method is not recommended unless lipid and nucleic acid contaminents are first removed.

Author Index

Numbers in parentheses are footnote reference numbers and indicate that an author's work is referred to although the name is not cited in the text.

A

Adelstein, R. S., 286, 288
Affolter, H. U., 369
Agard, D. A., 233, 234(3), 235(3), 241(2), 243
Aiba, H., 161
Aitken, A., 38(8), 60, 248, 249(18), 250(18), 337, 339(1), 344, 345(4, 5)
Alaba, J. O., 260, 267(12)
Alexander, M. C., 21, 23, 26(8), 27(8), 28(8)
Alexander, R. L., 231(15)
Alfano, J., 184
Allen, W. R., 392
Allfrey, V. G., 139
Allison, W. S., 141, 142
Alma, N. C. M., 105
Alper, S. L., 21
Ames, B. N., 19
Ames, G. F. L., 28
Amy, C. M., 21, 25, 26(12), 27(12), 35(12)
Anderson, L., 31
Anderson, N. G., 31, 33, 36
Anderson, W. B., 202, 204
Anderson, R. G. G., 289, 297(12)
Andree, P. J., 104
Anfinsen, C. B., 126, 150
Annamalai, A. E., 146
Antoniw, J. F., 338, 339
Antonoff, R. S., 154
Armstrong, R. N., 93, 97(6), 98(7), 99(7), 101(7), 102(7), 111(6, 7), 113(7, 17)
Aromatorio, D. K., 287
Ashby, C. D., 5, 77, 80, 82, 90(14), 92, 177, 300
Assimacopoulos, F. D., 228, 231(7), 299
Atmar, V. J., 367, 369, 370(9, 10, 16), 371(9, 10), 372
Avruch, J., 21, 23, 25, 26(8, 15), 27(8, 15), 28(8)

B

Baglioni, C., 347
Balaram, P., 106
Balkow, K., 346
Baltimore, D., 373, 374, 376, 378(2), 388, 389
Bamann, E., 391
Banargee, D., 205
Barany, G., 121, 122
Bardawill, C. J., 332
Barker, W. C., 146
Barrera, C. R., 336
Barrio, J. R., 162, 164(2, 3)
Bass, M. A., 243, 268, 275, 276(38), 300, 338, 339(10)
Baum, B. J., 21, 25, 26(11)
Bear, J. L., 207
Beavo, J. A., 8, 15, 16(9), 17(9), 20(9), 51, 55, 56, 63, 120, 127(9), 130(11), 137(8, 9, 10), 138(8, 10), 141, 151(1), 156, 158, 164, 166, 177, 186, 187, 188, 208, 212, 239
Bechtel, P. J., 15, 16(9), 17(9), 20(9), 55, 56, 63, 156, 164, 176, 177, 187, 188, 208, 239, 259, 267
Beckman E., 263, 263(18), 267(18)
Beebe, S. J., 80
Beemon, K., 388, 392, 394, 398(28)
Beham, R. A., 77
Bell, R. L., 289
Benjamini, E., 45, 120, 130(5), 137(5), 138(86), 275
Benjamin, W. B., 21
Bennett, H. P. J., 133
Bennett, W., 268
Benson, E. M., 328
Berandsen, H. J. C., 104
Berglund, L., 120, 130(4), 134(4), 215

Bernstein, I. A., 369
Berry, M. N., 22
Bethione, J. L., 150
Bilham, T., 38(8), 337, 345(4, 5)
Billah, M. M., 289
Bingham, E. W., 268, 308, 315(5), 317
Bishop, J. M., 388, 402
Bitte, L., 9
Blackmore, P. F., 299
Blake, J., 127, 128
Bloom, B. R., 197
Blumenthal, D., 286
Bodanszky M., 120, 129, 130(61)
Btohlen, P., 134, 398
Bolen, D. W., 94, 113(17), 139
Bolton, A. E., 357
Bonner, W. M., 392
Borland, M. K., 25, 26(14), 27(14), 299
Bornet, E. P., 268
Bornstein, P., 15, 387, 390
Bossinger, C. D., 126
Bossinger, J., 36, 242
Bothner-By, A. A., 103, 104(31, 32), 106
Bownds, D., 362
Brackett, N. L., 138, 208, 211(16), 212(16), 289, 297(12)
Bradford, M. M., 341, 383
Bramson, H. N., 93, 94, 99(26, 27), 100(27), 103(26, 27), 108(26, 27), 110(27), 111(10, 27), 112(9, 10), 113(10, 17), 114(10), 115(10), 120, 138, 139, 141
Bratvold, G. E., 252
Braun, R., 369
Brautigan, D. L., 15, 260, 267(12), 390
Bray, G. A., 272
Breakefield, X., 161
Brenner, M., 125
Breslow, E., 106
Brooker, G., 26
Brostrom, C. O., 81, 231(15, 17), 246, 253, 268, 273
Brown, B. M., 366
Brown, E., 93
Brown, J. R., 336
Browne, C. A., 133
Browning, E. T., 21, 23, 25(6), 26(6), 35(6)
Browning, M., 268
Brumley, F. T., 299
Brunner, J. R., 308
Brunton, L. L., 77

Buhrow, S. S., 379
Builder, S. E., 8, 158, 166, 188
Bukowski, M., 197, 198(8), 199(8), 200(8), 201(8), 202, 203(8), 205
Burchell, A., 250, 252(2), 260
Burgett, M. W., 335
Busch, M. A., 23, 26(8), 27(8), 28(8)
Buss, J. E., 9, 10(7), 19, 208, 387
Butler, J. R., 336
Bylund, D. B., 9, 272

C

Cabral, F., 202, 204
Caillibot, V., 198
Calkins, D., 5, 77, 80, 177, 300
Callis, B., 390
Carlson, G. M., 134, 259, 263, 267, 272
Carmichael, D. F., 319
Carnegie, P. R., 120, 134(2, 3)
Carpenter, G., 379, 380(7), 381(5), 384(7), 385(4), 388
Casillas, E. R., 372
Casnellie, J. E., 162, 387
Castagna, M., 297
Caudwell, F. B., 39, 337, 338
Cavadore, J.-C., 129
Cavallo, P. F., 124
Cavanagh, H. D., 362
Chalklay, R., 208, 290
Chambaz, E. M., 318, 328(5)
Chan, K.-F. J., 260, 261(7, 8), 262(7), 263(7, 8, 9, 10), 264(8), 265(7, 8), 266(8)
Channabasavaiah, K., 124
Chappell, J. B., 81, 253, 368
Chau, V., 138
Chen, J. I., 279
Chen, L. B., 389
Cheng, Y.-S. E., 389
Cherrington, A. D., 228, 231(7)
Cheung, W. Y., 280
Chin, B., 369
Chinkers, M., 379, 383(8), 384(8), 385(8), 387
Chiu, V. S., 207
Chock, P. B., 138
Chou, C. K., 184
Chow, R. C. L., 130
Chowdhry, V., 148

Chrambach, A., 28
Ciecuira, S. J., 198
Cimadevilla, G. M., 346
Clark, M. G., 229
Claus, T. H., 15, 16(11), 17(11), 18(11), 19(11), 20(11), 213, 214(2), 215, 217, 219
Cleland, W. W., 5, 94
Cochet, C., 318, 328(5)
Coffino, P., 60, 234, 235(5)
Cohen, P., 15, 38(8), 39, 43, 60, 187, 243, 244, 246, 247, 248, 249, 250, 252, 259, 260, 261(5, 6), 263(5), 267(2), 268, 269, 274, 276, 337, 338, 339, 340, 341(6), 342(6), 343(6, 15), 344, 345(4, 5, 6), 367
Cohen, P. T. W., 248, 249(14), 250, 252(2), 260, 261(6), 269
Cohen, S., 15, 146, 379, 380, 381, 383(8), 384(7, 8), 385, 386, 387, 388, 391, 398, 402
Cohn, M., 93, 94, 95(18, 19), 96, 98(1), 106
Cole, H. A., 268
Colescott, R. L., 126
Coletti-Previero, M. A., 129
Collett, M. S., 388
Colman, A., 225
Colman, R. F., 141, 142(7), 146
Condon, G. D., 337, 345(4)
Connolly, T. N., 96
Conti, M. A., 288
Cook, P. F., 3, 4(4), 6(4)
Cook, P. I., 126
Cooper, J. A., 389, 390, 401, 402, 403
Cooper, R. H., 250, 258, 259, 336
Corbin, J. D., 3, 15, 16, 17(16, 17, 18), 18, 19, 51, 52, 54, 55, 56, 57(6), 60, 63, 64(6), 65(5), 72, 82, 119, 135, 148, 151, 158, 159, 162(12), 164, 168, 169, 170(1, 2, 6), 171(1, 3), 172(3, 6), 173(3, 6), 175, 176, 178, 179(6), 186(6), 188, 203, 207, 208, 213, 227, 228, 231(5, 6, 7, 13, 14, 15, 17, 18), 232, 254, 301
Cori, G. T., 253, 261
Corley, L., 126
Cornelius, R. D., 94
Cornell, N. W., 24
Cosand, W. L., 121, 122
Costa, E., 80
Costa, M. R. C., 161
Coulter, B. P., 36
Cozzone, A. J., 398

Craven, G. R., 150
Crestfield, A. M., 150
Cuatrecasas, P., 289, 300
Cunningham, D. D., 392
Curtis, D. R., 366
Czernilofsky, A. P., 402

D

Dabrowska, R., 287
Dahmus, M. E., 313
Daile, P., 120, 134(2, 3)
Dairman, W., 11, 134
Daniel, V., 197
Daniels, G. R., 367, 369, 370(9, 16), 371(9), 372(9)
Das, A., 347, 359(30)
Das, H. K., 347, 359(30)
Dasgupta, A., 373, 378(2)
Datta, A., 347
David, M. M., 332
Davidson, E. A., 43
Davies, D. R., 217
Davies, P. J. A., 197
Davis, B., 89, 90
Davis, P. F., 336
Davis, R. H., 367
Dawes, J., 362
Dayhoff, M. O., 146
Dean, P. D. G., 312
Deaven, L. L., 197
Dedman, J. R., 92
DeGrado, W. F., 138
de Haro, C., 347
DeLange, R. L., 268
Demaille, J. G., 80, 83(2), 84(2), 90, 207, 286
DePaoli-Roach, A. A., 243, 268, 275
Desmeules, P., 98, 99(26), 100, 103(26), 108(26)
Desmond, J., 133
de Vellis, J., 26
DeWald, H. A., 131
DeWulf, H., 299
Dills, W. L., 16, 17(18), 56, 63, 64(6), 164, 175, 208, 239
DiSalvo, J., 286
Dixon, G. H., 336
Donella-Deana, A., 248, 249(18), 250(18)

Dorman, L. C., 124, 130
Doscher, M. S., 121
Doskeland, A., 218
Doskeland, S. O., 169, 170(6), 172(6), 173(6), 218
Douglas, W. W., 21
Dreyer, W. J., 362, 363(1), 365(1)
Drummond, G. I., 92
Dubois, M. F., 347
Dubs, A., 104, 105(38), 108(38)
Duewer, T., 244, 246(10)
Dumont, J. N., 224
Dunaway-Mariano, D., 94
du Vigneaud, V., 129, 130(61)

E

Ebner, K. E., 308
Eckhart, W., 388
Edwards, M. A., 328
Ehrenfeld, E., 346
Eigenbrodt, E., 215
Eisen, H., 389
Eksteen, R., 133
Eley, M. H., 336
El-Maghrabi, M. R., 15, 16(11), 17(11), 18(11), 19(11), 20(11), 213, 214(2), 215, 217, 219
Embi, N., 39, 243, 276, 337, 339, 344(7), 345(4)
Engelhard, M., 121, 122, 124
Engle, J. L., 93, 97(3), 98(3), 109(3), 113(3)
Engström, L., 120, 130(4), 134(4), 213, 215
Entman, M. L., 268
Erickson, B. W., 123, 127, 128, 129
Ericsoon, L. H., 271(28), 273, 276(28)
Ericsson, L. H., 243, 246(1), 276
Erikson, R. L., 388, 402
Erlichman, J., 15, 16, 17(10), 18(10), 20(10), 51, 55, 60, 156, 160, 162(22), 176, 177, 178, 179(14), 183(14), 184, 185, 186, 187, 188, 189, 190(5), 192, 193(4), 194, 195(10, 15), 196(15), 207
Ernst, V., 346, 347(8), 348, 349(6, 32), 350(32)
Esch, F. S., 141, 142, 402
Evain, D., 202, 204
Exton, J. H., 299

F

Fagard, R., 346
Fairbanks, G., 29
Farr, A. L., 56, 64, 86, 291, 313, 324
Farrell, H. M., Jr., 308
Farrell, P. J., 346, 347
Federman, P., 347
Fee, J. A., 94, 95(20)
Feinberg, R. S., 131
Feliu, J. E., 215
Felix, A. M., 126
Feramisco, J. R., 3, 120, 135, 138
Ferguson, J. J., 154
Ferraz, C., 90
Filburn, C. R., 21
Fillingame, R. H., 366, 367(1)
Finn, F. M., 120
Fischer, E. H., 5, 15, 77, 80, 83(2), 84(2), 90, 120, 177, 207, 243, 244, 246(1, 10), 252, 260, 267, 268, 269, 270, 271(28), 272(17), 273, 274, 276(28), 300
Fleischer, N., 186, 187, 189, 191, 192(4), 193(4), 194(4), 195(10, 15), 196(15)
Flink, I., 286
Flockerzi, V., 63, 64(7), 70(7), 71
Flockhart, D. A., 15, 16, 17(11, 17), 18, 19(11, 17), 20(11), 55, 56(6), 57(6), 60, 119, 151, 158, 162(12), 168, 169, 170(1, 6), 171(1, 6), 173(6), 176, 178, 179(6), 186, 188, 213, 227
Florio, V. A., 25, 26(14), 27(14), 299
Flug, B. M., 164
Foe, L. G., 216
Folch, J., 290
Fong, S.-L., 366
Forn, J., 21, 26, 27(18)
Foulkes, J. G., 15, 250, 252(2), 390
Fox, E., 217
Fox, J. E. B., 21, 25
Fraker, F. J., 190
Frank, R. N., 362
Frankhauser, P., 125
Freiberg, J. M., 21, 25, 26(11)
Freist, W., 18
Fridkin, M., 128
Friedman, R., 205
Friedrich, U., 60, 234, 235(5)
Friend, D. S., 22
Fries, P., 125

Friest, W., 141
Fryling, C., 402
Fujii, T., 288
Fujikura, T., 296
Fujitaki, J. M., 391, 398(23)
Furuya, E., 217

G

Gabriel, T. F., 133
Gallant, S., 21
Gallis, B., 15, 387
Ganapathi, M. K., 15
Gani, V., 141
Garbers, D. L., 15, 381, 387, 391, 398(24)
Garrels, J. I., 31, 36(31), 242
Garrison, J. C., 21, 22, 23(5), 25, 26(13, 14), 27(13, 14), 28(13), 35(13), 36, 299
Geahlen, R. L., 60, 162, 176, 235(8), 319
Geiduschek, E. P., 36, 242
Gerhart, J. C., 225
Ghosh-Dastidar, P., 347, 359(30)
Gilboe, D. P., 244
Gill, G. N., 9, 10(7), 63, 67(4), 153, 156, 206, 207, 208(1), 209, 210(2), 211(2), 212(1), 387
Gisin, B. F., 124, 125(30), 126
Glass, D. B., 3, 51, 120, 127(9), 130(11), 135, 137(9), 138, 139(99), 176, 212
Glynn, I. M., 81, 253, 368
Goltzman, D., 133
Gonzalez, C., 5, 77, 80, 177, 300
Gonzatti-Haces, M. I., 318, 328(12)
Good, N. E., 96
Gooding, K. M., 133
Goodwin, C. D., 164
Goris, J., 339, 345
Gornall, A. G., 332
Gottesman, M. M., 197, 198, 199(8), 200, 201(8), 202, 203, 204, 205
Gould, H., 397
Grace, M., 347
Grand, R. J., 248
Granot, J., 93, 94, 97, 98(7, 25), 99(7, 25), 101(7, 25), 102(7), 103, 111(6, 7, 10), 112(9, 10), 113(7, 10, 17, 25), 114(10), 115(10, 11), 116(11), 120, 138(7)

Graves, D. J., 45, 120, 130(5), 134, 135, 137(5), 248, 251, 259, 260, 261(7), 262(7), 263, 264(8), 265(7, 8), 266(8, 18), 267, 269, 270, 271(16), 272, 275, 276, 277(43), 278
Gray, E. G., 297
Greengard, P., 21, 23, 25(7), 26, 27(18), 62, 147, 154, 156, 157, 159, 160, 161, 162, 202, 203, 207
Groppi, V. E., Jr., 21, 23, 25(6), 26(6), 35(6)
Grosfeld, H., 8, 347
Gross, E., 120, 121, 127
Grove, J., 367
Gubler, U., 369
Guidotti, A., 80
Gupta, N. K., 347, 359(30)
Gupta, R. K., 93, 94(4), 98(4), 99(4), 100(4), 101(4), 103, 113(4)
Gutmann, N. S., 197
Gutte, B., 124, 129(31)

H

Hagiwara, D., 139
Hahn, P., 92
Haley, B. E., 147, 148, 154, 162
Hamilton, L., 336
Hamilton, R. T., 392
Happi, J., 141
Hardesty, B., 346
Hare, P. E., 404(10)
Hargrave, P. A., 366
Harmsen, B. J. M., 105
Harper, S. C., 299
Harrelson, W. G., Jr., 300
Harris, J. S., 207
Harris, T. J. R., 388, 390(14)
Hart, P. A., 94
Hartl, F. T., 3, 4(4), 6(4)
Hartley, B. S., 151
Hartshorne, D. J., 279, 286, 287, 288
Hashimoto, E., 146, 268
Haslam, R. J., 21, 25
Haston, W. S., 15, 16(11), 17(11), 18(11), 19(11), 20(11), 213, 214(2), 219(2)
Hathaway, G. M., 308, 309(1), 311, 314, 316(2, 11), 317, 318, 319(1), 325, 326(3, 19), 327(4, 7, 8, 9, 10), 328(9, 10), 329(3), 330(1), 331(1)

Hathway, D. R., 288
Hayakawa, T., 37, 250, 251, 252(1), 259, 261(4), 262(4), 263, 266(4), 267(18), 269
Haynes, R. C., Jr., 22
Hedrick, J. L., 90
Heilmeyer, L. M. G., 260
Heinrikson, R. L., 141, 216
Heldin, C. H., 402
Hemmings, B. A., 43, 60, 246, 248, 249(18), 250(18), 337, 339, 340(6), 341(6), 342, 343(6), 344, 345(6)
Hems, D. A., 299
Henderson, L. E., 133
Hermann, J., 243, 246(1), 271(28), 273, 276(28)
Hers, H.-G., 215, 217, 299
Hickinbottom, J. P., 273, 300
Hidaka, H., 212
Hilbers, C. W., 105
Hilz, H., 55, 158, 177, 189
Hinkins, S., 286
Hirsh, A. H., 51
Hixon, C. S., 141
Hiyama, K., 93, 115(11), 116(11)
Ho, E. S., 81
Hodges, R. S., 123, 125, 126
Hofmann, F., 15, 16(9), 17(9), 20(9), 55, 60, 62, 63, 64(7), 70(7), 71(7), 156, 161, 176, 187, 188, 344
Hofmann, K., 120
Hokin, L. E., 289
Hokin, M. R., 289
Holdy, K. E., 63, 67(4)
Holladay, L. A., 16, 17(16), 51, 54(5), 169
Hollenberg, M. D., 379
Hoppe, J., 18
Horvitz, R. A., 263, 266(18), 267(18)
Hostmark, A. T., 299
Howard, R. F., 15, 390
Hranisavljevic, J., 43
Hsie, A. W., 197
Hsu, C., 26
Hu, E. S., 253
Huang, T.-S., 9, 272
Hucho, F., 331, 332(2), 333(2), 335(2)
Hue, L., 215
Hughes, R. C., 273
Hull, W. E., 105
Humble, E., 120, 130(4), 134(4)
Humphreys, J., 336

Hunkeler, F. L., 246, 273, 300
Hunt, T., 346, 348, 349(3, 31)
Hunter, T., 8, 132, 346, 349(3), 388, 389, 390, 392, 394, 396, 398(28), 401, 402, 403
Hunter, W. M., 357
Hurst, M., 270
Hutcheson, E. T., 336
Hutchinson, M. A., 388
Hutson, N. J., 37, 299

I

Illingworth, B., 253, 261
Inamasu, M., 231(16)
Inglish, D., 26
Inoue, M., 288, 292(1, 2)
Isohashi, F., 27
Ito, H., 21, 25, 26(11)
Itoh, M., 139
Iwai, H., 231(16)
Iwasa, Y., 288, 296
Izawa, S., 96
Izumiya, N., 129

J

Jackson, R. J., 346, 349(3)
Jänne, J., 367
Jahn, R., 21
James, T. L., 106
Jamieson, G. A., Jr., 280
Janski, A. M., 24
Jardetsky, O., 107
Jennissen, H. P., 260
Jesse Chan, J. F., 248
Jett, M. F., 37
Jimenez, M. H., 126
Jiménez, J. S., 141
Job, D., 318, 328(5)
Joder-Ohlenbusch, A. M., 367
John, M. J., 277(44)
Johns, E. W., 290
Johnson, E. M., 139
Johnson, L. N., 275
Johnson, M. L., 36
Johnson, R., 300
Johnson, R. A., 52, 77, 230

Josephson, R. V., 308
Jovin, T. M., 28
Jungmann, R. A., 233
Juszczak, E., 366

K

Kabat, D., 9
Kaibuchi, K., 289, 291, 295(19), 296(14), 297
Kaiser, E. T., 93, 94, 97(6), 98, 99(7, 26, 27), 100(27), 101(7), 102(7), 103(26, 27), 108(26, 27), 110(27), 111(6, 7, 10, 27), 112(9, 10), 113(7, 10, 16), 114(10), 115(10, 11, 16), 116(11, 12), 117(12), 120, 138, 139, 141
Kalk, A., 104
Kamiya, T., 139
Kanof, P., 159, 161(19)
Kanstein, C. B., 63, 67(4)
Kanter, J. R., 77
Karger, B., 133
Kashelikar, D. V., 129
Katoh, N., 289, 297(12)
Katsoyannis, P. G., 120
Kawahara, Y., 288
Kawanishi, S., 131
Keely, S. L., 55, 207, 227, 228, 231(5, 14, 18), 232
Keller, P. J., 253
Keller, P. M., 300
Kemp, B. E., 3, 45, 120, 130(5), 135, 137(5), 138(86), 139, 229, 272, 275, 277(44)
Kemp, D. S., 134
Kemp, R. B., 268
Kemp, R. G., 58, 216
Kennerly, D. A., 289
Kent, S. B. H., 124
Kenyon, G. L., 98, 99(26), 100, 103(26), 108(26)
Kenyon, K. R., 362
Kerlavage, A. R., 141, 148, 149, 150, 151, 153, 159, 161(17), 208
Kernohan, J. C., 43, 337, 339(6), 340(6), 341(6), 342(6), 343(6), 344(6), 345(6)
Kerrick, W. G. L., 260, 267(12)
Khandelwal, R. L., 15, 17(8)
Khatra, B. S., 37, 43, 243, 268, 275, 276(38), 300, 338, 339(10)
Kikkawa, U., 289, 291(5, 6), 296, 297

Kim, K.-H., 218, 368
King, C. A., 231(17)
King, L., 379, 380(7), 381(5), 384(7), 385(4)
Kirshner, N., 21, 25, 26(12), 27(12), 35(12)
Kishida, Y., 128, 130(53), 131(53)
Kishimoto, A., 288, 292(1, 2, 5, 6), 296(5, 6)
Klausner, Y. S., 120
Klee, C. B., 250, 286, 288
Knight, B. L., 92
Knight, M., 128
Koide, A., 243, 246(1), 271(28), 273, 276(28)
Komike, K., 268
Kondo, H., 93, 97(6), 98(7), 99(7), 101(7), 102(7), 111(6, 7), 113(7), 115(11), 116(11)
Koroscil, T. M., 21
Krakow, J. S., 161
Kramer, G., 346
Krebs, E. G., 5, 8, 15, 16(9), 17(8, 9), 20(9), 37, 45, 51, 55, 56, 58, 60, 63, 77, 78, 80, 81, 82, 120, 127(9), 130(5, 11), 137(5, 8, 9), 138, 139, 141, 146, 151(1), 156, 158, 162, 164, 165, 166, 167(7), 176, 177, 186, 187, 188, 208, 212, 220, 231(15, 17), 235(8), 239, 244, 244, 250, 251(1), 252, 253, 259, 261, 262(4), 263, 266(4, 18), 267, 268, 269, 272, 273, 275, 300, 315, 319, 387
Kreuger, B. K., 21, 22, 26, 27(18), 35(2)
Krim, M., 372
Kitchevsky, G., 290
Kudlow, J. E., 9, 10(7), 387
Kuehn, G. D., 367, 369, 370(9, 10, 16), 371(9, 10), 372
Kühn, H., 362, 363(1), 365, 366(5)
Kumar, K. S., 46
Kumar, R., 318, 327(11)
Kumor, S., 243, 246(1)
Kuo, J. F., 62, 138, 208, 211(16), 212(16), 289, 297(12)
Kupfer, A., 141
Kupiec, J. W., 60, 160

L

Laemmli, U. K., 28, 63, 148, 175, 250, 251, 283, 326, 350, 381
Lane, C. D., 225
Lane, L. K., 268

Langan, T. A., 212
Lapetina, E. G., 289
Larner, J., 243, 268, 275
Larson, K. L., 244
Laskey, R. A., 392
Latshaw, S. P., 216
Layne, P. P., 14
LeCam, A., 197, 198, 199(8), 200(8), 201(8), 202, 203(8), 204, 205
Lee, E. Y. C., 15
Lee, K.-H., 218
Lee, R. H., 366
Lees, M., 290
Leimgruber, W., 11, 134
Leiter, A. B., 27
Leitschuh, M., 202
Lemarie, B., 106
Lemkin, P., 36
Lenard, J., 130, 131(63, 65)
Lengyel, P., 347
Lent, B., 368
Lenz, J. R., 347
Leonard, N. J., 162, 164(2, 3)
Leone, G. R., 21, 25, 26(15), 27(15)
Lerch, K., 15, 274
Leroux, A., 346
Levin, D. H., 346, 347, 348, 349(5, 6, 32), 350(32), 362(16, 27)
Levine, R. L., 9, 395, 397(33), 403
Levinson, A. D., 388
Levitsky, A., 200
Li, C. H., 127, 128, 129
Lichti, U., 197, 198(9), 205
Liebman, M. N., 97, 98(25), 99(25), 101(25), 113(25)
Lincoln, T. M., 16, 17(18), 55, 56(6), 57(6), 63, 64(6), 151, 158, 162(12), 164, 168, 169(1), 170(1), 171(1), 175, 188, 208, 227
Linn, T. C., 331, 332, 333, 335(2), 336
Lipmann, F., 15, 17(13)
Litwack, G., 197
Liu, A. Y.-C., 154, 159(4), 160(4), 161(4), 203
Ljungstrom, O., 215
Lloyd, C. W., 382
Lohmann, S. M., 157, 160, 162(11)
Lolley, R. N., 366
London, I. M., 346, 347, 348, 349(5, 6, 31, 32), 350(32), 362(16, 27)
Love, D. S., 252

Lowry, O. H., 56, 64, 86, 291, 313, 324
Lundak, T. S., 311, 325, 326(19)
Lutz, S. A., 80
Lynch, G., 268
Lynham, J. A., 21
Lyons, R. M., 21

M

McArthy, D., 16
McCann, P. P., 367
McCarthy, D., 55, 56(6), 57(6), 151, 158, 162(12), 168, 169(1), 170(1), 171(1), 178, 188, 227
McCrane, M. M., 213
McCullough, T. F., 250, 258(5), 259(5)
McDowell, J. H., 366
McFann, L. J., 138, 139(99)
MacKinlay, A. G., 207
McPherson, J. M., 80(6, 7), 81, 83(6), 88, 90(6, 7), 232
Mäenpää, P. H., 318, 328(6)
Majerus, P. W., 289
Majumdar, A., 347
Maller, J. L., 15, 17(8), 139, 220
Mamont, P. S., 367
Manai, M., 398
Manen, C. A., 367
Mansour, T. E., 141
March, S. C., 300
Margoliash, E., 192
Maita, T., 279
Markley, L. D., 124
Marshall, G. R., 124
Martensen, T. M., 9, 278, 395, 397(33), 403, 404(9), 405(9)
Martin, D. B., 21, 25, 26(15), 27(15)
Martinez, R., 401
Marutzky, R., 18
Mas, M. T., 366
Masaracchia, R. A., 3, 135
Matsubara, T., 289, 296(14)
Matsuda, G., 279
Matsueda, R., 141
Matthews, H. R., 397
Maudsley, D. B., 366
Means, A. R., 80, 92
Meienhofer, J., 120, 121, 127, 128(22), 133
Melenik, D. A., 260, 267(12)

Merlevede, W., 339, 345
Merrifield, R. B., 120, 121, 122, 123, 124, 125, 126, 127, 128, 129, 130(13, 23), 131, 139, 270
Merril, C., 36
Meshitsuka, S., 98
Meyer, W. L., 252
Meyers, K. R., 63
Michalewsky, J., 133
Michell, R. H., 289
Mildvan, A. S., 93, 94, 95(18), 96, 97, 98, 99(3, 4, 5, 7, 25, 26), 100, 101(4, 7, 25), 102(7), 103(26, 27), 108(26, 27), 109(3), 110(27), 111(6, 7, 10, 27, 42), 112(9, 10), 113(3, 4, 7, 10, 16, 17, 25), 114(10), 115(10, 11, 16), 116(11, 12), 117(12), 118, 120, 138(7)
Milhaud, P. G., 197, 205
Miller, J., 286, 362
Miller, M. D., 138, 139(99)
Miller, M. J., 36, 242
Millikin, D. M., 275
Mills, A. D., 392
Minakuchi, R., 289, 292(11), 296(13), 297
Mitchell, A. R., 123, 124
Mitsui, K. I., 347
Miyamoto, E., 207
Mockerlava, L., 289, 297(12)
Moeschler, H. J., 260, 267(17)
Mohberg, J., 369
Mohun, T. J., 225
Mojsov, S., 121, 122
Molla, A., 286
Monken, C. E., 207
Moore, S., 150
Mori, T., 288, 289, 291(5, 6), 296, 297
Morr, C. V., 308
Morris, D. R., 366, 367(1)
Morrissey, J. H., 29
Moylan, R. D., 26
Muir, L. W., 15
Munk, P., 336
Murtaugh, M., 202
Muschel, R. J., 197

N

Nairn, A. C., 248, 249(14), 260, 261(6), 269
Najjar, V. A., 14

Nakamura, K. D., 401
Nakomura, S., 268
Namihira, G., 336
Nelson, D. A., 130
Nelson, N. C., 141, 142, 146(14)
Neurath, H., 243, 246(1), 271(28), 273, 276(28)
Neville, M. E., 3, 4(4), 6(4)
Nicolaides, E. D., 131
Nicolas, J. C., 202, 204
Nilsen-Hamilton, M., 392
Nimmo, H. G., 187, 338, 339
Nishizuka, Y., 268, 288, 291, 292(1, 2, 11), 295(19), 296, 297
Nisonoff, A., 192
Noble, R. L., 129
Noggle, J. H., 103, 104(30), 107(30)
Noiman, E. S., 138
Nolan, C., 120, 269, 272
Nomoto, A., 388, 390(14)
Novoa, W. B., 120, 269, 272(17)
Nussli, M., 367
Nuttall, F. Q., 244

O

O'Brian, C., 94, 116(12), 117(12)
Ochoa, S., 8, 347
Oertel, W., 36
O'Farrell, P. H., 28, 30(21), 31(21), 33(21), 60, 233, 234, 235(5), 243(4)
Ogried, D., 158, 171, 218
Ohno, M., 129
Okada, M., 128, 130(53), 131(53)
Oliver, D., 208, 290
Ondetti, M. A., 120
Oppermann, H., 388
Ornstein, L., 28, 89, 90
Oroszlan, S., 133
Osborn, M., 56, 87, 295

P

Pace, M., 124
Paiva, A. C. M., 128
Paiva, T., 128
Pal, R. K., 141, 142(7), 146

Palmer, G., 94
Palmer, W. K., 232
Palmieri, S., 347
Panyin, S., 208, 290
Pardee, A. B., 392
Parikh, I., 300
Paris, H., 15
Park, C. R., 55, 207, 227, 228, 231(5, 6, 7, 13, 14, 18), 232
Parker, P. J., 39, 248, 249(18), 250(18), 337
Pastan, I., 197, 198, 199(8), 200(8), 201(8), 202, 203, 204, 205
Patschinsky, T., 401, 402
Paulus, T. J., 367
Pawelek, J. M., 197
Payne, M. E., 299, 300
Pelley, J. W., 331, 332(2), 333(2), 335, 336
Penny, C. L., 12
Perkins, J. P., 37, 81, 250, 251(1), 252(1), 253, 259, 261(4), 262(4), 266, 269
Perrin, M., 133
Perry, S. V., 248, 249(14), 260, 261(6), 268
Persechini, A., 279
Peters, K. A., 80, 83(2), 84(2), 90, 207
Peterson, D. F., 197
Peterson, G. L., 384
Petrovic, S., 15
Petryshyn, R., 346, 347, 362(16)
Pettit, F. H., 331, 332(2), 333, 335, 336
Petzold, G. L., 207
Picton, C., 38(8), 249, 250, 260, 267(12)
Pietta, P. G., 124
Pike, L. J., 387
Pilkis, S. J., 15, 16(11), 17(11), 18(11), 19(11), 20(11), 213, 214(2), 215, 217, 219
Pirollet, F., 318, 328(5)
Piscitello, J., 197
Plimmer, R. H. A., 388, 391(4), 403
Pocinwong, S., 260, 267(12)
Pomerantz, A. H., 139, 147
Poorman, R., 216
Patter, R. L., 148, 149(25), 186
Potuzak, H., 312
Powell, A. J., 382
Previero, A., 129
Pringle, J. R., 295
Proud, C. G., 338
Prywes, R., 373
Puck, T. T., 197, 198
Purchio, A. F., 388, 402

Q

Queen, C., 374, 376(3)

R

Rabbani, L. D., 106
Rabenowitz, M., 15, 17(13)
Rabin, M. S., 200
Rae, P. A., 197
Ragnarsson, U., 120, 130(4), 134(4)
Raina, A., 367
Rall, T. W., 20
Ralston, O., 347, 359(30)
Ralston, R., 347
Randal, P. J., 336
Randall, D. D., 331, 332(2), 333(2), 335(2)
Randall, R. J., 56, 64, 86, 291, 313, 324
Rangel-Aldao, R., 60, 156, 160, 162(22), 176, 177(8), 178(8), 180, 181, 182, 185, 186, 188, 195(10)
Rannels, S. R., 60, 148, 159, 168, 169, 170(1, 2, 6), 171(1, 3), 172(3, 6), 173(3, 6), 227
Ranu, R. S., 346, 347, 349(6), 359(29)
Ratner, L., 347
Ray, W. J., 94, 95(21)
Redfield, A. G., 103, 105, 108(40), 109(40)
Reed, G. H., 94, 95(21)
Reed, L. J., 331, 332(2), 333, 335, 336
Rees, D. A., 382
Regnier, F. E., 133
Reichlin, M., 191, 192
Reimann, E. M., 3, 16, 17(16), 19, 51, 52, 54, 72, 77, 79(3), 80, 82, 135, 169, 203, 213, 228, 243, 254, 261, 273, 275, 300, 301, 315, 319
Ressler, C., 129
Revel, M., 347
Reynolds, F. H., Jr., 402
Rich, D. H., 126
Richert, N., 204
Riley, W. D., 58, 268
Riniker, B., 132
Riordan, J. F., 139
Riou, J. P., 213
Rivier, J. E., 133
Roach, P. J., 37, 243, 268, 275
Roberts, G. C. K., 107
Robertson, H. D., 346, 349(3)

Robinson, A. B., 130, 131(63)
Robinson-Steiner, A. M., 169, 170(6), 172(6), 173(6)
Roche, T. E., 335, 336
Rodbard, D., 28
Rodrigues, L. M., 299
Rosebrough, N. J., 56, 64, 86, 291, 313, 324
Rosen, N., 197
Rosen, O. M., 15, 16, 17(10), 18(10), 20(10), 37, 51, 55, 60, 80, 156, 160, 176, 177, 178, 179(14), 180, 181, 182, 183(14), 184, 185, 187, 188, 190(5), 191, 197, 207
Rosenberg, N. E., 388
Rosenfeld, R., 60, 176
Rosevear, P. R., 94, 98, 99(26, 27), 100, 103(26, 27), 108(26, 27), 110(27), 111(27), 116(12), 117(12)
Roskoski, R., 3, 52, 134, 135(84)
Ross, A. H., 389
Ross, E., 64
Ross, H. H., 136
Roth, C., 202, 203, 204
Roth, G. S., 21, 25, 26(11)
Rothberg, P. G., 388, 390
Rouser, G., 290
Roy, R., 347, 359(30)
Rubin, C. S., 154, 156, 160, 162(22), 177, 178(9), 186, 188, 189, 190, 194, 195, 196
Rudolph, S. A., 21, 22, 23, 25(7), 35(2), 147
Rusch, H. P., 369
Russell, D. H., 233, 367
Ryabtsev, M. N., 123
Rylatt, D. B., 243, 276, 337, 339, 344(7)
Rymond, M., 60, 344

S

Saari, J. C., 270, 274
Sachs, D. H., 126
Sachs, L., 197
Sarkar, D., 160, 162(22), 186, 187, 189, 192(4), 193(4), 194(4), 195(10, 15), 196(15)
Sakai, L. J., 63
Sakakibara, S., 128, 130, 131(64), 131(53, 64)
Salomonsson, I., 289, 297(12)
Sampath Kumar, K. S. V., 391
Samuel, C., 347

Sano, K., 289, 296(13), 297
Sano, S., 131
Sansom, M. S. P., 275
Sarma, P. S., 46, 391
Savage, C. R., 380
Say, A. K., 25
Scandora, A. E., 36
Schaffer, N. K., 137
Schatz, G., 64
Schimmer, B. P., 197
Schirmer, R. E., 103, 104(30), 107(30)
Schlender, K. K., 52, 54, 77, 79(3), 80, 243, 275
Schlichter, D. J., 162
Schneck, J., 197
Schnegraf, A., 391
Schoner, W., 215
Schulman, L., 347
Schultz, M., 36
Schwartz, A., 268
Schwartz, G. P., 120
Schwechheimer, K., 161
Schwyzer, R., 132
Secrist, J. A., 162, 164
Seebeck, T., 369
Sefton, B. M., 8, 388, 392, 396, 398(28), 401, 402, 403
Seighart, W., 21
Sekar, V., 372
Sen, E. G., 347
Sevilla, C. L., 273
Shaltiel, S., 18, 128, 141
Shank, P. R., 34
Shapira, R., 138, 208, 211(16), 212(16)
Shapiro, B. M., 388
Sharief, F., 300
Sharma, R. K., 260
Sharp, S. B., 324
Sharts, C. M., 131
Shaw, J. O., 21
Shenolikar, S., 248, 249(14), 250, 260, 261(6), 269
Sheorian, V. S., 43, 276
Sheppard, W. A., 131
Sherry, J. M. F., 287
Shichi, H., 362, 365(4), 366
Shimonishi, Y., 128, 130(53), 131(53)
Shizuta, Y., 15, 16(9), 17(8, 9), 20(9)
Shoji, M., 138, 208, 211(16), 212, 289, 297(12)

Shulman, H., 159, 161(19)
Siekierka, J., 347
Silberman, S. R., 15
Simantov, R., 197
Siminovitch, L., 197, 198
Simpson, D. L., 43
Singer, I., 21
Singh, J., 126
Singh, R. M. M., 96
Singh, T. J., 202, 203, 204
Sjöquist, J., 402
Skala, J. P., 92
Skuster, J. R., 135, 248, 260, 261(7), 262, 263(7), 265(7), 270(16), 271(16), 272(16), 274(16)
Slattery, E., 347
Smart, J. E., 402
Smith, A. J., 90
Smith, D. P., 366
Smith, E., 36
Smith, G. M., 97, 98, 99(25), 101(25), 113(25)
Smith, R. A., 391, 398(23)
Smith, S. B., 60, 164, 165, 167(7)
Snyder, S. H., 367
Soderling, T. R., 37, 39, 43, 55, 227, 228, 231(5, 6, 13), 232, 243, 268, 273, 275, 276, 299, 300, 338, 339(10)
Sold, G., 62
Soling, H. D., 21
Solomon, I., 104, 105(34)
Solomon, S., 133
Somers, R. L., 362, 365(4), 366
Sommer, D. R., 208
Sommer, K. R., 290
Sowder, R., 133
Speck, J. C., 190
Speeg, K. V., Jr., 15
Speichermann, N., 63, 64(7), 70(7), 71(7)
Spiess, J., 133
Spiker, S., 208, 290
Srivastava, A. K., 243, 268, 275, 276(38), 300, 338, 339(10)
Stadtman, E. R., 388
Stahlman, M., 362
Standjord, T. P., 90
Stanley, G. H. S., 290
Stanford, N., 289
Stanley, P., 198
Staros, J. V., 146, 379

Steck, T. L., 29
Stedman, J. D., 392
Steers, E., 150
Steglich, W., 131
Stein, P. J., 111(42)
Stein, S., 11, 134
Stein, W. H., 150
Steinberg, R. A., 21, 28, 233, 234, 235(3, 5), 241(2), 243
Steiner, A., 202
Stengelin, J., 139
Stephenson, J. R., 402
Stewart, A. A., 339
Stewart, J. M., 121, 123(20), 124, 127(20), 128, 129(20), 130(20), 131(20), 132(20), 134(20)
Stingeln, J., 94, 113(17)
Storvick, C. A., 328
Stoscheck, C., 379, 383(8), 384(8), 385(8), 386(8), 387(8)
Strepp, L. R., 335
Strickland, W. G., 299
Stroud, R. M., 243
Studier, W. F., 349
Stull, J. T., 19, 208, 268
Stura, E. A., 275
Sugden, P. H., 16, 17(16), 51, 54(5), 55(6), 56, 57(6), 151, 153(31), 158, 162(12), 168, 169, 170(1), 171(1), 178, 188, 227
Sugihara, H., 128, 130(53), 131(53)
Sul, H. S., 250, 252, 258(5), 259
Sulakhe, P. V., 268
Sundararajan, T. A., 46, 391
Sutherland, E. W., 20
Svenden, P. J., 28
Swaisgood, H. E., 308
Swanstrom, R., 34
Swarup, G., 15, 381, 387, 391, 398(24)

T

Tabatabai, L. B., 135, 269, 270(16), 271(16), 272, 274, 275(34), 277(34)
Taborsky, G., 7, 8(1), 137, 388
Tabuchi, H., 268
Tahara, S. M., 311, 325, 326(19)
Takahashi, J., 296

Takai, Y., 288, 289, 291, 292(1, 2), 295(19), 296, 297
Takeyama, S., 231(16)
Takio, K., 60, 118, 146
Tallant, E. A., 280
Tam, S. W., 260
Tamaki, Y., 212
Tamura, A., 288
Tao, M., 207, 318, 327(11)
Tash, J. S., 80, 92
Taylor, J., 36
Taylor, S. S., 141, 142, 143, 144(2), 147, 148, 149(25), 151, 153, 159, 161(17), 186, 208
Tee, J., 208, 211(16), 212(16)
Templeton, D., 395
Terasaki, W. L., 26
Tessmer, G. W., 269, 270(16), 271(16), 272, 274(16), 278
Theoharidies, T., 21
Thom, D., 382
Thomas, N., 94, 113(17), 120, 138, 141
Tiao-Te Lo, R., 106
Tigani, M. C., 169, 170(6), 172(6), 173(6)
Titani, K., 60, 118, 243, 246(1), 271(28), 273, 276(28)
Todaro, G. J., 402
Tomkins, G. M., 197
Townsend, J., 94, 95(18)
Trachsel, H., 346, 347
Trapmann, H., 391
Traugh, J. A., 77, 308, 309(1), 311, 314, 315(5), 316, 317, 318, 319, 324, 325, 326(3, 19), 327(3, 7, 8, 9, 10), 328(9, 10, 12), 329(3), 330(1), 331(1, 2)
Trayser, K. A., 252
Tropp, J., 105, 108(40), 109(40)
Tse, J., 138
Tsukamoto, S., 129
Tu, J.-I., 278
Tuazon, P. T., 308, 315(5), 316, 317, 318, 319, 331(2)
Twible, D. A., 25, 26(14), 27(14), 299

U

Undenfriend, S., 11, 134
Uno, I., 154, 159(4), 160(4), 161(4), 203
Ushiro, H., 379, 383(8), 384(8), 385(8), 386(8), 387(8), 398, 402
Utter, M. F., 27
Uyeda, K., 217

V

Vale, W., 133
Vallee, B. L., 139
Vallet, B., 286
Vanaman, T. C., 250, 252(2), 280, 300
van Boom, J. H., 105
Vandenheede, J. R., 15, 17(8), 345
Vanderhoff, G., 348, 349(31)
van der Marel, G., 105
Van Keuren, M., 36
Van Schaftingen, E., 217
Van Wintle, W. B., 268
Vardanis, A., 71, 73(1, 2), 75, 76(2)
Varmus, H. E., 388
Villar-Palasi, C., 162, 164(3)
Viriya, J., 273
Vo, K.-P., 242
Vrana, K. E., 3, 4(4), 6(4)

W

Wade, M., 36
Wade, R. D., 243, 246(1)
Wagner, G., 104, 105(37), 108(37)
Wagner, J. D., 23, 28(9)
Waisman, D. M., 260
Wallace, R. W., 280
Wallach, D. F. H., 29
Walseth, T. F., 52, 77, 230
Walsh, D. A., 5, 37, 51, 77, 80, 81, 82, 83(6), 88(6), 90(6, 7, 14), 92, 177, 232, 250, 251(1), 252, 253, 258, 259, 261, 262(4), 266(4), 269, 273, 300, 315, 319
Walsh, K. A., 60, 243, 246(1)
Walsh, K. Y., 243
Walsh, M. P., 279, 286, 288
Walsh, R. X., 275
Walter, R., 106
Walter, U., 154, 156, 157, 159, 160, 161, 162, 203
Walton, G. M., 63, 67(4), 153, 156, 206, 207, 208(1), 209, 210(2), 211(2), 212(1)
Wang, J. H., 260
Wang, J. K., 366

Wang, J. Y. J., 374, 376
Wang, K.-T., 128
Wang, S.-S., 125
Wasteson, A., 388, 402
Watson, D. C., 336
Watterson, D. M., 16, 17(17), 19(17), 60, 119, 176, 179(6), 186(6), 300
Weber, G., 162, 164(2)
Weber, H., 80
Weber, I. T., 275
Weber, K., 56, 87, 295
Weber, M., 401
Weber, W., 55, 158, 177, 189
Wechter, W. J., 141
Weigeke, M., 11, 134
Weinberg, M., 27
Weller, M., 8, 9(4)
Welsh, M. J., 80
Weng, L., 141
West, L., 55, 56(6), 57(6), 151, 158, 162(12), 168, 169(1), 170(1), 171(1), 178, 188, 227
Westermark, B. Ek, B., 388, 402
Westheimer, F. H., 148
Weygand, F., 131
Whitaker, J. R., 14, 43
White, H. D., 164, 165, 167(7)
Whitehouse, S., 80(6, 7, 8), 81(6), 83(6), 88(6), 90(6, 7), 252, 259(9), 336
Whitney, R. M., 308
Whittaker, V. P., 297
Whitton, P. D., 299
Wilden, U., 365
Willey, J. C., 80
Wilson, K. S., 275
Wimmer, E., 388, 390(14)
Winget, G. D., 96
Winter, W., 96
Wise, B. C., 289, 297(12)
Witt, J. J., 3, 52, 135
Witte, O. N., 373, 378(2), 388
Witters, L. A., 21, 23, 26(8), 27(8), 28(8)
Woodring, M., 328
Wosilait, W. D., 20

Wrenn, R. W., 289, 297(12)
Wright, D. E., 138
Wu, M., 225
Wuthrich, K., 104, 105(37, 38), 108(37, 38)
Wyatt, J. L., 141, 142(7)
Wylie, C. C., 225

X

Xuong, N.-H., 242

Y

Yaghmai, B., 347, 359(30)
Yamamoto, A., 290
Yamamura, H., 268
Yamashiro, D., 127, 129, 132
Yang, J. C., 391, 398(23)
Yang, S. D., 345
Yeaman, S. J., 336, 339
Yeh, L.-A., 218
Yellowlees, D., 43, 248, 249(18), 250(18), 337, 339(1, 6), 340(6), 341(6), 342(6), 343(6), 344(1, 6), 345(6)
Yokoyama, M., 217
Young, J. D., 120, 121, 123(20), 127(20), 129(20), 130(20), 131(20), 132(20), 134(3, 20)
Yu, B., 289, 292(11), 296, 297
Yunes, A. A., 273

Z

Zabits, J. G., 207
Zeilig, C. E., 138, 139(99), 212
Zetterqvist, Ö., 120, 130(4), 134(4)
Ziemecki, A., 15, 390
Zilberstein, A., 347
Zoller, M. J., 141, 142, 143, 144(2), 146(14), 147, 148, 208, 318, 327(4)

Subject Index

A

A23187, 21
A-431 cells
  EGF receptor on, 379
  growth, 381
  preparation of membrane and vesicle fractions, 382–383
Abelson murine leukemia virus, 373
Acetic acid, in assay of protein kinase, 3, 5
Acetone, 27
Acetonitrile, 39, 40, 42, 45, 46, 47, 151
Acetyl-CoA carboxylase, 218
Acid hydrolysis procedure, for proteins in gels, 395–397
ACS, 230, 237
ACTH, see Adrenocorticotropic hormone
Acyl-bound phosphate, 8
  test for, 27
Adenosine, inhibitor, of rhodopsin kinase, 365
Adenosine 3',5'-cyclic phosphate, see also Dibutyryl cyclic AMP
  analogs
    selectivity studies, 169–171
    site-selective, cooperativity studies, 171–174
  contaminant, in protein-bound phosphate determination, 7–8
  fluorescent analog, see 1,N^6-Etheno-cAMP
  photosensitive analog, see 8-Azido-cAMP
  in purification of catalytic subunit of cAMP-dependent protein kinase, 51, 53
[^{3}H]Adenosine 3',5'-cyclic phosphate, in assay of R subunit, 56, 57
Adenosine 3',5'-cyclic phosphate-agarose, 73
Adenosine 3',5'-cyclic phosphate-Sepharose affinity column, in purification of R subunits, 58
Adenosine 5'-diphosphate, inhibitor, of pyruvate dehydrogenase kinase, 336
Adenosine 5'-diphosphate, magnesium salt, in reverse reaction of protein kinases, 14, 15
Adenosine 5'-monophosphate, inhibitor, of rhodopsin kinase, 365
Adenosine 5'-triphosphate
  analogs of, in affinity labeling studies, 141–146
  cobalt salt of, binding and conformation studies, 98–103
[γ-^{32}P]Adenosine 5'-triphosphate
  in assay of protein kinases, 3–4, 52, 63, 72, 78, 81, 228, 230, 246, 253, 261, 270–271, 279, 290–291, 300, 319, 332, 337–338, 348, 349, 362, 363, 368, 376
  in phosphorylation of glycogen synthase, 37
  specific activity determination, 26–27
[$adenine$-2,8-^{3}H]Adenosine 5'-triphosphate, in assay of rhodopsin kinase, 362, 363
ADP, see Adenosine 5'-diphosphate
Adrenocorticotropic hormone, 21
Affi-Gel 10, 380
Affi-Gel 15, 340
Affi-Gel Blue Sepharose, 282, 286, 313, 341
Affinity labeling, uses, 140–141
Agarose-poly (I)·poly(C), 356
Alanine, pyruvate kinase phosphorylation and, 214, 215
Alkaline hydrolysis
  of proteins in gels, 390
  of proteins in solution, 403–405
  in purification of phosphoproteins, 9–12
Alkaline β-elimination, of phosphoserine, 43, 45, 46
Amino acid
  phosphorylated, 8, 388
  synthesis of Boc derivatives, 139

Amino acid analysis, for
  phosphotgrosine, 402–405
Amino acid side chains, protection
  during peptide synthesis, 127–130
N^6-(2-Aminoethyl)-cAMP-Sepharose, 239
8-2-Aminoethylthio-cGMP-Sepharose 4B,
  63, 64, 65
Amino groups, quantitative tests for, 126
ε-Amino group, protection, 127–128
N^6-Aminohexylcarbamoylmethyl-cAMP,
  170
Ammonium molybdate, 12, 245
Ammonium sulfate precipitation, 17, 65,
  246–247, 333–334, 340–341
  reverse, 281–282
Anion-exchange column, in assay of
  protein kinases, 135, 136–137
Anisole, 131
ANSA, see Aminonitrosulfonic acid
Antiserum, preparation, 190
Aprotinin, 161
Arginine
  side chain deprotection, 132
  side chain protection, 127
  test for, 134
Ashing reagent, 12
Asparagine
  in peptide synthesis, 129–130
  in purification of pyruvate
    dehydrogenase kinase, 334
ATP, see Adenosine 5'-triphosphate
Autoradiography
  photography of autoradiographs, 36
  of slab gels, 34–36
8-Azido-cAMP, 147
  binding specificity, 148, 174–176
  use in crude extracts, 159–161
8-Azido-[^{32}P]cAMP, synthesis and
  characterization, 154

B

Barium sulfate, 132
Barth medium, modified, 222
Beckman Ready-Solv EP, 230
Benzamide, 189
Benzamidine, 25
Benzhydrylamine resin, 123, 124
Benzyl group, as protecting group, 128

Benzyloxycarbonyl moiety, as protecting
  group, 127–128
Bio-Rex 63, 270
Bio-Rex 70, 370
1,4-Bis[2-(5-phenyloxazolyl)]benzene,
  135, 208, 209
2,3-Bisphosphoglycerate, inhibitor, of
  casein kinase I, 314
  of casein kinase II, 318, 327–328
Blue-Sepharose 4B, 266
Blue-Sepharose CL-6B, 294, 364
Brain, rat, crude extract preparation,
  292
8-Bromo-cAMP, 170
  effect on R subunit, 241–242
  in isolation and characterization of
    cAMP-resistant mutants, 199,
    200–201, 205
Bromodecane:bromododecane, 24
Budget-Solve, 4
2,3-Butanedione, 207
1-Butanol:acetic acid:water solvent
  system, 154, 316

C

Calcium ion
  activator, of protein kinase C, 288,
    295
  β-elimination reaction and, 46
  myosin light chain kinase and, 279
  phosphorylase kinase and, 249
  removal from reagents, 290–291
  stimulant, of protein phosphorylation,
    21
Calmodulin
  activator, of glycogen synthase
    kinase, 299
  molecular weight, 248
  purification, 280–281
  requirement, of myosin light chain
    kinase, 279, 288
Calmodulin-Sepharose, in purification of
  glycogen synthase kinase, 305–306
cAMP, see Adenosine 3',5'-cyclic
  phosphate
Carbachol, 21
Casein, see also Dephasphocasein
  variants of, 314
κ-Casein, 268

SUBJECT INDEX

Casein kinase, 248
  partial purification, 304
Casein kinase I
  activity, 308–309, 314–316, 331
  assay, 309–310
  chymotryptic digestion, 316
  identification in cell lysates and homogenates, 316–317, 330–331
  inhibitors, 314
  properties, 308
  purification, 310–313
  recognition sequence, 308
  stability, 311
Casein kinase II
  activation by polyamines, 328
  activity, 316, 317–318, 319, 331
  assay, 319–320
  autophosphorylation, 325–326
  characteristics, 317–319, 320
  identification in cell lysates or extracts, 328–331
  inhibitors, 314, 318, 326–328
  in phosphorylation of glycogen synthase, 37, 40, 42
  purification, 324–325
  recognition site, 318
  storage, 325
Cell
  metabolic labeling, 234–237
  phosphate labeling procedure, 22–23, 392–393
  lysis procedure, 374, 393
Cell, cultured, response to cAMP, 197
Cell extract, determination of incorporated radioactivity, 237–238
Cell fractionation, with digitonin, 23–24
Cellulose phosphate paper, see Phosphocellulose paper
Cerebral cortex, subclasses of protein kinase in, 195–196
Charcoal, in purification of cGMP-dependent protein kinase, 70
Chelex 100, 291
Chinese hamster ovary fibroblast
  response to cAMP, 197–198
  cAMP-resistant mutants
    characterization, 201–206
    isolation, 199–201
    growth and handling, 198–199
    mutation procedure, 200–201

suspension cultures, 198–199
2-Chloro-10-(3-aminopropyl)phenothiazine-Sepharose 4B, 280
2-Chlorobenzyloxycarbonyl moiety, as protecting group, 128
Chloroform:methanol, 12, 27
Chloromethylated polystyrene-co-divinylbenzene resin, 123–124
Chlorpromazine, inhibitor, of protein kinase C, 297
Cholera toxin, 199, 205
Chymotrypsin
  digestion of casein kinases, 316–317
  effect on phosphorylase kinase, 249
  in preparation of peptide substrate, 269
CM-Sephadex C-50, 73
Collagen, 21
Collagenase method, for ioslation of oocytes, 224–225
Competitive displacement radioimmunoassay, 191
Compound 48/80, 21
Creatine kinase, conformation studies, 98
Cyanogen bromide fragmentation, in preparation of peptide substrate, 270
Cyclic AMP, see Adenosine 3',5'-cyclic phosphate
[^{35}S]Cysteic acid, preparation, 45
Cysteine, side chain protection, 129

D

DEAE-cellulose chromatography
  in purification of casein kinase I, 310
  of casein kinase II, 324
  of cGMP-dependent protein kinase, 64–65, 67
  of eIF-2α kinase, 353
  of glycogen synthase kinase-3, 341
  of insect protein kinase, 73
  of phosphorylase kinase, 257
  of protein kinase C, 293
  of protein kinase inhibitor, 78, 85
  of protein kinase subunit, 53–54, 57–58
  of pyruvate dehydrogenase kinase, 334
DEAE-Sephacel, 283, 302, 303

DEAE-Sephadex A-25, 145
DEAE-Sephadex A-50, 61–62
Dehydroalanine, conversion to [^{35}S]cysteic acid, 45
Densitometry of autoradiographs, 34–36
Deoxycholate, 374, 393
Deoxyribonucleic acid-cellulose chromatography, in purification of casein kinase I, 312
Dephosphocasein
  in assay of casein kinase I, 309
  preparation, 315
Diacylglycerol, activator, of protein kinase C, 288–289, 292, 295
Dibucaine, inhibitor, of protein kinase C, 297
Dibutyryl cyclic AMP, 21
  effect on R subunit, 241–242
  response of CHO cells to, 199
Dichloroacetate, inhibitor, of pyruvate dehydrogenase kinase, 336
O-2,6-Dichlorobenzyl group, as protecting group, 128
DIEA, see Diisopropylethylamine
Digitonin
  in cell fractionation, 23–24
  inhibitor, of rhodopsin kinase, 365
Dihydrolipoamide acetyltransferase, 331
Dihydrolipoamide reductase, 331
Diisopropylethylamine, 125
S-3,4-Dimethylbenzyl group, as protecting group, 129
Dimethylformamide:methylene chloride, 127, 128
Dimethyl sulfoxide, effect on thawed cells, 199
Dinitrophenyl moiety, as protecting group, 128
Diolein, in assay of protein kinase C, 290
2,5-Diphenyloxazole, 135, 208, 209
5,5′-Dithiobis(2-nitrobenzoic acid), inhibitor, of pyruvate dehydrogenase kinase, 336
Dounce homogenizer, 23
DPNH, see Nicotinamide adenine dinucleotide, reduced form
dsI, see Protein kinase, cAMP-independent, double-stranded RNA-dependent

E

EDTA, see Ethylenediaminetetraacetic acid
EGF, see Epidermal growth factor
EGRA, see Ethylene glycol bis(β-aminoethyl ether)-N,N,N',N'tetracetic acid
Electron paramagnetic resonance, Mn^{2+} binding studies and, 94–96
Emulphogene BC720, inhibitor, of rhodopsin kinase, 365
Epidermal growth factor, covalent linkage to gel, 380
Epidermal growth factor receptor, 379
  autophosphorylation assay, 380–381
  purification, 384–387
Epididymal fat pad, rat
  homogenization procedure, 228, 229
  protein kinase activity ratio in, 231
Epinephrine, 21, 231
EPR, see Electron paramagnetic resonance
Ethanolamine, 380
Ethanol:diethyl ether, 11
1,N^6-Etheno-cAMP, 162–167
2-Ethoxyethanol, 368
N^6-(Ethyl-2-diazomalonyl)-cAMP, 154
Ethylenediaminetetraacetic acid, in fractionation buffer, 23, 24, 25
Ethylene glycol bis(β-aminoethyl ether)-N,N,N',N'-tetraacetic acid, 286, 290
N-Ethylmaleimide, inhibitor, of pyruvate dehydrogenase kinase, 336
Ethylmethanesulfonate, 200
Eyes, bovine, crude extract preparation from, 364

F

Film, X-ray, characteristics, 34, 35
Fluorescamine, 134
Fluorescamine assay, sensitivity, 11
5′-p-Fluorosulfonylbenzoyladenosine, 379
Formic acid, 270, 381
Formyl group, as protecting group, 129
Freezing-thawing, effects of protein kinase activity, 228
Fructose 1,6-biphosphate, pyruvate kinase phosphorylation and, 214, 215

SUBJECT INDEX

G

Gel electrophoresis
  gel slicing and counting, 34
  one-dimensional, 28–30
  photography of gels and autoradiographs, 36
  staining, 29
  two-dimensional, 30–33, 238–242
Gizzard, turkey, crude extract preparation, 281
Glucagon, 21, 30, 32, 231
Glucose, as substrate-directed effector, 18
Glutamine, in peptide synthesis, 129–130
Glycerol
  for support storage, 70
  for enzyme storage, 70, 244
Glycerol gradient centrifugation, in kinase purification, 355–356
Glycerol:potassium phosphate:dithiothreitol storage solution, 16
Glycogen synthase, 15
  chromatographic analysis of labeled peptides of, 39–43
  phosphorylation, 37–38, 51, 268, 275–277
  proteolysis, 38–39
  purification, 84
  radiophosphorylation, 37–38
Glycogen synthase a
  in assay of glycogen synthase kinase, 300
  attachment to agarose, 340
  phosphorylation, 243, 337–339
Glycogen synthase kinase, calmodulin-dependent
  activator, 299
  assay, 300–301
  inhibitor, 299
  physicochemical properties, 299
  purification, 301–307
Glycogen synthase kinase, cAMP-independent, in phosphorylation of glycogen synthase, 37, 42
Glycogen synthase kinase-3
  activity, 337, 344–345
  assay, 337–340
  autophosphorylation, 343–344
  properties, 343–345

  purification, 340–343
  recognition site, 345
  specificity, 344
  stability, 341
Glycogen synthase kinase-5, 248, 249
Grasshopper, growth and dissection, 71–72
Guanidine hydrochloride, 149, 281
Guanidino moiety, protection of, 127
[γ-^{32}P]Guanosine 5′-triphosphate, in phosphorylation of glycogen synthase, 37

H

Heart, bovine
  advantages as source of protein kinase, 54–55
  crude extract preparation, 56–57, 255
Heart, rat
  homogenization procedure, 228, 229
  protein kinase activity ratio in, 231
Heat treatment, in purification of phosphoproteins, 8
  of protein kinase inhibitor, 84
HEDTA buffer, see N-(2-Hydroxyethyl)ethylenediamine-$N,N,'N'$-triacetic acid buffer
Heparin
  effect on casein kinase I, 314, 316
  inhibitor, of casein kinase II, 314, 316, 318, 327, 328–330
  of glycogen synthase kinase-5, 249
Hepatocyte, rat
  homogenization procedure, 228, 229
  protein kinase activity ratio in, 231
  protein phosphorylation in, 21, 22
Hexokinase, 16, 17
  in determination of [γ-^{32}P]ATP specific activity, 27
High-pressure liquid chromatography
  for determination of [^{32}P]ATP specific activity, 26–27
  in purification of basic, synthetic peptides, 133–134
    of modified peptides, 151
    of phosphopeptides, 43–48
    of phosphorylation sites, 39–43
Histidine
  side chain deprotection, 132

side chain protection, 128
Histone, effect on nucleotide binding, 207, 211
Histone H1, 268
Homogenization, 23
  choice of procedure, 228, 229
Hormone, in stimulation of protein phosphorylation, 20–22
Hydrogenation, catalytic, 132
Hydrogen bromide-trifluoroacetic acid, use in peptide synthesis, 130, 131–132
Hydrogen fluoride, use in peptide synthesis, 130–131
N'-(2-Hydroxyethyl)ethylenediamine-N,N',N'-triacetic acid buffer, 244
Hydroxy group, protection, 128
Hydroxylapatite chromatography, in purification of casein kinase I, 311, 313
  of protein kinase subunit, 53–54
p-Hydroxymercuriphenyl sulfonate, in purification of pyruvate dehydrogenase kinase, 334

I

IBMX, see 1-Isobutylmethylxanthine
Imidazole, in purification of pyruvate dehydrogenase kinase, 334
β-Imidozole ring, protection, 128
Imidazolium picrate, 126
Imipramine, inhibitor, of protein kinase C, 297
Immunoprecipitation, indirect, of protein kinases, 190, 191
β-Indole ring, protection, 128–129
Inhibitor protein, 338
Initiation factor 2, 318
Initiation factor eIF-2α, phosphorylation, 346
[^{3}H]Inosine 3',5'-cyclic phosphate, 170, 172
Inositol hexasulfate, inhibitor, of casein kinase I, 314
  of casein kinase II, 318, 326
Insect, see Grasshopper
Insulin, 21
Interferon, response of mutant CHO cells to, 205
Iodination procedure, 190–191

Ion-exchange chromatography, in purification of basic, synthetic peptides, 132–133
Ionomycin, 21
1-Isobutylmethylxanthine, 155, 161
Isoelectric focusing
  of phosphoproteins, 31
  of protein kinase C, 293–294
Isopropanol, 125, 151
Isoproterenol, 21, 200

K

Kemptide, 45

L

Label-chase procedure, for lymphoma cells, 237
Lactate dehydrogenase reaction, in assay of protein kinase, 4–6
Leupeptin, 25, 301, 303
Lithium bromide, in dissociation of phosphorylase kinase, 263
Liver, rabbit, crude extract preparation, 301–302
Lung, bovine, crude extract preparation, 64
Lysine, side chain protection, 127

M

Malachite green, 7, 12
Magnesium ion
  activator, of insect protein kinase, 76
  casein kinase II and, 321, 323
Manganese ion, binding studies, 94–98
MEK buffer, 164
Membrane fraction of A-431 cells
  preparation, 382
  solubilization, 383–384
MES, see 2-(N-Morpholino)ethanesulfonic acid
Methionine, side chain protection, 129
[^{35}S]Methionine
  counting procedure, 237–238
  in intracellular enzyme labeling, 234–236
p-Methylbenzhydrylamine resin, 123, 124

SUBJECT INDEX

S-4-Methylbenzyl, as protecting group, 129
Methyl ethyl sulfide, 129
2-Methylnaphthalene, 392
Mg·ADP, see Adenosine 5'-diphosphate, magnesium salt
Micromanipulator, choice of, 222
Micropipet, preparation, 222, 223
Millipore filtration binding assay, 56
Modified Barth medium, 222
Modulator proteins, 207
    effect on protein kinases, 209, 212
Moloney murine leukemia virus, 373
MOPS, see Morpholinopropanesulfonic acid
2-(N-Morpholino)ethanesulfonic acid, 16, 52, 77, 81, 82, 86, 89, 90, 155, 195, 239, 261
Morpholinopropanesulfonic acid, 3, 5, 164
MS222, 224
Murine leukemia virus, 373
Myosin, smooth muscle, purification, 281
Myosin light chain kinase, 3
    activity, 279
    amino acid composition, 287
    assay, 279–280
    phosphorylation of, by cAMP-dependent protein kinase, 288
    properties, 286
    purification, 281–285
    regulation, 287–288
    substrate specificity, 286

N

Nicotinamide adenine dinucleotide, reduced form, in assay of protein kinase, 4–6
Nicotine, 21
Nitrilotriacetate, 244
Nitro moiety, as protecting group, 127
Nonidet P-40, 236, 239, 393
Norepinephrine, 21, 30
Norleucine, use in peptide synthesis, 129
NTA, see Nitrilotriacetate
Nuclear magnetic resonance methods, 98–119
Nuclear Overhauser effect, principles, 103–107

Nucleotide, contaminant, in protein-bound phosphate determination, 7–8

O

Oligopeptide, see also Peptide
    applications, 137–139
    cleavage and deprotection, 130–132
    phosphorylation studies, 137
    purification and analysis, 132–134
    in substrate specificity studies, 137–138
    synthesis of protected form, 121–130
    use in protein kinase assay, 134–139
Oocytes
    isolation, 223–225
    labeling, 225
    maturation and protein kinases, 219–220
    microinjection
        equipment, 221–222, 223
        procedure, 225–226
        solutions, 222–223
OR2 medium, 222
Ornithine decarboxylase, 197, 205, 218
    regulation, 366–368

P

Palladium oxide, 132
PAM resin, see Phenylacetamidomethyl resin
Paramagnetic probe-T_1 method, 98–103
PEM buffer, 64
Pepstatin, 25, 301, 303
Peptide, see also Oligopeptide
    8-azido-[^{3}H]cAMP-modofied, isolation, 149–151
    FSO$_2$BzAdo-modified, isolation, 144–146
    synthetic, purification, 269–270
Peptide bond formation, quantitative test for, 126
Peptide sequencing, 37–38, 401–402
Peptide synthesis, solid-phase, 119–139, see also Oligopeptide
Phentolamine, inhibitor, of protein kinase C, 297
Phenylacetamidomethyl resin, 124–124
Phenylalanine hydroxylase, 218

Phenylmethylsulfonyl fluroide, 25, 213, 254, 324, 374
Phenyl-Sepharose, 294
Phosphate, inorganic, determination, 12–14, 19
Phosphate moiety
  labeling of, 392–393
  removal from proteins, 14–20
  protein-bound, measurement of, 7–14
Phosphate bond, free energy of hydrolysis, 16, 19–20
Phosphatidylinositol turnover, 289
Phosphoamidate, in protein, 8
Phosphoamino acids
  preparation of, 395–397
  quantitation, 401
  separation of, 397–401
Phosphoaspartate, 8
Phosphocellulose column chromatography
  in purification of casein kinase I, 310–311
    of casein kinase II, 325
    of eIF-2α kinase, 354
    of glycogen synthase kinase, 303, 341
    of polyamine-dependent protein kinase, 370, 371
Phosphocellulose paper, in assay of protein kinases, 3–4, 77, 135–136, 271
Phosphoenolpyruvate, pyruvate kinase phosphorylation and, 214, 215
Phosphofructokinase, 213
  purification, 84
  substrate-directed regulation of phosphorylation, 215–216
6-Phosphofructo-2-kinase/fructose-2,6-biphosphatase, 213, 217, 218
Phosphoglutamate, 8
Phosphohistidine, 8, 9
Phosphohistone, 3–4
Phospholipid
  activator, of protein kinase C, 288, 290
  removal, 8
  test for, 27
Phospholysine, 8, 9
O-Phosphomonoester, in proteins, 8

Phosphopeptides, purification using HPLC, 37–48
  electrophoretic separation
    one-dimensional, 28–30
    two-dimensional, 30–33
  identification, 33
  purification, 7–12
Phosphoprotein phosphatase, 183–185, 186, 212
Phosphoric acid, in assay of protein kinase, 3–4, 5
Phosphorylase a, 220, 221, 339
Phosphorylase b
  in assay of phosphorylase kinase, 245–246, 253, 260
  lyophilization, 253
  peptide preparation from, 269–270
  phosphorylation site, 243
  preparation, 244
  storage, 244, 253
Phosphorylase kinase, 3, 15, 17, 18
  activation, 269
  activity, 268
  assay, 244–246, 252–255, 260–261
  dissociation,
    monitoring of, 260–262
    procedure, 262–267
  molecular weight, 240, 259
  peptide inhibitors, 277–278
  in phosphorylation of glycogen synthase, 37, 42
  phosphorylation rates, determination of, 270–272
  phosphorylation sites, 51
  preparation of active $\alpha\gamma\delta$ complex, 263–265
  of active $\gamma\delta$ complex, 265–266
  properties, 248–250, 261
  purification, 84, 246–248, 255–258, 269
  radioassay, 246, 253, 260–261, 270–272
  ratio of activities at pH 6.8 and 8.2, 249, 261, 269
  substrate specificity, 272–277
  subunits, 248–249, 251–252, 259–260
Phosphorylase phosphatase, 15
Phosphorylatable region, 268

Phosphorylation, cAMP-dependent, substrate-directed regulation of, 212–219
Phosphoserine
  abundance, 8, 9, 388
  alkaline hydrolysis, 9, 10
  alkaline β-elimination, 45, 46–48
  determination, 315
  radiolabeled, selective purification, 43–48
  separation of, 398–402
Phosphoserine bond, test for, 27
Phosphothreonine
  abundance, 8, 9, 388
  determination, 315
  separation of, 398–402
Phosphotyrosine
  abundance, 8, 388–389
  alkaline hydrolysis, 9, 10
  amino acid analysis and, 402–405
  assay for, 375–376
  chemical properties, 391
  detection of, 391–402
  labeling of, 391–393
  physical properties, 391
  separation of, 398–402
  sequencing behavior, 401–402
  synthesis of, 390–391
[^{32}P]Phosphotyrosine, hydrolysis and identification of, 381
Phosvitin kinase, 15, 17
Photoaffinity probe, disadvantage, 147
Photography, of gels and autoradiographs, 36
Physarum polycephalum, isolation of nuclei from, 369
Picric acid, 126
1,4-Piperazinediethane sulfonic acid, 393
PMSF, *see* Phenylmethylsulfonyl fluoride
Polyamines, activators, of casein kinase II, 328
Polyarginine, 16, 206–207, 209, 210, 211, 212
Polyethylene glycol, in purification of phosphorylase kinase, 255
Potassium chloride
  effect on casein kinase I, 309
  on casein kinase II, 320–321, 323
Potassium ion, stimulant, of protein phosphorylation, 21

Progesterone, oocyte maturation and, 220, 224
1-Propanol, 39, 42, 44
Prostaglandin E_1, 21, 200
Protamine, 292
Protein
  ashing procedure, 12–13
  associated phosphate-containing compounds, 7–9
  cytosolic, extraction methods, 23–26
  measurement of chemical phosphate in, 7–14
  ^{32}P-labeled, separation, 393–395
  radiophosphorylation, 17
Protein kinase, *see also specific enzymes*
  phosphorylated substrate preparation, 17
  properties, 6
  radioassay, 3–4, 6
  reverse reaction, 14–20
  spectrophotometric assay, 4–5
  substrate specificity, 137–138, 387
Protein kinase, cAMP-dependent
  activation kinetic assay, 163
  activity, 51, 140–141
  activity ratio in intact tissues, determination of, 227–232
  affinity labeling of ATP-binding site, 141–146
    of cAMP-binding site, 147–153, 155–161
  autophosphorylation, determination of, 210
  conformation studies, 98–113
  as contaminant, removal, 69–70
  cooperativity studies, 166–168
  C subunit, 6, 51–55
    amino acid sequencing, 144, 146
    binding and conformation studies, 93–113
    kinetics of inhibition, 142–144
    kinetics of substrate exchange, 113–115
    magnetic resonance studies, 94–98
    storage, 82
  C titration assay, 164–167
  cyclic nucleotide binding activity, determination of, 209
  cyclic nucleotide exchange kinetic assay, 163

deactivation equilibrium assay, 164–167
deactivation kinetic assay, 163–164
dissociation, 151–153, 162–163, 168, 227
effect on oocyte maturation, 220
inhibitors, 77–93
interactions of R and C subunits, 115–116
isolation from insects, 73
kinase activity, determination of, 209–210
Mg^{2+} sensitivity, 6
modulator proteins of, 206–212
from mutant strains, characteristics, 201–206
in phosphorylation of glycogen synthase, 37, 40, 42
pH optimum, 6
photoaffinity labeling, 154–162
radioassay, 52, 56, 228–232
reconstitution, 168–169
reverse reaction, 15, 17–18
reversible autophosphorylation, 183–185
R subunit, 51
  assay, 56
  autophosphorylation, 155–157
  binding sites, 153, 158–159, 168
  binding studies, 168–175
  molecular weight, 60
  photoaffinity labeling, 147–153, 159–161
  physical characteristics, 60
  ^{31}P resonances in, 116–119
  purification, 55–62
  saturation with fluorescent cAMP analog, 165–166
  structural domains, 148–149
storage, 16
substrate-directed regulation of phosphorylation by, 212–219
substrate specificity, 137–138
tissue distribution, 298
Protein kinase, cAMP-dependent, type I
  R subunit
    properties, 238
    quantitation of radioactivity in, 242
    radiolabeling and detection of, 233–243
    subclasses of, 234–235
    two-dimensional gel analysis of, 238–242
Protein kinase, cAMP-dependent, type II
  autophosphorylation, 177–178, 179–185
  immunochemical analyses, 187–189, 191–196
  preparation, 176–177
  R subunit, 16
    autophosphorylation, 6
    phosphorylation, 176
    preparation, 177, 189
    radioiodination, 190–191
    structure and regulation studies, 185–186
  subclasses, 192–196
Protein kinase, (cAMP/cGMP)-dependent, 62–71, see also Protein kinase, insect
  assay, 72
Protein kinase, cAMP-independent, double-stranded RNA-dependent
  activation, 347, 350–351
  activity, 346–347
  assay, 349
  autophosphorylation, 359–361
  properties, 355, 357–361
  purification of latent form, 351–357
Protein kinase, cGMP-dependent, 15, 62–71
  assay, 63–64
  autophosphorylation, determination of, 210
  cGMP binding studies, 162
  modulator proteins of, 206–212
  nucleotide binding studies, 174–175
  purification, 64–71
  reverse reaction, 18
  storage, 70
Protein kinase, cyclic nucleotide-dependent, 3, see also specific enzymes
  assay using synthetic peptides, 134–139
Protein kinase, insect
  enzymatic properties, 76
  molecular weight, 76

purification, 73
stability, 75
Protein kinase, polyamine-dependent
   activity, 366
   assay, 368–369
   molecular weight, 372
   purification, 369–372
   specificity, 372
   stability, 372
Protein kinase, tyrosine-specific, *see also* Epidermal growth factor receptor
   functions, 388–389
Protein kinase, tyrosine-specific, viral
   assay, 374–375
   autophosphorylation assay, 374
   partial purification, 376
   source, 373
   specificity, 373–374
Protein kinase C
   activators, 297
   activity, 288–290
   assay, 290–292
   inhibitors, 297
   kinetic properties, 296
   mode of activation, 295–296
   physicochemical properties, 295
   purification, 292–294
   specificity, 296
   stability, 294–295
   subcellular distribution, 297–298
   tissue distribution, 297, 298
Protein kinase inhibitor, 5
   activity, 79–80
   assay, 77–78, 81–82, 83–84
   charge forms, 80–81, 90
   effect on protein kinase autophosphorylation, 179
   partial purification, 77–80
   properties, 90–91
   purification, 84–90, 177
   quantitation in crude extracts, 91
   regulation of tissue levels of, 92–93
   shape conformers, 80–81
      separation, 87–90
   stability, 79
   tissue distribution, 92
   unit of activity, 82
Protein kinase inhibitor, synthetic, 138, 139

Protein phosphatase, inhibition of, 389–390
Protein phosphatase, MgATP-dependent, 337
   activation, 339–340
Protein phosphatase inhibitor-2, 220, 221
Protein phosphorylation
   charge shift test for, 28
   in intact cells, 20–36
   posthomogenization, test for, 27–28
   quantitation of changes in, 33–36
   with radiolabel, 17
Protein synthesis, inhibition of, 346–347, 356
Proton relaxation rate, Mn^{2+} binding studies and, 96–98
PRR, *see* Proton relaxation rate
Putrescine, 367, 372
Pyridine:acetic acid system, 269–270, 381
Pyridine:acetic acid:water solvent system, 316
Pyridoxal 5'-phosphate, inhibitor, of casein kinase I, 314
   of casein kinase II, 318, 326, 328
Pyruvate, inhibitor, of pyruvate dehydrogenase kinase, 336
Pyruvate dehydrogenase complex, 331
Pyruvate dehydrogenase kinase
   activators, 336
   activity, 331, 336
   assay, 332
   inhibitors, 336
   properties, 335–336
   purification, 333–335
Pyruvate kinase, 15–16
   conformation studies, 98
   substrate-directed regulation of phosphorylation, 213–215
Pyruvate kinase reaction, in assay of protein kinase, 4–6

R

Resin, choice of, for solid-phase peptide synthesis, 123–124
Reticulocytes, rabbit
   casein kinase II from, 324
   lysis procedure, 324

Rhodopsin kinase
　activity, 362
　assay, 362–363
　autophosphorylation, 366
　distribution, 366
　inhibitors, 365
　phosphorylation sites, 365–366
　properties, 365–366
　purification, 364–366
　specificity, 365
　stability, 366
Ribonucleic acid, double-stranded, inhibitor, of protein synthesis initiation, 346
Ribosomal salt-wash, 351

S

Scint-A, 326
SDS-polyacrylamide gel electrophoresis
　of inhibitor protein, 87, 88
　of myosin light chain kinase, 285
　of R subunits, 59, 149
Sephacryl S-200, 364
Sephadex G-15, 270
Sephadex G-25, 149
Sephadex G-50, 150, 270
Sephadex G-75, 87, 89
Sephadex G-100, 344
Sephadex G-150, 263, 264, 293, 311
Sephadex G-200, 73, 94, 165, 358, 359
Sepharose 4B beads, used, storage, 70
Sepharose 6B, 66–67, 304, 305, 351
Sequencing, of phosphorylated peptides, 401–402
Serine
　side chain protection, 128
　in substrate of phosphorylase kinase, 269, 272
Side chain protecting groups, 127–130
Skeletal muscle, rabbit, crude extract preparation, 58, 84, 246
S49 mouse lymphoma cells, labeling procedure, 235–237
Sodium azide, for used support storage, 70
Sodium butyrate, mutant CHO cell response to, 205
Sodium dodecyl sulfate, in dissociation of phosphorylase kinase, 262

Sodium maleate, 245
Spermidine, 367, 368
Spermine, 367, 368
　activator, of casein kinase II, 328
SP-Sephadex, in purification of basic, synthetic peptides, 132–133
src gene product, 204
Standardization wedges, 238, 242
Substrate specificity, oligopeptides in studies of, 137–138
Sucrose density gradient centrifugation, in purification of phosphorylase kinase, 256–257, 261, 262
[^{35}S]Sulfite addition, 45, 47–48
Sulfopropyl-Sephadex, 311
Sulfoxide group, as protecting group, 129
Synthase kinase, calmodulin-dependent, in phosphorylation of glycogen synthase, 37, 42
Synthase kinase 3, in phosphorylation of glycogen synthase, 37, 41, 42
Synthase kinase F_A, in phosphorylation of glycogen synthase, 37, 41

T

Testes, bovine, crude extract preparation, 280
Tetracaine, inhibitor, of protein kinase C, 297
1,3,4,6-Tetrachloro-3α,6α-diphenyl glycouril, 190
12-O-Tetradecanoylphorbol 13-acetate, 205
Theophylline, 78
Thin-layer electrophoresis, two-dimensional, for separating phosphoamino acids, 316, 398–401
β-Thiol group, protection, 129
Threonine, side chain protection, 128
Thrombin, 21
TNC, see Troponin
Toluene scintillation fluid, 135
Tosyl moiety, as protecting group, 127, 128
N-Tosyl-L-phenylalanine trypsin, 37
TPA, see 12-O-Tetradecanoylphorbol 13-acetate
Transglutaminase, 197, 205
Trasylol, 393

Trichloroacetic acid precipitation
  in kinase assays, 254, 261, 280, 291, 301, 320, 332, 338, 363
  in purifications, 8, 9, 27, 77, 85, 237, 315
Trichloroacetic acid–tungstate solution, 11
Triethylamine, 125
Trifluoperazine, inhibitor, of protein kinase C, 297
Trifluoroacetic acid, 39, 40, 42, 44, 45, 151
Triton–toluene scintillation mixture, 326
Troponin, 209, 212, 252, 268, 319
Trypsin
  in dissociation of phosphorylase kinase, 266
  effect on phosphorylase kinase, 249
Trypsinization, of ^{32}P-labeled glycogen synthase, 38–39
Tryptophan, side chain protection, 128–129
Tumor, human, protein kinase type II isozymes in, 196
Tyrosine, see also Phosphotyrosine
  labeling of, 391–392
  side chain protection, 128
  specific phosphorylation, 373–374

U

Ultraviolet light, as mutagen, 200

Urea
  in preparation of peptide substrates, 270
  in protein separation, 31
  in purification of protein kinase R subunit, 60, 236

V

Vasopressin, 21
Verapamil, inhibitor, of protein kinase C, 297
Veratridine, 21
Vesicle fraction of A-431 membranes, preparation, 382–383
Vimentin, 204

W

Water, Ca^{2+}-free, preparation, 291

XYZ

Xanthenyl group, as protecting group, 130
Xenopus laevis oocytes, see Oocytes
Zinc ion, inhibitor, of rhodopsin kinase, 365